AF601886

Operator Theory: Advances and Applications

Volume 273

Founded in 1979 by Israel Gohberg

More information about this series at http://www.springer.com/series/4850

Valentin A. Zagrebnov

Gibbs Semigroups

Valentin A. Zagrebnov
Université d'Aix-Marseille
Institut de Mathématiques de Marseille
Marseille, France

ISSN 0255-0156 ISSN 2296-4878 (electronic)
Operator Theory: Advances and Applications
ISBN 978-3-030-18876-4 ISBN 978-3-030-18877-1 (eBook)
https://doi.org/10.1007/978-3-030-18877-1

Mathematics Subject Classification (2010): 47D03, 47D06, 46C05, 47B10

This book is published under the imprint Birkhäuser, www.birkhauser-science.com by the registered company Springer Nature Switzerland AG
The registered company address is: Gewerbestrasse 11, 6330 Cham, Switzerland

To the memory of

Robert A. Minlos

and

Hagen Neidhardt

Preface

The theory of one-parameter semigroups of operators emerged as an important branch of mathematics in the forties when it was realised that this theory has direct applications to partial differential equations, random processes, infinite-dimensional control theory, mathematical physics, etc. The theory of (strongly continuous) one-parameter semigroups is now generally accepted as an integral part of contemporary functional analysis.

In this framework the compact strongly continuous semigroups have been for a long time an important subject for research, as in almost all applications to partial differential equations with bounded domains the corresponding solution semigroups turn out to be families of compact operators. From this point of view, in the present book the emphasis is on a special subclass of compact semigroups. In fact, we mainly focus on the *strongly* continuous semigroups with values in the *trace-class* ideal $\mathcal{C}_1(\mathcal{H})$ of bounded operators acting on a Hilbert space $\mathcal{H}$.

Historically, this class of semigroups is closely related to the so-called *Density Matrix* in Quantum Statistical Mechanics. For this reason throughout the text for strongly continuous semigroups with values in the ideal $\mathcal{C}_1(\mathcal{H})$ we follow the terminology introduced in the first papers on this subject:[Uhl71], [ANB75], and call them the *Gibbs semigroups.*

The aim of this book is to give an accessible and complete introduction to different aspects of the theory of Gibbs semigroups. More than fifteen years has elapsed since the publication of the Leuven University Notes [Zag03b] on this topic. The present book constitutes a major expansion of those Notes with more details and new aspects, which almost doubles the number of chapters. It is halfway between a textbook and a monograph, which provides a systematic and a comprehensive up-to-date account of the Gibbs semigroup theory, addressed to students as well as to experienced researchers.

To this aim the first three chapters: Chapter 1 (*Semigroups and their generators*), Chapter 2 (*Classes of compact operators*), and Chapter 3 (*Trace inequalities*), contain preliminary material for the reading of the remaining main chapters. These preliminaries are accessible to graduate students with prerequisites corresponding to an introductory course in functional analysis. To recall certain elements of the spectral analysis, Appendix A, *Spectra of closed operators*, is also included in the preliminaries.

Each of the Chapters 1-3 concludes with Notes, which provide references to the literature, comments to discussions in each section of the corresponding chapter, and relevant historical remarks. The material of Chapters 1–3 and of Appendix A is standard and can be found in many sources. Therefore, the Notes are limited by knowledge and prejudices of the author.

The core of the book is constituted by Chapter 4 (*Gibbs semigroups*), Chapter 5 (*Product formulae for Gibbs semigroups*), Chapter 6 (*Symmetrically-normed ideals*), and Chapter 7 (*Product formulae in the Dixmier ideal*), where the principal topics are presented. Although the selection of material in these chapters is determined by the personal taste of the author, this part of the volume provides a self-consistent introduction to the theory of Gibbs semigroups for nonspecialists interested in this specific class of semigroups. The reader is expected to be familiar with basics of functional analysis and linear operator theory, but above this level, the necessary notations, definitions and propositions are provided in the text.

The Appendix B, *More inequalities*, serves to recall some particular inequalities useful for the exposition in Chapter 3 and in Chapters 5–7 of the book.

Appendix C, *Kato functions*, introduces and classifies different types of the Kato functions needed in the text. We also recall certain properties of these functions, that are used in Chapters 5–7.

The Notes at the end of each of Chapters 4–7 and Appendix B, Appendix C contain references and some historical remarks related to the corresponding sections. Sometimes we use the space of the Notes in order to elaborate on a discussion launched in a particular section.

Finally, Appendix D is dedicated to a review of results and of the literature closely related to the Gibbs semigroups. Mainly it is about evolution of our knowledge concerning convergence of the Trotter-Kato product formulae in the *operators-norm* and the *trace-norm* topologies. The exact formulation of results and applications to *lifting* the operator-norm estimates of rates of convergence to the trace-norm estimates are presented in Chapter 5 – Chapter 7. Recall that after the introduction of the self-adjoint Gibbs semigroups in the 1970s by [Uhl71] and [ANB75], the analysis of *bounded* [Uhl71] and *infinitesimally* Kato-small [ANB75] perturbations of generators was developed rather repidly. The analytic theory for the Kato-small perturbations of generators is more involved. It began developing in [Mai71] and then independently in [ZBT75], [Zag89]. Note that besides a purely mathematical interest, this activity was motivated by the study of the trace-class operator known in the Quantum Statistical Mechanics as a *density matrix*. We resume the analysis of strongly continuous semigroups with values in the trace-class ideal in Chapter 4.

My interest in Gibbs semigroups in the early 1980s was renewed after the following question posed to me by Robert A. Minlos: "Does the exponential Trotter product formula approximating these semigroups converge in the trace-norm topology?"

This question was motivated by our discussion of the very new at the time *infrared bounds* method in the mathematical theory of phase transitions proposed by J. Fröhlich, B. Simon and T. Spencer [FSS76]. In particular, we were intrigued by its *quantum* version ([FrLi78], [FILS78], [DLS78]) since in this case the proof of the infrared bounds involves the limit of the Trotter product formula *approximants* under the trace. Finally, to obtain the infrared bound one has to interchange the trace and the limit of approximants.

In fact this interchange is not problematic for quantum spin systems since the underlying Hilbert spaces are finite-dimensional, but it is so for, e.g., *unbounded* spins. A typical example is the problem of proving infrared bounds for the case of structural phase transitions in anharmonic quantum crystals [Fr76]. Then the arguments in the papers [DLP79], [PK87] (essentially motivated by [Fr76]) indicate that a combination of the Trotter product formula approximation and the path integral representation for the trace is sufficient for obtaining the *infrared bounds* and the *localisation* estimate [Kon94], [MVZ00].

A message delivered by Chapter 5 is that the answer to the question posed by Robert A. Minlos is *affirmative.* Consequently, this allows to *interchange* for Gibbs semigroups the limit in the Trotter product formula and the trace, which ensures the proof of the infrared bounds in general settings. Initially the convergence of the Trotter product formula in the *trace-norm* topology was proven in our paper [Zag88] for the Schrödinger (Gibbs) semigroups. The first results about the product formulae for abstract Gibbs semigroups in a Hilbert space, including a generalisation to convergence of the Trotter-Kato product formulae in the trace-norm, is due to our papers with Hagen Neidhardt [NZ90a], [NZ90b].

Moreover, Chapter 5 presents the trace-norm *error bounds* for the convergence rates of the Trotter-Kato product formulae. This stronger result is, in turn, due to the discovery that for strongly continuous semigroups the Trotter product formula converges in the *operator-norm* topology [Rog93]. This paper also suggested an error bound estimate for the rate of convergence.

Note that between 1959 (H. Trotter [Tro59]) and 1993 (Dzh. Rogava [Rog93]) it was common knowledge that for self-adjoint strongly continuous semigroups the Trotter product formula converges only in the *strong* operator topology. Although they did not provide error bound estimates, our papers about self-adjoint Gibbs semigroups [NZ90a], [NZ90b], published in 1990 were the first to challenge this state of affairs.

In [NZ98] the examination of the Rogava theorem allowed us to improve the estimate of the convergence rate in the operator-norm topology for the *Trotter-Kato* (i.e., *non-exponential*) generalisation of the product formulae under the same conditions as in [Rog93]

The first estimate of the trace-norm error bounds for the rate of convergence of the *exponential* Trotter-Kato product formula in the case of the self-adjoint Schrödinger (Gibbs) semigroups is due to [IT98b]. In Chapter 5 the trace-norm error bounds for the rate of convergence are obtained by *lifting* the operator-norm error bounds. This general method was invented in [NZ99d]. It allows to lift all, including the optimal, estimates of the convergence rates for the Trotter-Kato product formulae in the operator-norm topology to trace-norm error bounds estimates. The method is also able to cover the case of non-self-adjoint Gibbs semigroups [CZ01c] for a quite large class of Kato functions.

Chapters 6 and 7 are devoted to extending the Trotter-Kato product formulae to the more abstract framework of the *symmetrically-normed* ideals and to the case of the *Dixmier* ideal.

It turns out that results for the Gibbs semigroups (Chapter 5), and in general for the von Neumann-Schatten ideals, admit an extension to arbitrary symmetrically-normed ideals. Convergence of the Trotter-Kato product formulae in the norm of the symmetrically-normed ideals is proved together with estimates of the rates of convergence. These estimates correspond to the lifting of the operator-norm error bounds. To this end we essentially follow in Chapter 6 the results from [NZ99d] and our paper [Zag19].

The Trotter-Kato product formulae in the Dixmier ideal were considered for the first time in [NZ99d]. There under certain conditions the convergence of the (*singular*) Dixmier trace for the Trotter-Kato product formulae was conjectured. I made these arguments explicit in order to prove in Chapter 7 the convergence of the Trotter-Kato product formulae in the Dixmier ideal topology. The estimates of the rate of convergence are similar to those in the symmetrically-normed ideals [Zag19].

The Notes at the end of each chapter attempt to complete the exposition by comments and references that should make the contents more accessible to the readers, including graduate students in mathematics and mathematical physics.

Acknowledgements

I would like to thank many people, who have contributed to my understanding of the Gibbs semigroups and who have made the writing of this book possible.

First of all, I am grateful to my mentor Robert Adolfovich Minlos, who introduced me to functional analysis and sparked my interest in the problems surrounding the Trotter product formula. I must also mention Nicolae Angelescu, who many years ago brought to my attention the notion of the Gibbs semigroups. Never published with them on this particular topic, I benefited greatly from many discussions and collaborations with them in this and other projects.

Second, I am thankful to my co-authors in publications on the Gibbs semigroups theory and the Trotter-Kato product formulae: Hagen Neidhardt, Takashi Ichinose, Pavel Exner, Hiroshi Tamura, Hideo Tamura, Vincent Cachia and Artur Stephan. It was always a pleasure collaborating with them. With Hagen Neidhardt we obtained a complete answer to the Minlos question concerning the Trotter-Kato product formulae trace-norm convergence for abstract Gibbs semigroups.

In the beginning of 2000 Minlos strongly suggested that I should summarise the existing results on Gibbs semigroups as a detailed review. My first attempt to collect such results took the form the KU Leuven Notes [Zag03]. I am thankful to André Verbeure, the editor of this series, who initiated and supported this first enterprise, which was very helpful in the preparation of the present comprehensive volume.

It is my pleasure to thank Michael Th. Rassias for reading the manuscript and for making numerous suggestions and corrections.

Last but not least, I want to express my gratitude to my wife Galina. Her support over all these years has meant so much to me.

Marseilles, February 2018 – January 2019

Valentin A. Zagrebnov

Institut de Mathématiques de Marseille
and
Département de Mathématiques, AMU

Contents

Chapter 1

Semigroups and their generators

This chapter contains a brief account of basic results of the theory of generators of strongly continuous operator semigroups. We introduce here some notations and definitions indispensable for the following chapters and consider only some restricted type of semigroups and their perturbations. After we define of the strongly continuous exponential function, we consider contraction and quasi-bounded semigroups, emphasising the holomorphic semigroups on a complex separable Hilbert space. Preparing the introduction of the Gibbs semigroups we address the question of the continuity of semigroups in various topologies.

1.1 The exponential function

Let $\mathcal{B}$ be a Banach space. By $\mathcal{L}(\mathcal{B}) := \mathcal{L}(\mathcal{B}, \mathcal{B})$ we denote the Banach space of bounded linear transformations from $\mathcal{B}$ to itself. For any bounded operator $A \in \mathcal{L}(\mathcal{B})$ one can define the exponential function $t \mapsto \mathrm{e}^{-tA} = U_t(A)$, for complex $t \in \mathbb{C}$, by the series

$$U_t(A) := \sum_{n=0}^{\infty} \frac{t^n}{n!}(-A)^n, \tag{1.1}$$

which is convergent in the operator norm $\|\cdot\|$ on the Banach space $\mathcal{L}(\mathcal{B})$. Therefore, for any $A \in \mathcal{L}(\mathcal{B})$, the mapping $t \mapsto U_t(A)$ is an entire $\|\cdot\|$-holomorphic operator-valued function on the complex plane $\mathbb{C}$. The (semi)group property

$$U_{t_1+t_2}(A) = U_{t_1}(A)U_{t_2}(A), \quad t_1, t_2 \in \mathbb{C} \tag{1.2}$$

is a direct consequence of formula (1.1) as well as the equation

$$\|\cdot\|\text{-}\,\partial_t U_t(A) = (-A)U_t(A) = U_t(A)(-A), \tag{1.3}$$

V. A. Zagrebnov, *Gibbs Semigroups*, Operator Theory: Advances and Applications 273, https://doi.org/10.1007/978-3-030-18877-1_1

where $\|\cdot\|$-∂_t stands for differentiation in the sense of the operator norm on $\mathcal{B}$.

Now, let A be an unbounded linear operator in $\mathcal{B}$ with domain $\operatorname{dom} A \subset \mathcal{B}$. Since now the series (1.1) is ill-defined, the existence of the exponential function $U_{(\cdot)}(A) : D \to \mathcal{L}(\mathcal{B})$ for $D \subset \mathbb{C}$ is less obvious. One possibility of defining $U_t(A)$ is an alternative to (1.1), namely, via the famous *Euler formula*

$$U_t(A) = \lim_{n\to\infty}\left(\mathbb{1} + \frac{t}{n}A\right)^{-n}, \tag{1.4}$$

first for $t \in \mathbb{R}_0^+ := \{0\} \cup \mathbb{R}^+ = [0, +\infty)$. The following proposition gives sufficient conditions for that to be feasible.

Proposition 1.1. *Let $A \in \mathcal{C}(\mathcal{H})$ be a closed linear operator with a dense domain $\operatorname{dom} A \subset \mathcal{B}$ such that the following two conditions hold:*

(i) $\mathbb{R}^- := (-\infty, 0) = \mathbb{R}\backslash\mathbb{R}_0^+$ *belongs to the resolvent set $\rho(A)$ of A;*

(ii) $\|(A + \lambda\mathbb{1})^{-1}\| \leqslant \lambda^{-1}$ *for $\lambda > 0$.*

Then:

(a) *The operator A generates for $t \geqslant 0$ an operator-valued (exponential) function $t \mapsto U_t(A) \in \mathcal{L}(\mathcal{B})$.*

(b) $\|U_t(A)\| \leqslant 1$ *for all $t \geqslant 0$.*

(c) *The mapping $\mathbb{R}_0^+ \ni t \mapsto U_t(A)f$ is continuous for every $f \in \mathcal{B}$, i.e., strongly continuous in $\mathbb{R}_0^+$ with $\lim_{t\downarrow 0} U_t(A)f = f$.*

(d) *For $u_0 \in \operatorname{dom} A$ the function $u(t) := U_t(A)u_0$ is strongly differentiable in $\mathbb{R}^+$ and is a solution of the equation $\partial_t u(t) = -A\,u(t)$ with initial condition $u(t)|_{t=0} = u_0$.*

(e) $U_{t+s}(A) = U_t(A)U_s(A)$ *for all $t, s \geqslant 0$ and $U_0(A) = \mathbb{1}$.*

Proof. (a) Let us define for $t \geqslant 0$ the sequence

$$U_{t,n}(A) = \left(\mathbb{1} + \frac{t}{n}A\right)^{-n}, \quad n = 1, 2, \ldots \tag{1.5}$$

Then condition (ii) implies that $\|U_{t,n}(A)\| \leqslant 1$. Moreover, $t \mapsto U_{t,n}(A)$ is $\|\cdot\|$-holomorphic in the right half-plane $\mathbb{C}_+ := \{z \in \mathbb{C} : \Re z > 0\}$, inheriting this property from the resolvent

$$R_\zeta(A) := (A - \zeta\mathbb{1})^{-1} \tag{1.6}$$

with ζ in the resolvent set $\rho(A) := \{\zeta \in \mathbb{C} : R_\zeta(A) \in \mathcal{L}(\mathcal{B})\}$, see condition (i). Notice that

$$\|\cdot\|\text{-}\partial_t U_{t,n}(A) = (-A)\left(\mathbb{1} + \frac{t}{n}A\right)^{-(n+1)} \in \mathcal{L}(\mathcal{B}), \quad t > 0, \tag{1.7}$$

but $U_{t,n}(A)$ is right continuous at $t = 0$ only in the strong sense

$$\operatorname*{s-lim}_{t\downarrow 0} U_{t,n}(A) = U_{0,n}(A) = \mathbb{1}, \tag{1.8}$$

which means that

$$\lim_{t\downarrow 0} \|(U_{t,n}(A) - \mathbb{1})f\| = 0, \quad f \in \mathcal{B}.$$

Here $t \downarrow 0$ denotes $t \to +0$. By (1.8),

$$U_{t,n}(A)f - U_{t,k}(A)f = \lim_{\varepsilon\downarrow 0} \int_{\varepsilon}^{t-\varepsilon} \mathrm{d}\tau\, \partial_\tau \left[U_{t-\tau,k}(A)\, U_{\tau,n}(A)\right] f, \quad f \in \mathcal{B}.$$

Using (1.7) we get

$$\begin{aligned} &U_{t,n}(A)f - U_{t,k}(A)f \\ &= \lim_{\varepsilon\downarrow 0} \int_{\varepsilon}^{t-\varepsilon} \mathrm{d}\tau \left(\frac{\tau}{n} - \frac{t-\tau}{k}\right) A^2 \left(\mathbb{1} + \frac{t-\tau}{k} A\right)^{-(k+1)} \left(\mathbb{1} + \frac{\tau}{n} A\right)^{-(n+1)} f. \end{aligned}$$

Now let $f \in \operatorname{dom} A^2$. By the strong continuity of the integrand for $\tau \in [0,t]$ and condition (ii)

$$\|U_{t,n}(A)f - U_{t,k}(A)f\| \leqslant \frac{t^2}{2}\left(\frac{1}{n} + \frac{1}{k}\right)\|A^2 f\|, \tag{1.9}$$

which means that $\{U_{t,n}(A)f\}_{n\geqslant 1}$ is a Cauchy sequence for $f \in \operatorname{dom} A^2$. Since for $\lambda > 0$ $\operatorname{dom} A^2 = (A + \lambda\mathbb{1})^{-1}\operatorname{dom} A$ is dense in $\mathcal{B}$, and since $\|U_{t,n}(A)\| \leqslant 1$, it follows that the limit (1.4) exists in the strong sense

$$U_t(A) = \operatorname*{s-lim}_{n\to\infty} U_{t,n}(A), \quad t \geqslant 0. \tag{1.10}$$

(b) Thanks to the bound $\|U_{t,n}(A)\| \leqslant 1$, relation (1.8) and the estimate (1.9), we have

$$\|U_t(A)\| \leqslant 1 \quad \text{and} \quad U_0(A) = \mathbb{1}. \tag{1.11}$$

(c) Since $t \mapsto U_{t,n}(A)$ is strongly continuous for $t \geqslant 0$ and since the estimate (1.9) is uniform in t for any finite interval $0 \leqslant t \leqslant T$, the operator-valued function

$$t \mapsto U_t(A) : \mathbb{R}_0^+ \to \mathcal{L}(\mathcal{B}) \tag{1.12}$$

is strongly continuous in $\mathbb{R}^+$, including the right limit at $t = +0$, with $\operatorname{s-lim}_{t\downarrow 0} U_t(A) = \mathbb{1}$.

(d) From (1.7) and (1.8), it follows that

$$(U_{t,n}(A) - \mathbb{1})f = -\int_0^t \mathrm{d}\tau \left(\mathbb{1} + \frac{\tau}{n}A\right)^{-1} U_{\tau,n}(A)Af, \quad f \in \operatorname{dom} A. \tag{1.13}$$

Since

$$\operatorname*{s-lim}_{n\to\infty} \left(\mathbb{1} + \frac{\tau}{n}A\right)^{-1} U_{\tau,n}(A) = U_\tau(A)$$

uniformly for any finite interval $0 \leqslant \tau \leqslant T$, we get from (1.13) that

$$(U_t(A) - \mathbb{1})f = -\int_0^t \mathrm{d}\tau\, U_\tau(A)Af, \quad f \in \operatorname{dom} A. \tag{1.14}$$

On the other hand, due to (1.12) the vector-valued family $\{U_\tau(A)(Af)\}_{\tau \geqslant 0}$ is continuous in τ for any $f \in \operatorname{dom} A$. Therefore, (1.14) shows that the vector valued function $t \mapsto U_t(A)f$ is differentiable with respect to t for all $f \in \operatorname{dom} A$

$$\partial_t(U_t(A)f) = U_t(A)(-Af), \quad t > 0,\ f \in \operatorname{dom} A, \tag{1.15}$$

where by definition

$$\partial_t\, U_{t=0}(A)f := \lim_{t \downarrow 0} \frac{1}{t}(U_t(A) - \mathbb{1})f = \lim_{t \downarrow 0} U_t(A)(-Af).$$

In fact, (1.7) gives more. Since

$$\left(\mathbb{1} + \frac{t}{n}A\right)^{-1} \mathcal{B} \subseteq \operatorname{dom} A, \quad t > 0,$$

one has

$$\begin{aligned} \partial_t U_{t,n}(A) &= -U_{t,n}(A)A\left(\mathbb{1} + \frac{t}{n}A\right)^{-1} \\ &= -A\, U_{t,n}(A)\left(\mathbb{1} + \frac{t}{n}A\right)^{-1}, \end{aligned} \tag{1.16}$$

and

$$\lim_{n\to\infty} A\left(\mathbb{1} + \frac{t}{n}A\right)^{-1} u = \lim_{n\to\infty} \left(\mathbb{1} + \frac{t}{n}A\right)^{-1} Au = Au, \quad u \in \operatorname{dom} A,$$

because A is a closed operator. Hence, by virtue of (1.10) and the closedness of A, one gets that for $f \in \operatorname{dom} A$

$$\begin{aligned} \lim_{n\to\infty} U_{t,n}(A)A\left(\mathbb{1} + \frac{t}{n}A\right)^{-1} f &= U_t(A)Af \\ &= \lim_{n\to\infty} A\, U_{t,n}(A)\left(\mathbb{1} + \frac{t}{n}A\right)^{-1} f \\ &= A\, U_t(A)f. \end{aligned} \tag{1.17}$$

Therefore, $U_t(A)f \in \operatorname{dom} A$ and the differential equation (1.15) takes the form

$$\partial_t(U_t(A)f) = (-A)(U_t(A)f), \quad t > 0,\ f \in \operatorname{dom} A. \tag{1.18}$$

Summarising, $u(t) = U_t(A)u_0$ is a solution of the differential equation

$$\partial_t\, u(t) = -A\, u(t), \tag{1.19}$$

provided $u_0 := u(0) \in \operatorname{dom} A$.

(e) Let $u(t)$ be a solution of (1.19) with initial condition $u_0 = u(0) \in \operatorname{dom} A$. This means that the vector-valued function $u(\cdot) : \mathbb{R}_0^+ \to \operatorname{dom} A$ is strongly differentiable, i.e., the limit

$$\lim_{\delta \to 0} \delta^{-1} (u(t+\delta) - u(t)) = (\partial_t u)(t)$$

exists, and such that (1.19) holds for $t > 0$. Then by virtue of (1.15) and (1.19),

$$\partial_\tau (U_{t-\tau}(A) u(\tau)) = 0, \quad 0 < \tau < t,$$

with the left derivative

$$\lim_{\tau \downarrow 0} \partial_\tau (U_{t-\tau}(A) u(\tau)) = 0.$$

Thus, for any $0 \leqslant \tau \leqslant t$ one gets, for $t > 0$, that

$$U_{t-\tau}(A) u(\tau) = U_t(A) u(\tau = 0) = u(t) \in \operatorname{dom} A. \tag{1.20}$$

This means that any solution of (1.19) can be uniquely written as

$$u(t) = U_t(A) u_0, \tag{1.21}$$

that is, $U_{t-\tau}(A) U_\tau(A) u_0 = U_t(A) u_0$ for any $u_0 \in \operatorname{dom} A$. Since this holds on the dense set $\operatorname{dom} A \subset \mathcal{B}$ for $U_t(A) \in \mathcal{L}(\mathcal{B})$, $t \geqslant 0$, continuity yields the composition law:

$$U_s(A) U_\tau(A) = U_{s+\tau}(A) \ , \quad s, \tau \geqslant 0, \tag{1.22}$$

called the semigroup *functional equation*, on all elements of $\mathcal{B}$.

It is these observations that legitimate for $U_t(A)$ the *notation* of exponential function:

$$U_t(A) = \mathrm{e}^{-tA} \in \mathcal{L}(\mathcal{B}), \quad t \geqslant 0, \tag{1.23}$$

generated by the operator A. □

1.2 Strongly continuous semigroups

Proposition 1.1 motivates the following definition:

Definition 1.2. An operator-valued function of $t \in \mathbb{R}_0^+$, $t \mapsto U_t \in \mathcal{L}(\mathcal{B})$, or a family of bounded operators $\{U_t\}_{t \geqslant 0}$, on a Banach space $\mathcal{B}$ is called a *strongly continuous* one-parameter semigroup (or C_0-semigroup) if

(a) $U_0 = \mathbb{1}$,

(b) $U_t U_s = U_{t+s}$ for $t, s \in \mathbb{R}_0^+$,

(c) $t \mapsto U_t f \in \mathcal{B}$ is a continuous function of the parameter $t \in \mathbb{R}^+$ such that $\lim_{t \downarrow 0} U_t f = f$, for every $f \in \mathcal{B}$.

Therefore, in Section 1.1 we proved existence of a special class of strongly continuous semigroups with $\|U_t\| \leqslant 1$, generated by operators satisfying the conditions of Proposition 1.1.

Definition 1.3. An operator-valued function $t \mapsto U_t$ is called a strongly continuous *contraction semigroup* if it is a C_0-semigroup with the property that $\|U_t\| \leqslant 1$ for all $t \geqslant 0$.

It is important that the *converse* of Proposition 1.1 is also true. For this purpose we first briefly discuss integration of functions taking values in the Banach spaces $\mathcal{B}$ and $\mathcal{L}(\mathcal{B})$.

Remark 1.4. (a) Recall that a function $F : \mathbb{R} \to \mathcal{L}(\mathcal{B})$ is *norm continuous* (respectively *norm differentiable*) if it is continuous (respectively differentiable) with respect to the operator norm topology on the space $\mathcal{L}(\mathcal{B})$, and F is *strongly continuous* (respectively *strongly differentiable*) if $t \mapsto F(t)f$ is continuous (respectively differentiable) in $t \in \mathbb{R}$ with respect to the norm topology of the space $\mathcal{B}$ for all $f \in \mathcal{B}$.

(b) By the *uniform boundedness principle* (Proposition 1.6), one gets that if the function F is strongly continuous, then it is locally uniformly bounded:

$$M_I := \sup_{t \in I} \|F(t)\| < \infty,$$

for any compact $I \subset \mathbb{R}$. The local boundedness yields that the product of two strongly continuous mappings is *jointly* strongly continuous. In particular, the mapping $(t, s) \mapsto F(t)u(s) \in \mathcal{B}$ is norm continuous, for any $t \mapsto F(t)$, which is $\mathcal{L}(\mathcal{B})$-valued strongly continuous, and for $s \mapsto u(s)$, which is $\mathcal{B}$-valued norm continuous. One also sees that the product of strongly differentiable mappings is strongly differentiable in each variable.

(c) We recall that the strong continuity of $t \mapsto F(t)$ implies for all $f \in \mathcal{B}$ the continuity of the function $t \mapsto \|F(t)f\|$. Then, by the local uniform boundedness principle, the function $I \ni t \mapsto \|F(t)\|$ is bounded by M_I.

(d) We stress the point that $t \mapsto \|F(t)\|$ is not necessarily continuous, and may not even be measurable, unless $\mathcal{B}$ is *separable.*

(e) Below we systematically consider continuous one-parameter functions with values in Banach spaces $\mathcal{B}$ or $\mathcal{L}(\mathcal{B})$. Since the continuous image of a separable space (in our case $\mathbb{R}$), is separable, the corresponding integrals are well-defined (via the standard Darboux–Riemann scheme approximating this function by piecewise constant functions) and finite if and only if $\int_I \mathrm{d}s \|u(s)\| < \infty$ or $\int_I \mathrm{d}t \|F(t)\| < \infty$.

(f) Note that the space of bounded operators $\mathcal{L}(\mathcal{B})$ is *nonseparable*, but if $\mathcal{B}$ is separable it is convenient to use this fact to simplify some estimates. For example $\|F(t)\|$ is then a locally bounded measurable function, and hence locally integrable for $t \in I$.

(g) If $F(t) = U_t$ is a strongly continuous semigroup on a (*a priori* nonseparable) space $\mathcal{B}$, then this continuity and, hence, the local boundedness implies that

$$\| \lim_{t\to 0} \frac{1}{t}\int_0^t d\tau\, U_\tau f - f\| \leqslant \lim_{t\to 0} \sup_{0\leqslant s\leqslant t} \|U_\tau f - f\| = 0, \tag{1.24}$$

for any $f \in \mathcal{B}$.

Proposition 1.5. *Let* $\{U_t\}_{t\geqslant 0}$ *be a contraction semigroup on a Banach space* $\mathcal{B}$. *Then there exists a unique, closed, densely defined operator* A *with domain* dom A, *satisfying conditions* (i) *and* (ii) *of Proposition* 1.1, *such that*

$$U_t = U_t(A), \quad t \geqslant 0. \tag{1.25}$$

Proof. Set $A_{(t)} := t^{-1}(\mathbb{1} - U_t)$ and define

$$D(A) := \{u \in \mathcal{B} : \lim_{t\downarrow 0} A_{(t)}u \ \text{ exists}\}.$$

Then, for $u \in D(A)$, we can define the operator

$$Au := \lim_{t\downarrow 0} A_{(t)}u = -\partial_t\, U_t\big|_{t=+0}, \tag{1.26}$$

see (1.15), that is, $D(A) = \operatorname{dom} A$.

To verify that the closure $\overline{\operatorname{dom} A} = \mathcal{B}$, we follow Remark 1.4 and define for $t > 0$ and $u \in \mathcal{B}$ the Riemann integral

$$u^t := \int_0^t d\tau\, U_\tau u\,. \tag{1.27}$$

Then we obtain the representation

$$A_{(s)}u^t = \frac{1}{s}\int_0^s d\tau\, U_\tau u - \frac{1}{s}\int_t^{s+t} d\tau\, U_\tau u\,.$$

Since U_t is strongly continuous, (1.24) and (1.26) yield

$$\lim_{s\downarrow 0} A_{(s)}u^t = Au^t = u - U_t u, \quad u \in \mathcal{B}. \tag{1.28}$$

Therefore, $u^t \in \operatorname{dom} A$ for each $u \in \mathcal{B}$ and $t > 0$. Consequently, for any $s > 0$ the vector $t^{-1}u_t \in \operatorname{dom} A$. Since (1.24) yields $\lim_{s\downarrow 0} t^{-1}u_t = u$, the operator A is *densely* defined in $\mathcal{B}$.

To prove that A is *closed* we use (1.28) for $u \in \operatorname{dom} A$. Then by (1.27) one gets

$$\lim_{s\downarrow 0} A_{(s)}u^t = \int_0^t d\tau\, U_\tau A u = u - U_t u, \quad u \in \operatorname{dom} A. \tag{1.29}$$

Here we used that for $u \in \operatorname{dom} A$ the functions $\tau \mapsto U_\tau A_{(s)}$ converge uniformly on $[0, t]$ to the function $U_\tau Au$ when $s \downarrow 0$. Now we choose a sequence $\{u_n\}_{n\geqslant 1} \subset \operatorname{dom} A$, such that $u_n \to v$ and $Au_n \to w$. Then, by (1.29) for any $t > 0$,

$$\int_0^t d\tau\, U_\tau A u_n = u_n - U_t u_n\,,$$

and the uniform convergence for $n \to \infty$ of the functions $\tau \mapsto U_\tau A\, u_n$ on $[0,t]$ implies that

$$\int_0^t \mathrm{d}\tau\, U_\tau w = v - U_t v\,.$$

Therefore, one can take the limits:

$$\lim_{t\downarrow 0} \frac{1}{t} \int_0^t \mathrm{d}\tau\, U_\tau w = \lim_{t\downarrow 0} \frac{1}{t}(v - U_t v).$$

Since by (1.24) the limit in the left-hand side exists and is equal to w, we deduce the existence of the limit in the right-hand side. By (1.26), this implies $v \in \operatorname{dom} A$ and $w = Av$ and proves the *closedness* of A.

To check the uniqueness of A and the properties (i) and (ii) we first note that for any $\tau > 0$ one obviously has $A_{(\tau)}U_t = U_t A_{(\tau)}$. If $u \in \operatorname{dom} A$, the $\lim_{\tau\downarrow 0} U_t A_{(\tau)} u = U_t A u$ exists by (1.26). Then the limit $\lim_{\tau\downarrow 0} A_{(\tau)} U_t u$ also exists. This means that $U_t u \in \operatorname{dom} A$, i.e., $U_t : \operatorname{dom} A \to \operatorname{dom} A$, as well as that $AU_t u = U_t A u$. Since, by the definition of $A_{(\tau>0)}$,

$$\lim_{\tau\downarrow 0} U_t A_{(\tau)} u = -\lim_{\tau\downarrow 0} \frac{1}{\tau}(U_{t+\tau} u - U_t) = -\partial_t U_t u\ , \ u \in \operatorname{dom} A,$$

our arguments prove also the right-differentiability of $\{U_t u\}_{t\geqslant 0}$ for any $u \in \operatorname{dom} A$. The left derivative at $t > 0$ for $u \in \operatorname{dom} A$ is calculated by using representation (1.29) and the limit ($\tau < t$)

$$\partial_t U_t u = \lim_{\tau\downarrow 0} \frac{1}{\tau}(U_t u - U_{t-\tau} u) = -\lim_{\tau\downarrow 0} \frac{1}{\tau} \int_{t-\tau}^t \mathrm{d}\sigma\, U_\sigma A u = -U_t A u.$$

Therefore, if $u \in \operatorname{dom} A$, then one gets the equation

$$\partial_t U_t u = -AU_t u = -U_t A u. \tag{1.30}$$

Next we use (1.30) to verify (i) and (ii). To this end we analyse the Laplace transform $\hat{U}_z$ of the semigroup U_t:

$$\hat{U}_z = \lim_{T\to\infty} \int_0^T \mathrm{d}t\, \mathrm{e}^{-zt} U_t = \int_0^\infty \mathrm{d}t\, \mathrm{e}^{-zt} U_t\,. \tag{1.31}$$

The integral in the right-hand side of (1.31) is an improper strong Bochner integral, which is defined as the strongly convergent limit of the bounded operators $\{\hat{U}_{z,T} = \int_0^T \mathrm{d}t\, \mathrm{e}^{-zt} U_t\}_{T>0}$ for $z \in \mathbb{C}_+$. Here, the proper Bochner integrals are well-defined because the integrand is a strongly continuous function of t. Since $\|U_t\| \leqslant 1$, (1.31) defines a bounded linear operator with $\|\hat{U}_z\| \leqslant (\mathfrak{Re}\, z)^{-1}$. Now, let $u \in \operatorname{dom} A$. Then by virtue of (1.30),

$$\partial_t(\mathrm{e}^{-zt} U_t u) = -\mathrm{e}^{-zt} U_t (z\mathbb{1} + A)u, \quad \mathfrak{Re}\, z > 0.$$

Integration of this equality gives

$$u = \int_0^\infty \mathrm{d}t\, \mathrm{e}^{-zt}\, U_t(z\mathbb{1} + A)u,$$

or, with $v := (z\mathbb{1} + A)u$,

$$(z\mathbb{1} + A)^{-1}v = \int_0^\infty \mathrm{d}t\, \mathrm{e}^{-zt}\, U_t\, v, \quad \Re\mathrm{e}\, z > 0. \tag{1.32}$$

Since $\|\hat{U}_z\| \leqslant (\Re\mathrm{e}\, z)^{-1}$, the equality (1.32), means that

$$\hat{U}_z = (z\mathbb{1} + A)^{-1} = R_{-z}(A), \tag{1.33}$$

coincides with the resolvent of the operator A at the point $\zeta = -z$. Hence, the set $\Re\mathrm{e}\, \zeta < 0$ belongs to $\rho(A)$ (i) and we get the estimate (ii) of Proposition 1.1.

To prove *uniqueness*, let $U_t^{(1)}$ and $U_t^{(2)}$ be two different contraction semigroups, i.e., $U_{t_0}^{(1)} \neq U_{t_0}^{(2)}$ for some $t_0 > 0$. Then there exists a linear functional $\varphi \in \mathcal{B}^*$ and $v \in \mathcal{B}$ such that $\varphi(U_{t_0}^{(1)}v) \neq \varphi(U_{t_0}^{(2)}v)$, which, due to (1.32), implies that $\varphi((z\mathbb{1} + A_1)^{-1}v) \neq \varphi((z\mathbb{1} + A_2)^{-1}v)$ for some z with $\Re\mathrm{e}\, z > 0$. Thus $R_\zeta(A_1) \neq R_\zeta(A_2)$, so $A_1 \neq A_2$. Conversely, two different generators correspond, by (1.32), to different semigroups.

It remains to prove (1.25), i.e., that the semigroup $\{U_t\}_{t\geqslant 0}$ coincides with the one exhibited in Proposition 1.1 if we take as a generator the operator A just constructed above.

Since this operator A satisfies the conditions of Proposition 1.1, the contraction semigroup $U_t(A) = \mathrm{e}^{-tA}$ is well defined for $t \geqslant 0$. The Laplace transformation of this semigroup, calculated as in (1.31) and (1.32), gives

$$\int_0^\infty \mathrm{d}t\, \mathrm{e}^{-zt} U_t(A) = (z\mathbb{1} + A)^{-1}. \tag{1.34}$$

Therefore, (1.25) follows from the uniqueness we proved above. □

Since in this book we consider also other types of semigroup continuity (see Sections 1.4, 4.2) as well as *degenerate* semigroups (Chapters 5, 6), it is instructive to examine Definition 1.2 in more details and in a wider setting.

For example, although conditions (a) and (b) are important since they define the semigroup *functional equation* (1.22), the strong continuity (c) at $t = +0$ together with the topology of the *image* of the semigroup determine its analytic behaviour. A similar observation in Section 4.2 concerns the case of the *norm* and *trace-norm* continuity in $\mathbb{R}^+$. This demonstrates the important rôle playing by the condition at $t = 0$ for the notion of operator semigroup and for the entire theory, thus leading to a variety of semigroups, see Notes in Section 1.8.

To advance with analysis of behaviour at the point $t = 0$ we recall first the *uniform boundedness principle* for subsets of the set of bounded operators $\mathcal{L}(\mathcal{B})$ endowed with various topologies.

Proposition 1.6. *For a subset $\mathfrak{S} \subset \mathcal{L}(\mathcal{B})$ the following properties are equivalent:*

(a) *The subset $\mathfrak{S}$ is bounded in the strong operator topology, that is, the set $\{\|Bx\| : B \in \mathfrak{S}, x \in \mathcal{B}\}$ is bounded.*

(b) *The subset $\mathfrak{S}$ is bounded in the operator norm (uniform operator topology), i.e., the set $\{\|B\| : B \in \mathfrak{S}\}$ is bounded.*

Proposition 1.7. *Let $\{U_t\}_{t\geqslant 0}$ be a semigroup of operators on a Banach space $\mathcal{B}$, see Definition* 1.2 (a) *and* (b). *Then the following assertions are equivalent.*

(a) *The family $\{U_t\}_{t\geqslant 0}$ is strongly continuous.*

(b) *The family $\{U_t\}_{t\geqslant 0}$ is strongly right-continuous at $t = 0$.*

(c) *There exist $\delta > 0$, $M_\delta \geqslant 1$, and a dense $\mathcal{D} \subset \mathcal{B}$ such that*

  (1) $\|U_t\| \leqslant M_\delta$ *for* $0 \leqslant t \leqslant \delta$;

  (2) $\lim_{t\to +0} U_t x = x$ *for all* $x \in \mathcal{D}$.

Proof. Note that the implication (a)$\Rightarrow$(b) is trivial. Now, if (b) holds, then for any $t_0 > 0$ the sequence $\{\alpha_n(t_0) := \sup_{0\leqslant t\leqslant t_0/n} \|U_t\|\}_{n\in\mathbb{N}}$ is such that $\alpha_{n^*}(t_0) < \infty$ for some n^*. For otherwise there would exist $t_n \in [0, t_0/n]$ such that $\|U_{t_n}\| \geqslant n$. Therefore, by the uniform boundedness principle, Proposition 1.6, one gets for some $x \in \mathcal{B}$:

$$\limsup_{n\to\infty} \|U_{t_n} x\| = \infty\,,$$

which contradicts the strong right-continuity in (b) saying that $\lim_{n\to\infty} U_{t_n} x = x$ for any $x \in \mathcal{B}$. Since $\alpha_{n^*}(t_0)$ is bounded, the semigroup property yields on the interval $[0, t_0]$ the estimate

$$\|U_t\| \leqslant \sup_{0\leqslant t\leqslant t_0} \|U^n_{t/n}\| \leqslant \sup_{0\leqslant t\leqslant t_0} \|U_{t/n}\|^n \leqslant \alpha_{n^*}(t_0)^{n^*} =: \alpha(t_0). \tag{1.35}$$

Then for any $x \in \mathcal{B}$ and for any $t_0 > 0$ we obtain for $0 < \tau < t_0$

$$\begin{aligned} U_{t_0+\tau} x - U_{t_0} x &= U_{t_0}(U_\tau x - x)\,, \\ U_{t_0-\tau} x - U_{t_0} x &= U_{t_0-\tau}(x - U_\tau x)\,. \end{aligned}$$

These two identities together with (1.35) yield for $|h| < t_0$ the estimate

$$\|U_{t_0+h} x - U_{t_0} x\| \leqslant \alpha(t_0)\|(U_{|h|} x - x)\|\,, \quad x \in \mathcal{B},$$

which proves the strong continuity of the semigroup $\{U_t\}_{t\geqslant 0}$ for any $t_0 > 0$, and hence (b)$\Rightarrow$(a).

Since $\mathcal{D}$ is dense, the implication (a)$\Rightarrow$(c)-(2) is obvious. The implication (a)$\Rightarrow$ (c)-(1) follows by *reductio ad absurdum*. Indeed, suppose that for some sequence $\{\delta_n > 0\}_{n\in\mathbb{N}}$, such that $\lim_{n\to\infty} \delta_n = 0$, one has $\lim_{n\to\infty} \|U_{\delta_n}\| = \infty$. Then by the uniform boundedness principle there exists $x \in \mathcal{B}$ such that in turn one gets

$\lim_{n\to\infty} \|U_{\delta_n} x\| = \infty$, but this contradicts to the right-continuity at $t = 0$, which is a part of the properties in (a).

To prove that (c)⇒(b) we note that for $y \in \mathcal{B}$ and $x \in \mathcal{D}$ one gets

$$\|U_t y - y\| \leqslant \|U_t x - x\| + \|U_t\| \|y - x\| + \|y - x\| \,. \tag{1.36}$$

Since (c)-(2) yields $\lim_{t\to+0} \|U_t x - x\| = 0$ and (c)-(1) gives $\|U_t\| \leqslant M_\delta$ for $0 \leqslant t \leqslant \delta$, the denseness of $\mathcal{D}$ in $\mathcal{B}$ ensures that the right-hand side of (1.36) can be infinitesimally small in the limit $\lim_{t\to+0}$. This proves (c)⇒(b), and by the equivalence (a)⇔(b), also (c)⇒(a). □

Corollary 1.8. *By virtue of* (1.35) *one readily obtains the following estimate of* $\|U_t\|$ *for any* $t \geqslant 0$, *since the semigroup property yields*

$$\|U_t\| \leqslant (\alpha(t_0))^{[t/t_0]t_0+1} \leqslant (\alpha(t_0))^{t+1} \leqslant M e^{\omega t},$$

for some $M > 0$ *and* $\omega \geqslant 0$. *Here* $[t/t_0]$ *denotes the integer part of* $t/t_0 \geqslant 0$.

1.3 Generators and quasi-bounded semigroups

Propositions 1.1 and 1.5 show that the operator (*generator*) A plays a central rôle in the characterisation of a strongly continuous semigroup.

Definition 1.9. The infinitesimal *generator* A of a strongly continuous semigroup U_t on a Banach space $\mathcal{B}$ is defined by

$$Au = \lim_{t\downarrow 0} \frac{1}{t}(\mathbb{1} - U_t)u, \quad u \in \operatorname{dom} A, \tag{1.37}$$

where $\operatorname{dom} A$ is the set of all $u \in \mathcal{B}$ for which the limit (1.37) exists.

Since Definition 1.9 is (in a certain sense) a corollary of Proposition 1.5, below we collect explicitly some other fundamental properties of generators that follow from this proposition.

Corollary 1.10. *If* A *is the generator of a strongly continuous contraction semigroup* $\{U_t(A)\}_{t\geqslant 0}$ *on a Banach space* $\mathcal{B}$, *then* :

(a) *The dense subspace* $\operatorname{dom} A \subset \mathcal{B}$ *is invariant under* $U_t(A)$ *in the sense that* $U_t(A)(\operatorname{dom} A) \subseteq \operatorname{dom} A$ *for all* $t \geqslant 0$.

(b) *For all* $u \in \operatorname{dom} A$ *and all* $t > 0$ *one has* $AU_t u = U_t A u = -\partial_t U_t u$.

In fact, the proof of the Proposition 1.5 shows that the existence of the closed densely defined generator for a semigroup U_t is ensured by its strong continuity alone, i.e., it does not require that U_t is a contraction. Hence, in order that an operator A generates a strongly continuous semigroup, it is not necessary that the generator satisfies condition (ii) of Proposition 1.1.

Let U_t be a strongly continuous semigroup and $\tau > 0$. Since the vector-valued function $t \mapsto U_t f : [0,\tau] \to \mathcal{B}$ is continuous and $[0,\tau]$ is compact, the set $\{\|U_t f\| : 0 \leqslant t \leqslant \tau\}$ is bounded. Therefore, the uniform boundedness principle implies that there is an M such that $\|U_t\| \leqslant M$ for $t \in [0,\tau]$. Then (Corollary 1.8), by virtue of the semigroup property, one gets the estimate

$$\|U_t\| = \|(U_\tau)^n U_s\| \leqslant M e^{\omega t}, \tag{1.38}$$

for any $t = n\tau + s \in \mathbb{R}_0^+$, $n \in \{0\} \cup \mathbb{N}$, where $\mathbb{N} = \{1,2,\ldots\}$ and $\omega = \tau^{-1}\ln(M)$. Therefore, all strongly continuous semigroups are *exponentially* bounded.

Definition 1.11. Let ω_0 be the infimum of the numbers ω so that there is $M \geqslant 1$ satisfying the estimate (1.38). If $\omega_0 = 0$, we shall call U_t a *bounded semigroup.* If $\omega_0 \neq 0$, we shall call it *quasi-bounded* of *type* ω_0. We denote the corresponding class of generators by $Q(M,\omega_0)$.

Proposition 1.12. *Let A be a closed linear operator in a Banach space $\mathcal{B}$ with $\overline{\operatorname{dom} A} = \mathcal{B}$. Then, A is the generator of a strongly continuous quasi-bounded semigroup $U_t(A)$ of type ω_0, that is, $A \in Q(M,\omega_0)$, if and only if the following three conditions are satisfied:*

(i) *There is a real ω_0 such that $\{z \in \mathbb{C} : \Re z < -\omega_0\}$ belongs to the resolvent set $\rho(A)$.*

(ii) *There exists $M \geqslant 1$ such that $\|(z\mathbb{1} + A)^{-1}\| \leqslant M(\Re(z-\omega_0))^{-1}$, $z \in \mathbb{C}$.*

(iii) *One has*

$$\left\|(\lambda\mathbb{1} + A)^{-m}\right\| \leqslant \frac{M}{(\lambda-\omega_0)^m}\,, \tag{1.39}$$

for all $\lambda > \omega_0$ and $m \in \mathbb{N}$.

Proof. We denote $A_\zeta := A - \zeta\mathbb{1}$, $\zeta \in \mathbb{C}$. Then, (1.39) takes the form

$$\left\|(\xi\mathbb{1} + A_{-\omega_0})^{-m}\right\| \leqslant M\,\xi^{-m}, \quad \xi > 0,\ m \in \mathbb{N},$$

or, equivalently,

$$\left\|(\mathbb{1} + \alpha A_{-\omega_0})^{-m}\right\| \leqslant M, \quad \alpha \geqslant 0,\ m \in \mathbb{N}. \tag{1.40}$$

Inequality (1.40) implies that the operators $U_{t,n}(A_{-\omega_0})$ defined by (1.5) are uniformly bounded by M. Consequently, using (1.40) and the same reasoning as in the derivation of (1.9), one gets

$$\begin{aligned}
&\|U_{t,n}(A_{-\omega_0})f - U_{t,k}(A_{-\omega_0})f\| \\
&= \left\| \lim_{\varepsilon\downarrow 0}\int_\varepsilon^{t+\varepsilon} d\tau \left(\frac{\tau}{n} - \frac{t-\tau}{k}\right)\left(\mathbb{1} + \frac{t-\tau}{k}A_{-\omega_0}\right)^{-(k+1)} \frac{A^2_{-\omega_0}}{\left(\mathbb{1} + \frac{\tau}{n}A_{-\omega_0}\right)^{n+1}} f\right\| \\
&\leqslant \frac{t^2M^2}{2}\left(\frac{1}{n} + \frac{1}{k}\right)\left\|A^2_{-\omega_0}f\right\|,
\end{aligned} \tag{1.41}$$

for any $f \in \operatorname{dom} A^2$. Since $\overline{\operatorname{dom} A^2} = \mathcal{B}$ and $\|U_{t,n}(A_{-\omega_0})\| \leqslant M$, (1.41) implies the existence of the limit

$$U_t(A_{-\omega_0}) = \operatorname*{s-lim}_{n\to\infty} U_{t,n}(A_{-\omega_0}), \quad t \geqslant 0. \tag{1.42}$$

Moreover, since $U_{t,n}(A_{-\omega_0})$ is strongly continuous for $t \geqslant 0$, and since the estimate (1.41) is uniform in $t \in [0,T]$, the limit (1.42) is strongly continuous for $t \geqslant 0$, $U_{t=0}(A_{-\omega_0}) = \mathbb{1}$, and

$$\|U_t(A_{-\omega_0})\| \leqslant M. \tag{1.43}$$

Now, the arguments used to establish (1.13)–(1.23) carry over verbatim to verify that (1.42) defines a bounded semigroup

$$U_t(A_{-\omega_0}) = \mathrm{e}^{-tA_{-\omega_0}}, \quad t \geqslant 0. \tag{1.44}$$

Therefore, (1.43), (1.44) and the definition of A_ς imply the existence of the strongly continuous quasi-bounded semigroup

$$U_t(A) = \mathrm{e}^{t\omega_0} U_t(A_{-\omega_0}) = \mathrm{e}^{-tA}, \quad t \geqslant 0, \tag{1.45}$$

of type ω_0, with generator A and $\|U_t(A)\| \leqslant M\,\mathrm{e}^{t\omega_0}$.

To show the necessity of conditions (i), (ii) and (iii) we proceed as in the proof of Proposition 1.5. It follows from (1.45) that any quasi-bounded semigroup U_t of type ω_0 defines a strongly continuous bounded semigroup $\tilde{U}_t = \mathrm{e}^{-t\omega_0} U_t$. Then, we can define the family of operators $\tilde{A}_{(t)} = t^{-1}(\mathbb{1} - \tilde{U}_t)$, $t > 0$, and the limit operator $\tilde{A}u = \lim_{t\downarrow 0} \tilde{A}_{(t)} u$, $u \in \operatorname{dom} \tilde{A}$, see (1.37). Since the semigroup $\tilde{U}_t$ is bounded, its Laplace transform (1.31) exists for $\mathfrak{Re}\, z > 0$ and

$$\int_0^\infty \mathrm{d}t\, \mathrm{e}^{-zt} \tilde{U}_t f = (z\mathbb{1} + \tilde{A})^{-1} f, \quad f \in \mathcal{B}, \tag{1.46}$$

with $\|(z\mathbb{1} + \tilde{A})^{-1}\| \leqslant M(\mathfrak{Re}\, z)^{-1}$, or

$$\left\|(z\mathbb{1} + A)^{-1}\right\| \leqslant \frac{M}{\mathfrak{Re}(z - \omega_0)}, \quad \mathfrak{Re}(z - \omega_0) > 0, \tag{1.47}$$

which proves (i) and (ii). If $\mathfrak{Re}\, z = \lambda > \omega_0$, one gets from (1.46) that

$$\begin{aligned}\left\|(\lambda\mathbb{1} + A)^{-m} f\right\| &= \left\|\frac{(-1)^m}{m!} \partial_\lambda^m (A + \lambda\mathbb{1})^{-1} f\right\| \\ &\leqslant \frac{1}{m!} \|f\| \int_0^\infty \mathrm{d}t\, t^m\, \mathrm{e}^{-\lambda t}\, M\, \mathrm{e}^{\omega_0 t} = \frac{M\|f\|}{(\lambda - \omega_0)^m},\end{aligned}$$

which together with (1.47) proves (iii). □

We conclude this section by a result that is partially motivated by Proposition 1.7 (c). It is useful if one needs to determine a *core* for generator A. Recall that in general a set $\mathcal{D} := \operatorname{core} A \subset \operatorname{dom} A$ is called a core for operator A if the closure of the restriction of A to this set restores A, or explicitly if $\overline{(A \restriction \mathcal{D})} = A$. It is clear that $\operatorname{core} A$ is not unique.

Proposition 1.13. *Let $\mathcal{D} \subset \operatorname{dom} A$ be dense in $\mathcal{B}$ and invariant under the semigroup $\{U_t(A)\}_{t\geqslant 0}$. Then $\mathcal{D}$ is a core for the generator A.*

Proof. First we note that since the generator A of a strongly continuous semigroup is a closed operator, the space $\operatorname{dom} A$ is complete with respect to the *graph norm*

$$\|f\|_A := \|f\| + \|Af\|, \quad f \in \operatorname{dom} A. \tag{1.48}$$

Let $\overline{\mathcal{D}}$ denote the closure of $\mathcal{D}$ in the set $\operatorname{dom} A$ with respect to the norm (1.48). Let $f \in \operatorname{dom} A$. Then since $\mathcal{D}$ is dense in $\mathcal{B}$, there exists a sequence $\{f_n\}_{n\geqslant 1} \subset \mathcal{D}$ such that $\lim_{n\to\infty} \|f - f_n\| = 0$. Recall that the strongly continuous semigroup $\{U_t(A)\}_{t\geqslant 0}$ is continuous in topology (1.48). Therefore,

$$f_n^t := \int_0^t \mathrm{d}\tau \, U_\tau(A) f_n \in \overline{\mathcal{D}}.$$

Since $Af_n^t = f_n - U_t(A)f_n$, definition (1.48) and the boundedness property $\|U_t(A)\| \leqslant M\mathrm{e}^{\omega t}$ yield

$$\begin{aligned}\lim_{n\to\infty} \|f_n^t - f^t\|_A = {} & \lim_{n\to\infty} \left\| \int_0^t \mathrm{d}\tau \, U_\tau(A)(f - f_n) \right\| \\ & + \lim_{n\to\infty} \|f_n - U_t(A)f_n - f + U_t(A)f\| = 0,\end{aligned}$$

for any $t > 0$ and hence, $f^t \in \overline{\mathcal{D}}$. Here we used that $Af^t = f - U_t(A)f$. Now using this relation again we obtain that

$$\lim_{t\to 0} \left\| \frac{1}{t} f^t - f \right\|_A = \lim_{t\to 0} \left\| \frac{1}{t} f^t - f \right\| + \lim_{t\to 0} \left\| \frac{1}{t}(f - U_t(A)f) - Af \right\| = 0.$$

So, $f \in \overline{\mathcal{D}}$, and therefore $\operatorname{dom} A = \overline{\mathcal{D}}$. □

1.4 Norm and other continuity conditions

Here, we consider another continuity condition for semigroups. Suppose that instead of the strong continuity in Definition 1.2 (c), one has a semigroup $U_t(A)$ which is *norm* continuous in the sense that

$$(c)' \quad \lim_{t\downarrow 0} \|U_t(A) - \mathbb{1}\| = 0. \tag{1.49}$$

Since (1.49) implies (c) and consequently the quasi-boundedness (1.38) of $U_t(A)$, with generator A defined by (1.37), we also get

$$\lim_{\tau\to 0} \|U_{t+\tau}(A) - U_t(A)\| \leqslant \lim_{\tau\to 0} M\, \mathrm{e}^{\omega(t+|\tau|)} \left\|U_{|\tau|}(A) - \mathbb{1}\right\| = 0, \tag{1.50}$$

that is, the uniform continuity of the semigroup $U_t(A)$ at any $t > 0$.

If A is a bounded generator, then the semigroup $U_t(A)$ is well-defined by the $\|\cdot\|$-convergent series (1.1) for any $t \in \mathbb{C}$. It is obviously $\|\cdot\|$-continuous in the sense of (1.49) and (1.50).

Proposition 1.14. *A semigroup $\{U_t(A)\}_{t\geqslant 0}$ is norm continuous for $t \geqslant 0$ if and only if its generator $A \in \mathcal{L}(\mathcal{B})$.*

Proof. By the remark above and Propositions 1.1 and 1.2, we only have to prove that (1.49) (c)′ implies $A \in \mathcal{L}(\mathcal{B})$. Since

$$\left\|\int_0^\tau \mathrm{d}s\, (U_s(A) - \mathbb{1})\right\| \leqslant \tau \sup_{0\leqslant s\leqslant \tau} \|U_s(A) - \mathbb{1}\|\,, \tag{1.51}$$

taking $\tau > 0$ small enough, $\tau < \tau_0$, see (c)′, one gets

$$\left\|\tau^{-1}\int_0^\tau \mathrm{d}s\, U_s(A) - \mathbb{1}\right\| < 1.$$

Therefore, the operator $L_\tau := \int_0^\tau \mathrm{d}s\, U_s(A)$, $0 < \tau < \tau_0$, is invertible. Since

$$\begin{aligned}(U_t(A) - \mathbb{1})L_\tau &= \int_t^{t+\tau} \mathrm{d}s\, U_s(A) - \int_0^\tau \mathrm{d}s\, U_s(A)\\ &= \int_\tau^{t+\tau} \mathrm{d}s\, U_s(A) - \int_0^t \mathrm{d}s\, U_s(A)\\ &= (U_\tau(A) - \mathbb{1})\int_0^t \mathrm{d}s\, U_s(A)\end{aligned}$$

for $t \geqslant 0$, and, since all these operators commute, we obtain

$$U_t(A) - \mathbb{1} = -\int_0^t \mathrm{d}s\, U_s(A) K_\tau\,, \tag{1.52}$$

with $K_\tau := -(U_\tau(A) - \mathbb{1})L_\tau^{-1}$, a bounded operator. Relation (1.52) implies that the operator-valued function $t \mapsto U_t(A)$ is norm differentiable for $t > 0$, see (1.51),

$$\begin{aligned}&\lim_{\delta\to 0}\left\|\frac{1}{\delta}\left(U_{t+\delta}(A) - U_t(A)\right) + U_t(A)K_\tau\right\|\\ &= \lim_{\delta\to 0}\left\|\frac{1}{\delta}U_t(A)\int_0^\delta \mathrm{d}s\, (U_s(A) - \mathbb{1})K_\tau\right\|\\ &\leqslant \|U_t(A)\|\cdot\|K_\tau\|\left(\lim_{\delta\to 0}\sup_{0\leqslant s\leqslant|\delta|}\|U_s(A) - \mathbb{1}\|\right) = 0,\end{aligned}$$

with the operator-norm derivative

$$\|\cdot\|\text{-}\partial_t U_t(A) = -U_t(A)K_\tau. \tag{1.53}$$

The bounded operator K_τ generates a $\|\cdot\|$-continuous semigroup $\{U_t(K_\tau) = \mathrm{e}^{-tK_\tau}\}_{t\geqslant 0}$ defined by the corresponding $\|\cdot\|$-convergent series. The function $t \mapsto U_t(K_\tau)$ is in fact $\|\cdot\|$-holomorphic in the complex plane $\mathbb{C}$. Then, by (1.53), we get for $0 < s < t$ that

$$\begin{aligned}&\|\cdot\|\text{-}\partial_s \left(U_s(A)\mathrm{e}^{-(t-s)K_\tau}\right)\\ &= -U_s(A)K_\tau\,\mathrm{e}^{-(t-s)K_\tau} + U_s(A)K_\tau\,\mathrm{e}^{-(t-s)K_\tau} = 0.\end{aligned}$$

Hence, $U_t(A) = \mathrm{e}^{-tK_\tau}$, with $K_\tau \in \mathcal{L}(\mathcal{B})$. On the other hand, by (1.51), (1.52), and Definition 1.9, the infinitesimal generator A of the semigroup $U_t(A)$ can be calculated as the limit

$$Au = \lim_{t\downarrow 0} \frac{1}{t}\left(\mathbb{1} - U_t(A)\right)u = \left(\lim_{t\downarrow 0}\frac{1}{t}\int_0^t \mathrm{d}s\, U_s(A)\right) K_\tau u = K_\tau u,$$

for any $u \in \operatorname{dom} K_\tau = \mathcal{B}$. So, we conclude that $U_t(A) = \mathrm{e}^{-tA}$ with $A \in \mathcal{L}(\mathcal{B})$. □

It is relevant to mention here a special class of semigroups, which is of particular importance in the context of this chapter.

It is known (Corollary 1.10) that for any strongly continuous semigroup $U_t(A)$ on a Banach space $\mathcal{B}$ the subspace $\operatorname{dom} A$ is invariant under $U_t(A)$ in the sense that $U_t(A)$ maps $\operatorname{dom} A$ into $\operatorname{dom} A$, for all $t \geqslant 0$. But now we assume a bit *more*, namely that $U_t(A)$ maps $\mathcal{B}$ into $\operatorname{dom} A$, for all $t > 0$.

Proposition 1.15. *Let $\{U_t(A)\}_{t\geqslant 0}$ be a strongly continuous semigroup on $\mathcal{B}$ such that $U_t(A)\,\mathcal{B} \subseteq \operatorname{dom} A$ for $t > 0$. Then $\mathbb{R}^+ \ni t \mapsto U_t(A)$ is an infinitely $\|\cdot\|$-differentiable operator-valued function.*

Proof. Let $\delta > 0$. By the assumption of the proposition, the operator $A\,U_s(A)$ is closed with $\operatorname{dom}(A\,U_\delta(A)) = \mathcal{B}$, so that $\|A\,U_\delta(A)\| \leqslant C_\delta$. By (1.14),

$$U_t(A)f - f = -\int_0^t \mathrm{d}\tau\, U_\tau(A)\, Af,$$

for all $f \in \operatorname{dom} A$ and $t > 0$, so

$$\left(U_{t+\delta}(A) - U_\delta(A)\right)u = -\int_0^t \mathrm{d}\tau\, U_\tau(A)\, A\,U_\delta(A)\, u, \tag{1.54}$$

for $u \in \mathcal{B}$. Since every strongly continuous semigroup is exponentially bounded (see (1.38)), we have $\|U_t(A)\| \leqslant M\,\mathrm{e}^{\omega t}$, for some $M > 0$ and $\omega \geqslant 0$, for all $t \geqslant 0$. Then, (1.54) yields the estimate

$$\|U_{t+\delta}(A) - U_\delta(A)\| \leqslant C_\delta \int_0^t \mathrm{d}\tau\, M\,\mathrm{e}^{\omega\tau}. \tag{1.55}$$

Thus, the semigroup $\{U_t(A)\}_{t\geqslant 0}$ is $\|\cdot\|$-continuous at $t=\delta$ for all $\delta>0$. Moreover, by (1.54), (1.55) and by the closedness of the generator A, we get

$$\begin{aligned}&\left\|\frac{1}{t}\left(U_{t+\delta}(A)-U_\delta(A)\right)+A\,U_\delta(A)\right\|\\&=\left\|\frac{1}{t}\int_0^t \mathrm{d}\tau\ \left(U_{\tau+\delta/2}(A)-U_{\delta/2}(A)\right)A\,U_{\delta/2}(A)\right\|\\&\leqslant (C_{\delta/2})^2\int_0^t \mathrm{d}s\,M\,\mathrm{e}^{\omega s},\end{aligned}$$

which means that $U_t(A)$ is $\|\cdot\|$-differentiable with $\|\cdot\|$-$\partial_t U_t(A)=A\,U_t(A)$ for all $t>0$. For $0<\delta<t$, we can rewrite this as

$$\|\cdot\|\text{-}\,\partial_t U_t(A)=U_{t-\delta}(A)(-AU_\delta(A)), \tag{1.56}$$

and hence differentiate (1.56) repeatedly in the operator-norm topology. □

Having at our disposal semigroups generated by bounded operators we can provide another construction of strongly continuous semigroups, different from the one in Sections 1.1 and 1.2. The idea is to approximate them by sequences of the norm-continuous semigroups. Since the resolvent of the generator of a strongly continuous semigroup plays a key rôle in the theory, see Proposition 1.1 and (1.34), we first recall the following general convergence property.

Proposition 1.16. *Let A be a closed linear operator with dense domain* $\operatorname{dom} A\subseteq\mathcal{B}$. *If there exist a real ω and an $M>0$ such that $(-\infty,-\omega)\subset\rho(A)$ and $\|\lambda(A+\lambda\mathbb{1})^{-1}\|\leqslant M$ for all $\lambda\geqslant\omega$, then*

(i) $\text{s-lim}_{\lambda\to\infty}\lambda R_{-\lambda}(A)=\mathbb{1}$.

(ii) $\lim_{\lambda\to\infty}\lambda AR_{-\lambda}(A)u=\lim_{\lambda\to\infty}\lambda R_{-\lambda}(A)Au=Au,\quad u\in\operatorname{dom}A.$

Proof. If $u\in\operatorname{dom}A$, then $\lambda R_{-\lambda}(A)u=-R_{-\lambda}(A)Au+u$. The assumptions of the proposition imply that $\|R_{-\lambda}(A)Au\|\leqslant\lambda^{-1}M\|Au\|$, and we get (i) for any $u\in\operatorname{dom}A$. This result can be extended to all $u\in\mathcal{B}$, since by the same assumptions, the operators $\lambda R_{-\lambda}(A)$ are uniformly bounded for all $\lambda\geqslant\omega$ and $\operatorname{dom}A$ is dense in $\mathcal{B}$. Assertion (ii) is then a straightforward consequence of (i). □

Assertion (ii) suggests which bounded operator should be chosen to approximate the unbounded generator A. Since in Section 1.3 we showed that the results for the case of quasi-bounded semigroups can be deduced from those for contraction semigroups, the proof below is given only for the latter.

Proposition 1.17. *Under the assumptions of Proposition* 1.1, *that is, A is a closed linear operator with dense domain* $\operatorname{dom}A\subseteq\mathcal{B}$ *such that $(-\infty,0)\subset\rho(A)$ and $\|\lambda R_{-\lambda}(A)\|\leqslant 1$ for $\lambda>0$, the operator A generates a strongly continuous contraction semigroup $\{U_t(A)\}_{t\geqslant 0}$.*

Proof. With Proposition 1.16 in mind, we define a sequence of bounded operators

$$A_n := nAR_{-n}(A) = n\mathbb{1} - n^2 R_{-n}(A), \quad n \in \mathbb{N}, \tag{1.57}$$

which, by assertion (ii) therein, converges to A on $\operatorname{dom} A$. Consider then the corresponding sequence of norm-continuous semigroups

$$U_t(A_n) := \mathrm{e}^{-tA_n}, \quad t \geqslant 0,\ n \in \mathbb{N}. \tag{1.58}$$

By (1.57), each $\{U_t(A_n)\}_{t \geqslant 0}$ is a contraction semigroup:

$$\|U_t(A_n)\| \leqslant \mathrm{e}^{-tn} \mathrm{e}^{t\|n^2 R_{-n}(A)\|} \leqslant 1.$$

Therefore, by the uniform boundedness principle (Proposition 1.6), it is sufficient to prove the convergence of this approximating sequence of semigroups on the dense set $\operatorname{dom} A$.

For this purpose, we introduce the operator-valued functions

$$s \mapsto U_{t-s}(A_m)U_s(A_n)u, \quad u \in \operatorname{dom} A, \tag{1.59}$$

for $0 \leqslant s \leqslant t$ and $m, n \in \mathbb{N}$. The elements of the semigroups $\{U_t(A_n)\}_{t \geqslant 0}$ commute for all $n \in \mathbb{N}$ and are norm-continuous. Then, by virtue of (1.59), one gets

$$\begin{aligned} U_t(A_n)u - U_t(A_m)u &= \int_0^t \mathrm{d}s\, \partial_s \left(U_{t-s}(A_m)U_s(A_n)u\right) \\ &= \int_0^t \mathrm{d}s\, U_{t-s}(A_m)U_s(A_n)\left(A_m - A_n\right)u, \end{aligned}$$

which yields the estimate

$$\|U_t(A_n)u - U_t(A_m)u\| \leqslant t\left\|\left(A_n - A_m\right)u\right\|, \tag{1.60}$$

for any $u \in \mathcal{B}$. Since, by Proposition 1.16 (ii), $\{A_n u\}_{n \geqslant 1}$ is a Cauchy sequence for each $u \in \operatorname{dom} A$, estimate (1.60) implies that $\{U_t(A_n)u\}$ converges uniformly on each interval $[0, T]$. Hence, by the uniform boundedness of the sequence $\{U_t(A_n)\}_{n \geqslant 1}$, the limit

$$U_t u := \lim_{n \to \infty} U_t(A_n)u \tag{1.61}$$

exists for each $u \in \mathcal{B}$ and for each $t \geqslant 0$.

By virtue of (1.61), the limit $\{U_t\}_{t \geqslant 0}$ inherits the semigroup property of $\{U_t(A_n)\}$. Hence, it is a contraction semigroup, i.e., $\|U_t\| \leqslant 1$. Thanks to the estimate

$$\left\|\left(U_t - \mathbb{1}\right)u\right\| \leqslant \left\|\left(U_t - U_t(A_n)\right)u\right\| + \left\|\left(U_t(A_n) - \mathbb{1}\right)u\right\|,$$

and the uniform convergence of $\{U_t(A_n)u\}$ on each interval $[0, T]$ for any $u \in \operatorname{dom} A$, we get that

$$\lim_{t \downarrow 0} U_t u = u, \quad u \in \operatorname{dom} A. \tag{1.62}$$

The uniform boundedness of $\|U_t\| \leqslant 1$ allows us to extend (1.62) to $\mathcal{B}$ and conclude that the semigroup $\{U_t\}_{t\geqslant 0}$ is strongly continuous.

It remains to prove that the operator A generates this semigroup. Let B be the generator of $\{U_t\}_{t\geqslant 0}$. Fix $u \in \operatorname{dom} A$. Since the sequence $\{U_t(A_n)u\}_{n\geqslant 1}$ converges uniformly on each interval $[0, T]$, we get

$$\begin{aligned} u - U_t u &= u - \lim_{n\to\infty} U_t(A_n)u \\ &= \lim_{n\to\infty} \int_0^t \mathrm{d}s\, U_s(A_n)A_n u = \int_0^t \mathrm{d}s\, U_s A u, \end{aligned} \tag{1.63}$$

by virtue of (1.60), (1.61) and Proposition 1.16(ii). This implies that

$$Au = Bu, \quad u \in \operatorname{dom} A, \tag{1.64}$$

or $\operatorname{dom} A \subseteq \operatorname{dom} B$. Since $\{U_t\}_{t\geqslant 0}$ is a strongly continuous contraction semigroup, its Laplace transform (1.34) exists for the set $\{z : \Re\mathfrak{e}\, z > 0\}$. Hence, $(-\infty, 0) \subset \rho(B)$. Moreover, by (1.64), we get $R_\zeta(A) = R_\zeta(B)$ for any $\zeta \in (-\infty, 0)$, which implies $A = B$. □

We have already observed that inspecting the properties of semigroup one has first to focus on its behaviour at $t = 0$. After the strong and operator-norm continuity hypothesis in Definition 1.2(c) it is natural to discuss semigroups that are continuous at $t = +0$ in the *weak* operator topology. By observations in Proposition 1.7 and by (1.49) the continuity at $t = +0$ is decisive for the analysis of semigroups. So, we consider, instead of (c), or stronger (c)′, a weak continuity condition (c)″, see Definition 1.20. For this purpose we need some preliminaries.

Recall that on a Banach space $\mathcal{B}$ one can define a space of bounded linear functionals $\mathcal{L}(\mathcal{B}, \mathbb{C}) = \{\phi : \mathcal{B} \to \mathbb{C}\}$, where $\phi : u \mapsto \langle u|\phi\rangle$, $u \in \mathcal{B}$, and the norm $\|\phi\|^* := \sup\{|\langle u|\phi\rangle| : \|u\| \leqslant 1, u \in \mathcal{B}\}$ is finite. Then $\mathcal{B}^* := \mathcal{L}(\mathcal{B}, \mathbb{C})$ is the *dual* space of $\mathcal{B}$. The space $\mathcal{B}^*$ itself is Banach, with the norm $\|\cdot\|^*$.

Definition 1.18. The dual space $\mathcal{B}^*$ of $\mathcal{B}$ allows to define on $\mathcal{B}$ the *weak* topology, $\sigma(\mathcal{B}, \mathcal{B}^*)$, as the weakest topology on $\mathcal{B}$ in which all the functionals $\phi \in \mathcal{B}^*$ are continuous. This topology is generated on $\mathcal{B}$ by a collection of neighborhoods: $\{u \in \mathcal{B} : |\langle u|\phi\rangle| \leqslant \varepsilon,\ \phi \in \mathcal{B}^*,\ \varepsilon > 0\}$, of the vector $u = 0$. Then the equality

$$\lim_{n\to\infty} \langle u_n|\phi\rangle = \langle u|\phi\rangle, \quad \text{for all } \phi \in \mathcal{B}^*,$$

means that the sequence $\{u_n \in \mathcal{B}\}_{n\geqslant 1}$ converges *weakly* to the vector $u \in \mathcal{B}$, and we write $\text{w-}\lim_{n\to\infty} u_n = u$.

The *weak operator* topology on $\mathcal{L}(\mathcal{B})$ is the weakest topology such that the maps $\Phi_{u,\phi} : \mathcal{L}(\mathcal{B}) \to \mathbb{C}$ given by $\Phi_{u,\phi}(B) := \langle Bu|\phi\rangle$ are all continuous for all $u \in \mathcal{B}$, $\phi \in \mathcal{B}^*$. Then the convergence $\text{w-}\lim_{n\to\infty} B_n = B$ of the sequence $\{B_n\}_{n\geqslant 1}$ in the weak operator topology means that

$$\lim_{n\to\infty} \langle B_n u|\phi\rangle = \langle Bu|\phi\rangle,$$

for each $u \in \mathcal{B}$ and $\phi \in \mathcal{B}^*$.

Remark 1.19. (1) Although the topology $\sigma(\mathcal{B}, \mathcal{B}^*)$ is evidently weaker than the vector-norm topology on $\mathcal{B}$, the uniform boundedness principle is still valid for the weak topology on $\mathcal{B}$.

(2) One should not confuse the *weak operator* topology on $\mathcal{L}(\mathcal{B})$ with the *weak Banach space* topology $\sigma(\mathcal{L}(\mathcal{B}), \mathcal{L}(\mathcal{B})^*)$ on $\mathcal{L}(\mathcal{B})$ generated by the dual of $\mathcal{L}(\mathcal{B})$ space $\mathcal{L}(\mathcal{B})^*$.

(3) The weak Banach space and the operator-norm topologies are stronger than the weak operator topology on $\mathcal{L}(\mathcal{B})$, but the uniform boundedness principle is valid in this topology on $\mathcal{L}(\mathcal{B})$. So, we can add in Proposition 1.6 to the equivalent properties (a), (b), also:

(c) The subset $\mathfrak{S}$ is bounded in the weak operator topology, that is, the set $\{|\langle Bx \mid \phi\rangle| : B \in \mathfrak{S}, x \in \mathcal{B}, \phi \in \mathcal{B}^*\}$ is bounded.

Definition 1.20. A semigroup $\{U_t\}_{t\geqslant 0}$ on $\mathcal{B}$ is *weakly continuous* (or continuous in the weak operator topology) if in Definition 1.2 one substitutes (c) by the condition (cf. Definition 1.18):

(c)″ $t \mapsto U_t f \in \mathcal{B}$ is a *weakly* continuous vector-valued function of the parameter $t \in \mathbb{R}^+$ for every $f \in \mathcal{B}$ and weakly right-continuous at $t = 0$, that is,

$$\text{w-}\lim_{t\downarrow 0} U_t u = \lim_{t\downarrow 0} \langle U_t u|\phi\rangle = \langle u|\phi\rangle, \tag{1.65}$$

for each $u \in \mathcal{B}$ and $\phi \in \mathcal{B}^*$.

Proposition 1.21. *Let $\{U_t\}_{t\geqslant 0}$ be a semigroup on a Banach space $\mathcal{B}$, see Definition* 1.2 (a) *and* (b). *Then the following assertions are equivalent:*

(a) *The semigroup $\{U_t\}_{t\geqslant 0}$ is strongly continuous, Definition* 1.2(c).

(b) *The semigroup $\{U_t\}_{t\geqslant 0}$ is strongly right-continuous at $t = 0$.*

(a-w) *The semigroup $\{U_t\}_{t\geqslant 0}$ is weakly continuous, Definition* 1.20.

(b-w) *The semigroup $\{U_t\}_{t\geqslant 0}$ is weakly right-continuous at $t = 0$.*

The main impact of this surprising result is that it rules out possible non-trivial generalisations of the notion of semigroup by highlighting the relevance of the *strong* continuity at $t = +0$. The proof in a Banach space $\mathcal{B}$ needs a number of facts and tools, which are out of the scope of this book. Thus, the reader has to consult the Notes in Section 1.8 for the references.

Since the main objects studied in this book are the Gibbs semigroups, which are strongly continuous on a complex separable Hilbert space $\mathcal{H}$, we provide the proof of Proposition 1.21 only on such a space $\mathcal{H}$. Note that the *self-duality* of $\mathcal{H}$, which implies that the functionals, $\phi \in \mathcal{H}$, and gives their explicit form, (they are scalar products: $\langle u|\phi\rangle = (u, \phi)$), makes the proof simpler.

Proposition 1.22. *Let $\{U_t\}_{t\geqslant 0}$ be a contraction semigroup on a Hilbert space $\mathcal{H}$. Then the following assertions are equivalent:*

(a) *The semigroup $\{U_t\}_{t\geqslant 0}$ is strongly continuous, Definition* 1.2(c).

(b) *The semigroup $\{U_t\}_{t\geqslant 0}$ is strongly right-continuous at $t = 0$.*

(a-w) *The semigroup $\{U_t\}_{t\geqslant 0}$ is weakly continuous, Definition* 1.20.

(b-w) *The semigroup $\{U_t\}_{t\geqslant 0}$ is weakly right-continuous at $t = 0$.*

Proof. The equivalence: (a)⇔(b), follows from Proposition 1.6 since the strong right limit (b) implies operator-norm boundedness of semigroup in the vicinity of $t = 0$. By virtue of Remark 1.19(3) about the part (c) of the weak uniform boundedness principle, the proof of the equivalence (a-w) and (b-w) (1.65) follows the same line of reasoning as for (a)⇔(b). Since (b)⇒(b-w) is evident, it is sufficient to prove that (b-w)⇒(b). To this aim we note that since $\{U_t\}_{t\geqslant 0}$ is a contraction,

$$\begin{aligned}\lim_{t\to+0}\|U_t u - u\|^2 &= \lim_{t\to+0}\{\|U_t u\|^2 + \|u\|^2 - 2\,\mathfrak{Re}(U_t u, u)\}\\ &\leqslant \lim_{t\to+0}\{2\|u\|^2 - 2\,\mathfrak{Re}(U_t u, u)\} = 0,\end{aligned}$$

for any $u \in \mathcal{H}$, so $\{U_t\}_{t\geqslant 0}$ is strongly right-continuous at $t = 0$. □

Corollary 1.23. *If $\{U_t\}_{t\geqslant 0}$ is strongly continuous, then the adjoint semigroup $\{U_t^*\}_{t\geqslant 0}$ is also strongly continuous. To see this, note that by the definition $(U_t u, v) = (u, U_t^* v)$, $u, v \in \mathcal{H}$, the strong continuity of U_t yields the weak continuity of U_t^*, and hence its strong continuity.*

It is known that on a Banach space the dual semigroup U_t^* is *not* necessarily strongly continuous for a strongly continuous semigroup U_t. But it is always *weak**-continuous (in the topology $\sigma(\mathcal{L}(\mathcal{B}), \mathcal{L}(\mathcal{B})_*)$), which makes it still interesting to study, see comments to Section 1.4 in Notes in Section 1.8.

We conclude this section by describing a situation of failure of C_0-continuity, which is of a different nature than the modification of *topology* of continuity with respect to the parameter $t \in \mathbb{R}_0^+$. For that we modify (relax) the conditions (a) and (c) in Definition 1.2.

Definition 1.24. An operator-valued function of $t \in \mathbb{R}^+$, $t \mapsto U_t \in \mathcal{L}(\mathcal{B})$, is called a *degenerate* one-parameter semigroup on $\mathcal{B}$ if

(a′) $U_0 \neq \mathbb{1}$,

(b′) $U_t U_s = U_{t+s}$ for $t, s \in \mathbb{R}^+$,

(c′) $t \mapsto U_t f \in \mathcal{B}$ is a continuous function of the parameter $t \in \mathbb{R}^+$ for every fixed $f \in \mathcal{B}$.

Recall that for a degenerate semigroup $\{U_t\}_{t>0}$ the subspace

$$\mathcal{B}_0 := \{f \in \mathcal{B} : \lim_{t\downarrow 0} U_t f = f\} \subseteq \mathcal{B}\ , \tag{1.66}$$

is known as the *space of strong continuity*. Therefore, if $\mathcal{B}_0 = \mathcal{B}$, then the degenerate semigroup is a C_0-semigroup. Otherwise, the operator $P := \lim_{t\downarrow 0} U_t \neq \mathbb{1}$ violates condition (a) in Definition 1.2. By virtue of the semigroup property (b′) and by the continuity (c′), the operator P is a projection, which is orthogonal in the case of *self-adjoint* semigroups.

We comment here that the limits of the Trotter-Kato product formulae (away from zero) are often degenerate semigroups, see Chapter 5 and Chapter 6.

Proposition 1.25. *For a degenerate semigroup* $\{U_t\}_{t>0}$ *on* $\mathcal{H}$ *the following holds true:*

(i) $\{U_t\}_{t>0}$ *is exponentially bounded.*

(ii) $\ker(P)$ *and* $\operatorname{ran}(P)$ *are closed,* U_t*-invariant, and* $\mathcal{H} = \ker(P) \oplus \operatorname{ran}(P)$*, for self-adjoint semigroups.*

(iii) U_t *has the representation* $U_t = S_t\, P = P\, S_t$*, where* $\{S_t\}_{t\geqslant 0}$ *is a* C_0*-semigroup on* $\mathcal{H}$.

In Sections 5.4-5.6 and Sections 6.4-6.6 one encounters degenerate semigroups in the following context.

Let $P_0 : \mathcal{H} \to \mathcal{H}_0$ be an orthogonal projection from a Hilbert space $\mathcal{H}$ onto $\mathcal{H}_0$ and $\{e^{-tH}\}_{t\geqslant 0}$ be a C_0-semigroup on $\mathcal{H}_0$ with domain of generator $\operatorname{dom} H \subset \mathcal{H}_0$. Then by Definition 1.24, the operator family $\{W_t := e^{-tH} P_0\}_{t>0}$ is obviously a degenerate semigroup on $\mathcal{H}$ with the subspace $\mathcal{H}_0$ as its space of strong continuity (1.66).

1.5 Holomorphic semigroups

One can ask whether, instead of the construction of the strongly continuous semigroup $\{U_t(A) = \mathrm{e}^{-tA}\}_{t\geqslant 0}$ described in Sections 1.1–1.4, the functional calculus in the form of the *Riesz-Dunford* integral

$$U_t(A) = \frac{1}{2\pi i}\int_\Gamma \mathrm{d}z\, \mathrm{e}^{-tz}(z\mathbb{1} - A)^{-1}, \quad \Gamma \subset \rho(A)\ , \tag{1.67}$$

is appropriate for this purpose?

This formula is obviously valid if $A \in \mathcal{L}(\mathcal{B})$ and if Γ is a positively-oriented contour in the resolvent set $\rho(A)$, which encloses the spectrum $\sigma(A) := \mathbb{C}\backslash\rho(A)$ of the operator A in its interior. In fact, (1.67) is a way to construct semigroups if one assumes slightly more about generator A than in Section 1.1 or in Section 1.3.

Definition 1.26. Let $\theta \in (0, \pi/2]$ and denote by S_θ the open sector $S_\theta = \{z \in \mathbb{C}_+ : |\arg z| < \theta\}$. A strongly continuous semigroup $\{U_t\}_{t\geqslant 0}$ on a Banach space $\mathcal{B}$ is called a *bounded holomorphic semigroup of semi-angle* θ if:

(i) $\{U_t\}_{t>0}$ is the restriction to $\mathbb{R}^+$ of an analytic family of bounded operators $\{U_z\}_{z\in S_\theta}$, which obeys the semigroup property (functional equation): $U_{z+z'} = U_z U_{z'}$, for $z, z' \in S_\theta$.

(ii) For $\theta' < \theta$ and for $z \in \overline{S}_{\theta'}$, one has $\|U_z\| \leqslant M'$ and $\operatorname{s-lim}_{z\to 0} U_z = \mathbb{1}$.

If A is the generator of the semigroup $\{U_t\}_{t\geqslant 0}$, we write $U_z(A) = \mathrm{e}^{-zA}$.

We recall that for holomorphic families of bounded operators on $\mathcal{L}(\mathcal{H})$ there is no distinction between *uniform*, *strong*, or *weak* operator analyticity.

From Definition 1.26, one can deduce properties of the generator A. By (i) and (ii), for each $\varphi \in (-\theta, \theta)$ the family $\{U_z(A)\}_{z=r\mathrm{e}^{i\varphi}}$ is a bounded strongly continuous semigroup of parameter $r \in \mathbb{R}_0^+$ with generator $\mathrm{e}^{i\varphi}A$. Then, by Proposition 1.12, the spectrum $\sigma(\mathrm{e}^{i\varphi}A)$ of this generator lies in $\mathbb{C}_+$. Since this is true for all $|\varphi| < \theta$, the spectrum of A belongs to the closed sector $\overline{S}_{\pi/2-\theta}$:

$$\sigma(A) \subset \overline{S}_{\pi/2-\theta} = \Big\{z \in \mathbb{C}_+ \,:\, |\arg z| \leqslant \frac{\pi}{2} - \theta\Big\}. \tag{1.68}$$

By Definition 1.26(ii), for any $\varepsilon > 0$ and $|\varphi| \leqslant \theta' = \theta - \varepsilon/2$, the semigroup $\{U_{r\mathrm{e}^{i\varphi}}(A)\}_{r\geqslant 0}$ is uniformly bounded by M'_ε. Therefore, Proposition 1.12 yields the estimate $\|(z\mathbb{1} + \mathrm{e}^{i\varphi}A)^{-1}\| \leqslant M'_\varepsilon\,(\Re z)^{-1}$ for all z with $\Re z > 0$. For $|\arg z| \leqslant \pi/2 - \varepsilon/2$, this estimate implies

$$\big\|(z\,\mathrm{e}^{-i\varphi}\mathbb{1} + A)^{-1}\big\| \leqslant \frac{M'_\varepsilon}{|z|\sin\varepsilon}.$$

This means that for $\varepsilon > 0$

$$\big\|(\zeta\mathbb{1} + A)^{-1}\big\| \leqslant \frac{M_\varepsilon}{|\zeta|}, \quad \zeta \in \overline{S}_{\pi/2+\theta-\varepsilon}\backslash\{0\}, \tag{1.69}$$

with M_ε independent of ζ and $\varepsilon < \theta$.

In fact, the conditions (1.68) and (1.69) are also sufficient.

Proposition 1.27. *A closed linear operator A in a Banach space $\mathcal{B}$ is the generator of a bounded holomorphic semigroup $\{U_z(A)\}_{z\in S_\theta}$ of semi-angle $0 < \theta \leqslant \pi/2$ if and only if A satisfies conditions* (1.68) *and* (1.69).

Proof. The necessity part was already proved. To prove the sufficiency part we use the Riesz-Dunford integral representation (1.67). By virtue of (1.68) and (1.69), the integral is absolutely convergent for $t > 0$ in the *operator-norm* topology if Γ is chosen as a contour in $\rho(A)$ running from infinity with $\arg z = \pi/2 - \theta + \varepsilon$ and to infinity with $\arg z = -(\pi/2 - \theta + \varepsilon)$ inside the domain $D_{\theta-\varepsilon} = \mathbb{C}\backslash\overline{S}_{\pi/2-\theta+\varepsilon}$.

To verify the semigroup property, we define $U_s(A)$ by (1.67), with the contour of integration $\Gamma' \subset D_{\theta-\varepsilon}$ similar to Γ, but slightly shifted to the left. Then

$$U_t(A)U_s(A) = \frac{1}{(2\pi i)^2}\int_\Gamma \mathrm{d}z \int_{\Gamma'} \mathrm{d}z'\, \mathrm{e}^{-tz-sz'}(z\mathbb{1}-A)^{-1}(z'\mathbb{1}-A)^{-1}$$
$$= \frac{1}{(2\pi i)^2}\int_\Gamma \mathrm{d}z \int_{\Gamma'} \mathrm{d}z'\, \mathrm{e}^{-tz-sz'}\left[(z\mathbb{1}-A)^{-1}-(z'\mathbb{1}-A)^{-1}\right](z'-z)^{-1}$$
$$= \frac{1}{(2\pi i)^2}\int_\Gamma \mathrm{d}z\, \mathrm{e}^{-tz}(z\mathbb{1}-A)^{-1}\int_{\Gamma'} \mathrm{d}z'\, \mathrm{e}^{-sz'}(z'-z)^{-1}$$
$$+ \frac{1}{(2\pi i)^2}\int_{\Gamma'} \mathrm{d}z'\, \mathrm{e}^{-sz'}(z'\mathbb{1}-A)^{-1}\int_{\Gamma} \mathrm{d}z\, \mathrm{e}^{-tz}(z-z')^{-1}$$
$$= \frac{1}{2\pi i}\int_\Gamma \mathrm{d}z\, \mathrm{e}^{-(t+s)z}(z\mathbb{1}-A)^{-1} = U_{t+s}(A),$$

where we have used the resolvent equation

$$R_z(A)R_{z'}(A) = (R_z(A)-R_{z'}(A))(z-z')^{-1},$$

and properties of the Cauchy integral.

The integral (1.67) can be made absolutely convergent in the operator norm even for complex t by deforming the contour $\Gamma \subset D_{\theta-\varepsilon}$, if one ensures that $|\arg(tz)| < \pi/2$ for $z \in \Gamma$ and $|z| \to \infty$. Note that in the domain $D_{\theta-\varepsilon}$ one has $|\arg z| > \pi/2 - \theta$. Hence, the semigroup $U_t(A)$ can be extended to complex values of t in the sector $|\arg t| < \theta$. Since the Riesz-Dunford representation (1.67) is operator-norm differentiable *under* the integral sign, it follows that $U_t(A)$ is $\|\cdot\|$-holomorphic in the open sector $S_\theta = \{t \in \mathbb{C} : |\arg t| < \theta\}$. Within this sector, the Cauchy theorem gives

$$\|\cdot\|\text{-}\partial_t U_t(A) = \frac{1}{2\pi i}\int_\Gamma \mathrm{d}z\, \mathrm{e}^{-tz}(-z)(z\mathbb{1}-A)^{-1} \tag{1.70}$$
$$= \frac{(-1)}{2\pi i}\int_\Gamma \mathrm{d}z\, \mathrm{e}^{-tz}\left[\mathbb{1}+A(z\mathbb{1}-A)^{-1}\right]$$
$$= (-A)\frac{1}{2\pi i}\int_\Gamma \mathrm{d}z\, \mathrm{e}^{-tz}(z\mathbb{1}-A)^{-1} = (-A)U_t(A) \in \mathcal{L}(\mathcal{B}),$$

where the closedness of the operator A allows to take A out of the integral. Hence, we get

$$U_t(A)A \subset AU_t(A) = -\partial_t U_t(A) \in \mathcal{L}(\mathcal{B}), \quad t \in S_\theta. \tag{1.71}$$

Setting $z' = zt$, formula (1.67) becomes

$$U_t(A) = \frac{1}{2\pi i t}\int_{\Gamma'} \mathrm{d}z'\, \mathrm{e}^{-z'}\left(\frac{z'}{t}\mathbb{1}-A\right)^{-1}, \tag{1.72}$$

where the contour Γ' may be taken inside the domain D_θ independently of t. For $z' \in \Gamma'$, $\varepsilon > 0$ and $t \in \overline{S}_{\theta-\varepsilon}$, we have $\|(z'/t - A)^{-1}\| \leqslant M_\varepsilon |t/z'|$. Moreover, the condition $\Gamma' \subset D_\theta$ ensures the convergence of the integral (1.72), see estimate (1.69).

Therefore,

$$\|U_t(A)\| \leqslant M_\varepsilon \frac{1}{2\pi} \int_{\Gamma'} |\mathrm{d}z'|\, |z'|^{-1}\, |\mathrm{e}^{-z'}| = M'. \tag{1.73}$$

Thus, the semigroup $U_t(A)$ is uniformly bounded in the sector $\overline{S}_{\theta'}$ for $\theta' < \theta$. Similarly, we get the estimate, see (1.70),

$$\begin{aligned} \|\partial_t U_t(A)\| &= \left\| \frac{1}{2\pi i t^2} \int_{\Gamma'} \mathrm{d}z'\, z'\, \mathrm{e}^{-z'} \Big(\frac{z'}{t}\mathbb{1} - A\Big)^{-1} \right\| \\ &\leqslant \frac{M_\varepsilon}{|t|} \int_{\Gamma'} |\mathrm{d}z'|\, |\mathrm{e}^{-z'}| := \frac{M_1'}{|t|}, \quad t \in \overline{S}_{\theta' < \theta}. \end{aligned} \tag{1.74}$$

To verify the strong continuity of $U_t(A)$ when $t \to 0$ in the sector $\overline{S}_{\theta-\varepsilon}$, we use again the representation (1.72), writing

$$U_t(A) - \mathbb{1} = \frac{1}{2\pi i} \int_{\Gamma'} \mathrm{d}z' \, \frac{\mathrm{e}^{-z'}}{z'} \Big(\frac{z'}{t}\mathbb{1} - A\Big)^{-1} A.$$

Then, due to (1.69), for any $u \in \operatorname{dom} A$ we get

$$\|(U_t(A) - \mathbb{1})u\| \leqslant |t|\, M_\varepsilon\, \|Au\| \int_{\Gamma'} |\mathrm{d}z'|\, |z'|^{-2}\, |\mathrm{e}^{-z'}|, \quad t \in \overline{S}_{\theta-\varepsilon}. \tag{1.75}$$

Since $\overline{\operatorname{dom} A} = \mathcal{B}$ and $\|U_t(A)\| \leqslant M'$ for $t \in \overline{S}_{\theta-\varepsilon}$ and $\varepsilon > 0$, it follows that

$$\operatorname*{s-lim}_{t\to 0} U_t(A) = U_{t=0}(A) = \mathbb{1}, \quad t \in \overline{S}_{\theta-\varepsilon},$$

which finishes the proof. □

Corollary 1.28. *Since $\{U_t(A)\}_{t\geqslant 0}$ is holomorphic, a calculation similar to the one used in* (1.74) *gives that for any $t > 0$,*

$$\begin{aligned} \|\partial_t^n U_t(A)\| &= \|A^n U_t(A)\| \\ &= \left\| \frac{1}{2\pi i t^{n+1}} \int_{\Gamma'} \mathrm{d}z'\, (z')^n \mathrm{e}^{-z'} \Big(\frac{z'}{t}\mathbb{1} - A\Big)^{-1} \right\| \\ &\leqslant \frac{M_n'}{t^n}. \end{aligned} \tag{1.76}$$

The next corollary gives an alternative characterisation of the bounded holomorphic semigroups and their relation to the operator-norm continuity, see Section 1.4.

Corollary 1.29. *Let $\{U_z(A)\}_{z\in S_\theta}$ be a bounded holomorphic semigroup on $\mathcal{B}$ in the sector S_θ with $\theta < \pi/2$. Then*

$$U_z(A) : \mathcal{B} \to \operatorname{dom} A \tag{1.77}$$

for all $z \in S_\theta$. *Moreover, for any* $\varepsilon \in (0,\theta)$, *there is a* $C_\varepsilon > 0$ *such that*

$$\|A\,U_z(A)\| \leqslant \frac{C_\varepsilon}{|z|}, \tag{1.78}$$

for all $z \in S_{\theta-\varepsilon}$.

Proof. The mapping property (1.77) follows from (1.70) and (1.71). Estimate (1.78) follows from (1.70) and (1.74). □

Proposition 1.30. *Let* $\{U_t(A)\}_{t\geqslant 0}$ *be a strongly continuous semigroup on* $\mathcal{B}$ *with generator* A *such that for* $u \in \mathcal{B}$, $u \mapsto U_t(A)u \in \operatorname{dom} A$, $\|U_t(A)\| \leqslant M$ *and* $\|A\,U_t(A)\| \leqslant M_1' t^{-1}$ *for all* $t > 0$. *Then there exists an angle* $\theta > 0$, *which depends on* M_1', *such that* $U_t(A)$ *can be analytically continued to a bounded* $\|\cdot\|$*-holomorphic semigroup in the sector* S_θ.

Proof. Proposition 1.15 shows that $U_t(A)$ is an infinitely $\|\cdot\|$-differentiable operator-valued function on $\mathbb{R}^+$ with $\|\cdot\|$-$\partial_t^n U_t(A) = (-A\,U_{t/n}(A))^n$, see (1.56). Hence, by the assumptions of the proposition and by (1.76),

$$\|\partial_t^n U_t\| \leqslant (n\,M_1' t^{-1})^n.$$

This means that the function $U_t(A)$ can be analytically continued from $\mathbb{R}^+$ to the disc $D_t := \{z \in \mathbb{C}_+ : |z-t| < (e\,M_1')^{-1} t\}$ with $t > 0$ and $e\,M_1' > 1$, by setting

$$U_z(A) = \sum_{n=0}^{\infty} \frac{(z-t)^n}{n!} \partial_t^n U_t(A). \tag{1.79}$$

The union of these discs $\bigcup_{t>0} D_t = S_\theta$ is a sector with

$$\theta(M_1') = \arcsin\,(e\,M_1')^{-1} < \pi/2.$$

The rest of the proof that the operator-valued function (1.79) satisfies conditions (i) and (ii) of Definition 1.26 can be found in references provided in Notes in Section 1.8 □

Remark 1.31. For $t > 0$, the holomorphic semigroups enjoy an even *stronger* form of continuity than norm continuity. Since the family $\{U_t(A)\}_{t>0}$ can be differentiated any number of times, one gets by (1.76)

$$U_t(A)u \in \bigcap_{n=1}^{\infty} \operatorname{dom} A^n \quad \text{and} \quad \|A^n U_t(A)\| \leqslant M_n' t^{-n}, \tag{1.80}$$

for all $t > 0$. This in turn yields the estimate:

$$\begin{aligned}\|A^n(U_t(A) - U_s(A))\| &= \left\|\int_s^t d\tau\, A^n \partial_\tau U_\tau(A)\right\| \\ &\leqslant \frac{M_{n+1}'}{n}\left(\frac{1}{s^n} - \frac{1}{t^n}\right),\end{aligned} \tag{1.81}$$

for all $0 < s \leqslant t$ and $n \geqslant 1$.

The notion of a bounded holomorphic semigroup can be generalised in the same way as in Section 1.3 for strongly continuous quasi-bounded semigroups. This means that we can relax the boundedness condition in (ii) of Definition 1.26.

Definition 1.32. A strongly continuous semigroup $\{U_t(A)\}_{t\geqslant 0}$ on a Banach space $\mathcal{B}$ is called a quasi-bounded holomorphic semigroup of semi-angle $\theta \in (0, \pi/2)$ if $\{U_z(A)\}_{z\in S_\theta}$ is as in Definition 1.26, except that it is not required to be uniformly bounded in the sector $\overline{S}_{\theta'}$ for $\theta' < \theta$.

Proceeding as in Section 1.3, we can obtain an exponential estimate on the growth of $U_z(A)$ in the sector $S_{\theta'}$. If $U_z(A)$ is a holomorphic semigroup of angle θ, then $\|U_z(A)f\|$ is bounded in the domain $D_{\theta',\tau} = \{z \in \overline{S}_{\theta'} : |z| \leqslant \tau,\ \tau > 0\}$ for any $f \in \mathcal{B}$. Hence, by the uniform boundedness principle, $\|U_z(A)\| \leqslant M$ for $z \in D_{\theta',\tau}$. For any $z \in \overline{S}_{\theta'}$, the semigroup property (Definition 1.26 (i)) implies that $U_z(A) = (U_{z'}(A))^n U_\zeta(A)$, where $z = nz' + \zeta$ and $z', \zeta \in D_{\theta',\tau}$. Therefore, one concludes that there are constants $M, \omega > 0$ such that

$$\|U_z(A)\| \leqslant M\,\mathrm{e}^{\omega|z|}, \quad z \in \overline{S}_{\theta'}, \tag{1.82}$$

where M and ω depend on θ', and that $\tilde{U}_z(A) := \mathrm{e}^{-\omega z}U_z(A)$ is a bounded holomorphic semigroup of angle θ'.

The arguments above motivate a generalisation of Proposition 1.27 (cf. Proposition 1.12).

Proposition 1.33. *A closed operator A in a Banach space $\mathcal{B}$ is the generator of a quasi-bounded holomorphic semigroup $\{U_z(A)\}_{z\in S_\theta}$ of semi-angle $\theta \in (0, \pi/2]$ if and only if:*

(i) $\rho(A) = \mathbb{C}\backslash\{z \in \mathbb{C} : z + \omega_0 \in \overline{S}_{\pi/2-\theta}\}$ *for some $\omega_0 \in \mathbb{R}$, and*

(ii) *for $\varepsilon > 0$, there exists a constant $M_\varepsilon > 0$ such that*

$$\|(z\mathbb{1} + \omega_0\mathbb{1} + A)^{-1}\| \leqslant \frac{M_\varepsilon}{|z|}, \quad z \in \overline{S}_{\pi/2+\theta-\varepsilon}\backslash\{0\}\ ,\ \varepsilon < \theta\ . \tag{1.83}$$

We denote this class of generators by $\mathscr{H}(\theta, \omega_0)$, where ω_0 is the type of semigroup, cf. Definition 1.11.

Proof. The operator $A_0 := \omega_0\mathbb{1} + A$ satisfies the conditions of Proposition 1.27. Therefore, $U_z(A_0) = \mathrm{e}^{-\omega_0 z}U_z(A)$ is a bounded holomorphic semigroup of semi-angle $\theta' = \theta - \varepsilon$. Therefore, $\|U_z(A_0)\| \leqslant M'$ (1.73), and $\|U_z(A)\| \leqslant M'\,\mathrm{e}^{\omega_0|z|}$, which yields the proposition.

Note that $\mathscr{H}(\theta, \omega_0 < 0) \subset \mathscr{H}(\theta, 0)$, whereas $A \in \mathscr{H}(\theta, \omega_0 > 0)$ if and only if $A = A_0 - \omega_0\mathbb{1}$, where $A_0 \in \mathscr{H}(\theta, 0)$. □

Remark 1.34. To make evident that condition (1.83) (or (1.69) for $\omega_0 = 0$) is stronger than (1.39), it is sufficient to note that, since $t \mapsto U_t(A)$ is strongly continuous for $t \geqslant 0$, the condition (1.83) implies the existence of the operator-norm convergent Laplace transform

$$\hat{U}_\lambda(A) = \int_0^\infty dt\, e^{-\lambda t} U_t(A) = (A + \lambda \mathbb{1})^{-1} \in \mathcal{L}(\mathcal{B}),$$

for $\lambda > \omega_0$, cf.(1.31)–(1.33). Then, using Fubini's theorem, we get

$$\begin{aligned}(\lambda \mathbb{1} + A)^{-n} &= \int_0^\infty dt_1 \int_0^\infty dt_2 \ldots \int_0^\infty dt_n\, e^{-\lambda(t_1+\cdots+t_n)} U_{t_1+\cdots+t_n}(A) \\ &= \frac{1}{(n-1)!} \int_0^\infty d\tau\, \tau^{n-1} e^{-\lambda\tau} U_\tau(A). \end{aligned} \tag{1.84}$$

Since by Proposition 1.33 we have $\|U_\tau(A)\| \leqslant M' e^{\omega_0 \tau}$, we deduces the estimate (1.39) from the integral formula (1.84) for $M = M' \geqslant 1$.

Note that the condition (1.69) (or (1.83)) guarantees that the semigroup $U_t(A)$ is $\|\cdot\|$-continuous and that for $u \in \mathcal{B}$, $u \mapsto U_t(A)u \in \operatorname{dom} A$ for $t > 0$, see Section 1.4 and Corollary 1.29. This is in contrast with the strongly continuous semigroups, for which one has only for $u \in \operatorname{dom} A$, $u \mapsto U_t(A)u \in \operatorname{dom} A$ for $t > 0$, see Section 1.3.

1.6 Holomorphic semigroups on a Hilbert space

Since our main subject, the Gibbs semigroups, are defined on a Hilbert space $\mathcal{H}$, in this section we look closer at holomorphic semigroups on a separable complex space $\mathcal{H}$ with a sesquilinear inner product $(\cdot,\cdot) : \mathcal{H} \times \mathcal{H} \to \mathbb{C}$. We note that $(\alpha u, \beta v) = \alpha\overline{\beta}(u, v)$ for $u, v \in \mathcal{H}$ and $\alpha, \beta \in \mathbb{C}$.

First, we recall the notion of *numerical range* $\mathrm{Nr}A$ of an operator A. Let A be a linear operator in a Hilbert space $\mathcal{H}$. Then

$$\mathrm{Nr}A := \{(Au, u) : u \in \operatorname{dom} A,\ \|u\| = 1\}. \tag{1.85}$$

In general, $\mathrm{Nr}A$ is neither open nor closed, even if A is a closed operator in $\mathcal{H}$. A theorem of Hausdorff states only that $\mathrm{Nr}A$ is a convex subset in $\mathbb{C}$. Therefore $\overline{\mathrm{Nr}A}$ is a closed convex set, and the complement $\Delta := \mathbb{C}\backslash\overline{\mathrm{Nr}A}$ is either a connected open set, or it consists of two components Δ_1 and Δ_2. The latter possibility occurs whenever $\overline{\mathrm{Nr}A}$ is a strip bounded by two parallel straight lines, including the degenerate case when those lines coincide.

Next, we denote by $\operatorname{def} T := \dim(\operatorname{ran} T)^\perp$ the *deficiency* (or defect) of the closed operator T in $\mathcal{H}$.

Proposition 1.35. *Let A be a closed operator in $\mathcal{H}$. Then for any $\zeta \in \Delta$, the operator $A - \zeta\mathbb{1}$ is injective with a closed range* $\operatorname{ran}(A - \zeta\mathbb{1})$. *If, in addition, for* *each of those* ζ, *the deficiency* $\operatorname{def}(A - \zeta\mathbb{1})$ *is equal to zero, then* $\Delta \subset \rho(A)$ *or, equivalently,* $\sigma(A) \subset \overline{\operatorname{Nr}A}$. *Moreover,*

$$\|R_\zeta(A)\| \leqslant \frac{1}{\operatorname{dist}(\zeta, \overline{\operatorname{Nr}A})}. \tag{1.86}$$

Proof. For any $u \in \operatorname{dom} A$ with $\|u\| = 1$, one has

$$|(Au, u) - \zeta| = |((A - \zeta\mathbb{1})u, u)| \leqslant \|(A - \zeta\mathbb{1})u\|, \quad \zeta \in \mathbb{C}. \tag{1.87}$$

Then, if $\zeta \in \Delta = \mathbb{C}\backslash\overline{\operatorname{Nr}A}$ so that $\operatorname{dist}(\zeta, \overline{\operatorname{Nr}A}) = \delta > 0$, the estimate (1.87) gives $\|(A - \zeta\mathbb{1})u\| \geqslant \delta$, or

$$\|(A - \zeta\mathbb{1})v\| \geqslant \delta\|v\|, \quad v \in \operatorname{dom} A. \tag{1.88}$$

This implies that the operator $A - \zeta\mathbb{1}$ is injective, i.e., $\ker(A - \zeta\mathbb{1}) = \{0\}$. Moreover, since A is closed, the set $\operatorname{ran}(A - \zeta\mathbb{1})$ is also closed, i.e., it is a subspace. If the dimension of its orthogonal complement $(\operatorname{ran}(A - \zeta\mathbb{1}))^\perp$ is zero, then $\operatorname{ran}(A - \zeta\mathbb{1}) = \mathcal{H}$. Thus, the inverse operator $(A - \zeta\mathbb{1})^{-1} = R_\zeta(A)$ has domain $\mathcal{H}$, and $\|R_\zeta(A)\| \leqslant \delta^{-1}$ by virtue of (1.88). This proves that $\zeta \in \rho(A)$, or $\Delta \subset \rho(A)$, and the estimate (1.86). □

In fact, in Proposition 1.35 it is sufficient that $\operatorname{def}(A - \zeta\mathbb{1}) = 0$ for only one $\zeta \in \Delta$. As above, we set $A_\zeta := A - \zeta\mathbb{1}$ for $\zeta \in \mathbb{C}$.

Lemma 1.36. *The deficiency* $\operatorname{def} A_\zeta$ *is constant for* $\zeta \in \Delta$, *except for the above mentioned special case, in which the* $\operatorname{def} A_\zeta$ *is constant in both* Δ_1 *and* Δ_2. *These constants are called the deficiency of a* A *in* Δ, Δ_1 *and* Δ_2, *respectively.*

Proof. Let $\zeta \in \Delta$ so that $\operatorname{dist}(\zeta, \overline{\operatorname{Nr}A}) = \delta > 0$. We set $d_\zeta := \operatorname{def} A_\zeta$. For $a > 1$ and $\zeta' \in \mathbb{C}$ such that $|\zeta' - \zeta| \leqslant a\delta$, we have by (1.88) that

$$a\|A_\zeta u\| \geqslant \|(\zeta - \zeta')u\|, \quad u \in \operatorname{dom} A. \tag{1.89}$$

Notice that the operator $A_{\zeta'}$ is closed and $A_{\zeta'} = A_\zeta + (\zeta - \zeta')\mathbb{1}$.

Suppose that $d_{\zeta'} < d_\zeta$. Then there exists a $\varphi \in \mathcal{H} \ominus \operatorname{ran} A_\zeta$, $\varphi \neq 0$, such that $\varphi \perp (\mathcal{H} \ominus \operatorname{ran} A_{\zeta'})$. This means that $\varphi \in \operatorname{ran} A_{\zeta'}$, that is, $\varphi = A_{\zeta'}v$ for some $v \in \operatorname{dom} A$. Since $\varphi \perp \operatorname{ran} A_\zeta$, one has $(\varphi, A_\zeta v) = 0$, and hence,

$$(A_\zeta v, A_\zeta v) = -(\zeta - \zeta')v, A_\zeta v). \tag{1.90}$$

Suppose now that $d_{\zeta'} > d_\zeta$. Then, similarly, one can find $\psi \in \mathcal{H} \ominus \operatorname{ran} A_{\zeta'}$, $\psi \neq 0$, such that $\psi \in \operatorname{ran} A_\zeta$, i.e., $\psi = A_\zeta w$ for some $w \in \operatorname{dom} A$. Since $\psi \perp \operatorname{ran} A_{\zeta'}$, we get

$$(A_\zeta w, A_\zeta w) = -(A_\zeta w, (\zeta - \zeta')w). \tag{1.91}$$

Both (1.90) and (1.91) are impossible for $a < 1$, because, by (1.89)

$$\|A_\zeta v\|^2 \leqslant \|(\zeta - \zeta')v\|\|A_\zeta v\| \leqslant a\|A_\zeta v\|^2,$$

and

$$\|A_\zeta w\|^2 \leqslant \|(\zeta - \zeta')w\|\|A_\zeta w\| \leqslant a\|A_\zeta w\|^2.$$

Therefore, $d_{\zeta'} = d_\zeta$ for all ζ' in the disc $D_{r=\delta}(\zeta) := \{z \in \mathbb{C} : |z - \zeta| < \delta\} \subset \Delta$.

By the same reasoning, one can extend the equality $d_{\zeta''} = d_\zeta$ to all ζ'' in the disc $D_{r\leqslant\delta}(z) \subset \Delta$ for any centre $z \in D_{r=\delta}(\zeta)$. Covering Δ by these intersecting discs $D_{r\leqslant\delta}(z)$ we conclude that the deficiency of A is constant on Δ. The proof for the case with two components Δ_1 and Δ_2 is similar. □

The next statement gives a relation between the spectrum $\sigma(A)$ and the numerical range $\mathrm{Nr}A$ in the case of a bounded operator A.

Proposition 1.37. *If $A \in \mathcal{L}(\mathcal{H})$, then*

$$\sigma(A) \subset \overline{\mathrm{Nr}A}. \tag{1.92}$$

Proof. Since for $u \in \mathcal{H}$, $\|u\| = 1$, one has $|(Au, u)| \leqslant \|A\|$, we get $\overline{\mathrm{Nr}A} \subseteq \overline{D}_{r=\|A\|}(z = 0)$. Hence Δ is a connected open set containing the exterior of the disc $\overline{D}_{r=\|A\|}(z = 0)$. Since $A \in \mathcal{L}(\mathcal{H})$, this exterior belongs to the resolvent set $\rho(A)$, i.e., the resolvent $R_\zeta(A) \in \mathcal{L}(\mathcal{H})$, or $\operatorname{def} A_\zeta = 0$ for $\zeta \in \mathbb{C}\backslash\overline{D}_{\|A\|}(z = 0) \subset \Delta$. By Lemma 1.36, the same must be true for all $\zeta \in \Delta = \mathbb{C}\backslash\overline{\mathrm{Nr}A}$. Since the operator A is closed, the conditions of Proposition 1.35 are satisfied, and this implies that $\Delta \subset \rho(A)$, which is equivalent to (1.92). □

Remark 1.38. For a closed unbounded linear operator A in $\mathcal{H}$, one needs the additional condition

$$\operatorname{def} A_\zeta = \dim(\mathcal{H} \ominus \operatorname{ran} A_\zeta) = 0, \quad \text{for all } \zeta \in \Delta, \tag{1.93}$$

to ensure (1.92), cf. Proposition 1.35. Recall that z is in the spectrum $\sigma(A) = \{\zeta \in \mathbb{C} : R_\zeta(A) \notin \mathcal{L}(\mathcal{H})\}$ if z belongs to one of the following *disjoint* sets (Appendix A, Section A.3):

(i) The *point spectrum*

$$\sigma_{\mathrm{p}}(A) := \{\zeta \in \mathbb{C} : \ker A_\zeta \neq \{0\}\}.$$

In the case when $\ker A_z = \{0\}$, but $\operatorname{ran} A_z \neq \mathcal{H}$, i.e., $z \in \sigma(A)\backslash\sigma_{\mathrm{p}}(A)$, we have two other possibilities.

(ii) The *continuous spectrum*

$$\sigma_{\mathrm{cont}}(A) := \{\zeta \in \mathbb{C} : \ker A_\zeta = \{0\},\ \operatorname{ran} A_\zeta \neq \mathcal{H},\ \overline{\operatorname{ran} A_\zeta} = \mathcal{H}\}.$$

(iii) The *residual spectrum*

$$\sigma_{\mathrm{res}}(A) := \{\zeta \in \mathbb{C} : \ker A_\zeta = \{0\},\ \overline{\operatorname{ran} A_\zeta} \neq \mathcal{H}\}.$$

Therefore, in the case (iii), the deficiency def $A_z \neq 0$, i.e., the corresponding condition in Proposition 1.35 serves to exclude $\sigma_{\mathrm{res}}(A)$ from the set $\Delta = \mathbb{C}\backslash\overline{\mathrm{Nr}A}$. Otherwise, we would get instead of (1.92) that the *essential part* of the spectrum, $\sigma_{\mathrm{p}}(A)\cup\sigma_{\mathrm{cont}}(A)$, belongs to the closure of the numerical range: $\sigma_{\mathrm{p}}(A)\cup\sigma_{\mathrm{cont}}(A) \subset \overline{\mathrm{Nr}A}$.

The notion of numerical range is very useful for the classification of generators of contraction and holomorphic semigroups on a Hilbert space.

Definition 1.39. An operator A in a Hilbert space $\mathcal{H}$ is said to be *accretive*, if $\mathrm{Nr}A \subset \overline{\mathbb{C}}_+ = \{z \in \mathbb{C} : \mathfrak{Re}\, z \geqslant 0\}$. If, in addition, A is closed, then by the stability of the deficiency, Lemma 1.36, def $A_z = \dim(\mathrm{ran}\, A_z)^\perp$ is constant for $\mathfrak{Re}\, z < 0$. If def $A_z = 0$ for $\mathfrak{Re}\, z < 0$, then by Proposition 1.35, one has $\mathbb{C}_+ = \{z \in \mathbb{C} : \mathfrak{Re}\, z > 0\} \subset \rho(-A)$ with

$$(A + z\mathbb{1})^{-1} \in \mathcal{B}(\mathcal{H}) \quad \text{and} \quad \|(A + z\mathbb{1})^{-1}\| \leqslant \frac{1}{\mathfrak{Re}\, z}, \quad \mathfrak{Re}\, z > 0. \tag{1.94}$$

The operator A satisfying (1.94) is called *m-accretive.*

An m-accretive operator A is *maximally* accretive in the sense that A has no proper accretive extensions. Indeed, let A' be an accretive extension of A. Then, $(A + z\mathbb{1})^{-1} = (A' + z\mathbb{1})^{-1} \in \mathcal{B}(\mathcal{H})$ for $\mathfrak{Re}\, z > 0$, which implies $A = A'$.

Remark 1.40. Note that in Definition 1.39 we did not require that dom A be dense in $\mathcal{H}$. In fact, an m-accretive operator A is necessarily densely defined. Indeed, since dom $A = \mathrm{ran}(A + z\mathbb{1})^{-1}$ for $\mathfrak{Re}\, z > 0$, to prove this we have to show that $((A + z\mathbb{1})^{-1}u, v) = 0$ for all $u \in \mathcal{H}$ implies $v = 0$. Let $u = v$ and $w = (A + z\mathbb{1})^{-1}v$. Then,

$$0 = |(w, (A + z\mathbb{1})w)| \geqslant \mathfrak{Re}(Aw, w) + \mathfrak{Re}\, z\|w\|^2 \geqslant \mathfrak{Re}\, z\|w\|^2,$$

with $\mathfrak{Re}\, z > 0$, and hence, $w = 0$ and $v = 0$.

An operator A is called *quasi-accretive* if there is a $\gamma \in \mathbb{C}$ such that $A + \gamma\mathbb{1}$ is accretive. Similarly, we call A *quasi-m-accretive* if $A + \gamma\mathbb{1}$ is m-accretive for some $\gamma \in \mathbb{C}$.

Definition 1.41. An accretive operator A is called *sectorial with semi-angle* $\alpha \in (0, \pi/2)$, if

$$\mathrm{Nr}A \subseteq \overline{S}_\alpha := \{z \in \overline{\mathbb{C}}_+ : |\arg z| \leqslant \alpha\}. \tag{1.95}$$

If, in addition, A is m-accretive, then it is called *m-sectorial.* If A is quasi-m-accretive and

$$\mathrm{Nr}A \subseteq \overline{S}_{\alpha,\gamma} := \{z \in \overline{\mathbb{C}}_+ : |\arg(z - \gamma)| \leqslant \alpha\} \tag{1.96}$$

for some $\gamma \in \mathbb{C}$, then it is called *quasi-m-sectorial with vertex* γ *and semi-angle* α. Notice that γ and α are not uniquely determined.

Corollary 1.42. *Let A be a quasi-m-sectorial operator with vertex γ and semi-angle α. Then, by Proposition* 1.35, *we have*

$$\rho(A) \supseteq \mathbb{C}\backslash\overline{S}_{\alpha,\gamma}. \tag{1.97}$$

This is clear from Lemma 1.36, *since $\mathbb{C}\backslash\overline{\mathrm{Nr}A}$ is a connected set.*

Using the notions of m-accretive and of m-sectorial operators, we can identify m-accretive generators of the contraction, of the quasi-bounded, and of the holomorphic semigroups.

Proposition 1.43. *An operator A in $\mathcal{H}$ is a generator of a contraction semigroup $\{U_t(A)\}_{t\geqslant 0}$ on $\mathcal{H}$ if and only if A is m-accretive.*

Proof. If the operator A is m-accretive, then, by Definition 1.39 and Remark 1.40, it is densely defined, closed, and its properties (1.94) imply the conditions of Proposition 1.1. Hence, by Definition 1.3, the corresponding semigroup $\{U_t(A)\}_{t\geqslant 0}$ is a contraction semigroup. The converse follows from Proposition 1.1. □

Corollary 1.44. *An operator A is the generator of a quasi-bounded semigroup with $M = 1$ and of type ω_0 (i.e., $A \in Q(1, \omega_0)$, see Definition* 1.11*), if and only if it is quasi-m-accretive for some γ such that $\mathfrak{Re}(-\gamma) = \omega_0$.*

Remark 1.45. Let A be the generator of a quasi-bounded holomorphic semigroup with semi-angle $\theta \in (0, \pi/2]$, i.e., $A \in \mathscr{H}(\theta, \omega_0)$. In Section 1.5, Propositions 1.27 and Propositions 1.33, we have shown that $A \in \mathscr{H}(\theta, \omega_0 = 0)$ is equivalent to the existence, for each $\varepsilon > 0$, of a constant M'_ε such that $\mathrm{e}^{i\varphi}A \in Q(M'_\varepsilon, 0)$ for any angle $|\varphi| \leqslant \theta - \varepsilon/2$.

This remark gives a convenient condition for A to be the generator of a *contraction* holomorphic semigroup on $\mathcal{H}$.

Proposition 1.46. *Let A be an m-sectorial operator in a Hilbert space $\mathcal{H}$ with vertex $\gamma = 0$ and with semi-angle $\alpha \in [0, \pi/2)$. Then $A \in \mathscr{H}(\theta = \pi/2 - \alpha, 0)$ and the holomorphic semigroup $\{U_z(A)\}_{z\in S_\theta}$ is contraction semigroup since $\|U_z(A)\| \leqslant 1$ for $|\arg z| < \theta$.*

Proof. According to Definition 1.11 and Remark 1.45, it suffices to prove that $\mathrm{e}^{i\varphi}A \in Q(M = 1, 0)$ for any $|\varphi| \leqslant \pi/2 - \alpha$. Since, by (1.95) and (1.96), $\mathrm{Nr}A \subset \overline{S}_{\alpha,\gamma=0}$, one has $\mathrm{Nr}(\mathrm{e}^{i\varphi}A) = \mathrm{e}^{i\varphi}\mathrm{Nr}A \subset \overline{\mathbb{C}}_+$ for $|\varphi| \leqslant \pi/2 - \alpha$. Thus,

$$\rho(\mathrm{e}^{i\varphi}A) \supset \mathbb{C}\backslash\overline{\mathrm{Nr}(\mathrm{e}^{i\varphi}A)} \supset \mathbb{C}_- := \{z \in \mathbb{C} : \mathfrak{Re}\, z < 0\}\ ,$$

by Proposition 1.35. Therefore, if $\mathfrak{Re}\,\zeta < 0$, then by (1.86),

$$\|(-\zeta\mathbb{1} + \mathrm{e}^{i\varphi}A)^{-1}\| \leqslant \frac{1}{\mathrm{dist}(\zeta, \overline{\mathbb{C}}_+)}. \tag{1.98}$$

Since $\mathrm{dist}(\zeta, \overline{\mathbb{C}}_+)$ is actually the distance of ζ from the imaginary axis, (1.98) together with (1.94) and Corollary 1.44 imply that $\mathrm{e}^{i\varphi}A \in Q(M = 1, 0)$ for $|\varphi| \leqslant \pi/2 - \alpha$. Therefore, $\|U_z(A)\| = \|U_t(\mathrm{e}^{i\varphi}A)\| \leqslant 1$ where $z = \mathrm{e}^{i\varphi}t$ with $t \geqslant 0$. □

Corollary 1.47. *Let A be a positive self-adjoint operator: $A = A^* \geqslant 0$, or* $\mathrm{Nr}A \subset \mathbb{R}_0^+ = \{x \in \mathbb{R} : x \geqslant 0\}$. *Then $A \in \mathscr{H}(\pi/2, 0)$, which implies that, for $\mathfrak{Re}\, z > 0$, the semigroup $U_z(A)$ is holomorphic and a contraction, $\|U_z(A)\| \leqslant 1$.*

1.7 Perturbations of semigroups

Generally, bounded and *a fortiori* unbounded perturbations of quasi-bounded strongly continuous semigroups with generator $A \in Q(M, \omega_0)$ require a nontrivial analysis. See examples in Section 4.4. The situation is much simpler if the original semigroup has a generator $A \in Q(1, \omega_0)$.

Proposition 1.48. *Let $A \in Q(1, \omega_0)$ and $B \in \mathcal{L}(\mathcal{B})$. Then the operator $H := A + B$ with* $\mathrm{dom}\, A$ *is closed, and $H \in Q(1, \omega_0 + \|B\|)$.*

Proof. Since closedness is stable with respect to bounded perturbations, by virtue of Proposition 1.12 one only has to estimate $\|R_{-\lambda}(H)^m\|$ for $m = 1$. From the second Neumann series for the resolvent $R_{-\zeta}(H)$ one gets

$$(H + \zeta\mathbb{1})^{-1} = \sum_{k=0}^{\infty} (A + \zeta\mathbb{1})^{-1}[-B(A + \zeta\mathbb{1})^{-1}]^k. \tag{1.99}$$

Since $A \in Q(M = 1, \omega_0)$ implies that $\|(A + \zeta\mathbb{1})^{-1}\| \leqslant (\mathfrak{Re}\, \zeta - \omega_0)^{-1}$ for $\mathfrak{Re}\, \zeta > \omega_0$, from (1.99) we obtain for $\|B\|(\mathfrak{Re}\, \zeta - \omega_0)^{-1} < 1$ that

$$\|(H + \zeta\mathbb{1})^{-1}\| \leqslant (\mathfrak{Re}\, \zeta - \omega_0 - \|B\|)^{-1}. \tag{1.100}$$

By Definition 1.11, the estimate (1.100) yields $H \in Q(M = 1, \omega_0 + \|B\|)$. □

Remark 1.49. It is known that, in general, it is difficult to perturb a generator A of a strongly continuous semigroup by an unbounded operator B so that $A + B$ remains being a generator. For example, the operator $A+B$ need not be a generator in $Q(M', \omega_0')$ even if B is a *relatively bounded* perturbation of $A \in Q(M, \omega_0)$, that is,

$$\|Bu\| \leqslant a\|u\| + b\|Au\|, \quad u \in \mathrm{dom}\, A \subset \mathrm{dom}\, B, \tag{1.101}$$

for $A \geqslant 0$ and $a \geqslant 0$, $b > 0$. The *infimum* of all possible constants $b > 0$ in (1.101) is called the *relative bound* of the operator B (with respect to A), which we denote again by b.

Definition 1.50. Let A be a closed operator in $\mathcal{B}$. We denote by $\mathcal{P}_b(A)$ the class of closed operators verifying (1.101) with a relative bound $b > 0$. The class $\mathcal{P}_{b<1}(A)$ contains the so-called *Kato-small* perturbations of the operator A. On the other hand, if an *unbounded* operator B verifies (1.101) for *any* $b > 0$, i.e., the infimum is zero, we say that $b = +0$ (or $b := 0^+$) and that operator B belongs to the class $\mathcal{P}_{0^+}(A)$ of *infinitesimally* small unbounded perturbations of A.

Proposition 1.48 hints that one can make the class of admissible perturbations larger by enforcing conditions on the generator A. It is evident that if $B \in \mathcal{L}(\mathcal{B})$, then in (1.101) one can put $a = \|B\|$ and $b = 0$. Hence, the bounded operator $B \in \mathcal{P}_0$ for *any* unbounded operator A. So the class $\mathcal{P}_0$ of *bounded* perturbations considered in Proposition 1.48 is the smallest in the hierarchy:

$$\mathcal{P}_0 \subset \mathcal{P}_{0^+}(A) \subset \mathcal{P}_b(A) \ . \tag{1.102}$$

In Section 4.4 we shall study another class of infinitesimally small unbounded perturbations $\mathcal{P}$, which lies between $\mathcal{P}_0$ and $\mathcal{P}_{0^+}$, see (4.56).

Below we treat perturbations from the class $\mathcal{P}_{b<1}(A)$ under a rather strong condition on A. Namely, we request analyticity of the original unperturbed semigroup in a sector S_θ with semi-angle $\theta \in (0, \pi/2]$. The measure of the impact of perturbation is expressed by *decrement* $\theta - \varepsilon$, of the semi-angle of analyticity $(0 \leqslant \varepsilon < \theta)$, and by a *variation* of the semigroup type ω_0'.

Proposition 1.51. *Let A be the generator of a quasi-bounded holomorphic semigroup of semi-angle $\theta \in (0, \pi/2]$, i.e., $A \in \mathscr{H}(\theta, \omega_0)$. Then, for any non-negative $\varepsilon < \theta$, there exist constants $\omega_0' \geqslant 0$ and $\delta < 1$ such that, if $B \in \mathcal{P}_b(A)$ for some $a \geqslant 0$ and $b \leqslant \delta$, then the operator $H = A + B \in \mathscr{H}(\theta - \varepsilon, \omega_0')$. If, in particular, $\omega_0 = 0$ and $a = 0$, then $H \in \mathscr{H}(\theta - \varepsilon, 0)$ is the generator of a bounded holomorphic semigroup.*

Proof. Without loss of generality, we may assume that the type $\omega_0 = 0$, see Proposition 1.33. Let $A \in \mathscr{H}(\theta, 0)$. Then $B \in \mathcal{P}_b(A)$ implies that

$$\|B(A + \zeta\mathbb{1})^{-1}\| \leqslant a\|(A + \zeta\mathbb{1})^{-1}\| + b\|A(A + \zeta\mathbb{1})^{-1}\|. \tag{1.103}$$

If $|\arg\zeta| \leqslant \pi/2 + \theta - \varepsilon$, then by (1.69),

$$\|(A + \zeta\mathbb{1})^{-1}\| \leqslant \frac{M_\varepsilon}{|\zeta|},$$

and consequently,

$$\|A(A + \zeta\mathbb{1})^{-1}\| = \|\mathbb{1} - \zeta(A + \zeta\mathbb{1})^{-1}\| \leqslant 1 + M_\varepsilon.$$

Hence, (1.103) yields the estimate

$$\|B(A + \zeta\mathbb{1})^{-1}\| \leqslant a\, M_\varepsilon|\zeta|^{-1} + b\,(1 + M_\varepsilon), \tag{1.104}$$

which ensures the convergence of the Neumann series for $(H + \zeta\mathbb{1})^{-1}$ provided the right-hand side of (1.104) is smaller than one. The norm estimate of this series gives the additional condition

$$\begin{aligned}\|(H + \zeta\mathbb{1})^{-1}\| &\leqslant \frac{M_\varepsilon|\zeta|^{-1}}{1 - a\, M_\varepsilon|\zeta|^{-1} - b\,(1 + M_\varepsilon)} \\ &= \frac{M_\varepsilon(1 - b\,(1 + M_\varepsilon))^{-1}}{|\zeta| - a\, M_\varepsilon(1 - b\,(1 + M_\varepsilon))^{-1}}\,, \end{aligned} \tag{1.105}$$

for $|\arg\zeta| \leqslant \pi/2+\theta-\varepsilon$. If $b < (1+M_\varepsilon)^{-1} = \delta$, then to satisfy both conditions ζ should belong to a shifted sector $S_{\pi/2+\theta-\varepsilon,\ \gamma}$ with the vertex $\gamma := \gamma(a,b,M_\varepsilon) > 0$ such that the disc

$$\{\zeta : |\zeta| < aM_\varepsilon(1-b/\delta)^{-1}\} \subset \mathbb{C}\backslash S_{\pi/2+\theta-\varepsilon,\gamma}. \tag{1.106}$$

Equivalently, (1.105) and (1.106) mean that

$$\|(H+\gamma\mathbb{1}+\zeta\mathbb{1})^{-1}\| \leqslant \frac{M'}{|\zeta|}, \quad \zeta \in S_{\pi/2+\theta-\varepsilon}. \tag{1.107}$$

Therefore, by Proposition 1.33, the operator $H \in \mathscr{H}(\theta-\varepsilon,\omega_0')$ for $M' := M_\varepsilon(1-b/\delta)^{-1} > 0$ and for the type $\omega_0' \leqslant \gamma(a,b,M_\varepsilon)$. Note that then the vertex $\gamma_H \geqslant -\gamma(a,b,M_\varepsilon)$. From (1.105) we also deduce that one can put $\omega_0' = 0$ if $a = 0$, that is, $H \in \mathscr{H}(\theta-\varepsilon,0)$. □

Another case when one can extend the class of admissible perturbations to unbounded operators is perturbations of contraction semigroups, i.e., generators from the class $Q(1,0)$, cf. Proposition 1.48. Since this book is about the Gibbs semigroups, we consider only Hilbert spaces, see comments in Notes in Section 1.8.

Proposition 1.52. *Let $A \in Q(1,0)$ be the generator of a contraction semigroup on $\mathcal{H}$. If $B \in \mathcal{P}_{b<1}(A)$ is an accretive operator, then $H = A+B$ with* $\operatorname{dom} H = \operatorname{dom} A$ *is the generator of a contraction semigroup on $\mathcal{H}$, that is, $H \in Q(1,0)$.*

Proof. By Proposition 1.43, to prove the assertion it is sufficient to verify that H is an m-accretive operator in $\mathcal{H}$.

According to Definition 1.39, we have first to check that the operator $H = A+B$ with $\operatorname{dom} H = \operatorname{dom} A$ is closed. Indeed, since $B \in \mathcal{P}_{b<1}(A)$, (1.101) shows that

$$-a\|u\| + (1-b)\|Au\| \leqslant \|Hu\| \leqslant a\|u\| + (1+b)\|Au\|, \tag{1.108}$$

for $u \in \operatorname{dom} A$. Then by virtue of the inequalities (1.108), the operator H is closed since A is closed.

Next, by Proposition 1.43, A is an m-accretive operator because $A \in Q(1,0)$. Since B is an accretive operator, it follows that

$$\mathfrak{Re}(u,Hu) = \mathfrak{Re}(u,Au) + \mathfrak{Re}(u,Bu) \geqslant 0, \quad u \in \operatorname{dom} A\ .$$

Therefore, H is a closed accretive operator with $\mathrm{Nr}H \subseteq \overline{\mathbb{C}}_+$, and

$$\|(H+\zeta\mathbb{1})u\| \geqslant \mathfrak{Re}(u,(H+\zeta)u) \geqslant \mathfrak{Re}\,\zeta\|u\|^2, \quad u \in \operatorname{dom} A\ ,$$

for $\zeta \in \mathbb{C}_+$. This implies (Proposition 1.35) that the deficiency of the closed operator H is $\operatorname{def}(H+\zeta\mathbb{1}) = 0$. Hence, $\|(H+\zeta\mathbb{1})^{-1}\| \leqslant (\mathfrak{Re}\,\zeta)^{-1}$ and, by Definition 1.39, the operator H is m-accretive. Hence, it is a generator of the contraction semigroup $\{U_t(H)\}_{t\geqslant 0}$ on $\mathcal{H}$. □

Remark 1.53. Consider the operator $H(\kappa) := A + \kappa B$, $\operatorname{dom} H(\kappa) = \operatorname{dom} A$, where A and B are the same as in Proposition 1.52. Then one can not guarantee the result of this proposition for complex $\kappa \in \mathbb{C}$, even if $|\kappa|$ is small. However, if A is a generator of a holomorphic semigroup and $B \in \mathcal{P}_b(A)$, see Proposition 1.51, there is a nice theory for A perturbed by κB ($\kappa \in \mathbb{C}$) for $|\kappa\, b| < 1$. Recall that the linear function: $\{H(\kappa) := A + \kappa B\}_{\{\kappa:\ |\kappa|<b^{-1}\}}$, is a simplest example of a holomorphic family of *type* (A), see Notes in Section 5.6.

Below we treat perturbations from the class $\mathcal{P}_b(A)$ under the assumption that the original unperturbed semigroup is generated by an m-sectorial operator A with semi-angle $\alpha \in [0, \pi/2)$. The effect of the perturbation is expressed by the *increment* of the semi-angle for $H(\kappa)$. In contrast to Propositions 1.51 and 1.52, it is the analytic properties of the *family* of holomorphic semigroups $\{U_z(H(\kappa))\}_{\kappa\in\mathbb{C}}$ which are here the focus of our discussion. The case of a self-adjoint generator $A \geqslant 0$, that is, of an m-sectorial operator A with semi-angle $\alpha = 0$, is of a particular interest, see Proposition 1.46. Then some estimates, which are due to Proposition 1.51, become more explicit.

Proposition 1.54. *Let $A \geqslant 0$ be a densely defined self-adjoint operator in $\mathcal{H}$. If $B \in \mathcal{P}_b(A)$ for some $b \geqslant 0$, then the following statements hold:*

(i) *For any $|\kappa\, b| < 1$, the operator $H(\kappa) := A + \kappa B$ is m-sectorial and $H(\kappa) \in \mathscr{H}(\theta(\kappa, b), \omega_0(\kappa, a, b))$ for the semi-angle defined by equation*

$$\operatorname{ctg} \theta(\kappa, b) = \frac{|\kappa| b}{\sqrt{1 - |\kappa|^2 b^2}}, \tag{1.109}$$

and the semigroup type

$$\omega_0(\kappa, a, b) \leqslant \frac{a|\kappa|}{1 - b|\kappa|}. \tag{1.110}$$

(ii) *For a fixed $r < b^{-1}$, and for any z in the sector with vertex zero,*

$$S_{\theta(r,b)} := \bigcap_{|\kappa|<r} S_{\theta(\kappa,b)}, \tag{1.111}$$

the family of holomorphic semigroups $\{U_z(H(\kappa))\}_{\kappa\in D_r}$ is operator-norm holomorphic in the disc $D_r = \{\kappa \in \mathbb{C} : |\kappa| < r\}$.

Proof. (i) By Proposition 1.46, A is the generator of a holomorphic contraction semigroup: $A \in \mathscr{H}(\theta = \pi/2, 0)$. Since $\kappa B \in \mathcal{P}_{|\kappa b|<1}(A)$, Proposition 1.51 shows that $H(\kappa)$ is the generator of a bounded holomorphic semigroup: $H(\kappa) \in \mathscr{H}(\theta', \omega)$, for some $\theta' < \pi/2$ and type $\omega \geqslant 0$. To localise the numerical range of the operator $H(\kappa)$ we proceed as follows.

Note that $\{U_z(A)\}_{z\in S_{\pi/2}}$ is a contraction holomorphic semigroup in the open sector $S_{\theta=\pi/2}$. Since A is self-adjoint, Proposition 1.35 yields

$$\|R_{-\zeta}(A)\| \leqslant \begin{cases} |\zeta|^{-1}, & \text{if } |\arg\zeta| \leqslant \frac{\pi}{2}, \\ |\,\mathfrak{Im}\ \zeta|^{-1}, & \text{if } \frac{\pi}{2} < \pm\arg\zeta < \pi. \end{cases} \tag{1.112}$$

By the spectral representation we have

$$\begin{aligned} \|A(A+\zeta)^{-1}\| &= \left\| \int_0^\infty \mathrm{d}E_A(\lambda)\, \frac{\lambda}{\lambda+\zeta} \right\| \\ &\leqslant \sup_{\lambda\geqslant 0} \left| \frac{\lambda}{\lambda+\zeta} \right| \\ &= \begin{cases} 1, & \text{if } |\arg\zeta| \leqslant \frac{\pi}{2}, \\ |\zeta|\,|\,\mathfrak{Im}\ \zeta|^{-1}, & \text{if } \frac{\pi}{2} < \pm\arg\zeta < \pi. \end{cases} \end{aligned} \tag{1.113}$$

Then (1.103), (1.112) and (1.113) yield

$$\|B(A+\zeta\mathbb{1})^{-1}\| \leqslant \begin{cases} a|\zeta|^{-1} + b, & \text{if } |\arg\zeta| \leqslant \frac{\pi}{2}, \\ a|\,\mathfrak{Im}\ \zeta|^{-1} + b|\zeta|\,|\,\mathfrak{Im}\ \zeta|^{-1}, & \text{if } \frac{\pi}{2} < \pm\arg\zeta < \pi. \end{cases} \tag{1.114}$$

Therefore, the condition $\|\kappa\, B\, R_{-\zeta}(A)\| < 1$ for the operator-norm convergence of the Neumann series for the resolvent $R_{-\zeta}(H(\kappa))$ and the estimates (1.112), (1.114) show that the resolvent set $\rho(-H(\kappa))$ contains the domain

$$\begin{aligned} \mathcal{M}_{\kappa,b} := &\left\{ \zeta \in \mathbb{C} : \ |\arg\zeta| \leqslant \frac{\pi}{2} \ \wedge \ |\zeta| > \frac{|\kappa|a}{1-|\kappa|b} \right\} \\ \cup &\left\{ \zeta \in \mathbb{C} : \ \frac{\pi}{2} \leqslant |\arg\zeta| \leqslant \pi \ \wedge \right. \\ &\left. |\,\mathfrak{Im}\,\zeta| > \frac{|\kappa|a}{1-|\kappa|^2b^2} + \frac{|\kappa|b}{\sqrt{1-|\kappa|^2b^2}} \left[(\mathfrak{Re}\,\zeta)^2 + \frac{|\kappa|^2a^2}{1-|\kappa|^2b^2} \right]^{1/2} \right\}. \end{aligned} \tag{1.115}$$

By virtue of (1.115), for any $\kappa \in D_{b^{-1}}$ the asymptotic slopes of the border line of $\mathcal{M}_{\kappa,b}$, when $|\zeta| \to \infty$, are equal to

$$\lim_{|\zeta|\to\infty} \frac{\mathfrak{Im}\,\zeta}{|\,\mathfrak{Re}\,\zeta|} = \pm\,\mathrm{tg}\left(\frac{\pi}{2} + \theta(\kappa,b)\right) = \pm\frac{|\kappa|b}{\sqrt{1-|\kappa|^2b^2}}\,, \tag{1.116}$$

for, respectively, $\pi/2 < \pm\arg\zeta < \pi$.

Since $\mathcal{M}_{\kappa,b} \subset \rho(-H(\kappa))$, by (1.115) and (1.116) we obtain that $H(\kappa)$ is a sectorial operator (Definition 1.41) with semi-angle $\alpha = \pi/2 - \theta(\kappa,b)$ (1.109). Its numerical range lies in the sector:

$$\mathrm{Nr}(H(\kappa)) \subset \overline{S}_{\frac{\pi}{2}-\theta(\kappa,b),\,\gamma(\kappa,a,b)}, \tag{1.117}$$

with vertex $\gamma(\kappa,a,b) := -|\kappa|a/(1-|\kappa|b)$. Then for the vertex of m-sectorial operator $H(\kappa)$ one gets $\gamma_{H(\kappa)} \geqslant \gamma(\kappa,a,b)$. Since for the corresponding semigroup the type is $\omega_0 = -\gamma_{H(\kappa)}$, this proves (1.110) for $\omega_0(\kappa,a,b) \leqslant -\gamma(\kappa,a,b)$. Therefore, $H(\kappa) \in \mathscr{H}(\theta(\kappa,b),\omega_0(\kappa,a,b))$, by Proposition 1.33 and $\{U_z(H(\kappa))\}_{z\in S_{\theta(\kappa,b)}}$ is a quasi-bounded holomorphic semigroup for any $\kappa \in D_{b^{-1}}$.

(ii) For a fixed $r < b^{-1}$, there is a contour $\Gamma \subset \bigcap_{\kappa\in D_r}\mathcal{M}_{\kappa,b}$ such that, for any $\kappa \in D_r$, one gets the Riesz-Dunford representation (1.67),

$$U_z(H(\kappa)) = \frac{1}{2\pi i}\int_\Gamma d\zeta\, e^{z\zeta}(\zeta\mathbb{1} + H(\kappa))^{-1}, \tag{1.118}$$

for the family of holomorphic semigroups $\{U_z(H(\kappa))\}_{z\in S_{\theta(r,b)}}$ for the sector $S_{\theta(r,b)} = \bigcap_{\kappa\in D_r} S_{\theta(\kappa,b)}$ (1.111) and for the operator-norm convergent Bochner integral. Now we take into account that the resolvent identity gives the representation

$$\begin{aligned} &U_z(H(\kappa+\epsilon)) - U_z(H(\kappa)) \\ &= \frac{1}{2\pi i}\int_\Gamma d\zeta\, e^{z\zeta}\, R_{-\zeta}(H(\kappa+\epsilon))(-\epsilon B)R_{-\zeta}(H(\kappa)) \end{aligned} \tag{1.119}$$

for $\kappa, \kappa+\epsilon \in D_r$. Therefore, for any $z \in S_{\theta(r,b)}$, the function: $\kappa \mapsto U_z(H(\kappa))$ is operator-norm holomorphic in $\kappa \in D_r$. □

Corollary 1.55. *If $a = 0$, then by virtue of* (1.110), *we get $\omega_0(\kappa, a = 0, b) = 0$. Then operator $H(\kappa) \geqslant 0$ is the generator of a contraction holomorphic semigroup, see Proposition* 1.46.

Corollary 1.56. *If $b = 0$, that is, the operator $B \in \mathcal{P}_0(A)$, then $H(\kappa) \in \mathscr{H}(\pi/2, \omega(\kappa, a, b = 0))$ for $\kappa \in \mathbb{C}$, see* (1.109).

1.8 Notes

Notes to Section 1.1. The material of this chapter is standard. We give only some of the popular references for further reading and a few historical remarks.

One of the first definitions of the exponential function in infinite-dimensional spaces appeared probably in the paper [Gra10]. Here we follow Kato's book [Kat80], Ch. IX. A brief history of the exponential function (by T. Hahn and C. Parazzoli) can be found in a rather exhaustive treatise [EN00], Ch.VII.

Notes to Section 1.2. The continuity assumption at $t = 0$ is crucial in the semigroup theory. The right *strong* continuity for $t = 0$ seems to be the most appropriate for the basic classes of semigroups, see discussion in [HP57], [Dav80], [Dav07] and [EN00]. Several other types of semigroups are studied in the book of E. Hille and R. S. Phillips [HP57], the great classic on one-parameter semigroup theory, and in the more recent encyclopedia on semigroups by K.-J. Engel and R. Nagel [EN00].

Notes to Section 1.3. The material of this section is a part of standard courses on strongly continuous one-parameter semigroups. To complete the references quoted above, we only add here the books [Yos65], [Kre71] as well as [Paz83], [Gol85] and [Ves96].

Proposition 1.12 for $\omega_0 = 0$ and $M = 1$ is the famous Hille-Yosida theorem for contraction semigroups, whereas the general case is often called the Hille-Yosida-Phillips theorem.

Our proof of Proposition 1.12 is based on the classical *Euler formula* (or limit). Another popular way to prove this proposition resorts to the so-called Yosida approximants (1.57), that we consider in Section 1.4.

The criterion proved in Proposition 1.13 is due to E. Nelson. Here we followed [Dav07], Ch.6.1.

Notes to Section 1.4. Proposition 1.14 states that if a semigroup is right norm-continuous at $t = 0$, then it is in some sense *trivial.* Recall that it is not the case for the so-called *immediately* norm-continuous or *eventually* norm-continuous semigroups, that we discuss in Section 4.2, see also [EN00] Ch.II, Section 4. In the last case, the semigroup is norm-continuous for all $t > t_0$ and some $t_0 > 0$. For the immediately norm-continuous semigroup $t_0 = 0$. The both are strongly right-continuous at $t = 0$. Although such semigroups are known since [HP57], the complete characterisation of generators of these semigroups is (as far as I know) still an open problem, see [You92].

Proposition 1.17 is again the Hille-Yosida theorem. For the construction of the strongly continuous contraction semigroup, we used the Yosida approximants (1.57).

The main importance of the "no-go" Propositions 1.21 and 1.22 is that they rule out nontrivial extensions to *weakly* continuous semigroups. In a Hilbert space, Corollary 1.23 easily settles the question of continuity for the *adjoint* semigroup. In a Banach space the dual U_t^* (to a strongly continuous on $\mathcal{B}$ semigroup U_t) is not, in general, strongly continuous on $\mathcal{B}^*$. But $\mathbb{R}_0^+ \ni t \mapsto \langle u|U_t^*\phi\rangle$ is *continuous* for any $u \in \mathcal{B}, \phi \in \mathcal{B}^*$. The topology $\sigma(\mathcal{B}^*, \mathcal{B})$ generated on $\mathcal{B}^*$ by the set of functionals from $\mathcal{B}$ is the weak*-topology. This observation allows one to develop on the dual space $\mathcal{B}^*$ a consistent theory of dual weak*-continuous semigroups, see [EN00] Ch.II, Section 2.

Notes to Section 1.5. Holomorphic semigroups arise from parabolic partial differential equations and represent an important subclass of the strongly continuous semigroups, which are also immediately norm continuous in the above sense. In this section we followed essentially [Dav80], [Kat80] and [RS75].
A complete proof of the Proposition 1.30 one can find in [Dav07], Chapter 8.4, Theorem 8.4.9.

Notes to Section 1.6. The numerical range of a generator seems a very natural way to characterise the holomorphic semigroups on a Hilbert space. Our exposition follows [Kat80] and [RS75].

Notes to Section 1.7. We recall that there are essentially two ways to study perturbations of semigroups. They are the *semigroup based* methods and the *resolvent based* methods, Chapters 11.4 and 11.5 in [Dav07]. Since for contraction as well as for holomorphic semigroups one has to control only the resolvent (and not all its powers), we consider in this section only the resolvent based methods. Proposition 1.51 (Hille theorem, [HP57], p.418) and Proposition 1.52 give examples of such a method.

We note that the proof of Proposition 1.51 in a Banach space is more complicated. The standard Kato perturbation theory yields straightforwardly the proof for the relative bound $b < 1/2$, but the extension to $b < 1$ needs additional, although not very complicated arguments, see for example, 2.7 Theorem in [EN00], Ch.III, Section 2. If the Banach space is *reflexive*, the proof for $b < 1$ is simpler (Corollary III.2.9, [EN00]), and thus so is the proof of Proposition 1.51 for a Hilbert space.

We return to the perturbation theory in Chapter 4, where the semigroup based methods will be also presented.

The results of this section are contained in the monographs [HP57], [Kat80] and [Dav80]. In our proof of Propositions 1.54, we followed the line of reasoning of [Mai71] and [Zag89].

Chapter 2

Classes of compact operators

In this chapter, we deal with norm ideals (classes) of compact operators on a Hilbert space. After a rather standard presentation of compact operators, we introduce the von Neumann-Schatten ideals and discuss their properties making essential use of the notion of singular values. The following section is devoted to a detailed discussion of norm convergence theorems in these ideals. More inequalities involving the singular values of compact operators are featured in the last section, as they are indispensable for the study of Gibbs semigroups.

2.1 Compact operators on a Hilbert space

Let $\mathcal{H}$ be a complex, separable, infinite-dimensional Hilbert space with inner product $(\cdot,\cdot)$ as in Section 1.6. For $u, v \in \mathcal{H}$ and $\alpha, \beta \in \mathbb{C}$, $(\alpha u, \beta v) = \alpha\overline{\beta}\,(u, v)$.

Recall that a bounded operator $A \in \mathcal{L}(\mathcal{H})$ is called *non-negative* (or positive), if $(Au, u) \geqslant 0$ for all $u \in \mathcal{H}$. We then write $A \geqslant 0$, and correspondingly $A \geqslant B$ if $A - B \geqslant 0$.

Remark 2.1. Every non-negative operator $A \in \mathcal{L}(\mathcal{H})$ on a complex Hilbert space $\mathcal{H}$ is self-adjoint: $A^* = A$. This is a consequence of

$$0 \leqslant (Au, u) = (u, A^*u) = \overline{(u, Au)} = (u, Au)$$

and the *polarisation identity*

$$(u, v) = \frac{1}{4}\{(\|u+v\|^2 - \|u-v\|^2) - i(\|u+iv\|^2 - \|u-iv\|^2)\},$$

which leads to $(Au, v) = (u, Av)$ for $u, v \in \mathcal{H}$. Here $\|f\| := \sqrt{(f, f)}$ denotes the usual vector norm on $\mathcal{H}$.

For any $A \in \mathcal{L}(\mathcal{H})$, one has $A^*A \geqslant 0$ since $(A^*Au, u) = \|Au\|^2 \geqslant 0$. Let $E_K(\lambda)$ be the spectral measure corresponding to the self-adjoint operator $K = A^*A$. Then

V. A. Zagrebnov, *Gibbs Semigroups*, Operator Theory: Advances and Applications 273, https://doi.org/10.1007/978-3-030-18877-1_2

(the *square-root lemma*) the bounded operator

$$\sqrt{K} = \int_{\sigma(K)} \mathrm{d}E_K(\lambda)\,\sqrt{\lambda} \tag{2.1}$$

is well defined, self-adjoint and non-negative, cf. Section 3.1. Therefore, for any $A \in \mathcal{L}(\mathcal{H})$, one can define the absolute value of A by (2.1), i.e., $|A| := \sqrt{A^*A} = |A|^*$.

Remark 2.2. In spite of the fact that $|\alpha A| = |\alpha|\,|A|$ for $\alpha \in \mathbb{C}$, relations such as $|AB| = |A|\,|B|$, $|A| = |A^*|$ or $|A+B| \leqslant |A| + |B|$ are *false* in general.

Recall that an operator $U \in \mathcal{L}(\mathcal{H})$ such that $\|Uu\| = \|u\|$ for all $u \in \mathcal{H}$ is called an *isometry*. The operator U is a *partial isometry* if $U \restriction \ker U^\perp$, i.e., the restriction of U to the orthogonal complement of $\ker U := \{f \in \mathcal{H} : Uf = 0\}$, is an isometry. Hence, in this case

$$\mathcal{H} = \ker U \oplus \ker U^\perp = \operatorname{ran} U \oplus \operatorname{ran} U^\perp$$

and $U : \ker U^\perp \to \operatorname{ran} U$ is unitary. Note that $U^* : \operatorname{ran} U \to \ker U^\perp$ acts as the inverse of U. Therefore, $P_1 = U^*U$ and $P_2 = UU^*$ are the orthogonal projections onto $\ker U^\perp$ and $\operatorname{ran} U$, which are also called the *initial* and *final* spaces of U.

Proposition 2.3 (polar decomposition). *Let $A \in \mathcal{L}(\mathcal{H})$. There exists a unique partial isometry U such that $A = U|A|$ and* $\ker U = \ker A$. *Moreover,* $\operatorname{ran} U = \overline{\operatorname{ran} A}$.

Proof. Define $U : \operatorname{ran}|A| \to \operatorname{ran} A$ by $U(|A|u) := Au$. Since $\||A|u\| = \|Au\|$, one has $\ker|A| = \ker A$. Therefore, U is well defined, that is, if $Uf = g$, where $f = |A|u$ and $g = Au$, then $f = 0$ implies that $u \in \ker|A|$ and hence $g = 0$. Moreover, we get also that $\overline{\operatorname{ran}|A|} = \mathcal{H} \ominus \ker|A| = \mathcal{H} \ominus \ker A = \overline{\operatorname{ran} A^*}$. Hence, as $\|f\| = \|g\|$, the operator U is isometric, i.e. it extends to an isometry from $\overline{\operatorname{ran}|A|} = \overline{\operatorname{ran} A^*}$ to $\overline{\operatorname{ran} A}$. To extend U to all of $\mathcal{H}$, we define it to be zero on $\operatorname{ran}|A|^\perp$. Therefore, U is a partial isometry with initial space $\overline{\operatorname{ran} A^*}$ and final space $\operatorname{ran} U = \overline{\operatorname{ran} A}$. Since $|A|$ is self-adjoint, $\operatorname{ran}|A|^\perp = \ker|A| = \ker A$, thus $\ker U = \ker A$. □

Now we pass to the main subject of this section: the *compact* operators and norm ideals (classes) in this ring of operators on a Hilbert space $\mathcal{H}$.

Definition 2.4. A bounded operator A on $\mathcal{H}$ is *compact* if the image $\{Au_n\}_{n\geqslant 1}$ of any bounded sequence $\{u_n\}_{n\geqslant 1} \subset \mathcal{H}$ contains a Cauchy sequence. This is equivalent to the statement that A maps bounded subsets of $\mathcal{H}$ into precompact sets, i.e., into subsets with a compact closure. We shall denote the set of compact operators on $\mathcal{H}$ by $\mathcal{C}_\infty(\mathcal{H})$.

Recall that a sequence $\{u_n\}_{n\geqslant 1}$ on $\mathcal{H}$ converges *weakly* to u if for any $v \in \mathcal{H}$, $(u_n, v) \to (u, v)$ when $n \to \infty$. This will be denoted by $\text{w-lim}_{n\to\infty} u_n = u$. The corresponding topology is weaker than the topology defined by the vector norm on $\mathcal{H}$, and used in Definition 2.4.

Note that there is no difference between compact $\mathcal{C}_\infty(\mathcal{H})$ and *completely continuous* operators $\mathcal{L}_\infty(\mathcal{H})$ on $\mathcal{H}$, see Notes to Section 2.1 in Section 2.5.

Proposition 2.5. *A compact operator A maps any weakly convergent sequence $\{u_n\}_{n\geqslant 1}$ into a norm convergent sequence $\{Au_n\}_{n\geqslant 1}$.*

Proof. Suppose that $\text{w-lim}_{n\to\infty} u_n = u$. Then by the uniform boundedness principle, $\sup_n \|u_n\| < \infty$ and $Au = \text{w-lim}_{n\to\infty} Au_n$. If $\|Au_n - Au\| \nrightarrow 0$, for $n \to \infty$, then there exists a subsequence $\{u_{n'}\}_{n'\geqslant 1}$ such that $\inf_{n'} \|A(u_{n'} - u)\| = \varepsilon > 0$. As $A \in \mathcal{C}_\infty(\mathcal{H})$, the set $\{A(u_{n'} - u)\}$ is precompact. Therefore, one can find a subsequence $\{u_{n''}\}_{n''\geqslant 1}$ such that $\lim_{n''\to\infty} A(u_{n''} - u) = g$ with $\|g\| \geqslant \varepsilon$. However, $\text{w-lim}_{n\to\infty} Au_n = Au$ implies that $g = 0$, which is impossible if $\varepsilon > 0$. Thus, $\inf_{n\geqslant 1} \|A(u_n - u)\| = 0$, or $\lim_{n\to\infty} Au_n = Au$. □

Remark 2.6. Since any ball in a separable Hilbert space $\mathcal{H}$ is weakly compact, the converse of Proposition 2.5 also holds. Indeed, suppose that A maps any weakly convergent sequence $\{u_n\}_{n\geqslant 1}$ into a vector-norm convergent sequence $\{Au_n\}_{n\geqslant 1}$. Since by the uniform boundedness principle $\sup_{n\geqslant 1} \|u_n\| < \infty$, the sequence $\{u_n\}_{n\geqslant 1}$ belongs to a ball in $\mathcal{H}$. Then there is a subsequence $\{u_{n'}\}_{n'\geqslant 1}$ so that $u = \text{w-lim}_{n'\to\infty} u_{n'}$, and consequently $Au_{n'} \to Au$. Hence, the operator A is compact by Definition 2.4. This is not true in a Banach space if the space is not reflexive, and that makes there a difference between completely continuous and compact operators, see Notes to Section 2.1 in Section 2.5

Proposition 2.7. *The set of compact operators $\mathcal{C}_\infty(\mathcal{H})$ is a two-sided $*$-ideal in the algebra of bounded operators $\mathcal{L}(\mathcal{H})$:*

(a) *$\mathcal{C}_\infty(\mathcal{H})$ is a linear subspace of $\mathcal{L}(\mathcal{H})$.*

(b) *If $A \in \mathcal{C}_\infty(\mathcal{H})$ and $B \in \mathcal{L}(\mathcal{H})$, then $AB \in \mathcal{C}_\infty(\mathcal{H})$ and $BA \in \mathcal{C}_\infty(\mathcal{H})$.*

(c) *If $A \in \mathcal{C}_\infty(\mathcal{H})$, then $A^* \in \mathcal{C}_\infty(\mathcal{H})$.*

Proof. (a) Let $\{u_n\}_{n\geqslant 1}$ be a weakly convergent sequence in $\mathcal{H}$. For $A, B \in \mathcal{C}_\infty(\mathcal{H})$ and $\alpha, \beta \in \mathbb{C}$ one has $(\alpha A + \beta B)u_n = \alpha A u_n + \beta B u_n \to \alpha A u + \beta B u$. So, by Remark 2.6, $(\alpha A + \beta B) \in \mathcal{C}_\infty(\mathcal{H})$.

(b) Since a bounded operator B preserves the norm and weak topologies on $\mathcal{H}$, the assertion is a direct consequence of Proposition 2.5 and Remark 2.6.

(c) Since $\|A\| = \|A^*\|$, the operator A^* belongs to $\mathcal{L}(\mathcal{H})$ and, therefore, $AA^* \in \mathcal{C}_\infty(\mathcal{H})$. Let $u = \text{w-lim}_{n\to\infty} u_n$. Then $AA^*u_n \to AA^*u$, and since

$$\begin{aligned}\|A^*u_n - A^*u\|^2 &= (AA^*(u_n - u), (u_n - u))\\ &\leqslant \|AA^*(u_n - u)\|\|(u_n - u)\|,\end{aligned}$$

we conclude that $A^*u_n \to A^*u$, that is, $A^* \in \mathcal{C}_\infty(\mathcal{H})$. □

We caution the reader that $AB \in \mathcal{C}_\infty(\mathcal{H})$ does *not* imply that one of these operators is compact.

The next statement says that the ideal $\mathcal{C}_\infty(\mathcal{H})$ is closed in the operator norm topology. This type of convergence is denoted by

$$\|\cdot\|\text{-}\lim_{n\to\infty} A_n = A \quad \text{meaning} \quad \lim_{n\to\infty} \sup_{\substack{\varphi\in\mathcal{H}\\ \|\varphi\|=1}} \|(A - A_n)\varphi\| = 0.$$

Proposition 2.8. *If a sequence $\{A_n\}_{n\geqslant 1}$ of compact operators converges to A in the operator norm topology, then $A \in \mathcal{C}_\infty(\mathcal{H})$.*

Proof. Suppose that $\{u_m\}_{m\geqslant 1}$ is bounded. Then $\sup_{m\geqslant 1}\|u_m\| = \delta < \infty$ and for any subsequence $\{u_{m_k}\}_{k\geqslant 1}$ and $u \in \mathcal{H}$,

$$\|A(u_{m_k} - u)\| \leqslant \|(A - A_n)u_{m_k}\| + \|A_n(u_{m_k} - u)\| + \|(A_n - A)u\|. \tag{2.2}$$

For any $n'' > n' \geqslant 1$ there are subsequences $\{u_{n''_k}\}_{k\geqslant 1} \subset \{u_{n'_k}\}_{k\geqslant 1}$ such that $\{A_{n'}u_{n'_k}\}_{k\geqslant 1}$ and $\{A_{n''}u_{n''_k}\}_{k\geqslant 1}$ are Cauchy sequences. Then, by a diagonal trick, one finds a subsequence $\{u_{k_k}\}_{k\geqslant 1}$ such that $\{A_n u_{k_k}\}_{k\geqslant 1}$ is Cauchy for every $n \geqslant 1$. Hence, for any $\varepsilon > 0$ one can find $N(\varepsilon)$ and $K(\varepsilon)$ so that in (2.2)

$$\|(A - A_n)u_{k_k}\| \leqslant \|A - A_n\|\delta < \varepsilon,$$

also

$$\|(A_n - A)u_{k_k+p}\| \leqslant \|A_n - A\|\delta < \varepsilon,$$

for all $n > N(\varepsilon)$, and at the same time $\|A_n(u_{k_k} - u_{k_k+p})\| < \varepsilon$ for $k_k, k_k+p > K(\varepsilon)$. Thus $\{Au_{k_k}\}_{k\geqslant 1}$ is a Cauchy sequence. Hence, A is compact by Definition 2.4. □

An important subclass of compact operators is that of the *finite-rank* operators.

Definition 2.9. A bounded operator A is of finite rank if

$$\dim \operatorname{ran} A < \infty$$

or, equivalently, there exists an orthonormal basis $\{e_j\}_{j\geqslant 1} \subset \mathcal{H}$ such that

$$Au = \sum_{j=1}^{r}(Au, e_j)e_j, \quad u \in \mathcal{H} \tag{2.3}$$

for some finite r. The set of all finite-rank operators will be denoted by $\mathcal{K}(\mathcal{H})$. Let $\mathcal{K}_r(\mathcal{H}) := \{A \in \mathcal{L}(\mathcal{H}) : \dim \operatorname{ran} A = r\}$. Then $\mathcal{K}(\mathcal{H}) = \bigcup_{r\geqslant 0}\mathcal{K}_r(\mathcal{H})$.

Remark 2.10. The following properties of $\mathcal{K}(\mathcal{H})$ are a direct consequence of Definition 2.9:

(a) $\mathcal{K}(\mathcal{H}) \subset \mathcal{C}_\infty(\mathcal{H})$ because for any $A \in \mathcal{K}(\mathcal{H})$ the set $\{A\varphi : \|\varphi\| \leqslant 1\}$ is precompact in $\mathcal{H}$.

(b) By (2.3) one has

$$(Au, v) = \sum_{j=1}^{r}(u, A^*e_j)(e_j, v) = (u, A^*v).$$

Denoting $\tilde{e}_j := A^*e_j$, we get the representations

$$A = \sum_{j=1}^{r}(\cdot, \tilde{e}_j)e_j, \quad A^* = \sum_{j=1}^{r}(\cdot, e_j)\tilde{e}_j \tag{2.4}$$

for A and A^*. Hence, if $A \in \mathcal{K}_r(\mathcal{H})$, then $A^* \in \mathcal{K}_r(\mathcal{H})$.

(c) $\mathcal{K}(\mathcal{H})$ is a linear subspace of $\mathcal{C}_\infty(\mathcal{H})$. Moreover, if $A \in \mathcal{K}(\mathcal{H})$ and $B \in \mathcal{C}_\infty(\mathcal{H})$, then $AB \in \mathcal{K}(\mathcal{H})$ and $BA \in \mathcal{K}(\mathcal{H})$ by (2.4). Hence, $\mathcal{K}(\mathcal{H})$ is a two-sided $*$-ideal in $\mathcal{C}_\infty(\mathcal{H})$ and in $\mathcal{L}(\mathcal{H})$.

The following property is typical for compact operators, cf. Proposition 2.8.

Proposition 2.11. *Every compact operator $A \in \mathcal{C}_\infty(\mathcal{H})$ is the operator norm limit of a sequence $\{A_n\}_{n\geqslant 1} \subset \mathcal{K}(\mathcal{H})$.*

Proof. Let $\{e_j\}_{j\geqslant 1}$ be an orthonormal basis in $\mathcal{H}$ and let $\mathcal{H}_n^\perp$ be the orthogonal complement of the subspace $\mathcal{H}_n := [e_1, \dots, e_n]$ spanned by $\{e_j\}_{j=1}^n$. We set $\mathcal{H}_{n=0}^\perp = \mathcal{H}$ and define

$$s_{n+1} := \sup_{\substack{u \in \mathcal{H}_n^\perp \\ \|u\|=1}} \|Au\|. \tag{2.5}$$

Then the sequence $\{s_n\}_{n\geqslant 1}$ is monotonically decreasing, so it converges to a limit $s^* \geqslant 0$. Choose a normalised sequence $\{u_n \in \mathcal{H}_{n-1}^\perp\}_{n\geqslant 1}$ such that $\|Au_n\| \geqslant s^*/2$. Since for any $f \in \mathcal{H}$

$$\lim_{n\to\infty} (f, u_n) = \lim_{n\to\infty} \sum_{j=n+1}^{\infty} (f, e_j)(e_j, u_n) = 0,$$

w-lim $u_n = 0$, and, by Proposition 2.5, one gets $\lim_{n\to\infty} Au_n = 0$. Thus $s^* = 0$. Define now $A_n := \sum_{j=1}^n (\cdot, e_j)Ae_j \in \mathcal{K}_n(\mathcal{H})$. Then

$$\begin{aligned} \|A - A_n\| &= \sup_{\substack{v \in \mathcal{H} \\ \|v\|=1}} \|(A - A_n)v\| = \sup_{\substack{v \in \mathcal{H} \\ \|v\|=1}} \Big\| \sum_{j=n+1}^{\infty} (v, e_j)Ae_j \Big\| \\ &= \sup_{\substack{u \in \mathcal{H}_n^\perp \\ \|u\|=1}} \|Au\| = s_{n+1}. \end{aligned}$$

Hence, $\|\cdot\|$-$\lim_{n\to\infty} A_n = A$, as claimed. □

Remark 2.12. The above statement means that $\mathcal{C}_\infty(\mathcal{H})$ is the closure of $\mathcal{K}(\mathcal{H})$ in the operator norm. Let $P_n : \mathcal{H} \to \mathcal{H}_n$ be the orthogonal projection onto the subspace spanned by $\{e_j : j = 1, 2, \dots, n\}$, where $\{e_j\}_{j\geqslant 1}$ is an orthonormal basis in $\mathcal{H}$. Then, taking for any $B \in \mathcal{L}(\mathcal{H})$ the operator $B_n := P_n B \in \mathcal{K}_n(\mathcal{H})$, we have

$$\lim_{n\to\infty} \|(B - B_n)v\| \leqslant \|B\| \lim_{n\to\infty} \|(\mathbb{1} - P_n)v\| = 0, \quad v \in \mathcal{H}.$$

We say that B is the strong limit of the B_n: $B =$ s-lim $B_n = B$. Therefore, the *strong closure* of the class $\mathcal{K}(\mathcal{H})$ coincides with $\mathcal{L}(\mathcal{H})$. For example, s-$\lim_{n\to\infty} P_n = \mathbb{1} \notin \mathcal{C}_\infty(\mathcal{H})$.

2.2 The canonical form of a compact operator

Definition 2.13. Let A be a closed linear operator in $\mathcal{H}$ with $\operatorname{dom} A$. The *resolvent set* $\rho(A)$ of A is defined by

$$\rho(A) := \{\lambda \in \mathbb{C} : \ker A_\lambda = \{0\} \text{ and } \operatorname{ran} A_\lambda = \mathcal{H}\}.$$

Thus, the resolvent $R_\lambda(A) := A_\lambda^{-1}$ has $\operatorname{dom} R_\lambda(A) = \mathcal{H}$ and is a bounded operator with $\operatorname{ran} R_\lambda(A) = \operatorname{dom} A$ for any $\lambda \in \rho(A)$. The set $\sigma(A) = \mathbb{C}\backslash\rho(A)$ is called the *spectrum* of A. Here $A_\lambda = A - \lambda\mathbb{1}$.

We need to distinguish three subsets of the spectrum. For convenience they are recalled below, cf. Appendix A, Chapter 7.4.

Definition 2.14. Let A be a closed linear operator in $\mathcal{H}$. Then

(a) the set $\sigma_{\mathrm{p}}(A) := \{\lambda \in \mathbb{C} : \ker A_\lambda \neq \{0\}\}$ is called the *point spectrum* of A;

(b) the set $\sigma_{\mathrm{cont}}(A) := \{\lambda \in \mathbb{C} : \ker A_\lambda = \{0\} \;\; \wedge \; \operatorname{ran} A_\lambda \neq \overline{\operatorname{ran} A_\lambda} = \mathcal{H}\}$ is called the *continuous spectrum* of A. For $\lambda \in \sigma_{\mathrm{cont}}(A)$, the domain of the resolvent is a dense subspace of $\mathcal{H}$;

(c) the set $\sigma_{\mathrm{res}}(A) := \{\lambda \in \mathbb{C} : \ker A_\lambda = \{0\} \; \wedge \; \overline{\operatorname{ran} A_\lambda} \subset \mathcal{H}\}$ is called the *residual spectrum* of A. Note that the range of A_λ is not dense for $\lambda \in \sigma_{\mathrm{res}}(A)$.

Remark 2.15. If $\lambda = 0$ would belong to the resolvent set $\rho(A)$ of a compact operator A, then $A^{-1} \in \mathcal{L}(\mathcal{H})$ and, by Proposition 2.7, one would get $\mathbb{1} = A^{-1}A \in \mathcal{C}_\infty(\mathcal{H})$, which is impossible if $\mathcal{H}$ is infinite-dimensional. Hence, $\lambda = 0$ always belongs to $\sigma(A)$ for any $A \in \mathcal{C}_\infty(\mathcal{H})$.

Remark 2.16. According to Definition 2.14,

$$\sigma(A) = \sigma_{\mathrm{p}}(A) \cup \sigma_{\mathrm{cont}}(A) \cup \sigma_{\mathrm{res}}(A), \tag{2.6}$$

and these three subsets do not intersect.

Remark 2.17. Let $\lambda_0 \in \rho(A)$. By Definition 2.13, there is a $C(\lambda_0) > 0$ such that $\|A_{\lambda_0}u\| \geqslant C(\lambda_0)\|u\|$ for all $u \in \operatorname{dom} A$. This implies that for $|\lambda_0 - \lambda| < C(\lambda_0)$

$$\begin{aligned}\|[(\lambda_0 - \lambda)\mathbb{1} + A_{\lambda_0}]^{-1}u\| &= \|(A_{\lambda_0})^{-1}\sum_{n=0}^{\infty}(\lambda - \lambda_0)^n(A_{\lambda_0})^{-n}u\| \\ &\leqslant C^{-1}(\lambda_0)(1 - |\lambda_0 - \lambda|C^{-1}(\lambda_0))\|u\|.\end{aligned}$$

Therefore, the resolvent set $\rho(A)$ is open, i.e., the open neighbourhood $\{\lambda \in \mathbb{C} : |\lambda - \lambda_0| < C(\lambda_0)\}$ of a regular point $\lambda_0 \in \rho(A)$ belongs to the resolvent set. Hence, the spectrum $\sigma(A)$ is a closed subset of $\mathbb{C}$.

For a bounded operator $A \in \mathcal{L}(\mathcal{H})$, we can easily localise its spectrum $\sigma(A)$ in the complex plane. Let $|\lambda| > \|A\|$, then

$$\|R_\lambda(A)u\| = |\lambda|^{-1}\Big\|\sum_{n=0}^{\infty}(A/\lambda)^n u\Big\| \leqslant |\lambda|^{-1}\{1 - \|A\|/|\lambda|\}^{-1}\|u\|, \quad u \in \mathcal{H},$$

which means that $\sigma(A) \subset \{\lambda : |\lambda| \leqslant \|A\|\}$. For compact operators $\mathcal{C}_\infty(\mathcal{H})$ we can say more.

Proposition 2.18. *Let A be a compact operator on $\mathcal{H}$. Then $\sigma(A)$ is a discrete set with no other accumulation point than zero. Furthermore, each non-zero $\lambda \in \sigma(A)$ is an eigenvalue of A of finite multiplicity, i.e., the corresponding space of eigenvectors is finite-dimensional, and $\overline{\lambda}$ is an eigenvalue of A^* with the same multiplicity as λ.*

Lemma 2.19. *The eigenvalues of a compact operator A have no other accumulation points than zero.*

Proof. Suppose that the statement of the lemma is false. Then there exists a sequence $\{\lambda_n\}_{n\geqslant 1}$ of eigenvalues of A with corresponding eigenvectors u_n such that $|\lambda_n| \geqslant \delta > 0$. Suppose first that $\{\lambda_n\}$ contains only a finite number of distinct eigenvalues. Then there is a subsequence $\{\lambda_{n_j}\}_{n_j\geqslant 1}$ such that all λ_{n_j} are equal: $\lambda_{n_j} = \lambda^*$. Let $\mathcal{H}_{\lambda^*} \subset \mathcal{H}$ be the subspace spanned by the corresponding eigenvectors $\{u_{n_j}\}$. It is obviously invariant under A. In particular, the operator A maps the bounded set $\{v \in \mathcal{H}_{\lambda^*} : \|v\| \leqslant (\lambda^*)^{-1}\}$ into the unit ball $B = \{u \in \mathcal{H}_{\lambda^*} : \|u\| \leqslant 1\}$, which should be compact due to the compactness of A. This is possible only if $\dim \mathcal{H}_{\lambda^*} < \infty$. Hence, the subsequence $\{\lambda_{n_j}\}_{n_j\geqslant 1}$ is finite and λ^* is not a limit point, i.e., each eigenvalue has a finite multiplicity.

Next, suppose that $\{\lambda_n\}$ contains an infinite number $\{\lambda_\alpha\}$ of distinct eigenvalues with $|\lambda_\alpha| \geqslant \delta > 0$. Then, for any $m = 1, 2, \ldots$, the eigenvectors $\{u_\alpha : \alpha = 1, 2, \ldots, m\}$ are linearly independent. If this is not the case, then $u_m = \sum_{\alpha=1}^{m-1} C_\alpha u_\alpha$ and $\lambda_m u_m = \sum_{\alpha=1}^{m-1} C_\alpha \lambda_\alpha u_\alpha$, which implies that $\sum_{\alpha=1}^{m-1}(1 - \lambda_\alpha/\lambda_m)u_\alpha = 0$. Let $\mathcal{H}_m \subset \mathcal{H}$ now be the subspace spanned by $\{u_\alpha : \alpha = 1, 2, \ldots, m\}$; then $\mathcal{H}_n$ is invariant under A. Since the vectors $\{u_\alpha : \alpha = 1, 2, \ldots, m\}$ are linearly independent, $\mathcal{H}_{m-1}$ is a proper subspace of $\mathcal{H}_m$ and, by the *Gram-Schmidt orthogonalisation* procedure, there is a $v_m \in \mathcal{H}_m$ with $\|v_m\| = 1$ such that $v_m \perp \mathcal{H}_{m-1}$. Hence, the sequence $\{v_\alpha\}_{\alpha\geqslant 1}$ is orthonormal and the set $\{\lambda_\alpha^{-1} v_\alpha\} \subset \mathcal{H}$ is bounded. Consider $A(\lambda_n^{-1}v_n - \lambda_m^{-1}v_m) = v_n - [\lambda_m^{-1}Av_m - \lambda_n^{-1}(A - \lambda_n)v_n]$ with $m < n$. Then $Av_m \in \mathcal{H}_{n-1}$ because $v_m \in \mathcal{H}_{n-1}$ and $\mathcal{H}_{n-1}$ is invariant under A.

By the construction of $\mathcal{H}_n$ the operator $(A - \lambda_n \mathbb{1})$ maps $\mathcal{H}_n$ into $\mathcal{H}_{n-1}$. Hence, $(A - \lambda_n\mathbb{1})v_n \in \mathcal{H}_{n-1}$ and $[\lambda_m^{-1}Av_m - \lambda_n^{-1}(A - \lambda_n\mathbb{1})v_n] \in \mathcal{H}_{n-1} \perp v_n$. Therefore, $\|A(\lambda_n^{-1}v_n - \lambda_m^{-1}v_m)\| \geqslant \|v_n\| = 1$ for any $m < n$, showing that the set $\{A(\lambda_\alpha^{-1}v_\alpha)\}$ is not compact. This contradicts to compactness of A. □

Lemma 2.20. *Suppose that $\lambda \neq 0$ is not an eigenvalue of the compact operator A. Then* ran A_λ *is closed.*

Proof. Suppose that a sequence $\{A_\lambda u_n\}_{n\geqslant 1} \subset \operatorname{ran} A_\lambda$ converges to v. We need to show that $v \in \operatorname{ran}(\lambda\mathbb{1} - A)$.

If $\{u_n\}$ is bounded, then by the compactness of A, there is a subsequence $\{u_{n'}\}_{n'\geqslant 1}$ such that $Au_{n'} \to w$. Hence, $\lambda u_{n'} = (\lambda\mathbb{1} - A)u_{n'} + Au_{n'} \to v + w$ and $\lambda Au_{n'} \to A(v+w)$. Thus, $\lambda w = A(v+w)$, or $v = \lambda^{-1}(\lambda\mathbb{1} - A)(v+w) \in \operatorname{ran} A_\lambda$.

Now, if $\{u_n\}$ is unbounded, then there is a subsequence $\{u_{n'}\}_{n'\geqslant 1}$ such that $\|u_{n'}\| \to \infty$. Set $\tilde{u}_{n'} := u_{n'}/\|u_{n'}\|$; then $\{\tilde{u}_{n'}\}_{n'\geqslant 1}$ is a bounded sequence and $\lim_{n'\to\infty} A_\lambda u_{n'}/\|u_{n'}\| = 0$. By the same argument as above, there is a subsequence $\{\tilde{u}_{n''}\}_{n''\geqslant 1}$ such that $A\tilde{u}_{n''} \to \tilde{w}$, $\lambda\tilde{u}_{n''} = (\lambda\mathbb{1} - A)\tilde{u}_{n''} + A\tilde{u}_{n''} \to \tilde{w}$ and $\lambda A\tilde{u}_{n''} \to A\tilde{w}$, or $\lambda\tilde{w} - A\tilde{w} = 0$, where $\|\tilde{w}\| = \lim_{n''\to\infty} \|\lambda\tilde{u}_{n''}\| = |\lambda| > 0$. So, $\tilde{w}$ is an eigenvector of A corresponding to an eigenvalue $\lambda \neq 0$, which is contrary to the assumption of the lemma. □

Lemma 2.21. *Let $\lambda \neq 0$. If A is a compact operator and* $\ker A_\lambda = \{0\}$, *then* $\operatorname{ran} A_\lambda = \mathcal{H}$.

Proof. Define a sequence of subspaces of $\mathcal{H}$ by $\mathcal{H}^{(n)} := \operatorname{ran}(R_\lambda^{-n}(A))$ for $n = 0, 1, 2, \ldots$, where $\mathcal{H}^{(0)} := \mathcal{H}$. Then $\mathcal{H} \supseteq \mathcal{H}^{(1)} \supseteq \cdots$ and, by Lemma 2.20, each $\mathcal{H}^{(n+1)} = A_\lambda\mathcal{H}^{(n)}$ is a closed subspace of $\mathcal{H}^{(n)}$. Moreover, there is an n^* so that $\mathcal{H}^{(n+1)} = \mathcal{H}^{(n)}$ for all $n \geqslant n^*$.

Indeed, suppose to the contrary that all $\{\mathcal{H}^{(n)}\}$ are distinct. Then by the Gram-Schmidt orthogonalisation procedure there exists an orthonormal sequence $\{v_n\}_{n\geqslant 1}$ such that $v_{n+1} \in \mathcal{H}^{(n+1)}$ and $v_{n+1} \perp \mathcal{H}^{(n)}$. Then $Av_n - Av_m = -\lambda v_m + (\lambda v_n + R_\lambda^{-1}(A)v_n - R_\lambda^{-1}(A)v_m)$ and the vector $\lambda v_n + R_\lambda^{-1}(A)v_n - R_\lambda^{-1}(A)v_m$ belongs to $\mathcal{H}^{(m+1)}$ for $m < n$. Therefore, $\|Av_n - Av_m\| \geqslant |\lambda|\|v_m\| = |\lambda|$, showing that the set $\{Av_n\}$ is not compact. This contradicts the assumption that A is compact.

Assume that $\operatorname{ran} A_\lambda \subset \mathcal{H}$, i.e., $\mathcal{H} \supset \mathcal{H}^{(1)}$. Since $\ker A_\lambda = \{0\}$, the mapping $A_\lambda : \mathcal{H} \to \mathcal{H}^{(1)}$ is a bijection and consequently $A_\lambda : \mathcal{H}^{(1)} \to \mathcal{H}^{(2)} \subset \mathcal{H}^{(1)}$. If this would not be the case, then $\mathcal{H}^{(2)} = \mathcal{H}^{(1)}$ but, as $A_\lambda^{-1}\mathcal{H}^{(2)} = \mathcal{H}^{(1)}$ and $A_\lambda^{-1}\mathcal{H}^{(1)} = \mathcal{H}$, it would follow that $\mathcal{H}^{(1)} = \mathcal{H}$. The same reasoning applies to all $n \geqslant 1$: $A_\lambda : \mathcal{H}^{(n)} \to \mathcal{H}^{(n+1)} \subset \mathcal{H}^{(n)}$, which contradicts to the existence of n^*. Hence, $\operatorname{ran} A_\lambda = \mathcal{H}$. □

Lemma 2.22. *Let $A \in \mathcal{L}(\mathcal{H})$. Then $\sigma(A^*) = \{\lambda \in \mathbb{C} : \overline{\lambda} \in \sigma(A)\}$ and $R_\lambda(A^*) = R_{\overline{\lambda}}(A)^*$.*

Proof. If $\overline{\lambda} \in \rho(A)$, then $R_{\overline{\lambda}}(A)A_{\overline{\lambda}} = \mathbb{1} = A_{\overline{\lambda}}R_{\overline{\lambda}}(A)$. Therefore, $R_{\overline{\lambda}}(A)^*A_\lambda^* = \mathbb{1} = A_\lambda^*R_{\overline{\lambda}}(A)^*$, which proves the lemma. □

Proof of Proposition 2.18. By Remark 2.15, $\sigma(A) \neq \varnothing$ and contains at least the point $\lambda = 0$. Lemma 2.20 excludes that the continuous spectrum of A contains any $\lambda \neq 0$, and the same holds for the residual spectrum by Lemma 2.21. Hence, according to Lemma 2.19 and (2.6), we have $\sigma(A)\backslash\{0\} = \sigma_{\mathrm{p}}(A)\backslash\{0\}$, and the point spectrum $\sigma_{\mathrm{p}}(A)\backslash\{0\}$ is a discrete set of eigenvalues of finite multiplicities.

By Remark 2.15 and Lemma 2.22, $0 \in \sigma(A^*)$ and $\sigma(A^*)\backslash\{0\} = \sigma_{\mathrm{p}}(A^*)\backslash\{0\} = \{\lambda \neq 0 : \overline{\lambda} \in \sigma_{\mathrm{p}}(A)\}$. □

Remark 2.23. The only point in the spectrum $\sigma(A)$ that we did not yet classify is $\lambda = 0$. In general, it may belong to $\sigma_{\text{cont}}(A) \cup \sigma_{\text{res}}(A)$ as well as to $\sigma_{\text{p}}(A)$, see Appendix A (Section A.6). In the latter case it can be an eigenvalue of finite or infinite multiplicity. By Remark 2.15 the point $\lambda = 0$ always belongs to $\sigma(A)$, if the Hilbert space $\mathcal{H}$ is infinite-dimensional. Compact operators for which $\sigma(A) = \{0\}$ are known as abstract *Volterra* operators. For more details about the classification of the point $\lambda = 0$ and for examples of compact operators, see Sections A.6 and A.7 of Appendix A.

A particularly interesting case is that of the *compact self-adjoint* operators, Section A.6 of Appendix A.

Proposition 2.24. *Let $A = A^* \in \mathcal{C}_\infty(\mathcal{H})$, then*

(a) $\sigma(A) \subset \mathbb{R}$ *and* $\sigma_{\text{res}}(A) = \varnothing$;

(b) *eigenvectors corresponding to distinct eigenvalues of A are orthogonal;*

(c) *there is an orthonormal basis $\{e_n\}_{n\geqslant 1}$ in $\mathcal{H}$ such that $Ae_n = \lambda_n e_n$. The set $\{\lambda_n \neq 0\}$ coincides with $\sigma_{\text{p}}(A)\backslash\{0\}$, and each eigenvalue λ_n has a finite multiplicity. Finally, any $u \in \mathcal{H}$ has a unique representation $u = \sum_{n\geqslant 1} u_n e_n + \xi$, where $\xi \in \ker A$.*

Proof. (a) Notice that in this item we do not use the compactness of the operator A. Let $\lambda = \alpha + i\beta$. By self-adjointness,

$$\|A_\lambda u\|^2 = \|(A_\lambda)^* u\|^2 \geqslant |\beta|^2 \|u\|^2, \quad u \in \mathcal{H}.$$

Therefore, $\ker A_\lambda = \ker(A_\lambda)^* = \{0\}$ and

$$\mathcal{H} = \overline{\operatorname{ran} A_\lambda} \oplus \ker(A_\lambda)^* = \operatorname{ran} A_\lambda, \quad \Im\mathrm{m}\, \lambda \neq 0.$$

Then by Definition 2.13 we obtain that $\sigma(A) \subset \mathbb{R}$. Now, since

$$\mathcal{H} = \overline{\operatorname{ran} A_\alpha} \oplus \ker(A_\alpha)^* = \overline{\operatorname{ran} A_\alpha} \oplus \ker A_\alpha,$$

Definition 2.14 implies that $\sigma_{\text{res}}(A) = \varnothing$, cf. Remark 2.23.

(b) Let $\lambda_m, \lambda_{m'} \in \sigma_{\text{p}}(A)$ with $\lambda_m \neq \lambda_{m'}$. By Proposition 2.18, there exist two eigenvectors u_m and $u_{m'}$ corresponding to λ_m and $\lambda_{m'}$. But then, $(u_m, Au_{m'}) = \lambda_{m'}(u_m, u_{m'}) = (Au_m, u_{m'}) = \lambda_m(u_m, u_{m'})$, which is only possible if $u_m \perp u_{m'}$.

(c) For each non-zero eigenvalue λ_m one can choose an orthonormal basis in the finite-dimensional subspace $\mathcal{H}_{\lambda_m}$ spanned by eigenvectors corresponding to λ_m. Since, by (a), eigenvectors corresponding to distinct $\{\lambda_n\}_{n\geqslant 1}$ are mutually orthogonal, the collection of all these vectors $\{e_n\}_{n\geqslant 1}$ is an orthonormal set in $\mathcal{H}$. Let $\mathcal{M}$ be the closure of the span of $\{e_n\}_{n\geqslant 1}$. Since $A = A^*$ and $\mathcal{M}$ is invariant under A by construction, one has also that $A : \mathcal{M}^\perp \to \mathcal{M}^\perp$. If $A_\perp := A \upharpoonright \mathcal{M}^\perp$ denotes the restriction of A to $\mathcal{M}^\perp$, then $A_\perp = A_\perp^* \in \mathcal{C}_\infty(\mathcal{H})$ since $A = A^* \in \mathcal{C}_\infty(\mathcal{H})$. Therefore, by Proposition 2.18, any non-zero $\lambda \in \sigma(A_\perp)$ belongs to $\sigma_{\rm p}(A_\perp)$, and thus to $\sigma_{\rm p}(A)$. However, since all eigenvectors corresponding to non-zero eigenvalues of A are in $\mathcal{M}$, the spectral radius $r = \sup_{\lambda\in\sigma(A_\perp)}|\lambda| = 0$. Since $\sigma(A_\perp)$ is a closed set, $\sigma(A_\perp) = \{0\}$. By the spectral theorem for self-adjoint operators, one gets $A_\perp = \int_{\sigma(A_\perp)=\{0\}} \mathrm{d}E_{A_\perp}(\lambda)\,\lambda = 0$, that is, $A_\perp$ is the zero operator on $\mathcal{M}^\perp$ or, equivalently, $\mathcal{M}^\perp = \ker A$.

Decomposing $\mathcal{H}$ as $\mathcal{M}\oplus\mathcal{M}^\perp$, each vector $u\in\mathcal{H}$ has the unique representation

$$u = \sum_{n\geqslant 1} u_n e_n + \xi, \tag{2.7}$$

with $u_n = (u, e_n)$ and $\xi\in\ker A$. □

Corollary 2.25. *We know that the subspace* $\ker A$, *at least if it is not reduced to* $\{0\}$, *corresponds to the point* $0\in\sigma(A)$, *see Remark* 2.23. *Let* $\{e'_m\}_{m\geqslant 1}$ *be an orthonormal basis in* $\mathcal{M}^\perp = \ker A$. *Then* (2.7) *allows us to write* $u = \sum_{n\geqslant 1} u_n e_n + \sum_m u'_m e'_m$, *i.e.*, $\{e_n\}_{n\geqslant 1}\cup\{e'_m\}_{m\geqslant 1}$ *is an orthonormal basis in* $\mathcal{H}$ *corresponding to the compact self-adjoint operator* A. *Therefore,* $0\in\sigma_{\rm p}(A)$ *is an eigenvalue of finite or infinite multiplicity according to whether* $\dim\mathcal{M}^\perp < \infty$ *or* $\dim\mathcal{M}^\perp = \infty$.

Corollary 2.26 (Canonical form for compact operators). *If* A *is a compact operator on* $\mathcal{H}$, *then* A^*A *is compact and self-adjoint, see Proposition* 2.7. *By Proposition* 2.24 *and Corollary* 2.25 *there is an, not necessarily complete, orthonormal set* $\{\varphi_n\}_{n\geqslant 1}$ *such that* $A^*A\varphi_n = \lambda_n(A^*A)\varphi_n$ *with* $\lambda_n(A^*A)\neq 0$. *The numbers* $\lambda_n(A^*A)$ *are the non-zero eigenvalues of* A^*A, *and for any* $u\in\mathcal{H}$ *one has* $A^*Au = \sum_{n\geqslant 1}\lambda_n(A^*A)u_n\varphi_n$, *with* $u_n = (u,\varphi_n)$. *Moreover,* $\ker(A^*A) = [\varphi_1,\varphi_2,\ldots]^\perp$ *and, since* $A^*A\geqslant 0$, *each* $\lambda_n(A^*A) > 0$. *Let* $s_n(A) := \sqrt{\lambda_n(A^*A)}$ *and set* $\psi_n = s_n(A)^{-1}A\varphi_n$. *Then the system of vectors* $\{\psi_n\}_{n\geqslant 1}$ *is orthonormal and we obtain the canonical representation of the operator* A:

$$Au = \sum_{n\geqslant 1} s_n(A)(u,\varphi_n)\psi_n, \quad A^*u = \sum_{n\geqslant 1} s_n(A)(u,\psi_n)\varphi_n, \quad u\in\mathcal{H}. \tag{2.8}$$

Remark 2.27. The canonical representation (2.8) entails

(a) that the orthonormal sets $\{\varphi_n\}_{n\geqslant 1}$ and $\{\psi_n\}_{n\geqslant 1}$ are related by $\psi_n = s_n(A)^{-1}A\varphi_n$, and that they are complete in $\overline{\operatorname{ran} A}$ and in $\overline{\operatorname{ran} A^*}$ respectively;

(b) that the numbers $\{s_n(A)\}_{n\geqslant 1}$, called the *singular values* of A, are the eigenvalues $\{\lambda_n(|A|)\}_{n\geqslant 1}$ of the self-adjoint operator $|A| = \sqrt{A^*A}$, see Proposition 2.3), with corresponding eigenvectors $\{\varphi_n\}_{n\geqslant 1}$. Therefore, if $A = U|A|$ is the polar decomposition of A, then $\lambda_n(|A|) = s_n(A)$ and $\psi_n = U\varphi_n$;

(c) that although $|A|\neq|A^*|$, one has $\{s_n(A) = s_n(A^*)\}_{n\geqslant 1}$.

For details, see Section 3.1.

The representation (2.8) of a compact operator is called *canonical* because the inverse to Corollary 2.26 is also true.

Proposition 2.28. *Let $\{\varphi_n\}_{n\geqslant 1}$ be an orthonormal (not necessarily complete) set and let U be a partial isometry on $\mathcal{H}$. Let $\{s_n\}_{n\geqslant 1}$ be a non-increasing sequence of positive real numbers such that $s_n \to 0$ when $n \to \infty$. Then (2.8), with $\psi_n = U\varphi_n$, defines a compact operator $A = U|A|$, such that $\{s_n(A) = s_n\}_{n\geqslant 1}$.*

Proof. Consider the rank-r operator

$$A_r := \sum_{n=1}^{r} s_n(\cdot, \varphi_n)U\varphi_n. \tag{2.9}$$

Since

$$\begin{aligned}\|(A_{r+N} - A_r)u\|^2 &= \sum_{n=r+1}^{r+N} s_n^2 |(u, \varphi_n)|^2 \\ &\leqslant s_{r+1}^2 \sum_{n=r+1}^{r+N} |(u, \varphi_n)|^2 \leqslant s_{r+1}^2 \|u\|^2,\end{aligned}$$

the sequence $\{A_r\}_{r\geqslant 1}$ is norm convergent when $r \to \infty$. By Proposition 2.11, the limit operator

$$\|\cdot\|\text{-}\lim_{r\to\infty} A_r = A \tag{2.10}$$

is compact with $\ker A = \ker U$ and $s_n = s_n(A)$. □

Remark 2.29. In general the operators

$$A^*A = \sum_{n\geqslant 1} s_n(A)(\cdot, \varphi_n)A^*\psi_n = \sum_{n\geqslant 1} s_n(A)^2(\cdot, \varphi_n)\varphi_n,$$

and

$$AA^* = \sum_{n\geqslant 1} s_n(A)(\cdot, \psi_n)A\varphi_n = \sum_{n\geqslant 1} s_n(A)^2(\cdot, \psi_n)\psi_n,$$

are different. Since $|A| \neq |A^*|$, their eigenfunctions $\{\varphi_n \neq \psi_n\}_{n\geqslant 1}$, although the singular values coincide $\{s_n(A) = s_n(A^*)\}_{n\geqslant 1}$. If the operator A is normal, i.e., $A^*A = AA^*$, then $\{\varphi_n = \psi_n\}_{n\geqslant 1}$. Therefore, operators A and A^* have a common set of eigenfunctions, but with different eigenvalues: $\lambda(A^*) = \overline{\lambda(A)}$, cf. Section 3.1.

2.3 Trace class and $\mathcal{C}_p(\mathcal{H})$-ideals

Above, we associated with any compact operator A its singular values $\{s_n(A) = \sqrt{\lambda_n(A^*A)}\}_{n\geqslant 1}$, where $\{\lambda_n(A^*A)\}_{n\geqslant 1}$ is the set of decreasing eigenvalues of the positive self-adjoint compact operator $A^*A > 0$. It is clear that $s_n(A) = \sqrt{\lambda_n(|A|^2)} = s_n(|A|)$ and that

$$s_n(A) = \sqrt{\lambda_n(AA^*)} = s_n(A^*), \tag{2.11}$$

see Remark 2.29. With singular values at our disposal, we can describe the *von Neumann-Schatten classes* $\mathcal{C}_p(\mathcal{H})$ $(1 \leqslant p < \infty)$ of linear operators on a Hilbert space $\mathcal{H}$. We start with the trace class $\mathcal{C}_1(\mathcal{H})$.

Definition 2.30. A compact operator $A \in \mathcal{C}_\infty(\mathcal{H})$ belongs to the *trace class* $\mathcal{C}_1(\mathcal{H})$ if

$$\|A\|_1 := \sum_{n \geqslant 1} s_n(A) < \infty. \tag{2.12}$$

By (2.8), Proposition 2.28 and Definition 2.12,

$$\|A\| = s_1(A) \leqslant \|A\|_1 , \tag{2.13}$$

where equality holds only for *rank-one* operators: $A \in \mathcal{K}_1(\mathcal{H})$. Moreover,

$$\|A\|_1 = \||A|\|_1 \quad \text{and} \quad \|A\|_1 = \|A^*\|_1 , \tag{2.14}$$

by virtue of (2.11).

Proposition 2.31. *$\mathcal{C}_1(\mathcal{H})$ is a two-sided $*$-ideal in the ring of bounded operators $\mathcal{L}(\mathcal{H})$:*

(a) *If $A \in \mathcal{C}_1(\mathcal{H})$, then $A^* \in \mathcal{C}_1(\mathcal{H})$.*

(b) *If $A \in \mathcal{C}_1(\mathcal{H})$ and $B \in \mathcal{L}(\mathcal{H})$, then $AB \in \mathcal{C}_1(\mathcal{H})$ and $BA \in \mathcal{C}_1(\mathcal{H})$.*

(c) *$\mathcal{C}_1(\mathcal{H})$ is a vector space.*

Proof. (a) Follows from (2.11), see (2.14).
(b) By the canonical representation (2.8) of a compact operator with non-increasing singular values $\{s_n(A)\}_{n \geqslant 1}$, one readily gets $s_1(A) = \sup_{u \in \mathcal{H}:\|u\|=1} \|Au\|$ and

$$s_n(A) = \sup_{\substack{\|u\|=1 , \\ u \in [\varphi_1,\ldots,\varphi_{n-1}]^\perp}} \|Au\|, \quad \text{for } n \geqslant 2. \tag{2.15}$$

If $B \in \mathcal{L}(\mathcal{H})$, then $\|BAu\| \leqslant \|B\|\|Au\|$. Together with (2.15), this gives

$$s_{n+1}(BA) \leqslant \|B\| s_{n+1}(A). \tag{2.16}$$

Since by (2.11) $s_{n+1}(AB) = s_{n+1}(B^* A^*)^*) = s_{n+1}(B^* A^*)$, we also get

$$s_{n+1}(AB) \leqslant \|B^*\| s_{n+1}(A^*) = \|B\| s_{n+1}(A). \tag{2.17}$$

Then (2.12) and the estimates (2.16) and (2.17) give

$$\|BA\|_1 \leqslant \|B\|\|A\|_1 \quad \text{and} \quad \|AB\|_1 \leqslant \|B\|\|A\|_1 . \tag{2.18}$$

(c) Notice that by (2.8), (2.9) and (2.15) we obtain:

$$\|A - A_r\| = \sup_{\substack{\|u\|=1 \ \wedge \\ u \in [\varphi_1,\ldots,\varphi_{n-1}]^\perp}} \|Au\| = s_{r+1}(A). \tag{2.19}$$

On the other hand, there is a characterisation of singular values, which generalises (2.15) or (2.19), namely, the *minimax principle*. For the case of unbounded operators, see Proposition 4.23.

Remark 2.32 (Minimax principle). Let $s_1(A) = \sup_{u\in\mathcal{H}:\|u\|=1} \|Au\|$ and

$$s_{n+1}(A) = \inf_{\mathcal{M}_n\subset\mathcal{H}} \sup_{\substack{\|u\|=1\ \wedge \\ u\in\mathcal{M}_n^\perp}} \|Au\|, \quad \text{for } n \in \mathbb{N} . \tag{2.20}$$

Here $\mathcal{M}_n \subset \mathcal{H}$ are linear subspaces of dimension n and $\mathcal{M}_n^\perp$ are their orthogonal complements. Indeed, suppose that there exists a subspace $\widetilde{\mathcal{M}}_n$ such that $\|Au\| < s_{n+1}(A)$ for all $u \in \widetilde{\mathcal{M}}_n^\perp$. Let $\mathcal{F}_{n+1} = [\varphi_1, \ldots, \varphi_{n+1}]$ be the subspace spanned by the eigenvectors $\{\varphi_k\}$ of the operator A. Then $\mathcal{F}_{n+1} \cap \widetilde{\mathcal{M}}_n^\perp \neq \{0\}$, since $\dim \widetilde{\mathcal{M}}_n < \dim \mathcal{F}_{n+1}$ and there is a unit vector $\tilde{u} \in \mathcal{F}_{n+1} \cap \widetilde{\mathcal{M}}_n^\perp$ such that $\|A\tilde{u}\| < s_{n+1}(A)$. However, $\|A\tilde{u}\|^2 = \|A\sum_{k=1}^{n+1}(\tilde{u}, \varphi_k)\varphi_k\|^2 = \sum_{k=1}^{n+1} |(\tilde{u}, \varphi_k)|^2 s_k^2(A) \geqslant s_{n+1}^2(A)$ since the $\{s_k(A)\}_{k\geqslant 1}$ are non-increasing. This contradiction shows that

$$\sup_{\substack{\|u\|=1\ \wedge \\ u\in\mathcal{M}_n^\perp}} \|Au\| \geqslant s_{n+1}(A). \tag{2.21}$$

It remains only to observe that the equality in (2.21) is attained for $\mathcal{M}_n = \mathcal{F}_n$.

Now, let $A, B \in \mathcal{C}_1(\mathcal{H})$ and let $[e_1, \ldots, e_{n+m}]$ be a subspace spanned by linearly independent vectors $\{e_j : j = 1, 2, \ldots, n+m\}$. Then

$$\|(A+B)u\| \leqslant \|Au\| + \|Bu\|$$

and

$$\sup_{\substack{\|u\|=1, \\ u\in[e_1,\ldots,e_{n+m}]^\perp}} \|(A+B)u\| \leqslant \sup_{\substack{\|u\|=1, \\ u\in[e_1,\ldots,e_{n+m}]^\perp}} \|Au\| + \sup_{\substack{\|u\|=1, \\ u\in[e_1,\ldots,e_{n+m}]^\perp}} \|Bu\|. \tag{2.22}$$

Minimising the left- and right-hand sides of (2.22) over $\{e_j : j = 1, 2, \ldots, n+m\}$, one gets by (2.20) that

$$s_{n+m+1}(A+B) \leqslant s_{n+1}(A) + s_{m+1}(B), \tag{2.23}$$

which shows that $A + B \in \mathcal{C}_1(\mathcal{H})$. □

Corollary 2.33. (a) *Let U_1 and U_2 be unitary operators. By (2.16) and (2.17) we get $s_n(A) = s_n(U_1^{-1}U_1AU_2U_2^{-1}) \leqslant s_n(U_1AU_2) \leqslant s_n(A)$ for any $A \in \mathcal{C}_1(A)$. Therefore,*

$$s_n(A) = s_n(U_1AU_2) \quad \textit{and} \quad \|A\|_1 = \|U_1AU_2\|_1. \tag{2.24}$$

(b) *Let $C \in \mathcal{K}_n(\mathcal{H})$ be a rank-n operator, see Definition* 2.9. *Then combining* (2.19) *and* (2.20), *we obtain another characterisation of the singular values of an operator $A \in \mathcal{C}_\infty(\mathcal{H})$:*

$$\begin{aligned} s_{n+1}(A) &\leqslant \sup_{\substack{\|u\|=1,\\ u\in\operatorname{ran}(C)^\perp}} \|Au\| = \sup_{\substack{\|u\|=1,\\ u\in\operatorname{ran}(C)^\perp}} |(Au,u)| \\ &= \sup_{\substack{\|u\|=1,\\ u\in\operatorname{ran}(C)^\perp}} |((A-C)u,u)| \leqslant \sup_{\|u\|=1} |((A-C)u,u)| \\ &= \|A-C\|. \end{aligned} \tag{2.25}$$

On the other hand, we know that the equality in (2.25) *is attained for $C = A_n$, see* (2.19). *Therefore*

$$s_{n+1}(A) = \inf_{\{C:\, C\in\mathcal{K}_n(\mathcal{H})\}} \|A-C\|. \tag{2.26}$$

Corollary 2.34. *Suppose that the compact operators A and B satisfy $0 \leqslant A \leqslant B$. Then $A = |A|$, $B = |B|$, and $\|A^{1/2}u\| \leqslant \|B^{1/2}u\|$. The minimax principle* (2.20) *and $s_n(A^{1/2}) = s_n^{1/2}(A)$, $s_n(B^{1/2}) = s_n^{1/2}(B)$ imply that*

$$s_n(A) \leqslant s_n(B), \quad n \geqslant 1. \tag{2.27}$$

See Section 3.1 (e) *for details.*

Lemma 2.35. *Let $A \in \mathcal{L}(\mathcal{H})$ and suppose that $\sum_{j=1}^\infty \|Ae_j\|^2 < \infty$ for an orthonormal basis $\{e_j\}_{j\geqslant1}$ in $\mathcal{H}$. Then $A \in \mathcal{C}_\infty(\mathcal{H})$.*

Proof. Consider a bounded sequence $\{u_n\}_{n\geqslant1}$ with $\|u_n\| \leqslant a$. Then

$$\|A^*u_n\|^2 = \sum_{j=1}^\infty |(A^*u_n, e_j)|^2 = \sum_{j=1}^\infty |(u_n, Ae_j)|^2.$$

Now, let $u = \text{w-lim}\, u_n$. Then $(u_n - u, Ae_j) \to 0$, and one has

$$\begin{aligned} \|A^*(u_n-u)\|^2 &= \sum_{j=1}^\infty |(u_n-u, Ae_j)|^2 \\ &\leqslant \|(u_n-u)\|^2 \sum_{j=1}^\infty \|Ae_j\|^2 < \infty. \end{aligned}$$

This means that $\|A^*(u_n-u)\| \to 0$, or $A^* \in \mathcal{C}_\infty(\mathcal{H})$, see Proposition 2.5. Hence, $A \in \mathcal{C}_\infty(\mathcal{H})$, by Proposition 2.7. □

Lemma 2.36. *Let $A \in \mathcal{L}(\mathcal{H})$ and $A \geqslant 0$. If $\sum_{j=1}^\infty (e_j, Ae_j) < \infty$ for an orthonormal basis $\{e_j\}_{j\geqslant1}$ in $\mathcal{H}$, then $A \in \mathcal{C}_1(\mathcal{H})$ and*

$$\sum_{j=1}^\infty (\tilde{e}_j, A\tilde{e}_j) = \sum_{n=1}^\infty s_n(A) = \|A\|_1$$

for any orthonormal basis $\{\tilde{e}_j\}_{j\geqslant 1}$ *in* $\mathcal{H}$.

Proof. Recall that $A \geqslant 0$ implies $A = A^*$, see Remark 2.1. Therefore, the operator $A^{1/2}$ is well defined and, by the assumption of the lemma, $\sum_{j=1}^{\infty} \|A^{1/2}e_j\|^2 < \infty$. By Lemma 2.35, $A^{1/2} \in \mathcal{C}_\infty(\mathcal{H})$, and by Proposition 2.7, $A \in \mathcal{C}_\infty(\mathcal{H})$. Notice that for $A \geqslant 0$ the polar decomposition has the form $|A| = A$. Hence, in the canonical representation of A, see (2.8), $\psi_n = \varphi_n$ and we get

$$\sum_{j=1}^{\infty}(\tilde{e}_j, A\tilde{e}_j) = \sum_{j=1}^{\infty}\sum_{n=1}^{\infty} s_n(A)|(\tilde{e}_j,\varphi_n)|^2 = \sum_{n=1}^{\infty} s_n(A)\|\varphi_n\|^2 = \sum_{n=1}^{\infty} s_n(A). \qquad \square$$

Proposition 2.37. *Let* $A \in \mathcal{L}(\mathcal{H})$ *and suppose that* $\sum_{j=1}^{\infty}(e_j, A\tilde{e}_j) < \infty$ *for any two orthonormal bases* $\{e_j\}_{j\geqslant 1}$ *and* $\{\tilde{e}_j\}_{j\geqslant 1}$. *Then* $A \in \mathcal{C}_1(\mathcal{H})$.

Proof. By the polar decomposition, $A = U|A|$, see Proposition 2.3. Let $\{\tilde{e}_j\}_{j\geqslant 1}$ be an orthonormal basis in $\overline{\operatorname{ran}|A|}$ and $e_j := U\tilde{e}_j$. Since the restriction of U to $\overline{\operatorname{ran}|A|}$ is an isometry, the system $\{e_j\}$ is also orthonormal and

$$\sum_{j=1}^{\infty}(\tilde{e}_j, |A|\tilde{e}_j) = \sum_{j=1}^{\infty}(U\tilde{e}_j, U|A|\tilde{e}_j) = \sum_{j=1}^{\infty}(e_j, A\tilde{e}_j) < \infty, \tag{2.28}$$

by the assumption of the proposition. We can complete the system $\{\tilde{e}_j\}_{j\geqslant 1}$ to an orthonormal basis for $\mathcal{H}$ without altering the sum (2.28) because $\mathcal{H} \ominus \overline{\operatorname{ran}|A|} = \ker|A|$. Then by Lemma 2.36 we get that $|A| \in \mathcal{C}_1(\mathcal{H})$, and hence, $A = U|A| \in \mathcal{C}_1(\mathcal{H})$. $\square$

The next statement says that the converse is also true.

Proposition 2.38. *Let* $A \in \mathcal{C}_1(\mathcal{H})$ *and* $\{e_j\}$, $\{\tilde{e}_j\}$ *be two arbitrary orthonormal bases for* $\mathcal{H}$. *Then*

$$\sum_{j=1}^{\infty}|(e_j, A\tilde{e}_j)| \leqslant \|A\|_1\,. \tag{2.29}$$

Equality is reached for $e_j = \psi_j$ *and* $\tilde{e}_j = \varphi_j$, *where* $\{\psi_j\}$ *and* $\{\varphi_j\}$ *are defined by the canonical representation* (2.8).

Proof. Since $A \in \mathcal{C}_\infty(\mathcal{H})$, one can use the canonical representation (2.8) to get the following estimate:

$$\begin{aligned}\sum_{j=1}^{\infty}|(e_j, A\tilde{e}_j)| &\leqslant \sum_{j,n=1}^{\infty} s_n(A)|(\tilde{e}_j,\varphi_n)(e_j,\psi_n)| \\ &\leqslant \sum_{n=1}^{\infty} s_n(A)\Big[\sum_{j=1}^{\infty}|(\tilde{e}_j,\varphi_n)|^2\Big]^{1/2}\Big[\sum_{j=1}^{\infty}|(e_j,\psi_n)|^2\Big]^{1/2} \\ &= \sum_{n=1}^{\infty} s_n(A)\|\varphi_n\|\|\psi_n\| = \sum_{n=1}^{\infty} s_n(A) = \|A\|_1\,. \end{aligned} \tag{2.30}$$

Again by (2.8) we get that equality in (2.30) is attained for $e_j = \psi_j$ and $\tilde{e}_j = \varphi_j$. □

Instead of Proposition 2.37, the following property may be useful:

Proposition 2.39. *If $A \in \mathcal{L}(\mathcal{H})$ and $\sum_{j=1}^{\infty} \|Ae_j\| < \infty$ for an orthonormal basis $\{e_j\}_{j\geqslant 1}$ of $\mathcal{H}$, then $A \in \mathcal{C}_1(\mathcal{H})$.*

Proof. By the polar decomposition, we have $\|Au\| = \||A|u\|$ and

$$\sum_{j=1}^{\infty}(e_j, |A|e_j) \leqslant \sum_{j=1}^{\infty} \||A|e_j\| = \sum_{j=1}^{\infty} \|Ae_j\| < \infty.$$

Therefore, by Lemma 2.36, $|A| \in \mathcal{C}_1(\mathcal{H})$ and, consequently, $A \in \mathcal{C}_1(\mathcal{H})$. □

Remark 2.40. Let $\{e_j\}_{j\geqslant 1}$ and $\{\tilde{e}_j\}_{j\geqslant 1}$ be two orthonormal bases and $A \in \mathcal{C}_1(\mathcal{H})$. Then

$$\begin{aligned}\sum_{j=1}^{\infty}(\tilde{e}_j, A\tilde{e}_j) &= \sum_{j=1}^{\infty} \sum_{n,n'=1}^{\infty} (\tilde{e}_j, e_n)(e_n, Ae_{n'})(e_{n'}, e_j) \\ &= \sum_{n,n'=1}^{\infty} (e_n, Ae_{n'}) \sum_{j=1}^{\infty}(\tilde{e}_j, e_n)(e_{n'}, \tilde{e}_j) = \sum_{n=1}^{\infty}(e_n, Ae_n),\end{aligned}$$

where the interchanging of sums is allowed by absolute convergence, see Proposition 2.38. This motivates the following definition.

Definition 2.41. The map $\operatorname{Tr} : \mathcal{C}_1(\mathcal{H}) \to \mathbb{C}$ given by

$$A \mapsto \operatorname{Tr} A := \sum_{j=1}^{\infty}(e_j, Ae_j), \quad A \in \mathcal{C}_1(\mathcal{H}),$$

where $\{e_j\}_{j\geqslant 1}$ is *any* orthonormal basis in $\mathcal{H}$, is called the *trace*.

To bolster this concept we calculate the result of this map explicitly.

Proposition 2.42. *Let $A \in \mathcal{C}_1(\mathcal{H})$. Then for any orthonormal basis $\{e_j\}_{j\geqslant 1} \subset \mathcal{H}$ the sum $\sum_{j=1}^{\infty} |(e_j, Ae_j)| < \infty$ and the matrix trace $\operatorname{Tr} A$ is independent of the choice of a basis:*

$$\sum_{j=1}^{\infty}(e_j, Ae_j) = \sum_{n=1}^{\infty} s_n(A)(\varphi_n, \psi_n). \tag{2.31}$$

The map $A \mapsto \operatorname{Tr} A$ is a bounded linear functional on the space $\mathcal{C}_1(\mathcal{H})$ with $|\operatorname{Tr} A| \leqslant \|A\|_1$.

Proof. By the canonical representation (2.8) of compact operators and the absolute convergence of the double sum, which justifies their interchange, one obtains the representation (2.31):

$$\sum_{j=1}^{\infty}(e_j, Ae_j) = \sum_{j,n=1}^{\infty} s_n(A)(\varphi_n, e_j)(e_j, \psi_n) = \sum_{n=1}^{\infty} s_n(A)(\varphi_n, \psi_n) \,.$$

Moreover, since $s_n(A) \geqslant 0$ and $|(\varphi_n, \psi_n)| \leqslant 1$, this representation yields the $\|\cdot\|_1$-estimate $|\operatorname{Tr} A| \leqslant \|A\|_1$ for the linear functional $\operatorname{Tr}(\cdot)$. □

Corollary 2.43. *The (matrix) trace* $\operatorname{Tr}(\cdot)$ *is* $\|\cdot\|_1$*-continuous.*

Remark 2.44. Let $A \in \mathcal{C}_1(\mathcal{H})$. By the estimate in Corollary 3.4, the sum of eigenvalues $\Lambda(A) := \sum_{n=1}^{\infty} \lambda_n(A)$ (*spectral trace*) converges absolutely and defines on $\mathcal{C}_1(\mathcal{H})$ another bounded linear functional. Whether the spectral and the matrix traces *coincide* is a subtle question, the proof of which is out of the scope of this book. See Notes in Section 2.5 for references.

Proposition 2.45 (Lidskiĭ trace theorem). *If the operator* $A \in \mathcal{C}_1(\mathcal{H})$, *then* $\operatorname{Tr} A = \Lambda(A)$.

This positive answer for $A \in \mathcal{C}_1(\mathcal{H})$ allows us to establish a number of trace and $\|\cdot\|_1$-norm inequalities, Section 3.3. To appreciate the nontriviality of equality $\operatorname{Tr} A = \Lambda(A)$ we discuss in Appendix A (Section A.7) the *Volterra* operator.

The next assertion confirms that the domain $\mathcal{C}_1(\mathcal{H})$ for the functional $\Lambda(A)$ is also reasonable, because it is closed under the mapping $A \mapsto |A|$.

Proposition 2.46. *An operator* $A \in \mathcal{L}(\mathcal{H})$ *belongs to the trace class* $\mathcal{C}_1(\mathcal{H})$ *if and only if* $\operatorname{Tr}|A| < \infty$, *and in this case*

$$\operatorname{Tr}|A| = \|A\|_1. \tag{2.32}$$

Proof. If $A \in \mathcal{C}_1(\mathcal{H})$, then by (2.14) $|A| \in \mathcal{C}_1(\mathcal{H})$. By (2.8), the canonical representation of $|A|$ reads $|A| = \sum_{n\geqslant 1} s_n(A)(\cdot, \varphi_n)\varphi_n$, where the eigenvalues $\lambda_n(|A|)$ of $|A|$ coincide with the singular values $s_n(A)$ of A and $\{\varphi_n\}_{n\geqslant 1}$ is the orthonormal set of the corresponding eigenvectors. By the *Hilbert-Schmidt* theorem (Corollary 2.25) we can complete $\{\varphi_n\}_{n\geqslant 1}$ to an orthonormal basis $\{e_j\}_{j\geqslant 1}$ of $\mathcal{H}$, such that

$$\sum_{j=1}^{\infty}(e_j, |A|e_j) = \sum_{n\geqslant 1}^{\infty}(\varphi_n, |A|\varphi_n) = \sum_{n\geqslant 1} s_n(|A|). \tag{2.33}$$

Since $|A| \in \mathcal{C}_1(\mathcal{H})$, the sum (2.33) converges. Then by Lemma 2.36 it converges for any orthonormal basis of $\mathcal{H}$, i.e., by Definitions 2.30 and 2.41 one gets

$$\operatorname{Tr}|A| = \sum_{n\geqslant 1} s_n(|A|) = \|A\|_1 < \infty\,. \tag{2.34}$$

Now, let $A \in \mathcal{L}(\mathcal{H})$ be such that $\operatorname{Tr}|A| < \infty$. Then for any orthonormal basis $\{\tilde{e}_j\}_{j\geqslant 1}$ one has $\sum_{j=1}^{\infty}(\tilde{e}_j, |A|\tilde{e}_j) = \sum_{j=1}^{\infty}\||A|^{1/2}\tilde{e}_j\|^2 < \infty$. Therefore, by Lemma 2.35, $|A|^{1/2} \in \mathcal{C}_\infty(\mathcal{H})$. Hence, $|A| \in \mathcal{C}_\infty(\mathcal{H})$. Then, calculating the trace in the complete orthonormal basis $\{e_j\}_{j\geqslant 1}$ corresponding to the compact self-adjoint operator $|A|$, see Corollaries 2.25 and 2.26, we get

$$\operatorname{Tr}|A| = \sum_{j=1}^{\infty}(e_j, |A|e_j) = \sum_{j=1}^{\infty}\lambda_j(|A|) = \sum_{n\geqslant 1} s_n(A) = \|A\|_1. \qquad \square$$

Remark 2.47. It is not true that if $\sum_{j=1}^{\infty}|(\tilde{e}_j, A\tilde{e}_j)| < \infty$ for some orthonormal basis $\{\tilde{e}_j\}_{j\geqslant 1}$, then $A \in \mathcal{C}_1(\mathcal{H})$, cf. Propositions 2.37–2.39 and Remark 2.40.

Some basic properties of the *trace* are:

Proposition 2.48. *Let $A, B \in \mathcal{C}_1(\mathcal{H})$. Then:*

(a) $\operatorname{Tr}(A + B) = \operatorname{Tr} A + \operatorname{Tr} B$.

(b) $\operatorname{Tr}(\lambda A) = \lambda \operatorname{Tr} A,\ \lambda \in \mathbb{C}$.

(c) $\operatorname{Tr}(UAU^{-1}) = \operatorname{Tr} A$ *for any unitary operator* U.

(d) $\operatorname{Tr}(AC) = \operatorname{Tr}(CA)$ *for any* $C \in \mathcal{L}(\mathcal{H})$.

(e) *If* $0 \leqslant A \leqslant B$, *then* $\operatorname{Tr} A \leqslant \operatorname{Tr} B$.

(f) *If a sequence $\{A_k\}_{k\geqslant 1}$ from the cone $\mathcal{C}_{1,+}(\mathcal{H})$ of positive trace-class operators converges weakly to a trace-class operator A:* $\text{w-lim}_{k\to\infty} A_k = A$, *then* $\operatorname{Tr} A \leqslant \liminf_{k\to\infty} \operatorname{Tr} A_k$, *that is, the functional* $\operatorname{Tr}(\cdot)$ *is weakly lower semi-continuous.*

Proof. Properties (a) and (b) follow directly from Definition 2.41.

(c) Note that if $\{e_j\}_{j\geqslant 1}$ is an orthonormal basis, then so is $\{\tilde{e}_{j'}\}_{j'\geqslant 1} = \{Ue_j\}_{j'\geqslant 1}$. Thus $\operatorname{Tr}(UAU^{-1}) = \sum_{j=1}^{\infty}(\tilde{e}_j, UAU^{-1}\tilde{e}_j) = \operatorname{Tr} A$.

(d) Choose an orthonormal basis $\{e_j\}_{j\geqslant 1} \supset \{\psi_n\}_{n\geqslant 1}$, where $\{\psi_n\}_{n\geqslant 1}$ is the orthonormal system corresponding to the canonical form of A^*, (2.8), then

$$\operatorname{Tr}(AC) = \sum_{j=1}^{\infty}(e_j, ACe_j) = \sum_{n\geqslant 1}(A^*\psi_n, C\psi_n) = \sum_{n\geqslant 1} s_n(A)(\varphi_n, C\psi_n).$$

Similarly, using the orthonormal system $\{\varphi_n\}_{n\geqslant 1}$ corresponding to A (2.8), one gets

$$\operatorname{Tr}(CA) = \sum_{n\geqslant 1}(\varphi_n, CA\varphi_n) = \sum_{n\geqslant 1} s_n(A)(\varphi_n, C\psi_n) = \operatorname{Tr}(AC).$$

(e) Since $A \leqslant B$ is equivalent to $(u, Au) \leqslant (u, Bu)$ for any $u \in \mathcal{H}$, the inequality $\operatorname{Tr} A \leqslant \operatorname{Tr} B$ follows from the definition of trace.

(f) Note that $\operatorname{Tr} A_k = \sum_{n=1}^{\infty}(e_n, A_k e_n)$ for any $k \geqslant 1$ independently of the orthonormal basis $\{e_n\}_{n\geqslant 1}$, where each positive term $(e_n, A_k e_n)$ has limit:

$\lim_{k\to\infty}(e_n, A_k e_n) = (e_n, Ae_n)$. Then using a discrete version of Fatou's lemma we get inequality

$$\operatorname{Tr} A = \sum_{n=1}^{\infty} \lim_{k\to\infty} (e_n, A_k e_n) \leqslant \liminf_{k\to\infty} \sum_{n=1}^{\infty} (e_n, A_k e_n), \tag{2.35}$$

which proves the weak lower semi-continuity of the trace when the weak limit $A \in \mathcal{C}_1(\mathcal{H})$. □

Note that the weak operator convergence of the trace-class operators $\{A_k\}_{k\geqslant 1}$ above does not guarantee that the positive bounded operator $A \in \mathcal{C}_{1,+}(\mathcal{H})$, since both sides of inequality (2.35) could be infinite.

We can summarise the properties of the functional $\|\cdot\|_1 := \operatorname{Tr}(|\cdot|)$ on $\mathcal{C}_1(\mathcal{H})$ as

Proposition 2.49. *The functional $\|\cdot\|_1$ is a norm on the two-sided $*$-ideal $\mathcal{C}_1(\mathcal{H})$:*

(a) *If $A \in \mathcal{C}_1(\mathcal{H})$, then $\|A\|_1 \geqslant 0$, and $\|A\|_1 = 0$ implies $A = 0$.*

(b) *$\|\alpha A\|_1 = |\alpha|\|A\|_1$ for $\alpha \in \mathbb{C}$ and $A \in \mathcal{C}_1(\mathcal{H})$.*

(c) *$\|A + B\|_1 \leqslant \|A\|_1 + \|B\|_1$ for $A, B \in \mathcal{C}_1(\mathcal{H})$.*

(d) *$\|AB\|_1 \leqslant \|A\|_1\|B\|_1$ for $A, B \in \mathcal{C}_1(\mathcal{H})$.*

(e) *The functional $\|\cdot\|_1$ is weakly lower semi-continuous.*

Proof. (a) This follows from (2.12) by the positivity of the singular values and by the canonical representation (2.8) of compact operators.

(b) This results from the polar decomposition $|\alpha A| = |\alpha||A|$ and Proposition 2.53: $\|\alpha A\|_1 = \operatorname{Tr}|\alpha A| = |\alpha| \operatorname{Tr}|A| = |\alpha|\|A\|_1$.

(c) Let $\{e_j\}_{j\geqslant 1}$ and $\{\tilde{e}_j\}_{j\geqslant 1}$ be two orthonormal bases in $\mathcal{H}$. Then

$$\begin{aligned}\sum_{j=1}^{\infty} |(e_j, (A+B)\tilde{e}_j)| &\leqslant \sum_{j=1}^{\infty} |(e_j, A\tilde{e}_j)| + \sum_{j=1}^{\infty} |(e_j, B\tilde{e}_j)| \\ &\leqslant \|A\|_1 + \|B\|_1,\end{aligned} \tag{2.36}$$

by (2.29). Applying Proposition 2.38 to the left-hand side of (2.36), we get $\|A + B\|_1 \leqslant \|A\|_1 + \|B\|_1$.

(d) This follows directly from (2.13) and (2.18): $\|AB\|_1 \leqslant \|A\|_1\|B\| \leqslant \|A\|_1\|B\|_1$.

(e) The proof is similar to (f) in Proposition 2.48. □

In fact, the $*$-ideal $\mathcal{C}_1(\mathcal{H})$ equipped with the *trace-norm* $\|\cdot\|_1$ is a Banach space. This is similar to the $*$-ideal $\mathcal{C}_\infty(\mathcal{H})$, which is a Banach space with respect to the operator norm $\|\cdot\|$. We shall consider topological properties of the $*$-ideals $\mathcal{C}_p(\mathcal{H})$, $1 \leqslant p < \infty$, in the next section.

Definition 2.50. An operator $A \in \mathcal{C}_\infty(\mathcal{H})$ is said to belong to the *von Neumann-Schatten class* $\mathcal{C}_p(\mathcal{H})$, $1 \leqslant p < \infty$, if

$$\|A\|_p := \left\{ \sum_{n \geqslant 1} s_n^p(A) \right\}^{1/p} < \infty. \tag{2.37}$$

Since by (2.37) $p \mapsto \|A\|_p$ is a non-increasing function of $p \geqslant 1$, one immediately gets for $1 < p < q < \infty$ that

$$\mathcal{C}_1(\mathcal{H}) \subset \mathcal{C}_p(\mathcal{H}) \subset \mathcal{C}_q(\mathcal{H}) \subset \mathcal{C}_\infty(\mathcal{H}), \tag{2.38}$$

which corresponds to the estimates

$$\|A\|_1 \geqslant \|A\|_p \geqslant \|A\|_q \geqslant \|A\|. \tag{2.39}$$

The norms in (2.39) coincide only for rank-one operators $A \in \mathcal{K}_1(\mathcal{H})$.

Another immediate consequence of (2.37) is an analogue of Proposition 2.46. If $A \in \mathcal{C}_p(\mathcal{H})$, then by (2.38) $A \in \mathcal{C}_\infty(\mathcal{H})$ and $|A| \in \mathcal{C}_\infty(\mathcal{H})$. By the spectral representation of the self-adjoint operator $|A|$, we can define, for any $1 \leqslant p < \infty$, the operator

$$|A|^p = \int_{\sigma(|A|)} \mathrm{d}E_{|A|}(\lambda)\, \lambda^p = \sum_{n \geqslant 1} \lambda_n^p(|A|)(\cdot, \varphi_n)\varphi_n, \tag{2.40}$$

where $\{\varphi_n\}$ and $\{\lambda_n(|A|) = s_n(A)\}$ are the eigenvectors and eigenvalues of $|A|$. Therefore, $|A|^p \in \mathcal{C}_1(\mathcal{H})$ and (2.37) reads

$$\|A\|_p = (\mathrm{Tr}\, |A|^p)^{1/p}. \tag{2.41}$$

The class $\mathcal{C}_p(\mathcal{H})$ is a linear space. If $A \in \mathcal{C}_p(\mathcal{H})$, then obviously $\|\alpha A\|_p = |\alpha| \|A\|_p$, i.e., $\alpha A \in \mathcal{C}_p(\mathcal{H})$. Now let $A, B \in \mathcal{C}_p(\mathcal{H})$. Then by (2.38) we get $(A + B) \in \mathcal{C}_\infty(\mathcal{H})$. For the singular values of this compact operator we have inequalities

$$s_{2n}(A + B) \leqslant s_{2n-1}(A + B) \leqslant s_n(A) + s_n(B),$$

see (2.23). Then by using the elementary estimates

$$(s_n(A) + s_n(B))^p \leqslant \Big[\max\{2s_n(A), 2s_n(B)\}\Big]^p \leqslant (2s_n(A))^p + (2s_n(B))^p,$$

we obtain

$$\begin{aligned} \|A + B\|_p &= \sum_{n \geqslant 1} \{(s_{2n-1}(A + B))^p + (s_{2n}(A + B))^p\} \\ &\leqslant 2^{p+1} \Big[\sum_{n \geqslant 1} (s_n(A))^p + \sum_{n \geqslant 1} (s_n(B))^p \Big] \\ &= 2^{p+1}(\|A\|_p + \|B\|_p), \end{aligned} \tag{2.42}$$

that is, $(A+B) \in \mathcal{C}_p(\mathcal{H})$.

As $s_n(A^*) = s_n(A)$, see (2.11), we get that $A^* \in \mathcal{C}_p(\mathcal{H})$ whenever $A \in \mathcal{C}_p(\mathcal{H})$, and that

$$\|A\|_p = \|A^*\|_p. \tag{2.43}$$

Inequality (2.17) implies

$$\|AB\|_p \leqslant \|B\|\|A\|_p \quad \text{and} \quad \|BA\|_p \leqslant \|B\|\|A\|_p, \tag{2.44}$$

which means that $\mathcal{C}_p(\mathcal{H})$ is a two-sided $*$-ideal in $\mathcal{L}(\mathcal{H})$.

Notice that the inequalities (2.39) and (2.44) imply

$$\|AB\|_p \leqslant \|A\|_p\|B\|_p, \tag{2.45}$$

and that $\|A\|_p = 0$ implies $A = 0$, see (2.37) and (2.40). Therefore, we can summarise the above observations about $\|\cdot\|_p$ as

Proposition 2.51. *For $p \geqslant 1$ the functional $\|\cdot\|_p$ is a norm on the $*$-ideal $\mathcal{C}_p(\mathcal{H})$.*

To prove this statement it remains to verify the *triangle inequality*

$$\|A+B\|_p \leqslant \|A\|_p + \|B\|_p \tag{2.46}$$

for $A, B \in \mathcal{C}_p(\mathcal{H})$. Note that $A+B \in \mathcal{C}_p(\mathcal{H})$ by (2.42) and that we have already established (2.46) for $p = 1$ in Proposition 2.49 (c). We shall first establish some auxiliary results before proving (2.46) in Proposition 2.55.

Lemma 2.52. *Let $W = [w_{nm}]$ be a square matrix of dimension r satisfying*

$$\sum_{n=1}^{r} |w_{nm}| \leqslant 1 \quad \text{and} \quad \sum_{m=1}^{r} |w_{nm}| \leqslant 1. \tag{2.47}$$

If $x = (x_1, x_2, \ldots, x_r)$ and $y = (y_1, y_2, \ldots, y_r)$ are two vectors with ordered components $0 \leqslant x_r \leqslant x_{r-1} \leqslant \cdots \leqslant x_1$ and $0 \leqslant y_r \leqslant y_{r-1} \leqslant \cdots \leqslant y_1$, then

$$|(Wx, y)| \leqslant (x, y). \tag{2.48}$$

Proof. Let $f_k = (1, \ldots, 1, 0, \ldots, 0)$ be the vector with the first k components equal to 1. Then according to (2.47),

$$|(Wf_n, f_m)| = \Big|\sum_{i \leqslant n, j \leqslant m} w_{ij}\Big| \leqslant \min\{n, m\} = (f_n, f_m), \tag{2.49}$$

for $1 \leqslant n, m \leqslant r$. It is clear that the vectors x and y can be represented as $x = \sum_{n=1}^{r} \alpha_n f_n$ and $y = \sum_{m=1}^{r} \beta_m f_m$, where $\alpha_n \geqslant 0$ and $\beta_m \geqslant 0$. By (2.49) we get

$$|(Wx, y)| = \Big|\sum_{n,m=1}^{r} \alpha_n \beta_m (Wf_n, f_m)\Big| \leqslant \sum_{n,m=1}^{r} \alpha_n \beta_m (f_n, f_m) = (x, y). \qquad \square$$

Lemma 2.53. *Let $A, B \in \mathcal{C}_\infty(\mathcal{H})$. The singular values of A, B and AB satisfy the inequalities*

$$\sum_{n=1}^{r} s_n(AB) \leqslant \sum_{n=1}^{r} s_n(A)s_n(B), \quad r \in \mathbb{N}. \tag{2.50}$$

Proof. By the polar decomposition, $AB = U|AB|$. Hence, $|AB| = U^*AB$. Let P_r be a projection onto the subspace $[\varphi_1, \varphi_2, \dots, \varphi_r]$ spanned by the eigenvectors $\{\varphi_j : j = 1, 2, \dots, r\}$ of the compact operator $|AB|$, then

$$\begin{aligned} \sum_{n=1}^{r} s_n(AB) &= \sum_{n=1}^{r} s_n(|AB|) = \mathrm{Tr}(|AB|P_r) \\ &= \mathrm{Tr}(P_r U^* ABP_r) = \mathrm{Tr}(\tilde{A}\tilde{B}), \end{aligned} \tag{2.51}$$

where $\tilde{A} := P_r U^* A \in \mathcal{K}_r(\mathcal{H})$ and $\tilde{B} := BP_r \in \mathcal{K}_r(\mathcal{H})$. Using the canonical decomposition (2.8) one gets

$$\tilde{A} = \sum_{n=1}^{r} s_n(\tilde{A})(\cdot, \varphi_n(\tilde{A})\psi_n(\tilde{A}), \quad \tilde{B} = \sum_{n=1}^{r} s_n(\tilde{B})(\cdot, \varphi_n(\tilde{B})\psi_n(\tilde{B}). \tag{2.52}$$

Therefore, (2.51) and (2.52) yield

$$\begin{aligned} \sum_{n=1}^{r} s_n(AB) &= \sum_{n=1}^{r} s_n(\tilde{A})(\tilde{B}\psi_n(\tilde{A}), \varphi_n(\tilde{A})) \\ &= \sum_{n,m=1}^{r} s_n(\tilde{A})s_m(\tilde{B})(\psi_n(\tilde{A}), \varphi_m(\tilde{B}))(\psi_m(\tilde{B}), \varphi_n(\tilde{A})) \\ &=: (Wx, y), \end{aligned} \tag{2.53}$$

where matrix $W := [w_{nm}]$ with

$$w_{nm} := (\psi_n(\tilde{A}), \varphi_m(\tilde{B}))(\psi_m(\tilde{B}), \varphi_n(\tilde{A})),$$

and

$$x := (s_1(\tilde{A}), s_2(\tilde{A}), \dots, s_r(\tilde{A})), \; y := (s_1(\tilde{B}), s_2(\tilde{B}), \dots, s_r(\tilde{B})).$$

The matrix W satisfies the conditions of Lemma 2.52

$$\begin{aligned} \sum_{n=1}^{r} |w_{nm}| &\leqslant \Big[\sum_{n=1}^{r} |(\psi_n(\tilde{A}), \varphi_m(\tilde{B}))|^2\Big]^{1/2} \Big[\sum_{n=1}^{r} |(\psi_m(\tilde{B}), \varphi_n(\tilde{A}))|^2\Big]^{1/2} \\ &\leqslant \|\varphi_m(\tilde{B})\| \|\psi_m(\tilde{B})\| = 1, \end{aligned}$$

and similarly for the row sums

$$\begin{aligned} \sum_{m=1}^{r} |w_{nm}| &\leqslant \Big[\sum_{m=1}^{r} |(\psi_n(\tilde{A}), \varphi_m(\tilde{B}))|^2\Big]^{1/2} \Big[\sum_{m=1}^{r} |(\psi_m(\tilde{B}), \varphi_n(\tilde{A})|^2\Big]^{1/2} \\ &\leqslant \|\psi_n(\tilde{A})\| \|\varphi_n(\tilde{A})\| = 1. \end{aligned}$$

Therefore, by the estimate (2.48) and (2.53) one gets

$$\sum_{n=1}^{r} s_n(AB) = (Wx, y) \leqslant (x, y) = \sum_{n=1}^{r} s_n(\tilde{A})s_n(\tilde{B}). \tag{2.54}$$

Then (2.50) follows from the inequalities (2.16), (2.17) and (2.54), since

$$s_n(\tilde{A}) \leqslant \|P_r\|\|U^*\|s_n(A) = s_n(A) \text{ and } s_n(\tilde{B}) \leqslant \|P_r\|s_n(B) = s_n(B). \qquad \square$$

Proposition 2.54 ($\mathcal{C}_p$-Hölder inequality). *Let $A \in \mathcal{C}_p(\mathcal{H})$ and $B \in \mathcal{C}_q(\mathcal{H})$, where $p > 1$ and $p^{-1} + q^{-1} = 1$. Then $AB \in \mathcal{C}_1(\mathcal{H})$ and*

$$\|AB\|_1 \leqslant \|A\|_p\|B\|_q\,. \tag{2.55}$$

Proof. By Definition 2.30

$$\|AB\|_1 = \sum_{n\geqslant 1} s_n(AB). \tag{2.56}$$

Applying the classical Hölder inequality to the estimate (2.50) for the truncated sum corresponding to (2.56), one gets

$$\begin{aligned}\sum_{n=1}^{r} s_n(AB) &\leqslant \sum_{n=1}^{r} s_n(A)s_n(B)\\ &\leqslant \Big[\sum_{n=1}^{r} s_n^p(A)\Big]^{1/p}\Big[\sum_{n=1}^{r} s_n^q(B)\Big]^{1/q}.\end{aligned} \tag{2.57}$$

Taking the limit $r \to \infty$, we obtain (2.55). $\square$

Proposition 2.55 ($\|\cdot\|_p$-triangle inequality). *Let $A \in \mathcal{C}_p(\mathcal{H})$ and $B \in \mathcal{C}_p(\mathcal{H})$, where $1 < p < \infty$. Then inequality* (2.46) *holds:*

$$\|A + B\|_p \leqslant \|A\|_p + \|B\|_p\,.$$

Proof. Let $C \in \mathcal{C}_1(\mathcal{H})$. Then, by Proposition 2.38,

$$|\operatorname{Tr} C| \leqslant \|C\|_1\,. \tag{2.58}$$

For $C = TQ$, with $T \in \mathcal{C}_p(\mathcal{H})$, $Q \in \mathcal{C}_q(\mathcal{H})$, $p > 1$ and $p^{-1} + q^{-1} = 1$, the estimate (2.58) implies

$$|\operatorname{Tr}(TQ)| \leqslant \|T\|_p\|Q\|_q\,, \tag{2.59}$$

see Proposition 2.54. For a fixed $T \in \mathcal{C}_p(\mathcal{H})$ the inequality (2.59) is saturated for $Q = Q_{\max} := |T|^{p-1}U^*$, where we used the polar decomposition $T = U|T|$. Indeed,

$$\operatorname{Tr}(TQ_{\max}) = \operatorname{Tr}(U|T|^pU^*) = (\|T\|_p)^p,$$

see (2.32), (2.37) and Proposition 2.48 (d)) and

$$\|Q_{\max}\|_q = \Big[\sum_{n\geqslant 1}(s_n(T))^{(p-1)q}\Big]^{1/q} = \Big[\sum_{n\geqslant 1} s_n^p(T)\Big]^{1/q} = (\|T\|_p)^{p-1}.$$

Therefore,

$$\|T\|_p = \sup_{Q\in\mathcal{C}_q(\mathcal{H})} \frac{|\operatorname{Tr}(TQ)|}{\|Q\|_q} = \sup_{\|Q\|_q=1} |\operatorname{Tr}(TQ)|, \tag{2.60}$$

where $p > 1$ and $p^{-1} + q^{-1} = 1$.

For $A, B \in \mathcal{C}_p(\mathcal{H})$, $Q \in \mathcal{C}_q(\mathcal{H})$, $p > 1$ and $p^{-1} + q^{-1} = 1$ we have the obvious inequality

$$|\operatorname{Tr}(A + B)Q| \leqslant |\operatorname{Tr}(AQ)| + |\operatorname{Tr}(BQ)|. \tag{2.61}$$

Taking here the supremum over $\{Q : \|Q\|_q = 1\}$, we get by (2.60) the $\|\cdot\|_p$-triangle inequality

$$\|A + B\|_p \leqslant \|A\|_p + \|B\|_p. \tag{2.62}$$

□

Proof of Proposition 2.51. We have already checked all necessary properties of the functional $\|\cdot\|_p$, including the triangle inequality for $p = 1$, see Proposition 2.49 and for $p > 1$, see Proposition 2.55. □

In fact, the $\mathcal{C}_p(\mathcal{H})$-ideals are Banach spaces with respect to the norms $\|\cdot\|_p$, $p \geqslant 1$. We shall discuss this in the next section.

2.4 Convergence theorems for $\mathcal{C}_p(\mathcal{H})$

This section is devoted to topological properties of the space $\mathcal{C}_p(\mathcal{H})$. We begin with the following lemma.

Lemma 2.56. *If $C_1, C_2 \in \mathcal{C}_\infty(\mathcal{H})$, then*

$$|s_\ell(C_1) - s_\ell(C_2)| \leqslant \|C_1 - C_2\|, \quad \ell \geqslant 1. \tag{2.63}$$

Proof. Put in the estimate (2.23) $A + B = C_1$, $A = C_2$ and $m = 0$. Then, taking into account (2.8),

$$s_{n+1}(C_1) - s_{n+1}(C_2) \leqslant s_1(C_1 - C_2) = \|C_1 - C_2\|, \quad n \geqslant 0. \tag{2.64}$$

Now, let $A + B = C_2$ and $A = C_1$. Then for $m = 0$ the estimate (2.23) gives

$$s_{n+1}(C_2) - s_{n+1}(C_1) \leqslant \|C_1 - C_1\|, \quad n \geqslant 0,$$

which together with (2.64) proves (2.63). □

Lemma 2.57. *The space $\mathcal{L}(\mathcal{H})$ of bounded operators equipped with the operator-norm topology is a Banach space.*

Proof. We only have to prove the completeness of $\mathcal{L}(\mathcal{H})$. Let $\{A_k\}_{k\geqslant 1}$ be a Cauchy sequence of elements of $\mathcal{L}(\mathcal{H})$. Then, as $\|A_{k+n}u - A_k u\| \leqslant \|A_{k+n} - A_k\|\|u\|$, $\{A_k u\}_{k\geqslant 1}$ is a Cauchy sequence in $\mathcal{H}$ for each fixed $u \in \mathcal{H}$. Since the space $\mathcal{H}$ is complete, there is a $v \in \mathcal{H}$ such that $\lim_{k\to\infty} \|A_k u - v\| = 0$. Let A be an operator defined by $Au := v$. Then A is obviously linear and bounded, with $\|A\| \leqslant \sup_{k\geqslant 1} \|A_k\| \leqslant M$. The latter holds because $\{A_k\}_{k\geqslant 1}$ is a Cauchy sequence, which implies that $\{\|A_k\|\}_{k\geqslant 1}$ is also a Cauchy sequence. Hence,

$$\|(A_k - A)u\| \leqslant \lim_{n\to\infty} \|A_k - A_{k+n}\|\|u\|,$$

or

$$\lim_{k\to\infty} \|A_k - A\| \leqslant \lim_{k,n\to\infty} \|A_k - A_{k+n}\| = 0. \qquad \square$$

Corollary 2.58. *The space* $\mathcal{C}_\infty(\mathcal{H})$ *is a* $*$*-ideal in* $\mathcal{L}(\mathcal{H})$*, see Proposition* 2.7*. By Proposition* 2.8*, it is a closed manifold of the Banach space* $\mathcal{L}(\mathcal{H})$*. Thus* $\mathcal{C}_\infty(\mathcal{H})$ *is itself a Banach space with respect to the operator norm topology. Moreover, since the finite-rank operators* $\mathcal{K}(\mathcal{H})$ *are dense in* $\mathcal{C}_\infty(\mathcal{H})$ *in the operator norm topology, see Proposition* 2.11*, the Banach space* $\mathcal{C}_\infty(\mathcal{H})$ *is separable.*

Similar properties hold for the von Neumann-Schatten classes $\mathcal{C}_p(\mathcal{H})$, $1 \leqslant p < \infty$.

Proposition 2.59. *For any* $1 \leqslant p < \infty$*, the space* $\mathcal{C}_p(\mathcal{H})$ *has the following properties:*

(a) $\mathcal{C}_p(\mathcal{H})$ *is closed in the* $\|\cdot\|_p$*-norm.*

(b) *The finite-rank operators are* $\|\cdot\|_p$*-dense in* $\mathcal{C}_p(\mathcal{H})$*.*

(c) $\mathcal{C}_p(\mathcal{H})$ *is a separable Banach space with respect to the norm* $\|\cdot\|_p$*.*

Proof. (a) Suppose that a sequence $\{A_k \in \mathcal{C}_p(\mathcal{H})\}_{k\geqslant 1}$ converges in the $\|\cdot\|_p$-norm to operator A, i.e., $\|A - A_k\|_p < \varepsilon$ for $k > k_\varepsilon$. By the $\|\cdot\|_p$-triangle inequality (2.62), $\|(A - A_k) + A_k\|_p \leqslant \|A - A_k\|_p + \|A_k\|_p$. Hence, $A \in \mathcal{C}_p(\mathcal{H})$.

(b) Using the canonical representation (2.8) one can construct for any $A \in \mathcal{C}_p(\mathcal{H})$ a finite-rank operator $A^{(N)} = \sum_{n=1}^N s_n(A)(\cdot, \varphi_n)\psi_n$ such that $\|A - A^{(N)}\|_p = \left(\sum_{n\geqslant N+1} s_n^p(A)\right)^{1/p} < \varepsilon$ for $N > N_\varepsilon$.

(c) Let $\{A_k\}_{k\geqslant 1}$ be a Cauchy sequence of elements of $\mathcal{C}_p(\mathcal{H})$, i.e., $\|A_{k+n} - A_k\|_p < \varepsilon$ for $k > k_\varepsilon$ and $n \geqslant 1$ arbitrary. Since $\mathcal{C}_p(\mathcal{H}) \subset \mathcal{C}_\infty(\mathcal{H})$ and $\|\cdot\| < \|\cdot\|_p$, $\{A_k\}_{k\geqslant 1}$ is a Cauchy sequence in $\mathcal{C}_\infty(\mathcal{H})$. Then by Corollary 2.58 there exists an operator $A \in \mathcal{C}_\infty(\mathcal{H})$ such that $A = \|\cdot\|\text{-}\lim_{k\to\infty} A_k$. Because $\{A_k\}_{k\geqslant 1}$ is a Cauchy sequence in the $\|\cdot\|_p$-norm

$$\sum_{\ell\geqslant 1} s_\ell(A_{k+n} - A_k) \leqslant \varepsilon, \quad k > k_\varepsilon\,, \quad n \geqslant 1. \tag{2.65}$$

By Lemma 2.56, the singular values are $\|\cdot\|$-continuous functions on the compact operators. Therefore, we can let $n \to \infty$ in (2.65) to obtain

$$\sum_{\ell\geqslant 1} s_\ell(A - A_k) < \varepsilon, \quad k > k_\varepsilon\,. \tag{2.66}$$

The estimate (2.66) means that $\lim_{k\to\infty} \|A - A_k\|_p = 0$, see (2.37). Hence, by (a), $A \in \mathcal{C}_p(\mathcal{H})$, that is, $\mathcal{C}_p(\mathcal{H})$ is a Banach space. Since finite-rank operators $\mathcal{K}(\mathcal{H})$ are $\|\cdot\|_p$-dense in $\mathcal{C}_p(\mathcal{H})$, the space $\mathcal{C}_p(\mathcal{H})$ is separable. □

Remark 2.60. The above arguments show that, for $1 \leqslant p < \infty$, the space $\mathcal{C}_p(\mathcal{H})$ is not closed in the operator norm topology. By Remark 2.12, the $\|\cdot\|$-closure of $\mathcal{C}_p(\mathcal{H})$ coincides with $\mathcal{C}_\infty(\mathcal{H})$.

Next we collect and discuss a number of convergence theorems in $\mathcal{C}_p(\mathcal{H})$, which are indispensable for the ensuing exposition of the theory of Gibbs semi-groups.

- The first type of the convergence theorems is a direct analogue in $\mathcal{C}_p(\mathcal{H})$ of the classical *dominated convergence* theorem in L^p.
- The second type of them is a kind of *lifting* propositions, stating that convergence of norms of operators in $\mathcal{C}_p(\mathcal{H})$ *bolstered* by a *weak* convergence of the operators themselves yields their *norm* convergence.
- The third type concerns a *local uniformity* of convergence in $\mathcal{C}_p(\mathcal{H})$ of one-parameter operator-valued functions $\{t \mapsto \Phi_n(t) \in \mathcal{C}_p(\mathcal{H})\}_{n\in\mathbb{N}}$ for $t \in [a, b] \subset \mathbb{R}$.

The following proposition provides analogue in $\mathcal{C}_p(\mathcal{H})$ of the dominated convergence theorem for L^p-spaces. Note that analogue of the *pointwise* convergence here is convergence in the weak operator topology.

Proposition 2.61. *Let $0 \leqslant B \in \mathcal{L}(\mathcal{H})$ and suppose that $\{A_k\}_{k\geqslant 1}$ is a sequence of bounded operators such that $|A_k| \leqslant B$, $|A_k^*| \leqslant B$ for all $k \geqslant 1$. Suppose also that the weak operator limit exists: $A = \text{w-}\lim_{k\to\infty} A_k$, with $|A| \leqslant B$, $|A^*| \leqslant B$.*

If $B \in \mathcal{C}_p(\mathcal{H})$ for $1 \leqslant p < \infty$, then $A = \|\cdot\|_{\text{p}}\text{-}\lim_{k\to\infty} A_k \in \mathcal{C}_p(\mathcal{H})$.

Proof. First, note that $0 \leqslant B \in \mathcal{C}_\infty(\mathcal{H})$ implies that $B^{1/2} \in \mathcal{C}_\infty(\mathcal{H})$, see Corollary 2.26. Moreover, $s_n(B^{1/2}) = s_n^{1/2}(B)$ and $B^{1/2}$ has the same eigenvectors as B. Since $|A_k| \leqslant B$ is equivalent to $\||A_k|^{1/2}u\| \leqslant \|B^{1/2}u\|$, the operator $|A_k|^{1/2}$ maps any weakly convergent sequence $\{u_n\}_{n\geqslant 1}$ into a convergent sequence $\{|A_k|^{1/2}u_n\}_{n\geqslant 1}$. Therefore, $|A_k|^{1/2} \in \mathcal{C}_\infty(\mathcal{H})$ and, consequently, $A_k = U_k|A_k| \in \mathcal{C}_\infty(\mathcal{H})$ for each $k \geqslant 1$. By the same argument, $A \in \mathcal{C}_\infty(\mathcal{H})$.

Now, for any fixed $\varepsilon > 0$, we can find a finite-rank projection P such that $\|QBQ\| < \varepsilon$, where $Q = \mathbb{1} - P$ and $p \leqslant \infty$, see Proposition 2.28. Note that $|A_k| \leqslant B$ implies $Q|A_k|Q \leqslant QBQ$. Hence, by Corollary 2.34, $s_n(Q|A_k|Q) \leqslant s_n(QBQ)$ and, consequently, $\|Q|A_k|Q\|_p \leqslant \|QBQ\|_p \leqslant \varepsilon$, or $\||A_j|^{1/2}Q\|_{2p} \leqslant \varepsilon^{1/2}$. Note that $\|\cdot\|_\infty = \|\cdot\|$. The estimate

$$\|PA_kQ\|_p \leqslant \|A_kQ\|_p = \|U|A_k|Q\|_p \leqslant \||A_k|Q\|_p$$

follows from (2.44), $\|P\| \leqslant 1$, $\|U\| = 1$, the Hölder inequality (2.55) and the fact that $|A_k| \leqslant B$:

$$\||A_k|^{1/2}|A_k|^{1/2}Q\|_p \leqslant \||A_k|^{1/2}\|_{2p}\||A_k|^{1/2}Q\|_{2p} \leqslant \|B\|_p\varepsilon^{1/2}.$$

Similar estimates for A_k^*, A and A^* hold by the assumptions of the proposition. Next, we write

$$\begin{aligned}\|A - A_k\|_p &= \|(P+Q)(A-A_k)(P+Q)\|_p \\ &\leqslant \|P(A-A_k)P\|_p + \|P(A-A_k)Q\|_p + \|Q(A-A_k)P\|_p \\ &\quad + \|Q(A-A_k)Q\|_p. \end{aligned} \tag{2.67}$$

We then estimate the different terms in (2.67). For the first term we get,

$$\|Q(A-A_k)Q\|_p \leqslant \||A|Q\|_p + \||A_k|Q\|_p \leqslant 2\varepsilon^{1/2}\|B\|. \tag{2.68}$$

The second and the third terms can be estimated in a similar way, see (2.43),

$$\begin{aligned}\|P(A-A_k)Q\|_p &\leqslant \|(A-A_k)Q\|_p \leqslant 2\varepsilon^{1/2}\|B\|_p, \\ \|Q(A-A_k)P\|_p &\leqslant \|(A-A_k)^*Q\|_p \leqslant 2\varepsilon^{1/2}\|B\|_p.\end{aligned} \tag{2.69}$$

Therefore, (2.67)–(2.69) yield

$$\|A - A_k\|_p \leqslant 6\varepsilon^{1/2}\|B\|_p + \|P(A-A_k)P\|_p. \tag{2.70}$$

Since P is a finite-rank operator, $\|P(A-A_k)P\|_p \to 0$ as $k \to \infty$ as $\text{w-lim}_{k\to\infty} A_k = A$. This proves the proposition. □

Remark 2.62. The condition $|A_k^*| \leqslant B$ cannot be dropped. Indeed, let $\{e_k\}_{k\geqslant 1}$ be an orthonormal set and $A_k := (\cdot, e_1)e_k \in \mathcal{C}_1(\mathcal{H})$. Then $\text{w-lim}_{k\to\infty} A_k = 0$ and $|A_k| \leqslant (\cdot, e_1)e_1 =: B \in \mathcal{C}_1(\mathcal{H})$, but $\lim_{k\to\infty} \|A_k\| \neq 0$. In this example, $A_k^* = (\cdot, e_k)e_1$ and $|A_k^*| = (\cdot, e_k)e_k \in \mathcal{C}_1(\mathcal{H})$, but the estimate $|A_k^*| \leqslant B$ is not valid.

If in the assumption of Proposition 2.61 we replace the weak convergence by the strong convergence, then we get the following statement for the case when $B \in \mathcal{C}_\infty(\mathcal{H})$.

Proposition 2.63. *Let B be a positive compact operator and let $\{A_k\}_{k\geqslant 1}$ be a sequence of bounded self-adjoint operators such that*

$$-B \leqslant A_k \leqslant B \quad \textit{and} \quad \operatorname*{s-lim}_{k\to\infty} A_k = A. \tag{2.71}$$

Then $A_k \in \mathcal{C}_\infty(\mathcal{H})$, $A = A^* \in \mathcal{C}_\infty(\mathcal{H})$, $-B \leqslant A \leqslant B$, and also $\lim_{k\to\infty} \|A - A_k\| = 0$.

Proof. The two-side bound (2.71) by compact operator B imply that $A_k \in \mathcal{C}_\infty(\mathcal{H})$ for any $k \geqslant 1$. Since $A_k^* = A_k$ and since the strong convergence (2.71) yields $\text{w-lim}_{k\to\infty} A_k^* = A^*$, we get $A^* = A$. Moreover, this limit and (2.71) give $-(Bu,u) \leqslant \lim_{k\to\infty}(A_k u, u) \leqslant (Bu,u)$, or $-B \leqslant A \leqslant B$.

For the rest of the proof, we can restrict ourselves to the case $A = 0$. Suppose that for a subsequence $\{A_\ell\}_{\ell\geqslant 1}$ one has $\|A_\ell\| \geqslant a > 0$. Since $A_\ell \in \mathcal{C}_\infty(\mathcal{H})$, there is a sequence $\{u_\ell\}_{\ell\geqslant 1}$ of normalised vectors such that $A_\ell u_\ell = b_\ell u_\ell$ with $|b_\ell| \geqslant a$.

Then (2.71) implies $|(u_\ell, A_\ell u_\ell)| \leqslant (u_\ell, Bu_\ell)$, or $\|B^{1/2}u_\ell\| \geqslant a^{1/2}$. As $B^{1/2} \in \mathcal{C}_\infty(\mathcal{H})$, there exists a subspace M of $\mathcal{H}$, with $\dim M < \infty$ such that

$$\|B^{1/2}v\| \leqslant \delta a^{1/2}\|v\|, \quad v \in M^\perp, \tag{2.72}$$

for some $0 < \delta < 1$, see (2.20) and (2.26). Let P_M be the projection onto M. Then (2.72) and the estimate

$$\|B^{1/2}u_\ell\| \leqslant \|B^{1/2}P_M u_\ell\| + \|B^{1/2}(\mathbb{1} - P_M)u_\ell\|$$

show that

$$\|B^{1/2}\|\|P_M u_\ell\| \geqslant \|B^{1/2}P_M u_\ell\| \geqslant (1-\delta)a^{1/2}.$$

Hence, $\|P_M u_\ell\| \geqslant (1-\delta)a^{1/2}\|B^{1/2}\|^{-1}$, $\|u_\ell\| = 1$ and

$$\|P_M A_\ell u_\ell\| \geqslant (1-\delta)a^{3/2}\|B^{1/2}\|^{-1}, \quad \|u_\ell\| = 1. \tag{2.73}$$

On the other hand, $\text{s-lim}_{\ell\to\infty} A_\ell = 0$ implies $\|A_\ell P_M\| \to 0$ and hence $\lim_\ell \|P_M A_\ell\| = 0$, which contradicts the estimate from below in (2.73). □

Corollary 2.64. *Propositions* 2.61 *and* 2.63 *are valid assuming that* $B \in \mathcal{C}_p(\mathcal{H})$ *with* $1 \leqslant p < \infty$ *and* $A_k = A_k^*$. *The conclusion is that* $A_k \in \mathcal{C}_p(\mathcal{H})$ *and* $\|\cdot\|_{\mathrm{p}}\text{-}\lim_{k\to\infty} A_k = A \in \mathcal{C}_p(\mathcal{H})$. *The proofs carry through verbatim.*

In applications the following version of the *lifting* proposition for convergence of compact operators $\mathcal{C}_\infty(\mathcal{H})$ is useful.

Proposition 2.65. *Let B be a positive compact operator and let $\{A_k\}_{k\geqslant 1}$ be a sequence of bounded self-adjoint operators such that*

$$0 \leqslant A_k \leqslant B \quad \textit{and} \quad \underset{k\to\infty}{\text{w-lim}}\, A_k = A. \tag{2.74}$$

Then $A_k \in \mathcal{C}_\infty(\mathcal{H})$, $A = A^* \in \mathcal{C}_\infty(\mathcal{H})$, $0 \leqslant A \leqslant B$, *and* $\lim_{k\to\infty} \|A - A_k\| = 0$.

Proof. The first part of this assertion including the inequalities: $0 \leqslant A \leqslant B$, is a corollary of the arguments of the proof for $\{A_k\}_{k\geqslant 1}$ in Propositions 2.63.

To prove the result on the lifting of the weak operator convergence (2.74) to the operator-norm convergence we first note that operators $B^{1/2} \in \mathcal{C}_\infty(\mathcal{H})$, $A^{1/2} \in \mathcal{C}_\infty(\mathcal{H})$, ${A_k}^{1/2} \in \mathcal{C}_\infty(\mathcal{H}), k \in \mathbb{N}$, and that

$$0 \leqslant A^{1/2} \leqslant B^{1/2} \quad \text{and} \quad 0 \leqslant {A_k}^{1/2} \leqslant B^{1/2}, \quad k \in \mathbb{N}. \tag{2.75}$$

Now, we use the canonical representation for the compact operators $A^{1/2}$, $B^{1/2}$ (2.8) (Corollary 2.26) and the minimax principle (Corollary 2.34), which gives $s_n(A^{1/2}) \leqslant s_n(B^{1/2})$, $n \geqslant 1$. They allow to construct a *contraction* $\Gamma_{A,B}$ such that

$$\Gamma_{A,B} : \overline{\operatorname{ran} B^{1/2}} \to \mathcal{H} \quad \text{and} \quad A^{1/2} = \Gamma_{A,B}B^{1/2}. \tag{2.76}$$

Similarly one constructs the contractions $\{\Gamma_{A,B}(k)\}_{k\geqslant 1}$ such that

$$\Gamma_{A,B}(k) : \overline{\operatorname{ran} B^{1/2}} \to \mathcal{H} \quad \text{and} \quad A_k^{1/2} = \Gamma_{A,B}(k)B^{1/2}. \tag{2.77}$$

Since $A = (A^{1/2})^* A^{1/2} = B^{1/2}\Gamma_{A,B}^*\Gamma_{A,B}B^{1/2}$, and similarly $A_k = (A_k^{1/2})^* A_k^{1/2} = B^{1/2}\Gamma_{A,B}(k)^*\Gamma_{A,B}(k)B^{1/2}$, the weak operator limit in (2.74) yields for any $u, v \in \mathcal{H}$ that

$$\begin{aligned} &\lim_{k\to\infty} ((A - A_k)u, v) \\ &= \lim_{k\to\infty} ((\Gamma_{A,B}^*\Gamma_{A,B} - \Gamma_{A,B}(k)^*\Gamma_{A,B}(k))B^{1/2}u, B^{1/2}v) = 0. \end{aligned} \tag{2.78}$$

Note that by virtue of the canonical form of self-adjoint compact operators (2.8) (Corollary 2.26), $\overline{\operatorname{ran} B^{1/2}} = \mathcal{H}$. Therefore, (2.78) implies $\text{w-lim}_{k\to\infty} \Gamma_{A,B}(k)^*\Gamma_{A,B}(k) = \Gamma_{A,B}^*\Gamma_{A,B}$, and thus

$$\lim_{k\to\infty} \|B^{1/2}(\Gamma_{A,B}^*\Gamma_{A,B} - \Gamma_{A,B}(k)^*\Gamma_{A,B}(k))B^{1/2}\| = 0, \tag{2.79}$$

since $B^{1/2} \in \mathcal{C}_\infty(\mathcal{H})$. □

The use of Proposition 2.61 may pose a problem because the triangle inequality *fails* for absolute values of operators, see Remark 2.2. For this reason, the following notion of the operator (s)-*order* and the next statement are helpful.

Definition 2.66. We write $A \overset{(s)}{<} B$, if $s_n(A) \leqslant s_n(B)$ for $n \geqslant 1$. We say that $(s)\text{-}\lim_{k\to\infty} A_k = A$, if $\lim_{k\to\infty} s_n(A_k) = s_n(A)$ for all $n \geqslant 1$.

Remark 2.67. Since $s_1(A) = \|A\| \geqslant s_n(A)$, $(s)\text{-}\lim_{k\to\infty} A_k = 0$ if and only if $\lim_{k\to\infty} \|A_k\| \to 0$.

Proposition 2.68. *Let $\{A_k\}_{k\geqslant 1}$ be a sequence of bounded operators such that $\{A_k \overset{(s)}{<} B\}_{k\geqslant 1}$ for an operator $B \in \mathcal{C}_p(\mathcal{H})$. If $\|\cdot\|\text{-}\lim_{k\to\infty} A_k = A$, then $\|\cdot\|_{\mathrm{p}}\text{-}\lim_{k\to\infty} A_k = A$.*

Proof. Since by Remark 2.67 $\|A_k - A\| \to 0$ implies $s_n(A_k - A) \to 0$, it is sufficient to show that $s_n(A_k - A) \leqslant \alpha_n$ with $\{\alpha_n\}_{n\geqslant 1} \in l^p$, because then we can appeal to the dominated convergence theorem in the l^p-space. By inequality (2.23) and $s_n(A_k) \leqslant s_n(B)$, one gets

$$\begin{aligned} &s_{2n+1}(A_k - A) \leqslant s_{n+1}(A_k) + s_{n+1}(A) \leqslant 2s_{n+1}(B), \\ &s_{2n}(A_k - A \leqslant s_{n+1}(A_k) + s_n(A) \leqslant s_n(A_k) + s_n(A) \leqslant 2s_n(B). \end{aligned}$$

Hence, we can take $\{\alpha_n\}_{n\geqslant 1} = 2\{s_n(B)\}_{n\geqslant 1} \in l^p$. □

Therefore, Proposition 2.68 is an alternative device to Proposition 2.61 and Corollary 2.64.

The next result is of the *second* type. Here the lifting of the convergence of norms of operators is bolstered by conditions of strong convergence of operators and their adjoints.

Proposition 2.69. *Let* $\{A_k\}_{k\geqslant 1}$ *be a sequence of bounded operators such that* $\text{s-lim}_{k\to\infty} A_k = A$ *and* $\text{s-lim}_{k\to\infty} A_k^* = A^*$. *If there is* $1 \leqslant p < \infty$ *such that* $A_k, A \in \mathcal{C}_p(\mathcal{H})$ *and* $\lim_{k\to\infty} \|A_k\|_p = \|A\|_p$, *then*

$$\|\cdot\|_p\text{-}\lim_{k\to\infty} A_k = A. \tag{2.80}$$

To carry on we first have to establish two lemmata.

Lemma 2.70 (Strong continuity of multiplication). *Let* $\text{s-lim}_{k\to\infty} A_k = A \in \mathcal{L}(\mathcal{H})$ *and* $\text{s-lim}_{k\to\infty} B_k = B \in \mathcal{L}(\mathcal{H})$. *Then* $\text{s-lim}_{k\to\infty} A_k B_k = AB$.

Proof. For any $u \in \mathcal{H}$ one has

$$A_k B_k u - ABu = A_k(B_k - B)u + (A_k - A)Bu. \tag{2.81}$$

Then the first term in (2.81) converges to zero because

$$\|A_k(B_k - B)u\| \leqslant \sup_{k\geqslant 1} \|A_k\| \|(B_k - B)u\|,$$

where $\sup_{k\geqslant 1} \|A_k\| < \infty$ by the uniform boundedness principle (Proposition 1.6). The second term in (2.81) converges to zero because $\text{s-lim}_{k\to\infty} A_k = A$. □

Lemma 2.71. *Suppose that* $\text{s-lim}_{k\to\infty} A_k = A \in \mathcal{L}(\mathcal{H})$ *and that also* $\text{s-lim}_{k\to\infty} A_k^* = A^*$. *Then* $\text{s-lim}_{k\to\infty} |A_k| = |A|$ *and* $\text{s-lim}_{k\to\infty} |A_k^*| = |A^*|$.

Proof. First, by Lemma 2.70,

$$\text{s-lim}_{k\to\infty} |A_k|^2 = \text{s-lim}_{k\to\infty} A_k^* A_k = |A|^2,$$

and similarly $\text{s-lim}_{k\to\infty} |A_k^*|^2 = |A^*|^2$. Next, let $C_k := A_k/M$ and $C := A/M$, where $M = \sup_{k\geqslant 1} \|A_k\|$. Then $\|C_k\|, \|C\| \leqslant 1$. We now observe that $|C_k|$ can be computed using the series

$$|C_k| = \sqrt{\mathbb{1} - (\mathbb{1} - C_k^* C_k)} = \mathbb{1} + \sum_{m=1}^{\infty} \gamma_m (\mathbb{1} - C_k^* C_k)^m, \tag{2.82}$$

which converges in the operator norm. Indeed, $\sqrt{1-z}$ is analytic for $|z| < 1$, and as all $\gamma_{m\geqslant 1}$ are negative, we have for all N the estimate

$$\sum_{m=0}^{N} |\gamma_m| = 2 - \sum_{m=0}^{N} \gamma_m = 2 - \lim_{x\uparrow 1} \sum_{m=0}^{N} \gamma_m x^m \leqslant 2 - \lim_{x\uparrow 1} \sqrt{1-x} = 2.$$

This implies that the power series for $\sqrt{1-z}$ is absolutely convergent for $|z| \leqslant 1$. Since $0 \leqslant C_k^* C_k \leqslant \mathbb{1}$, we have

$$\|\mathbb{1} - C_k^* C_k\| = \sup_{\substack{u \in \mathcal{H} \\ \|u\|=1}} (u, (\mathbb{1} - C_k^* C_k)u) \leqslant 1,$$

and consequently the series (2.82) yields $|C_k|$. Since the convergence in (2.82) is uniform, we can use the strong continuity of multiplication, see Lemma 2.70, to take the limit in each term of the sum, therefore

$$|C| = \operatorname*{s-lim}_{k\to\infty} |C_k| \quad \text{and similarly} \quad |C^*| = \operatorname*{s-lim}_{k\to\infty} |C_k^*|. \qquad \square$$

Proof of Proposition 2.69. Since $A \neq 0$, then by virtue of the estimate

$$\begin{aligned} \|A_k - A\|_p &= \left\| \frac{A_k}{\|A_k\|_p} \|A_k\|_p - \frac{A}{\|A\|_p} \|A\|_p \right\|_p \\ &\leqslant \big| \|A_k\|_p - \|A\|_p \big| + \left\| \frac{A_k}{\|A_k\|_p} - \frac{A}{\|A\|_p} \right\|_p \|A\|_p, \end{aligned} \tag{2.83}$$

we can assume with no loss of generality that $\|A_k\|_p = \|A\|_p = 1$. Fix $\varepsilon > 0$. Then we can find a projection $P \in \mathcal{K}(\mathcal{H})$ such that $\|P|A|P\|_p \geqslant 1 - \varepsilon$ and $\|P|A^*|P\|_p \geqslant 1 - \varepsilon$, see Remark 2.12 and Proposition 2.59 (b). For example, P can be the projection onto the span of the first few eigenvectors of $|A|$ and $|A^*|$. Since

$$\operatorname*{s-lim}_{k\to\infty} |A_k| = |A| \quad \text{and} \quad \operatorname*{s-lim}_{k\to\infty} |A_k^*| = |A|, \tag{2.84}$$

there exists an $N(\varepsilon)$ such that for $k \geqslant N(\varepsilon)$ we have $\|P|A_k|P\|_p \geqslant 1 - 2\varepsilon$ and $\|P|A_k^*|P\|_p \geqslant 1 - 2\varepsilon$. Put $Q := \mathbb{1} - P$. By Definition 2.50 of $\|\cdot\|_p$ and by the canonical representation of $|A_k|$,

$$1 = \sum_{n=1}^{\dim P} s_n^p(A_k) + \sum_{n=1+\dim P}^{\infty} s_n^p(A_k) = \|P|A_k|P\|_p^p + \|Q|A_k|Q\|_p^p,$$

i.e., we get

$$\|Q|A_k|Q\|_p \leqslant \{1 - (1-2\varepsilon)^p\}^{1/p} =: \delta(2\varepsilon), \tag{2.85}$$

and the analogous for $\|Q|A_k^*|Q\|$. Hence, $\||A_k|^{1/2}Q\|_{2p} \leqslant \delta^{1/2}(\varepsilon)$ and

$$\||A_k|Q\|_p \leqslant \||A_k|^{1/2}\|_{2p} \||A_k|^{1/2}Q\|_{2p} \leqslant \delta^{1/2}(2\varepsilon), \tag{2.86}$$

where we used $\||A_k|^{1/2}\|_{2p} = \|A_k\|_p^{1/2} = 1$ and the Hölder inequality (2.55). Similarly, we derive the inequalities $\||A_k^*|Q\|_p \leqslant \delta^{1/2}(2\varepsilon)$ and

$$\||A|Q\|_p \leqslant \delta^{1/2}(\varepsilon) \quad \text{and} \quad \||A^*|Q\|_p \leqslant \delta^{1/2}(\varepsilon). \tag{2.87}$$

Now we can present (2.83) using inequality (2.67)

$$\|A_k - A\|_p \leqslant \|P(A_k - A)P\|_p + \|P(A_k - A)Q\|_p + \|Q(A_k - A)P\|_p + \|Q(A_k - A)Q\|_p .$$

The different terms can be estimated as follows: the last as

$$\|Q(A_k - A)Q\|_p \leqslant \||A_k|Q\|_p + \||A|Q\|_p \leqslant \delta^{1/2}(2\varepsilon) + \delta^{1/2}(\varepsilon), \tag{2.88}$$

and the second and third ones in a similar manner as

$$\begin{aligned} \|P(A_k - A)Q\|_p &\leqslant \|(A_k - A)Q\|_p \leqslant \delta^{1/2}(2\varepsilon) + \delta^{1/2}(\varepsilon), \\ \|Q(A_k - A)P\|_p &\leqslant \|(A_k^* - A^*)Q\|_p \leqslant \delta^{1/2}(2\varepsilon) + \delta^{1/2}(\varepsilon). \end{aligned} \tag{2.89}$$

Therefore, (2.88) and (2.89) yield

$$\|A_k - A\|_p \leqslant 3\delta^{1/2}(2\varepsilon) + 3\delta^{1/2}(\varepsilon) + \|P(A_k - A)P\|_p . \tag{2.90}$$

Since P is a finite-rank operator, $\lim_{k\to\infty} \|P(A_k - A)P\|_p = 0$, because $A =$ s-$\lim_{k\to\infty} A_k$. This proves the proposition. □

Corollary 2.72. *Examining the proof given above one finds that taking into account results and arguments of dominated convergence Proposition* 2.61, *the conditions of the strong operator convergence in Proposition* 2.69 *can be relaxed to the weak operator convergence.*

Consequently, the assertion of Proposition 2.69 follows in fact from the next statement.

Proposition 2.73. *Fix* $1 \leqslant p < \infty$. *Suppose that* w-$\lim_{k\to\infty} A_k = A$, w-$\lim_{k\to\infty} |A_k| = |A|$, w-$\lim_{k\to\infty} |A_k^*| = |A^*|$. *If operators* $A_k, A \in \mathcal{C}_p(\mathcal{H})$ *and* $\lim_{k\to\infty} \|A_k\|_p = \|A\|_p$, *then* $\|\cdot\|_p$-$\lim_{k\to\infty} A_k = A$.

Remark 2.74. We note that if $1 < p < \infty$, then due to the *uniform convexity* of the Banach spaces $\mathcal{C}_{1<p<\infty}(\mathcal{H})$, Proposition 2.73 can be strengthened to a "perfect" assertion about *lifting* the weak operator convergence to the $\|\cdot\|_p$-convergence, namely:

- *Suppose that* w-$\lim_{k\to\infty} A_k = A$ *and that* $\lim_{k\to\infty} \|A_k\|_p = \|A\|_p$. *Then* $\|\cdot\|_p$-$\lim_{k\to\infty} A_k = A$.

Moreover, this statement is valid even for the more difficult case $p = 1$. The reader is addressed to the Notes in Section 2.5 for a discussion and for references.

Therefore, in the case $p = 1$ it is instructive to consider first the positive cone of trace-class operators $\mathcal{C}_{1,+}(\mathcal{H}) \subset \mathcal{C}_1(\mathcal{H})$, when $\|A\|_1 = \operatorname{Tr} A \geqslant 0$. Then weak conditions of Proposition 2.73 reduce to only one condition: w-$\lim_{k\to\infty} A_k = A$. On the cone of positive trace-class operators $\mathcal{C}_{1,+}(\mathcal{H})$ Proposition 2.73 takes the following "imperfect", cf. Remark 2.74, yet useful form.

Proposition 2.75. *Let the sequence $\{A_k \in \mathcal{C}_{1,+}(\mathcal{H})\}_{k\geqslant 1}$ converge in the weak operator topology to $A \in \mathcal{C}_{1,+}(\mathcal{H})$. If the sequence $\{\operatorname{Tr} A_k\}_{k\geqslant 1}$ is uniformly bounded, then $\operatorname{Tr} A \leqslant \liminf_{k\to\infty} \operatorname{Tr} A_k$. Moreover, $\lim_{k\to\infty} \|A_k - A\|_1 = 0$ if and only if $\lim_{k\to\infty} \operatorname{Tr} A_k = \operatorname{Tr} A$.*

Proof. Note that the first part of the proposition is simply the weak lower semi-continuity of the trace, Proposition 2.48(f). The second part needs arguments similar to those for of Proposition 2.73. On the positive cone $\mathcal{C}_{1,+}(\mathcal{H})$ they are straightforward.

The proof in one direction is evident. Thus, let w-$\lim_{k\to\infty} A_k = A$ and $\lim_{k\to\infty} \|A_k\|_1 = \|A\|_1$. Since for $A = 0$ the statement is trivial, let $A \neq 0$. Then by virtue of the inequality

$$\begin{aligned}\|A_k - A\|_1 &= \left\| \frac{A_k}{\|A_k\|_1}\|A_k\|_1 - \frac{A}{\|A\|_1}\|A\|_1 \right\|_1 \\ &\leqslant \big|\|A_k\|_1 - \|A\|_1\big| + \left\| \frac{A_k}{\|A_k\|_1} - \frac{A}{\|A\|_1} \right\|_1 \|A\|_1 ,\end{aligned}$$

we can consider the case $\|A_k\|_1 = \|A\|_1 = 1$. Then for any finite-rank orthogonal projection P one gets

$$\begin{aligned}\|A_k - A\|_1 \leqslant{}& \|P(A_k - A)P\|_1 + 2\|PA_k(\mathbb{1} - P)\|_1 \\ &+ 2\|PA(\mathbb{1} - P)\|_1 + \|(\mathbb{1} - P)A_k(\mathbb{1} - P)\|_1 + \|(\mathbb{1} - P)A(\mathbb{1} - P)\|_1 .\end{aligned} \tag{2.91}$$

Note that on the positive cone $\mathcal{C}_{1,+}(\mathcal{H})$ one gets

$$\|(\mathbb{1} - P)A_k(\mathbb{1} - P)\|_1 = \operatorname{Tr}(\mathbb{1} - P)A_k = 1 - \operatorname{Tr} PA + \operatorname{Tr} P(A - A_k),$$

$$\|(\mathbb{1} - P)A(\mathbb{1} - P)\|_1 = 1 - \operatorname{Tr} PA, \quad \|P(A_k - A)P\|_1 = |\operatorname{Tr} P(A - A_k)|.$$

If $PA_k(\mathbb{1} - P) = U_k|PA_k(\mathbb{1} - P)|$ is the polar decomposition then $\|PA_k(\mathbb{1} - P)\|_1 = \operatorname{Tr} U_k^* PA_k^{1/2} A_k^{1/2}(\mathbb{1} - P)$ and the Cauchy-Schwarz inequality yields the estimate

$$\|PA_k(\mathbb{1} - P)\|_1 \leqslant (\operatorname{Tr} PA_k)^{1/2}(1 - \operatorname{Tr} PA_k)^{1/2} . \tag{2.92}$$

One obviously obtains the similar estimate for the term with A:

$$\|PA(\mathbb{1} - P)\|_1 \leqslant (\operatorname{Tr} PA)^{1/2}(1 - \operatorname{Tr} PA)^{1/2} . \tag{2.93}$$

Now, by weak convergence: w-$\lim_{k\to\infty} A_k = A$ of trace-class operators, for any (small) $\varepsilon > 0$ there exist a projection P_ε such that $1 - \operatorname{Tr} P_\varepsilon A < \varepsilon^2$, and a number N_ε such that $|\operatorname{Tr} P_\varepsilon(A - A_k)| < \varepsilon^2$ for all $k \geqslant N_\varepsilon$. Since $\operatorname{Tr} PA_k \leqslant 1$, $\operatorname{Tr} PA \leqslant 1$, then collecting (2.92), (2.93) and the identities above, we deduce from (2.91) the estimate

$$\|A_k - A\|_1 \leqslant \varepsilon^2 + 2\,(2\varepsilon^2)^{1/2} + 2\varepsilon + 2\varepsilon^2 + \varepsilon^2 < 10\,\varepsilon,$$

for all $k \geqslant N_\varepsilon$, which proves the assertion. □

Summarising these facts let us list conditions that are *sufficient* for the application of Proposition 2.69 and Propositions 2.73.

Corollary 2.76. *Let* $1 \leqslant p < \infty$. *Suppose that* $\lim_{k\to\infty} \|A_k - A\| = 0$ *and* $\lim_{k\to\infty} \|A_k\|_p = \|A\|_p$. *Then* $\lim_{k\to\infty} \|A_k - A\|_p = 0$. *Indeed, since* $\|A_k - A\| = \|A_k^* - A^*\|$, *one gets* $\text{s-}\lim_{k\to\infty} A_k = A$ *and* $\text{s-}\lim_{k\to\infty} A_k^* = A^*$. *Hence,* $\|\cdot\|_{\mathrm{p}}\text{-}\lim_{k\to\infty} A_k = A$ *by Proposition* 2.69.

We know that the absolute value $|\cdot|$ is strongly continuous in the sense of Lemma 2.71. It follows that in fact, Proposition 2.69 implies $\|\cdot\|_p$-*continuity* of $|\cdot|$.

Proposition 2.77. *Let* $1 \leqslant p < \infty$. *Suppose that* $\{A_k\}_{k\geqslant 1} \subset \mathcal{C}_p(\mathcal{H})$ *and* $\|\cdot\|_{\mathrm{p}}\text{-}\lim_{k\to\infty} A_k = A$, *then* $\|\cdot\|_{\mathrm{p}}\text{-}\lim_{k\to\infty} |A_k| = |A|$.

Proof. So far as $\lim_{k\to\infty} \|A_k - A\|_p = 0$, we have both $\text{s-}\lim_{k\to\infty} A_k = A$ and $\text{s-}\lim_{k\to\infty} A_k^* = A^*$. Then by Lemma 2.71 we get $\text{s-}\lim_{k\to\infty} |A_k| = |A|$ and $\text{s-}\lim_{k\to\infty} |A_k^*| = |A^*|$. Next, by the triangle inequality (2.46),

$$\big|\|A_k\|_p - \|A\|_p\big| \leqslant \|A_k - A\|_p\,. \tag{2.94}$$

Since $\|A_k\|_p = \||A_k|\|_p$ and $\|A\|_p = \||A|\|_p$, the estimate (2.90) implies

$$\|\cdot\|_{\mathrm{p}}\text{-}\lim_{k\to\infty} \||A_k|\| = \||A|\|. \tag{2.95}$$

Therefore, the proof follows from Proposition 2.69. □

The next proposition shows that one can lift the strong continuity of multiplication in Lemma 2.70, up to $\|\cdot\|_p$-continuity for $1 \leqslant p < \infty$ if one of the factors converges in the $\|\cdot\|_p$-norm.

Proposition 2.78. *Suppose that* $\text{s-}\lim_{k\to\infty} A_k = A \in \mathcal{L}(\mathcal{H})$ *and that* $\|\cdot\|_{\mathrm{p}}\text{-}\lim_{k\to\infty} B_k = B \in \mathcal{C}_p(\mathcal{H})$ *for a* $1 \leqslant p < \infty$. *Then*

$$\|\cdot\|_{\mathrm{p}}\text{-}\lim_{k\to\infty} A_k B_k = AB. \tag{2.96}$$

If, in addition, $\text{s-}\lim_{k\to\infty} A_k^* = A^*$, *then also*

$$\|\cdot\|_{\mathrm{p}}\text{-}\lim_{k\to\infty} B_k A_k = BA. \tag{2.97}$$

Proof. Writing

$$\|A_k B_k - AB\|_p \leqslant \|A_k\|\|B_k - B\|_p + \|(A_k - A)B\|_p\,, \tag{2.98}$$

and using the uniform boundedness principle

$$\sup_{k\geqslant 1} \|A_k\| \leqslant M < \infty,$$

for a strongly convergent sequence $\{A_k\}_{k\geqslant 1}$, we see that it suffices to show that in (2.98) $\lim_{k\to\infty}\|(A_k - A)B\|_p = 0$. Given $\varepsilon > 0$, we can find a projection $P \in \mathcal{K}(\mathcal{H})$ such that $\|(\mathbb{1} - P)B\|_p < \varepsilon$. Then the last term in (2.98) can be estimated as

$$\|(A_k - A)B\|_p \leqslant \|(A_k - A)P\|\|B\|_p + (\|A_k\| + \|A\|)\|(\mathbb{1} - P)B\|_p. \tag{2.99}$$

Then, since the strong convergence of $\{A_k\}_{k\geqslant 1}$ implies $\lim_{k\to\infty}\|(A_k - A)P\| = 0$, (2.99) gives

$$\limsup_{k\to\infty}\|(A_k - A)B\|_p \leqslant (M + \|A\|)\varepsilon.$$

Since ε is arbitrary, the result (2.96) is proven.

The proof of (2.97) follows along the same line of reasoning, with the specificity of the norm $\|\cdot\|_p$ taken into account. □

Remark 2.79. This result does not hold if in the statement of the proposition the strong convergence is replaced by the weak convergence $\text{w-}\lim_{k\to\infty} A_k = A$. Take for example $A_k = (\cdot, e_1)e_k$ and $B_k = (\cdot, e_1)e_1 \in \mathcal{C}_1(\mathcal{H})$. Then $\|\cdot\|_p\text{-}\lim_{k\to\infty} A_k B_k = \|\cdot\|_p\text{-}\lim_{k\to\infty}(\cdot, e_1)e_k \neq 0$, while $\text{w-}\lim_{k\to\infty} A_k = 0$. Here, $\{e_k\}_{k\geqslant 1}$ is an orthonormal basis in $\mathcal{H}$.

Remark 2.80. This result also fails for the sequence $\{B_k A_k\}_{k\geqslant 1}$ instead of $\{A_k B_k\}_{k\geqslant 1}$. Take $A_k = (\cdot, e_k)e_1$. Then $\text{s-}\lim A_k = 0$ and $B_k = (\cdot, e_1)e_1 \in \mathcal{C}_1(\mathcal{H})$. However,

$$\|\cdot\|_p\text{-}\lim_{k\to\infty} A_k B_k = \|\cdot\|_p\text{-}\lim_{k\to\infty}(e_1, e_k)(\cdot, e_1)e_1 = 0\,, \text{ whereas,}$$
$$\|\cdot\|_p\text{-}\lim_{k\to\infty} B_k A_k = \|\cdot\|_p\text{-}\lim_{k\to\infty}(\cdot, e_k)e_1 \neq 0.$$

We conclude this section by extending the convergence results of Proposition 2.61 and Proposition 2.69 to sequences of operator-valued in $\mathcal{C}_p(\mathcal{H})$ *functions*. This is the *third* type of convergence theorems that we need, in particular, for the Trotter-Kato product formulae in Chapters 5–7.

Definition 2.81. Let $\{t \mapsto \Phi_n(t)\}_{n\in\mathbb{N}}$ be a sequence of operator-valued functions in $\mathcal{C}_p(\mathcal{H})$ for $t \in \mathbb{R}$. We say that the sequence $\{\Phi_n(\cdot)\}_{n\in\mathbb{N}}$, converges in $\|\cdot\|_p$-topology to the function $\Phi(\cdot) \in \mathcal{C}_p(\mathcal{H})$ *locally uniformly* in $D \subseteq \mathbb{R}$ if

$$\lim_{n\to\infty}\sup_{t\in[a,b]}\|\Phi_n(t) - \Phi(t)\|_p = 0,$$

for any compact $[a, b] \subset D$.

Lemma 2.82. *Let $\{X_n(\cdot)\}_{n\in\mathbb{N}}$ and $\{Y_n(\cdot)\}_{n\in\mathbb{N}}$ be sequences of operator-valued functions respectively in $\mathcal{L}(\mathcal{H})$ and in $\mathcal{C}_p(\mathcal{H})$ for $t \in [a, b] \subset \mathbb{R}$, and let $\sup_{t\in[a,b],n\geqslant 1}\|X_n(t)\| < \infty$.*

Let respectively, $X(\cdot) : [a,b] \to \mathcal{L}(\mathcal{H})$ *and* $Y(\cdot) : [a,b] \to \mathcal{C}_p(\mathcal{H})$ *be such that*

$$\sup_{t\in[a,b]} \|Y(t)\| < \infty, \quad and \quad \|\cdot\|_p\text{-}\lim_{n\to\infty} Y_n(t) = Y(t),$$

uniformly in $t \in [a,b]$.

(i) *If* $\text{s-lim}_{n\to\infty} X_n(t) = X(t)$ *uniformly on* $[a,b]$ *and if for some sequence of finite-rank orthogonal projections* $\{P_k\}_{k\geqslant 1}$ *such that* $\text{s-lim}_{k\to\infty} P_k = \mathbb{1}$ *one has*

$$\lim_{k\to\infty} \sup_{t\in[a,b]} \|(\mathbb{1} - P_k)Y(t)\|_p = 0,$$

then $\|\cdot\|_p\text{-}\lim_{n\to\infty} X_n(t)Y_n(t) = X(t)Y(t)$ *uniformly on* $[a,b]$.

(ii) *If* $\text{s-lim}_{n\to\infty} X_n(t)^* = X(t)^*$ *uniformly in* $t \in [a,b]$ *and if for some sequence of finite-rank orthogonal projections* $\{Q_k\}_{k\geqslant 1}$ *such that* $\text{s-lim}_{k\to\infty} Q_k = \mathbb{1}$ *one has*

$$\lim_{k\to\infty} \sup_{t\in[a,b]} \|Y(t)(\mathbb{1} - Q_k)\|_p = 0 ,$$

then $\|\cdot\|_p\text{-}\lim_{n\to\infty} Y_n(t)X_n(t) = Y(t)X(t)$ *uniformly on* $[a,b]$.

Proof. Modulo the evident t-uniformity, the proof essentially mimics the line of reasoning of the proof of Proposition 2.78. □

Proposition 2.83. *Let* $\{X_n(t)\}_{n\in\mathbb{N}}$ *be a sequence of operator-valued functions in* $\mathcal{C}_p(\mathcal{H})$ *for* $t \in [a,b]$ *and* $1 \leqslant p < \infty$, *such that* $\sup_{t\in[a,b],n\geqslant 1} \|X_n(t)\| < \infty$.

Let $X(\cdot) : [a,b] \to \mathcal{C}_p(\mathcal{H})$ *be such that there are two sequences of finite-rank orthogonal projections* $\{P_k\}_{k\geqslant 1}$ *and* $\{Q_k\}_{k\geqslant 1}$ *obeying* $\text{s-lim}_{k\to\infty} P_k = \text{s-lim}_{k\to\infty} Q_k = \mathbb{1}$, *with the property:*

$$\lim_{k\to\infty} \sup_{t\in[a,b]} \|(\mathbb{1} - P_k)X(t)\| = \lim_{k\to\infty} \sup_{t\in[a,b]} \|X(t)(\mathbb{1} - Q_k)\| = 0, \tag{2.100}$$

and in addition

$$\lim_{k\to\infty} \sup_{t\in[a,b]} \|(\mathbb{1} - P_k)X(t)\|_p = 0 \vee \lim_{k\to\infty} \sup_{t\in[a,b]} \|X(t)(\mathbb{1} - Q_k)\|_p = 0. \tag{2.101}$$

If $\text{s-lim}_{n\to\infty} X_n(t) = X(t)$, $\text{s-lim}_{n\to\infty} X_n(t)^* = X(t)^*$, *and*

$$\lim_{n\to\infty} \|X_n(t)\|_p = \|X(t)\|_p \ , \ 1 \leqslant p < \infty, \tag{2.102}$$

uniformly on $[a,b]$, *then* $\|\cdot\|_p\text{-}\lim_{n\to\infty} X_n(t) = X(t)$ *uniformly in* $t \in [a,b]$ *for* $p \in \mathbb{N}$.

Proof. Note that the conditions of this assertion for $t \in [a,b]$ coincide with the conditions of Proposition 2.69 *pointwise*, and so does the proof. The extension to uniformity follows from the conditions (2.100)–(2.102) that should be incorporated in the argument. However, the latter is straightforward. □

Corollary 2.84. *Let $\{X_n(\cdot)\}_{n\in\mathbb{N}}$ and $\{Y_n(\cdot)\}_{n\in\mathbb{N}}$ be sequences of operator-valued functions $[a,b] \subset \mathbb{R}$ with values in $\mathcal{C}_p(\mathcal{H})$, $1 \leqslant p < \infty$, such that $\sup_{t\in[a,b],n\geqslant 1} \|X_n(t)\| < \infty$.*

Let $X(\cdot) : [a,b] \to \mathcal{C}_p(\mathcal{H})$ be a function obeying (2.101) and (2.102). If

$$\operatorname*{s-lim}_{n\to\infty} X_n(t) = X(t), \quad \operatorname*{s-lim}_{n\to\infty} X_n(t)^* = X(t)^*, \quad \|\cdot\|_p\text{-}\lim_{n\to\infty} Y_n(t) = Y(t),$$

uniformly on $[a,b]$ and $\|X_n(t)\|_p \leqslant \|Y_n(t)\|_p$ for $t \in [a,b]$, then

$$\|\cdot\|_p\text{-}\lim_{n\to\infty} X_n(t) = X(t), \quad 1 \leqslant p < \infty ,$$

uniformly on $[a,b]$.

Furthermore, we shall use a certain generalisation of the Lebesgue dominated convergence theorem for one-parameter operator-valued families in $\mathcal{C}_p(\mathcal{H})$-ideals. First, we need the following lemma.

Lemma 2.85. *Let $\{X_n\}_{n\in\mathbb{N}}$ and $\{Y_n\}_{n\in\mathbb{N}}$ be sequences of non-negative self-adjoint operators strongly converging, respectively, to the operators X and Y. Suppose that $\{Y_n\}_{n\in\mathbb{N}} \subset \mathcal{C}_p(\mathcal{H})$, $Y \in \mathcal{C}_p(\mathcal{H})$.*

If $X_n \leqslant Y_n$ for all $n \in \mathbb{N}$, then $\{X_n\}_{n\in\mathbb{N}} \subset \mathcal{C}_p(\mathcal{H})$ and $\text{s-lim}_{n\to\infty} X_n = X \in \mathcal{C}_p(\mathcal{H})$, $1 \leqslant p < \infty$.

The proof is straightforward and essentially the same as in L^p spaces.

Proposition 2.86. *Let $\{X_n(\cdot)\}_{n\in\mathbb{N}}$, $X(\cdot)$ and $\{Y(\cdot)\}$ be operator-valued functions on $[a,b] \subset \mathbb{R}$ such that for each $t \in [a,b]$ the operators satisfy the conditions of Lemma 2.85.*

If in addition $\text{s-lim}_{n\to\infty} X_n(t) = X(t)$, $\|\cdot\|_p\text{-}\lim_{n\to\infty} Y_n(t) = Y(t)$ uniformly on $[a,b]$, $\sup_{t\in[a,b],n\geqslant 1} \|X_n(t)\| < \infty$ and $Y(t)$ satisfy conditions (2.100):

$$\lim_{k\to\infty} \sup_{t\in[a,b]} \|(\mathbb{1} - P_k)Y(t)\| = \lim_{k\to\infty} \sup_{t\in[a,b]} \|Y(t)(\mathbb{1} - Q_k)\| = 0,$$

then $\|\cdot\|_p\text{-}\lim_{n\to\infty} X_n(t) = X(t)$ uniformly on $[a,b]$.

Proof. Pointwise the conditions of this assertion for $t \in [a,b]$ coincide with the conditions of Lemma 2.85. The extension to uniformity follows from the uniform boundedness $\sup_{t\in[a,b],n\geqslant 1} \|X_n(t)\| < \infty$ and conditions (2.100). □

2.5 Notes

Notes to Section 2.1. The basic definitions and properties of compact operators are standard and can be found in many books on operators on Hilbert spaces. We were inspired by [BS87], [KF99], [Sim05], [BEH94], [RS80], and [HN01].

Note that Definition 2.4 proposes a general definition for compact operators. One can repeat it verbatim for a Banach space $\mathcal{B}$.

From Proposition 2.5 and Remark 2.6 it follows that the more general concept of the *completely continuous* operators $\mathcal{L}_\infty(\mathcal{B})$ is useful, see [KF99](Ch.IV, §6), [Horn50], [Fan51].

Definition. A bounded operator A on a Banach space $\mathcal{B}$ is called *completely continuous* if it maps any weakly convergent *sequence* $\{x_n\}_{n\geqslant 1} \subset \mathcal{B}$ into a norm-convergent sequence $\{Ax_n\}_{n\geqslant 1}$.

Note that if A is a compact operator on $\mathcal{B}$, then A is completely continuous, that is, $\mathcal{C}_\infty(\mathcal{B}) \subseteq \mathcal{L}_\infty(\mathcal{B})$.

If A is completely continuous on $\mathcal{B}$ and the Banach space $\mathcal{B}$ is *reflexive*, then A is compact. In particular, $\mathcal{C}_\infty(\mathcal{H}) = \mathcal{L}_\infty(\mathcal{H})$ when this concept is applied for operators in Hilbert spaces. The following example shows that in general for Banach spaces the difference $\mathcal{L}_\infty(\mathcal{B})\backslash\mathcal{C}_\infty(\mathcal{B})$ is nontrivial.

Example. Recall that *every* bounded operator on the Banach space $\mathcal{B}= l^1$ is completely continuous, but among them there are many, that are noncompact, for example, the identity operator.

Notes to Section 2.2. There are many good references on the material of this section. It is only question of the place in the spectrum of the accumulation point 0 which is a rather subtle matter. See also the comments in Appendix A.

Here we followed essentially [KF99], [BS87], [RS80] and [Sad91].

Notes to Section 2.3. The great classics on the $\mathcal{C}_p(\mathcal{H})$-ideals are [GK69], [Sch70] and [Sim05].

H. Weyl and R. Courant understood the importance of the minimax principle for the study of singular values. Most textbooks on the theory of linear operators deal with this topic, the treatments in [BS87] and in [BEH94] are especially well written.

Proposition 2.45 is the celebrated Lidskiĭ *Trace Theorem*, see [Lid59] and [Sim05] for more comments.

Notes to Section 2.4. This section contains the most important convergence theorems in $\mathcal{C}_p(\mathcal{H})$-ideals that we need. The proof of the $\mathcal{C}_\infty(\mathcal{H})$-*dominated convergence theorem*, Proposition 2.61, and Remark 2.62 are due to [Sim05], Theorem 2.16. Proposition 2.63 and Corollary 2.64 are proved in [Zag80]. Proposition 2.65 can be found in [NZ99b]. The device in Definition 2.66 as well as Proposition 2.68 were proposed by B. Simon [Sim05], Theorem 2.17. Propositions 2.69 and 2.78 are proved in [Grü73], but Remarks 2.79 and 2.80 are again due to [Sim05].

The observation that Grümm's convergence theorem (Propositions 2.69) for $1 \leqslant p < \infty$ can be strengthened to Proposition 2.73 is in fact Theorem 2.20 in [Sim05].

Note that for $1 < p < \infty$ there exists an improved version of these observations that we formulated in Remark 2.74 as a "perfect" lifting assertion. It can be stated as follows:

- *The weak convergence of operators enhanced by the convergence of their operator norms implies the norm convergence of these operators.*

For $1 < p < \infty$ this follows from the *uniform convexity* of Banach spaces $\mathcal{C}_p(\mathcal{H})$, see [Sim05], Theorem 2.21.

The long standing question: whether this assertion holds true also for the "flat" trace-class ideal $\mathcal{C}_1(\mathcal{H})$, was solved affirmatively in the general setting of symmetrically normed ideals by the Arazy-Simon theorem, see e.g., [Sim05] (Addendum H), Theorem A6.

The affirmative answer (Proposition 2.75) for the special case of the positive cone $\mathcal{C}_{1,+}(\mathcal{H})$ is due to G. F. dell'Antonio [dellA67] and E. B. Davis [Dav69].

For the extension of convergence theorems to *sequences* of operator-valued functions in $\mathcal{C}_p(\mathcal{H})$ (Lemmata 2.82, 2.85 and Propositions 2.83, 2.86) we followed here [NZ90b].

Chapter 3

Trace inequalities

To study the Gibbs semigroups (including the semigroups in *symmetrically-normed* ideals) we need more $\mathcal{C}_p$-norm estimates than those provided in Section 2.3. Here we present inequalities involving eigenvalues and singular values of operators from the ideals $\{\mathcal{C}_p(\mathcal{H})\}_{p\geqslant 1}$.

3.1 Singular values of compact operators

Here we collect some properties of singular values of compact operators that we systematically use throughout the book. Some of them were already mentioned in Chapter 2.

(*a*) Recall that, by the *square-root lemma*, any positive (hence self-adjoint) bounded operator $K \in \mathcal{L}(\mathcal{H})$ has a unique positive *square root* $\sqrt{K} \in \mathcal{L}(\mathcal{H})$, such that $\sqrt{K}$ commutes with all the operators $C \in \mathcal{L}(\mathcal{H})$ verifying $[C, K] = 0$, and that $(\sqrt{K})^2 = K$, cf. Section 2.1. Since the operator $K = A^*A \geqslant 0$, this observation motivates the definition of the *absolute value* of A by $|A| := \sqrt{A^*A} \geqslant 0$. Then $\||A|u\| = \|Au\|$ since $|A|^2 = A^*A$. For some other properties of the operator absolute value, see Remark 2.2.

By the *polar decomposition* theorem (Proposition 2.3), for any $A \in \mathcal{L}(\mathcal{H})$ there exists a unique *partial isometry* $U : (\ker U)^{\perp} \to \operatorname{ran} U$, such that one has the representation $A = U\,|A|$. Here $\operatorname{ran} U = \overline{\operatorname{ran} U}$, the absolute value is $|A| = U^*\,A$, and U is uniquely determined by the condition $\ker U = \ker A$. Then $U^*\,A = A^*\,U$, but also by definition $|A| \neq |A^*|$, if A is not normal. If $A \geqslant 0$, then $A = |A|$, i.e., $U = \mathbb{1}$.

(*b*) Let $A \in \mathcal{C}_\infty(\mathcal{H})$ be a compact operator. Then by Proposition 2.7(b) its absolute value $|A| = U^*\,A$ is also a compact operator. Proposition 2.24 states that:
- The operator $|A|$ has an orthonormal set of eigenfunctions $\{\varphi_n\}_{n\geqslant 1}$ with eigenvalues $\{\lambda_n(|A|)\}_{n\geqslant 1}$

$$|A|\,\varphi_n = \lambda_n(|A|)\varphi_n\;, \quad \lambda_n(|A|) =: s_n(A), \quad n \in \mathbb{N}. \tag{3.1}$$

V. A. Zagrebnov, *Gibbs Semigroups*, Operator Theory: Advances and Applications 273, https://doi.org/10.1007/978-3-030-18877-1_3

- The positive eigenvalues $\{\lambda_n(|A|)\}_{n\geqslant 1}$ of $|A|$ convergent to zero and have finite multiplicity. They are called the *singular values* (or s-numbers) $s(A) := \{s_n(A)\}_{n\geqslant 1}$ of the operator A.
- The eigenvectors $\{\phi_n\}_{n\geqslant 1}$ of the operator $|A|$ form an orthonormal basis in $\mathcal{H}$.

The importance of the s-numbers as compared with the eigenvalues follows from the fact that it is the s-numbers that express the canonical form of compact operators as well as define the norms on the ideals $\mathcal{C}_p(\mathcal{H})$, $p \geqslant 1$, of bounded operators in $\mathcal{L}(\mathcal{H})$.

(*c*) By virtue of (3.1) $\{\varphi_n\}_{n\geqslant 1}$ are also eigenfunctions of the operator $|A|^2 = A^*A$ with eigenvalues $\{\lambda_n(|A|^2) = s_n^2(A)\}_{n\geqslant 1}$ and similarly for $|A^*|^2 = AA^*$. Therefore

$$\lambda_n(A^*A) = s_n^2(A) \quad \text{and} \quad \lambda_n(AA^*) = s_n^2(A^*), \quad n \in \mathbb{N}. \tag{3.2}$$

Since the identity $\lambda(\lambda\mathbb{1} - A^*A)^{-1} = \mathbb{1} + A^*(\lambda\mathbb{1} - AA^*)^{-1}A$ implies that $\sigma_{\mathrm{p}}(A^*A)\backslash\{0\} = \sigma_{\mathrm{p}}(AA^*)\backslash\{0\}$, we infer that $s_n(A) = s_n(A^*)$ in (3.2). So, if A is not *normal*, then $|A| \neq |A^*|$, yet $\lambda_n(|A|) = \lambda_n(|A^*|)$.

(*d*) If A is a normal compact operator, then its singular values are $\{s_n(A) = |\lambda_n(A)|\}_{n\geqslant 1}$, cf. (3.1). To see this we note that by normality, i.e., $AA^* = A^*A$, the operators A and A^* have a common set of eigenfunctions $\{u_n\}_{n\geqslant 1}$ and $\lambda_n(A^*) = \overline{\lambda_n(A)}$. Indeed, $A^*Au_n = \lambda_n(A^*A)u_n$ yields $\{\lambda_n(A^*A) = |\lambda_n(A)|^2 = s_n^2(A)\}_{n\geqslant 1}$ for $n \in \mathbb{N}$.

(*e*) If $A \geqslant 0$ is a compact operator, then $|A| = A$ and the singular values $s_n(A) = \lambda_n(A) \geqslant 0$. By (3.1) this yields $s_n(A^{1/2}) = s_n^{1/2}(A)$, cf. Corollary 2.34, for $n \in \mathbb{N}$.

(*f*) We recall the following *Weyl-Horn* chain of inequalities that involves s-numbers, which will be needed in the sequel. Let $A \in \mathcal{C}_\infty(\mathcal{H})$. Then for *any* system of vectors $\{u_j\}_{j=1}^k$ one has inequalities:

$$\det[(Au_i, Au_j)]_1^k \leqslant \prod_{n=1}^k s_n^2(A) \det[(u_i, u_j)]_1^k, \quad k \in \mathbb{N}.$$

Here $[(\xi_i, \eta_j)]_1^k$ denotes the $k \times k$ matrix with entries $m_{ij} := (\xi_i, \eta_j)$, for vectors $\{\xi_i : i = 1, 2, \ldots, k\}$ and $\{\eta_j : j = 1, 2, \ldots, k\}$ in $\mathcal{H}$.

3.2 Inequalities for s-numbers and eigenvalues

First we consider inequalities relating the s-numbers and eigenvalues of compact operators.

Lemma 3.1. *Let A be a compact operator. Then for any $k \in \mathbb{N}$*

$$\left|\prod_{j=1}^{k} \lambda_j(A)\right| \leqslant \prod_{j=1}^{k} s_j(A) ,$$

and equalities hold if and only if A is normal.

The proof is based on the Weyl-Horn inequalities and on the relations between eigenvalues and singular values for normal operators, Section 3.1 (d) and the Notes to Section 3.2.

To continue we need the following abstract lemma.

Lemma 3.2. *Let $\Phi : \mathbb{R} \to \mathbb{R}$ be a convex function that vanishes at $-\infty$ (which implies $\Phi(x) \geqslant 0$). Let $\{a_j\}_{j\geqslant 1}$ and $\{b_j\}_{j\geqslant 1}$ be non-increasing sequences of real numbers such that*

$$\sum_{j=1}^{k} a_j \leqslant \sum_{j=1}^{k} b_j \quad k = 1, 2, \dots . \tag{3.3}$$

Then

$$\sum_{j=1}^{k} \Phi(a_j) \leqslant \sum_{j=1}^{k} \Phi(b_j), \quad k = 1, 2, \dots . \tag{3.4}$$

If in addition Φ is strictly convex, then the equalities

$$\sum_{j=1}^{k} \Phi(a_j) = \sum_{j=1}^{k} \Phi(b_j), \quad k = 1, 2, \dots , \tag{3.5}$$

imply $a_j = b_j$.

Proof. Since Φ is convex, the left derivative

$$\Phi_l'(x) := \lim_{\epsilon \downarrow 0} \frac{\Phi(x) - \Phi(x - \epsilon)}{\epsilon}$$

exists for any $x \in \mathbb{R}$ and it is a non-negative nondecreasing function. Let us now consider the Stieltjes integral

$$\begin{aligned}\int_{-N}^{\infty} \mathrm{d}\Phi_l'(u)\,(x-u)_+ &= \int_{-N}^{x} \mathrm{d}\Phi_l'(u)\,(x-u)\\ &= -\Phi_l'(-N)(x+N) + \int_{-N}^{x} \mathrm{d}u\,\Phi_l'(u),\end{aligned} \tag{3.6}$$

where $y_+ := \max\{y, 0\}$ and $N > 0$. Since the integral (3.6) is positive, we obtain

$$(x+N)\,\Phi_l'(-N) \leqslant \int_{-N}^{x} \mathrm{d}u\,\Phi_l'(u) = \Phi(x) - \Phi(-N) \leqslant \Phi(x), \tag{3.7}$$

and hence

$$\limsup_{N\to\infty} N\,\Phi_l'(-N) < \infty \quad \text{and} \quad \lim_{N\to\infty} \Phi_l'(-N) = 0. \tag{3.8}$$

By assumption, $\lim_{x\to-\infty} \Phi(x) = 0$. Therefore, (3.7) and (3.8) imply that

$$\lim_{N\to\infty} (x+N)\,\Phi_l'(-N) = \lim_{N\to\infty} N\,\Phi_l'(-N) = 0. \tag{3.9}$$

Now, taking the limit $N \to \infty$ in (3.6) and using (3.9), we get the representation

$$\Phi(x) = \int_{-\infty}^{\infty} \mathrm{d}\Phi_l'(u)\,(x-u)_+, \quad x \in \mathbb{R}. \tag{3.10}$$

Let

$$A_k(u) := \sum_{j=1}^{k} (a_j - u)_+ \quad \text{and} \quad B_k(u) := \sum_{j=1}^{k} (b_j - u)_+\,. \tag{3.11}$$

Then the representation (3.10) yields

$$\begin{aligned} \sum_{j=1}^{k} \Phi(a_j) &= \int_{-\infty}^{\infty} \mathrm{d}\Phi_l'(u)\,A_k(u), \\ \sum_{j=1}^{k} \Phi(b_j) &= \int_{-\infty}^{\infty} \mathrm{d}\Phi_l'(u)\,B_k(u). \end{aligned} \tag{3.12}$$

We now derive inequalities between the functions A_k and B_k defined in (3.11). First, let $u \geqslant b_1$. Then, obviously $A_k(u) = B_k(u) = 0$. If $u \leqslant \min\{a_k, b_k\}$, then by (3.3) one gets

$$A_k(u) = \sum_{j=1}^{k} a_j - ku \leqslant \sum_{j=1}^{k} b_j - ku = B_k(u).$$

Now let

$$a_{q+1} \leqslant u < a_q \quad \text{and} \quad b_{p+1} \leqslant u < b_p$$

for some $q, p \leqslant k$. Then for $p \geqslant q$ we have

$$A_k(u) = \sum_{j=1}^{q} a_j - qu \leqslant \sum_{j=1}^{q} b_j - qu + (b_{q+1} - u) + \cdots + (b_p - u) = B_k(u).$$

Similarly, if $p < q$, then

$$A_k(u) = \sum_{j=1}^{q} a_j - qu \leqslant \sum_{j=1}^{q} a_j - qu - (b_q - u) - \cdots - (b_{p+1} - u) \leqslant B_k(u).$$

These estimates yield

$$A_k(u) \leqslant B_k(u), \quad k = 1, 2, \dots, \quad u \in \mathbb{R}. \tag{3.13}$$

The representations (3.12) show that (3.4) follows from the inequality (3.13).

In the case of a strictly convex function $\Phi(x)$ the equality (3.5) and representation (3.12) imply $A_k(u) = B_k(u)$ for $k = 1, 2, \dots$ and for any $u \in \mathbb{R}$, which in turn is only possible if $a_j = b_j$ for $j = 1, 2, \dots$ □

Proposition 3.3. *Let $A \in \mathcal{C}_\infty(\mathcal{H})$ and let $f : \mathbb{R}_0^+ \to \mathbb{R}$ be a function such that $f(0) = 0$ and $\mathbb{R} \ni t \mapsto f(\mathrm{e}^t)$ is convex. Then*

$$\sum_{j=1}^{k} f(|\lambda_j(A)|) \leqslant \sum_{j=1}^{k} f(s_j(A)), \quad k = 1, 2, \dots . \tag{3.14}$$

Proof. We use Lemmata 3.1 and 3.2. Notice that the function $t \mapsto f(\mathrm{e}^t)$ verifies the assumptions on the function $\Phi(t)$ in Lemma 3.2. If we define

$$a_j := \ln |\lambda_j(A)| \quad \text{and} \quad b_j := \ln s_j(A),$$

then the inequalities (3.3) follow from the estimate given by Lemma 3.1. Consequently, (3.14) results from (3.4) for $\Phi(t) = f(\mathrm{e}^t)$. □

Corollary 3.4. *For any compact operator $A \in \mathcal{C}_\infty(\mathcal{H})$ and $p > 0$,*

$$\sum_{j=1}^{k} |\lambda_j(A)|^p \leqslant \sum_{j=1}^{k} s_j(A)^p, \quad k = 1, 2, \dots . \tag{3.15}$$

If the right-hand side has a finite limit for $p \geqslant 1$ when $k \to \infty$, then $A \in \mathcal{C}_p(\mathcal{H})$.

Now we consider relations between singular values when more than *one* operator A is involved. They complement the inequalities provided by Lemma 2.53 and by (2.57) and will unable us to establish in Section 3.3 generalised $\mathcal{C}_p$-Hölder inequalities.

Lemma 3.5. *Let $A, B \in \mathcal{C}_\infty(\mathcal{H})$. The singular values $s_n(A)$ and $s_n(B)$ satisfy the relations*

$$\prod_{j=1}^{k} s_j(AB) \leqslant \prod_{j=1}^{k} s_j(A) \prod_{j=1}^{k} s_j(B), \quad k = 1, 2, \dots , \tag{3.16}$$

and

$$\sum_{j=1}^{k} s_j(A + B) \leqslant \sum_{j=1}^{k} s_j(A) + \sum_{j=1}^{k} s_j(B), \quad k = 1, 2, \dots . \tag{3.17}$$

Proof. Recall that for any $C \in \mathcal{C}_\infty(\mathcal{H})$ and any system of vectors $\{\phi_l : l = 1, 2, \ldots, k\} \subset \mathcal{H}$ one has the Weyl-Horn inequalities in the form

$$\det[(C\phi_i, C\phi_j)] \leqslant \prod_{l=1}^{k} s_l^2(C) \det[(\phi_i, \phi_j)]. \tag{3.18}$$

Here $[(\xi_i, \eta_j)]$ denote the $k \times k$ matrix with elements $m_{ij} := (\xi_i, \eta_j)$, for vectors $\{\xi_i : i = 1, 2, \ldots, k\}$ and $\{\eta_j : j = 1, 2, \ldots, k\}$.

Applying the inequalities (3.18) to the product $C = AB$ one gets

$$\begin{aligned} \det[(ABe_i, ABe_j)] &\leqslant \prod_{l=1}^{k} s_l^2(A) \det[(Be_i, Be_j)] \\ &\leqslant \prod_{l=1}^{k} s_l^2(A) \prod_{l=1}^{k} s_l^2(B). \end{aligned} \tag{3.19}$$

It follows from the canonical representation of the compact self-adjoint operator B^*A^*AB that there exists a complete system of eigenvectors $\{e_j : j = 1, 2, \ldots, k\}$ such that, by Remark 2.29, we have

$$\det[(ABe_i, ABe_j)] = \prod_{l=1}^{k} s_l^2(AB). \tag{3.20}$$

The inequalities (3.16) then follow from (3.19) and (3.20).

By Propositions 2.28 and 2.38, we can choose an orthonormal system of vectors $\{\psi_j : j = 1, 2, \ldots, k\}$ and a partial isometry U such that

$$\sum_{j=1}^{k} |(U(A+B)\psi_j, \psi_j)| = \sum_{j=1}^{k} s_j(A+B). \tag{3.21}$$

Since, again by Proposition 2.38, we have

$$\begin{aligned} \sum_{j=1}^{k} |(U(A+B)\psi_j, \psi_j)| &\leqslant \sum_{j=1}^{k} |(UA\psi_j, \psi_j)| + \sum_{j=1}^{k} |(UB\psi_j, \psi_j)| \\ &\leqslant \sum_{j=1}^{k} s_j(A) + \sum_{j=1}^{k} s_j(B), \end{aligned} \tag{3.22}$$

the inequalities (3.17) follow from (3.21) and (3.22). $\square$

With the help of the lemmata above, we can establish a generalisation of Proposition 3.3 for s-numbers of a pair of compact operators.

Proposition 3.6. *Let $A, B \in \mathcal{C}_\infty(\mathcal{H})$ and let $f : \mathbb{R}_0^+ \to \mathbb{R}$ be a function such that $f(0) = 0$ and for $t \in \mathbb{R}$, $t \mapsto f(\mathrm{e}^t)$ is convex. Then*

$$\sum_{j=1}^{k} f(s_j(AB)) \leqslant \sum_{j=1}^{k} f(s_j(A)s_j(B)), \quad k = 1, 2, \dots . \tag{3.23}$$

Proof. Notice that the function $t \mapsto f(\mathrm{e}^t)$ verifies the assumptions on the function $\Phi(t)$ in Lemma 3.2. Setting

$$a_j := \ln s_j(AB) \quad \text{and} \quad b_j := \ln(s_j(A)s_j(B)), \tag{3.24}$$

the inequalities (3.3) follow from (3.16). Consequently, (3.23) results from (3.4) with $\Phi(t) = f(\mathrm{e}^t)$. □

Corollary 3.7. *Choosing $f(x) = x$ for $x \geqslant 0$, (3.23) reproduces the result of Lemma 2.53. On the other hand, since the inequality (3.16) can be extended to a product of operators $\{A_s \in \mathcal{C}_\infty(\mathcal{H}) : s = 1, \dots, r\}$, we get the following generalisation of (2.50):*

$$\sum_{j=1}^{k} s_j(A_1 A_2 \cdots A_r) \leqslant \sum_{j=1}^{k} s_j(A_1) \cdots s_j(A_r). \tag{3.25}$$

Corollary 3.8. *Let $A \in \mathcal{C}_\infty(\mathcal{H})$. For any $n = 1, 2, \dots$ and any $\alpha > 0$ we get*

$$\sum_{j=1}^{k} s_j^{\alpha/n}(A^n) \leqslant \sum_{j=1}^{k} s_j^{\alpha}(A), \quad k = 1, 2, \dots . \tag{3.26}$$

Indeed, by (3.23) we have

$$\sum_{j=1}^{k} f\Big(s_j\big(\prod_{s=1}^{r} A_s\big)\Big) \leqslant \sum_{j=1}^{k} f\Big(\prod_{s=1}^{r} s_j(A_s)\Big), \quad k = 1, 2, \dots . \tag{3.27}$$

Then for $A_s = A$ and $r = n$ in (3.27) and with the function $f(x) = x^{\alpha/n}$, for $x \geqslant 0$, the inequalities (3.27) imply (3.26).

3.3 Trace and $\mathcal{C}_p(\mathcal{H})$-norm estimates

The inequalities for singular values that we established in Section 3.2 yield a number of important relations involving traces and $\|\cdot\|_p$-norms on $\mathcal{C}_p(\mathcal{H})$, $p \geqslant 1$. Here we present some of them.

To this aim we consider first trace and norm estimates motivated by upper bounds from Proposition 3.3 and Corollary 3.4. Recall that for a trace-class operator $X \in \mathcal{C}_1(\mathcal{H})$ its *spectral trace* $\Lambda(X) = \sum_{j=1}^{k} \lambda_j(X)$ (Proposition

2.45) and its *matrix trace* $\Lambda(X) = \mathrm{Tr}(X)$ coincide. Since $\lambda_j(X^p) = \lambda_j(X)^p$ and $s_j(X) = \lambda_j(|X|)$, we can rewrite (3.15) for $p \geqslant 1$ as follows:

$$|\,\mathrm{Tr}(A^p)| = |\sum_{j=1}^{\infty} \lambda_j(A)^p| \leqslant \sum_{j=1}^{\infty} |\lambda_j(A)|^p \tag{3.28}$$
$$\leqslant \sum_{j=1}^{\infty} s_j(A)^p = \sum_{j=1}^{\infty} \lambda_j(|A|)^p = \mathrm{Tr}(|A|^p)\,.$$

Remark 3.9. Note that a direct corollary of (3.28) for the trace-class operators ($p = 1$) is the estimate $|\,\mathrm{Tr}(A)| \leqslant \mathrm{Tr}(|A|) = \|A\|_1$. It implies the $\|\cdot\|_1$-norm continuity of the map $\mathrm{Tr} : \mathcal{C}_1(\mathcal{H}) \to \mathbb{C}$ given by

$$A \mapsto \mathrm{Tr}(A) = \sum_{j=1}^{\infty} (e_j, Ae_j),$$

where $\{e_j\}_{j\geqslant 1}$ is *any* orthonormal basis in $\mathcal{H}$, cf. Corollary 2.43.

Corollary 3.10. *Let $X \geqslant 0$ and $Y \geqslant 0$ be self-adjoint trace-class operators. Then $\mathrm{Tr}(XY) = \mathrm{Tr}(X^{1/2}YX^{1/2}) \geqslant 0$ and $\mathrm{Tr}((XY)^p) = \mathrm{Tr}((X^{1/2}YX^{1/2})^p) \geqslant 0$. Since $|XY|^2 = (XY)^*(XY) = YXXY$, one also gets $\mathrm{Tr}((|XY|)^2) = \mathrm{Tr}(X^2\,Y^2)$. These observations together with inequality (3.28) for the trace-class operator $A := XY$ and for integer $p = 2q$ yield*

$$0 \leqslant \mathrm{Tr}((XY)^{2q}) \leqslant \mathrm{Tr}((X^2\,Y^2)^q), \quad q \in \mathbb{N}. \tag{3.29}$$

If $q = 2^{n-1}$, $n \in \mathbb{N}$, then iterating the estimate (3.29) we get

$$0 \leqslant \mathrm{Tr}((XY)^{2^n}) \leqslant \mathrm{Tr}((X^2\,Y^2)^{2^{n-1}}) \leqslant \ldots \leqslant \mathrm{Tr}(X^{2^n}\,Y^{2^n}). \tag{3.30}$$

An important applications of the estimates (3.30) concerns the *operator-valued exponential* functions: $X = \mathrm{e}^{-A}$, $Y = \mathrm{e}^{-B}$.

Proposition 3.11. *Let A and B be self-adjoint non-negative operators such that $B \in \mathcal{P}_{b<1}(A)$ is a Kato-small perturbation of A (Definition 1.50). Let $\mathrm{e}^{-A} \in \mathcal{C}_s(\mathcal{H})$, $\mathrm{e}^{-B} \in \mathcal{C}_r(\mathcal{H})$ for $1/s + 1/r = 1$, $s, r \in \mathbb{N}$, including the case $s = 1$, when we set $\mathrm{e}^{-B} \in \mathcal{L}(\mathcal{H})$. Then the operator sum $A + B$ is a well-defined non-negative self-adjoint operator on $\mathrm{dom}\,(A + B) = \mathrm{dom}\,A$, and*

$$\mathrm{Tr}(\mathrm{e}^{-(A+B)}) \leqslant \mathrm{Tr}(\mathrm{e}^{-A}\mathrm{e}^{-B}). \tag{3.31}$$

Proof. The first part of the assertion follows from the Kato-smallness of the perturbation B, see Proposition 1.51, which gives a C_0-semigroup with generator $A + B$.

By the assumptions made on the operator exponentials the product $e^{-A}e^{-B}$ belongs to $\mathcal{C}_1(\mathcal{H})$. Hence, the trace in the left-hand side of (3.31) exists. Then for $X := e^{-A/2^n}$ and $Y := e^{-B/2^n}$ the inequalities (3.30) yield

$$0 \leqslant \operatorname{Tr}((e^{-A/2^n}e^{-B/2^n})^{2^n}) \leqslant \operatorname{Tr}(e^{-A}e^{-B}). \tag{3.32}$$

Note that the generators A and B satisfy the conditions of Proposition 5.8. Then by the Lie-Trotter *product formula*, there exists the strong limit $\text{s-lim}_{n\to\infty}(e^{-A/2^n}e^{-B/2^n})^{2^n} = e^{-(A+B)}$. Since the trace (as well as the trace norm) is weakly lower semi-continuous (Proposition 2.48 (f) and Corollary 2.75) and the left-hand side of (3.32) is bounded from above, this strong limit gives the estimate

$$\operatorname{Tr}(e^{-(A+B)}) \leqslant \liminf_{n\to\infty} \operatorname{Tr}(e^{-A/2^n}e^{-B/2^n})^{2^n}. \tag{3.33}$$

Therefore, taking the limit $n \to \infty$ in the left-hand side of (3.32) and using the lower semi-continuity (3.33) we obtain the inequality (3.31). $\square$

Since by the definition of the $\|\cdot\|_p$-norm one has $\|A\|_p = (\operatorname{Tr}(|A|^p))^{1/p}$, $p \geqslant 1$, the trace inequality (3.31) gives the inequality for norms

$$\|e^{-(A+B)}\|_1 \leqslant \|e^{-A/2}e^{-B}e^{-A/2}\|_1 \,. \tag{3.34}$$

In the next assertion we relax the conditions of Proposition 3.11 on the generators A and B. Since in (3.33) the estimate from below comes from the weak lower semi-continuity of the trace, the topology of convergence for the Lie-Trotter product formula was not important for the proof of inequality (3.31). However, at least one of the generators A or B has to produce a Gibbs semigroup, see Chapter 4.

Proposition 3.12. *Let $A \geqslant 0$ be generator of the self-adjoint Gibbs semigroup $\{G_t(A) = e^{-tA}\}_{t\geqslant 0}$. Then Trotter-Kato product formula converges away from zero in the trace-norm topology for Kato functions: $f(x) = g(x) = e^{-x}$, and for any self-adjoint operator $B \geqslant 0$ to a degenerate Gibbs semigroup*

$$\|\cdot\|_1\text{-}\lim_{n\to\infty} \left(e^{-tA/n}e^{-tB/n}\right)^n = e^{-tH}P_0, \quad H = A \dot{+} B, \tag{3.35}$$

where orthogonal projection $P_0 : \mathcal{H} \to \overline{\operatorname{dom} H}$ and $t > 0$.

For the proof one has to consult Section 5.4, Proposition 5.53.

Corollary 3.13. *For the operators $X = e^{-A/2^n}$, $Y = e^{-B/2^n}$ and for inequalities (3.30) we obtain the same estimate (3.32). However, because of (3.35), the limit in the left-hand side of (3.33) takes now the different form*

$$\operatorname{Tr}(e^{-tH}P_0) \leqslant \liminf_{n\to\infty} \operatorname{Tr}(e^{-A/2^n}e^{-B/2^n})^{2^n}, \tag{3.36}$$

which gives instead of (3.31) the estimate

$$\operatorname{Tr}(e^{-tH}P_0) \leqslant \operatorname{Tr}(e^{-A}e^{-B}), \quad t > 0. \tag{3.37}$$

This inequality involves the degenerate Gibbs semigroup $\{e^{-tH}P_0\}_{t>0}$.

Proposition 3.14. *Let* $\{A_s \in \mathcal{C}_{p_s}(\mathcal{H}) : s = 1, 2, \ldots, r\}$ *and suppose that* $\sum_{s=1}^{r} p_s^{-1} \leqslant 1$. *Then*

$$A^{(r)} := \prod_{s=1}^{r} A_s \in \mathcal{C}_p(\mathcal{H}) \quad \text{with } p^{-1} := \sum_{s=1}^{r} p_s^{-1} \tag{3.38}$$

and

$$\Big\| \prod_{s=1}^{r} A_s \Big\|_p \leqslant \prod_{s=1}^{r} \|A_s\|_{p_s}. \tag{3.39}$$

Proof. Take $f(x) = x^p$, $x \geqslant 0$, in (3.27). By the definition (2.37) of the $\|\cdot\|_p$-norm, we obtain

$$\|A^{(r)}\|_p \leqslant \Big[\sum_{j=1}^{\infty} \prod_{s=1}^{r} s_j^p(A_s)\Big]^{1/p}. \tag{3.40}$$

Since by the Hölder inequality

$$\begin{aligned} \|A^{(r)}\|_p &\leqslant \Big[\sum_{j=1}^{\infty} \prod_{s=1}^{r} s_j^p(A_s)\Big]^{1/p} \\ &\leqslant \Big[\sum_{j=1}^{\infty} s_j^{p_1}(A_1)\Big]^{1/p_1} \cdots \Big[\sum_{j=1}^{\infty} s_j^{p_r}(A_r)\Big]^{1/p_r} \\ &= \|A_1\|_{p_1} \cdots \|A_r\|_{p_r}, \quad \text{for } p^{-1} = \sum_{s=1}^{r} p_s^{-1}, \end{aligned} \tag{3.41}$$

(3.40) and (3.41) imply (3.38) and (3.39). □

Remark 3.15. This result is in fact a generalisation of the $\mathcal{C}_p$-Hölder inequality proved in Proposition 2.54.

Corollary 3.16. *If* $A \in \mathcal{C}_p(\mathcal{H})$ *for some* $p > 0$, *then* $A^\alpha \in \mathcal{C}_{p/\alpha}(\mathcal{H})$, *for any* $\alpha > 0$, *and*

$$\|A^\alpha\|_{p/\alpha} \leqslant (\|A\|_p)^\alpha. \tag{3.42}$$

Proof. This result follows directly from the Definition 2.50 of $\|\cdot\|_p$ and from (2.41), (3.28). We note that the quantity $\|\cdot\|_p$ defined for $0 < p < 1$ by (2.37) has no longer the properties of a norm. □

Now we consider *non-self-adjoint* operator exponentials that belong to the trace-class $\mathcal{C}_1(\mathcal{H})$. In fact they are related to the non-self-adjoint Gibbs semigroups that we study in Section 5.5.

Proposition 3.17. *For any trace-class operator* X *and* $p \in \mathbb{N}$ *one has inequalities*

$$\mathrm{Tr}((X^*)^p X^p) \leqslant \mathrm{Tr}(X^* X)^p. \tag{3.43}$$

Proof. For matrices the result (3.43) follows directly from the Weyl estimate (3.15) in Corollary 3.4. The extension to $\mathcal{C}_1(\mathcal{H})$ is carried out by trace-norm finite-rank approximation of X. $\square$

We apply (3.43) for the Gibbs exponential $X = e^{-A} \in \mathcal{C}_1(\mathcal{H})$ generated by m-sectorial operator A with vertex $\gamma = 0$ and semi-angle α, Definition 4.26. Note that by Corollary 4.33 (or Corollary 5.79) we obtain

$$\|\cdot\|_1\text{-}\lim_{n\to\infty} (e^{-tA^*/n} e^{-tA/n})^n = e^{-t(A^* \dot{+} A)}, \quad t \in S_\theta\,, \tag{3.44}$$

where $\theta = \pi/2 - \alpha$ and $A^* \dot{+} A$ is the *form-sum* of the operators A^* and A. Since $K \mapsto \operatorname{Tr} K$ for $K \in \mathcal{C}_1(\mathcal{H})$ is continuous in the the trace-norm topology (Remark 3.9), the inequality (3.43) and the Trotter formula (3.44) yield the following assertion.

Proposition 3.18. *For the Gibbs exponential $X = e^{-A}$ generated by the m-sectorial operator A one has the inequality*

$$\operatorname{Tr}(e^{-A^*} e^{-A}) \leqslant \operatorname{Tr} e^{-(A^* \dot{+} A)}. \tag{3.45}$$

Comparing this result with Proposition 3.11 we see that the inequality (3.45) is *opposite* in sign to the inequality (3.31) for self-adjoint exponentials. We also note that (3.45) becomes equality if and only if operator A is *normal.*

3.4 Monotonicity, convexity and inequalities

Similarly to the Gibbs semigroups, the inequalities that we study in this section have significant applications in quantum statistical mechanics. For this reason we essentially focus here on the Gibbs exponential function and leave the discussion regarding general convex functions for the end. In order to be able to provide proofs we shall need to refer to some results of Chapter 4 (Sections 4.4 and 4.5).

Proposition 3.19 (Peierls-Bogoliubov inequality). *The Gibbs exponential $X = e^{-A} \in \mathcal{C}_{1,+}(\mathcal{H})$ generated by a self-adjoint non-negative operator $A \geqslant 0$ satisfies the inequality*

$$(e^{-A}u, u) \geqslant e^{-(Au,u)}, \tag{3.46}$$

for any unit vector $u \in \operatorname{dom} A$. Equality is attained for eigenvectors of the operator A.

Proof. Since $X^* = X$ belongs to the positive cone $\mathcal{C}_{1,+}(\mathcal{H})$, there exists a complete orthonormal set of eigenvectors $\{\varphi_j\}_{j\geqslant 1}$ of the operator $X \geqslant 0$ that forms an *orthonormal basis* (ONB) in $\mathcal{H}$. By the spectral mapping theorem (Section A.9), one gets that also $A\varphi_j = \lambda_j(A)\varphi_j$ for non-negative eigenvalues $\{\lambda_j(A)\}_{j\geqslant 1}$. Then

$u = \sum_{j\geqslant 1} u^{(j)}\varphi_j$ and since $\sum_{j\geqslant 1} |u^{(j)}|^2 = 1$, the *Jensen inequality* for (convex) exponential function yields

$$(\mathrm{e}^{-A}u, u) = \sum_{j\geqslant 1} |u^{(j)}|^2 \, \mathrm{e}^{-\lambda_j(A)} \geqslant \mathrm{e}^{-\sum_{j\geqslant 1} |u^{(j)}|^2 \, \lambda_j(A)} = \mathrm{e}^{-(Au,u)}, \tag{3.47}$$

which proves (3.46). Note that equality is attainded in (3.47) when $u = \varphi_j$, for any $j \geqslant 1$. □

Corollary 3.20. *Let $\{e_j\}_{j\geqslant 1}$ be an ONB in the Hilbert space $\mathcal{H}$. Then (3.46) yields for the Gibbs exponential $X = \mathrm{e}^{-A} \geqslant 0$ the inequality*

$$\sum_{i\geqslant 1} (\mathrm{e}^{-A}e_j, e_j) \geqslant \sum_{j\geqslant 1} \mathrm{e}^{-(Ae_j,e_j)}.$$

Since the left-hand side is the trace (Definition 2.41), which is independent of the choice of ONB, for any basis $\{e_j\}_{j\geqslant 1}$ one has the inequality

$$\operatorname{Tr} \mathrm{e}^{-A} \geqslant \sum_{j\geqslant 1} \mathrm{e}^{-(Ae_j,e_j)}, \tag{3.48}$$

also known as the Peierls-Bogoliubov inequality.

Moreover, taking into account the last assertion of Proposition 3.19 and inequality (3.48) we obtain that

$$\operatorname{Tr} \mathrm{e}^{-A} = \sum_{j\geqslant 1} \mathrm{e}^{-(A\varphi_j,\varphi_j)} = \sup_{\{e_j\}_{j\geqslant 1}} \sum_{j\geqslant 1} \mathrm{e}^{-(Ae_j,e_j)}. \tag{3.49}$$

Proposition 3.21. *Let $Op_{1,+} := \{A \geqslant 0 : X = \mathrm{e}^{-A} \in \mathcal{C}_{1,+}(\mathcal{H})\}$. Then the real-valued function*

$$A \mapsto \ln \operatorname{Tr} \mathrm{e}^{-A} \tag{3.50}$$

is convex and monotonically decreasing on the cone of positive operators $Op_{1,+}$.

Proof. Let $A_1, A_2 \in Op_{1,+}$. Taking into account (3.49) and the *Hölder inequality* one gets

$$\begin{aligned}
\operatorname{Tr} \mathrm{e}^{-(\alpha A_1+(1-\alpha)A_2)} &= \sup_{\{e_j\}_{j\geqslant 1}} \sum_{j\geqslant 1} \mathrm{e}^{-((\alpha A_1+(1-\alpha)A_2)e_j,e_j)} \\
&= \sup_{\{e_j\}_{j\geqslant 1}} \sum_{j\geqslant 1} (\mathrm{e}^{-(A_1e_j,e_j)})^{\alpha} \, (\mathrm{e}^{-(A_2e_j,e_j)})^{1-\alpha} \\
&\leqslant \sup_{\{e_j\}_{j\geqslant 1}} \Big[\Big(\sum_{j\geqslant 1} \mathrm{e}^{-(A_1e_j,e_j)}\Big)^{\alpha} \Big(\sum_{j\geqslant 1} \mathrm{e}^{-(A_2e_j,e_j)}\Big)^{1-\alpha} \Big],
\end{aligned} \tag{3.51}$$

for $\alpha \in [0,1]$. Then elementary estimate and again (3.49) yield

$$\sup_{\{e_j\}_{j\geqslant 1}} \left[\Big(\sum_{j\geqslant 1} \mathrm{e}^{-(A_1 e_j, e_j)}\Big)^{\alpha} \Big(\sum_{j\geqslant 1} \mathrm{e}^{-(A_2 e_j, e_j)}\Big)^{1-\alpha} \right] \tag{3.52}$$

$$\leqslant \sup_{\{e_j\}_{j\geqslant 1}} \Big(\sum_{j\geqslant 1} \mathrm{e}^{-(A_1 e_j, e_j)}\Big)^{\alpha} \sup_{\{e_j\}_{j\geqslant 1}} \Big(\sum_{j\geqslant 1} \mathrm{e}^{-(A_2 e_j, e_j)}\Big)^{1-\alpha} = (\mathrm{Tr}\, \mathrm{e}^{-A_1})^{\alpha} (\mathrm{Tr}\, \mathrm{e}^{-A_2})^{1-\alpha}.$$

The inequalities (3.51) and (3.52) imply

$$\ln \mathrm{Tr}\, \mathrm{e}^{-(\alpha A_1 + (1-\alpha) A_2)} \leqslant \alpha \ln \mathrm{Tr}\, \mathrm{e}^{-A_1} + (1-\alpha) \ln \mathrm{Tr}\, \mathrm{e}^{-A_2}, \tag{3.53}$$

which proves the convexity of the mapping (3.50).

Let $A_1, A_2 \in Op_{1,+}$ and $A_1 \leqslant A_2$. Then by (3.49)

$$\mathrm{Tr}\, \mathrm{e}^{-A_1} = \sup_{\{e_j\}_{j\geqslant 1}} \sum_{j\geqslant 1} \mathrm{e}^{-(A_1 e_j, e_j)} \geqslant \sup_{\{e_j\}_{j\geqslant 1}} \sum_{j\geqslant 1} \mathrm{e}^{-(A_2 e_j, e_j)} = \mathrm{Tr}\, \mathrm{e}^{-A_2},$$

and consequently

$$\ln \mathrm{Tr}\, \mathrm{e}^{-A_1} \geqslant \ln \mathrm{Tr}\, \mathrm{e}^{-A_2}, \quad A_1 \leqslant A_2 . \tag{3.54}$$

That is the mapping (3.50) is monotonically decreasing on the positive cone $Op_{1,+}$. □

Corollary 3.22. *We set $A_1 := A$ and $A_2 := A + B$, where $B \geqslant 0$ is also self-adjoint. Then*

$$\Omega(\alpha) := \ln \mathrm{Tr}\, \mathrm{e}^{-(\alpha A_1 + (1-\alpha) A_2)} = \ln \mathrm{Tr}\, \mathrm{e}^{-(A + (1-\alpha) B)} \tag{3.55}$$

is a convex monotonically increasing function $\Omega : [0,1] \to [\Omega(0), \Omega(1)]$. Recall that convex functions are almost everywhere differentiable and possess everywhere right- and left-derivatives. Hence, by convexity one gets the inequalities

$$\partial_\alpha \Omega(+0) \leqslant \Omega(1) - \Omega(0) \leqslant \partial_\alpha \Omega(1-0) . \tag{3.56}$$

Remark 3.23. The calculation in (3.56) of the right-derivative at $\alpha = +0$ and the left-derivative at $\alpha = 1 - 0$ needs an additional analysis of *differentiability* of the mapping (3.50) on the cone of positive operators $Op_{1,+}$. This analysis is presented in full generality in Sections 4.5 and 4.6. Here we recall only some necessary facts concerning the calculation of derivatives.

Heuristically one gets for the derivative $\partial_\alpha \Omega(\alpha)$ of the function (3.55) the following expression

$$\partial_\alpha \Omega(\alpha) = \frac{1}{\mathrm{Tr}\, \mathrm{e}^{-(A+(1-\alpha)B)}} \mathrm{Tr}\, B\, \mathrm{e}^{-(A+(1-\alpha)B)} =: \langle B \rangle_{A+(1-\alpha)B} \, . \tag{3.57}$$

Then the convexity inequality (3.56) takes on the form

$$\langle B \rangle_{A+B} \leqslant \ln \mathrm{Tr}\, \mathrm{e}^{-A} - \ln \mathrm{Tr}\, \mathrm{e}^{-(A+B)} \leqslant \langle B \rangle_{A} \, , \tag{3.58}$$

which is known as the two-sided *Bogoliubov (convexity) inequality*.

Remark 3.24. A key step towards justification of (3.57) is including the Gibbs exponential under the trace, into a C_0-semigroup family, which for $t > 0$ takes values in the trace class $\mathcal{C}_1(\mathcal{H})$ for any $\alpha \in [0,1]$. This yields $\{G_t(H(\kappa)) = \mathrm{e}^{-t\,H(\kappa)}\}_{t\geqslant 0}$, which is a Gibbs semigroup with self-adjoint generator $H(\kappa) = A + \kappa B$ in the sense of Definition 4.1. Here we set $\kappa := 1 - \alpha$.

Then to proceed further we use that by Corollary 4.32, the function: $t \mapsto \mathrm{e}^{-t\,H(\kappa)}$, is trace-norm differentiable for $t > 0$ and any fixed $\alpha \in [0,1]$. Then for the derivative of the function $W_{t,s}(\kappa,\epsilon) := G_{t-s}(H(\kappa))G_s(H(\kappa+\epsilon))$ we obtain

$$\|\cdot\|_1\text{-}\partial_s W_{t,s}(\kappa,\epsilon) = G_{t-s}(H(\kappa))(-\epsilon B)G_s(H(\kappa+\epsilon)) \in \mathcal{C}_1(\mathcal{H}). \tag{3.59}$$

Thus, definition of $W_{t,s}(\kappa,\epsilon)$ and (3.59) yield

$$G_t(H(\kappa+\epsilon)) - G_t(H(\kappa)) = \int_0^t ds\, G_{t-s}(H(\kappa))(-\epsilon B)G_s(H(\kappa+\epsilon)), \tag{3.60}$$

where the integral is well defined and belongs to $\mathcal{C}_1(\mathcal{H})$.

Note that if there exists $\xi > 0$, such that operator $B(A+\xi\,\mathbb{1})^{-1} \in \mathcal{L}(\mathcal{H})$, then (3.60) gives $\lim_{\epsilon\to 0}\|G_t(H(\kappa+\epsilon)) - G_t(H(\kappa))\|_1 = 0$. Consequently, by (3.60) we find that the $\|\cdot\|_1$-derivative

$$\partial_\kappa G_t(H(\kappa)) = -\int_0^t ds\, G_{t-s}(H(\kappa))\, B\, G_s(H(\kappa)) \in \mathcal{C}_1(\mathcal{H}), \tag{3.61}$$

exists and is also an operator in $\mathcal{C}_1(\mathcal{H})$, Proposition 4.52.

Since $\mathrm{Tr}(\cdot)$ is continuous in the $\|\cdot\|_1$-topology, $\mathrm{Tr}(\partial_\kappa G_t(H(\kappa))) = \partial_\kappa \mathrm{Tr}\, G_t(H(\kappa))$. Similarly, since the integrals in (3.60) and (3.61) are constructed as Bochner integrals in $\mathcal{C}_1(\mathcal{H})$, one can interchange $\mathrm{Tr}(\cdot)$ and the integration. Therefore, application of the $\mathrm{Tr}(\cdot)$ to derivative (3.61) yields

$$\begin{aligned}\partial_\kappa \mathrm{Tr}\, G_t(H(\kappa)) &= -\int_0^t ds\ \mathrm{Tr}\, G_{t-s}(H(\kappa))\, B\, G_s(H(\kappa)) = \\ &\quad -\ t\ \mathrm{Tr}\, B\, G_t(H(\kappa)),\end{aligned} \tag{3.62}$$

by the cyclicity of the trace. Recall that $G_t(H(\kappa = 1-\alpha)) = \mathrm{e}^{-t\,(A+(1-\alpha)B)}$. Then (3.62) gives for the derivative of $\Omega(\alpha)$ the heuristical value (3.57), provided that B is *relatively small* with respect to the generator A in the sense that $B(A+\xi\,\mathbb{1})^{-1} \in \mathcal{L}(\mathcal{H})$ for some $\xi > 0$. This proves the two-sided Bogoliubov (convexity) inequality (3.58).

We conclude this section by some remarks on trace functions and operator trace inequalities, which are similar to what we obtained for the Gibbs exponential $f(x) = \mathrm{e}^{-x}$ denoted by $\widehat{f}(A) = \mathrm{e}^{-A}$.

Let $f : \mathbb{R} \to \mathbb{R}$ be a monotonically decreasing convex function. Recall that $Op_{1,+}$ denotes cone of positive self-adjoint operators, see Proposition 3.21. By the spectral functional calculus, the mapping $Op_{1,+} \ni A \mapsto \widehat{f}(A)$, corresponding to the function f, is well defined and we assume that $\widehat{f} : Op_{1,+} \to \mathcal{C}_{1,+}(\mathcal{H})$. Then by the spectral mapping theorem (Section A.9) $\sigma(\widehat{f}(A)) = f(\sigma(A))$, and the non-negative spectrum $\sigma(A) = \{\lambda_j(A)\}_{j\geqslant 1}$ is pure point with accumulation at $+\infty$.

Proposition 3.25 (general Peierls-Bogoliubov inequality). *Let $f : \mathbb{R} \to \mathbb{R}_0^+$ be a monotonically decreasing convex function. Then for $A \in Op_{1,+}$ and for the corresponding operator $\widehat{f}(A) \in \mathcal{C}_{1,+}(\mathcal{H})$ one has the inequalities*

$$(\widehat{f}(A)u, u) \geqslant f((Au, u)) \quad \text{and} \quad \operatorname{Tr} \widehat{f}(A) \geqslant \sum_{j\geqslant 1} f((Ae_j, e_j)), \tag{3.63}$$

for any unit vector $u \in \operatorname{dom} A$ and for any ONB $\{e_j\}_{j\geqslant 1} \subset \operatorname{dom} A$. Equalities are attained on eigenvectors $\{\varphi_j\}_{j\geqslant 1}$ of the operator A.

Proof. The proof closely mimics that of Proposition 3.19 and Corollary 3.20. Since it is sufficient to check only the first inequality in (3.63), we note that for $\widehat{f}(A)\varphi_j = f(\lambda_j(A))\varphi_j$ and $u = \sum_{j\geqslant 1} u^{(j)}\varphi_j$, $\sum_{j\geqslant 1} |u^{(j)}|^2 = 1$, the *Jensen inequality* for the convex function f yields

$$(\widehat{f}(A)u, u) = \sum_{j\geqslant 1} |u^{(j)}|^2 \; f(\lambda_j(A)) \geqslant f(\sum_{j\geqslant 1} |u^{(j)}|^2 \; \lambda_j(A)) = f((Au, u)),$$

which proves (3.63). Note that the inequalities (3.63) become equalities, when $u = \varphi_j$, for any $j \geqslant 1$, and for $\{e_j = \varphi_j\}_{j\geqslant 1}$. □

Corollary 3.26. (a) *The monotonicity of the function f implies the monotonicity of the trace function $A \mapsto \operatorname{Tr} \widehat{f}(A)$.*

(b) *The convexity of the function f implies the convexity of the trace function $A \mapsto \operatorname{Tr} \widehat{f}(A)$.*

Proof. (a) Let $A_1, A_2 \in Op_{1,+}$ and $A_1 \leqslant A_2$, i.e., $\operatorname{dom} A_1 \supset \operatorname{dom} A_2$. Then by Proposition 3.25 and the monotonicity of f,

$$\begin{aligned} \operatorname{Tr} \widehat{f}(A_1) &= \sup_{\{e_j\}_{j\geqslant 1}} \sum_{j\geqslant 1} f((A_1 e_j, e_j)) \\ &\geqslant \sup_{\{e_j\}_{j\geqslant 1}} \sum_{j\geqslant 1} f((A_2 e_j, e_j)) = \operatorname{Tr} \widehat{f}(A_2), \end{aligned}$$

that is, the trace function is monotone.

(b) Using Proposition 3.25 and the convexity of f we estimate the trace function for a convex combination of operators $A_1, A_2 \in Op_{1,+}$ and $\alpha \in [0, 1]$ on the common domain $\mathcal{D} = \operatorname{dom} A_2$:

$$\begin{aligned}\operatorname{Tr}\widehat{f}(\alpha A_1 + (1-\alpha)A_2) &= \sup_{\{e_j\}_{j\geqslant 1}} \sum_{j\geqslant 1} f(((\alpha A_1 + (1-\alpha)A_2)e_j, e_j)) \\ &\leqslant \sup_{\{e_j\}_{j\geqslant 1}} \sum_{j\geqslant 1} (\alpha f((A_1 e_j, e_j)) + (1-\alpha) f((A_2 e_j, e_j))) \\ &\leqslant \alpha \sup_{\{e_j\}_{j\geqslant 1}} \sum_{j\geqslant 1} f((A_1 e_j, e_j)) + (1-\alpha) \sup_{\{e_j\}_{j\geqslant 1}} \sum_{j\geqslant 1} f((A_2 e_j, e_j)) \\ &= \alpha \operatorname{Tr}\widehat{f}(A_1) + (1-\alpha)\operatorname{Tr}\widehat{f}(A_2)\,.\end{aligned}$$

Hence, the trace function $A \mapsto \operatorname{Tr}\widehat{f}(A)$ is convex. □

Corollary 3.27 (Klein inequality). *Let the function*

$$\alpha \mapsto \omega(\alpha) := \operatorname{Tr}\widehat{f}(\alpha A_1 + (1-\alpha)A_2)$$

be convex for $\alpha \in [0,1]$, i.e., we get that

$$\omega(1) - \omega(0) \geqslant \frac{\omega(\alpha) - \omega(0)}{\alpha}\,. \tag{3.64}$$

So, the right derivative $\partial_\alpha \omega(\alpha)|_{\alpha=+0}$ exists and we obtain the Klein inequality in the form:

$$\operatorname{Tr}\{\widehat{f}(A_1) - \widehat{f}(A_2)\} \geqslant \partial_\alpha \operatorname{Tr}\widehat{f}(A_2 + \alpha(A_1 - A_2))\big|_{\alpha=+0}\,, \tag{3.65}$$

regardless of what the explicit expression in the right-hand side is.

Let $A(\alpha) := A_2 + \alpha(A_1 - A_2)$. To proceed with the calculation of derivative in (3.65) for small α, the function $\alpha \mapsto \widehat{f}(A(\alpha))$ should be trace-norm differentiable.

Since for $\alpha < 1/2$ the operator $\alpha(A_1 - A_2)$ is Kato-small with respect to A_2, with a relative bound less than one, the linear function $\alpha \mapsto A(\alpha)$ can be extended to a holomorphic family $\{A(z)\}_{z\in D_{1/2}}$ of *type* (A) in the disc of radius 1/2, Section 5.5, (subsection 5.5.1). Now we can use the Riesz-Dunford functional calculus (1.67) to express the operator $\widehat{f}(A(\alpha))$ as

$$\widehat{f}(A(\alpha)) = \frac{1}{2\pi i}\int_\Gamma d\zeta\, \frac{f(\zeta)}{\zeta\mathbb{1} - A(\alpha)}\,. \tag{3.66}$$

Here $\Gamma \subset \rho(A(\alpha))$ is a positively oriented contour in the resolvent set of $A(\alpha)$ that encircles the spectrum $\sigma(A(\alpha))$. Note that $\rho(A(\alpha)) \subset \mathbb{C}\backslash\mathbb{R}_0^+$ for any $\alpha \in [0,1]$.

To ensure that the representation (3.66) exists and gives trace-class operators, we require that

- the resolvent $R_\lambda(A_2) = (A_2 - \lambda\mathbb{1})^{-1} \in \mathcal{C}_1(\mathcal{H})$ for $\lambda < 0$;
- for $\Re\mathfrak{e}\,\zeta \to +\infty$, the function $f(\zeta) \to 0$ fast enough to guarantee that the integral is $\|\cdot\|_1$-convergent.

Then the left-hand side of (3.66) is $\|\cdot\|_1$-differentiable, which implies that the derivative ∂_α now commutes with the integral and also with $\mathrm{Tr}(\cdot)$. Consequently,

$$\partial_\alpha \operatorname{Tr} \widehat{f}(A_2 + \alpha(A_1 - A_2))\big|_{\alpha=+0} = \operatorname{Tr}(A_1 - A_2)\widehat{f}'(A_2)\,,$$

where

$$\widehat{f}'(A_2) = \frac{1}{2\pi i}\int_\Gamma \mathrm{d}\zeta\, \frac{f(\zeta)}{(\zeta\mathbb{1} - A_2)^2}\ .$$

Hence, we obtain the Klein inequality (3.65) in its well-known *canonical* form:

$$\operatorname{Tr}[\widehat{f}(A_1) - \widehat{f}(A_2) - (A_1 - A_2)\widehat{f}'(A_2)] \geqslant 0\ . \tag{3.67}$$

3.5 Notes

Notes to Section 3.1. The description of singular values in (a)–(e) is standard and can be found in [GK69] (Ch.II, §1–§2) and in [Sim05], Ch.1.

For the *square-root lemma* in (a), see [RS80] Ch.VI.4. The usual way to prove that singular values satisfy $\{s_n(A) = s_n(A^*)\}_{n\geqslant 1}$ is a reference to the canonical form of the compact operators. In (c) we outline another way, based on the observation that the compact operators AB and BA have the same non-zero eigenvalues, with the same multiplicities, see [Sim05], Ch.1.

Section 3.1 contains only preliminaries about s-numbers; other properties of singular values of compact operators are explained in other parts of the book, where they are indispensable.

The *Weyl-Horn* chain of inequalities in (f) is due to H. Weyl [Wey49] and A. Horn [Horn50]. For a short proof, see [GK69] (Ch.II, §3) or [Sad91] (Ch.5.2.3).

Notes to Section 3.2. Lemma 3.1 is due to H. Weyl [Wey49] and A. Horn [Horn50], for the proof see, e.g., [Sim05] (Theorem 1.13 and Theorem 1.14). The important Lemma 3.2 was proved by H. Weyl [Wey49], here we followed [GK69] (Ch.II, §3.2). Lemma 3.5 is due to A. Horn [Horn50] and K. Fan [Fan51]. Our proof here is based on the Weyl-Horn chain of inequalities. Proposition 3.6 is due to A. Horn [Horn50].

Notes to Section 3.3. For additional information about *spectral* and *matrix* traces, including historical remarks, see [Piet14].

Proposition 3.11 proves the *Golden-Thompson* inequality (3.31) for self-adjoint exponentials with values in the trace ideals for a Hilbert space. For matrices it was proved in [Gold65] and [Thom65]. Comprehensive surveys of related trace inequalities with historical remarks can be found in [Sim05] (Chapter 8) and in [Petz83], [Car09].

Corollary 3.13 shows the *Golden-Thompson* inequality for the case when the Trotter product limit is a *degenerate* Gibbs semigroup. Similar to the case of C_0-semigroups the weak lower semi-continuity of the trace is sufficient to infer the result in this case.

Proposition 3.6, together with Corollaries 3.7 and 3.8, yield fundamental estimates for $\mathcal{C}_p(\mathcal{H})$-ideals, see [GK69], [Sch70]. These estimates are collected in Proposition 3.14 and Corollary 3.16. They are useful in applications to non-self-adjoint Gibbs semigroups.

Proposition 3.17 is a generalisation to the Hilbert space case of the Ky Fan inequalities for matrices [Fan49]. Then Proposition 3.18 is an extension to non-self-adjoint Gibbs exponentials of the matrix *Bernstein* inequality [Bern88]. Note that inequality (3.45) is opposite in sign to the Golden-Thompson inequality (3.31) for self-adjoint exponentials.

Notes to Section 3.4. The Peierls-Bogoliubov inequality appeared originally in [Peie38]. Apparently it was rediscovered in fifties and intensively exploited by Bogoliubov's school as a quantum variational principle, especially in the theory of magnetism [Tya67]. To keep a contact with the main topic of this book, we provide here the proof of Proposition 3.19 and Corollary 3.20 in a Hilbert space for Gibbs exponentials. It is a *direct* proof with reference to the Jensen inequality, cf. [Sim05] and compare with e.g., [Car09].

Proposition 3.21 is also proved in a Hilbert space and for Gibbs exponentials, cf. [Rue69], Proposition 2.5.5. The proof avoids derivatives, which are easy to treat for matrices (see [Car09], [Petz83]) but difficult in infinite-dimensional spaces.

Corollary 3.22 (3.55) serves as a preparation of the two-side Bogoliubov convexity inequality (3.58). Note that the right-hand side of (3.58) appeared first as a new variational principle in the paper of one of Bogoliubov's student [Kva56]. Then it became popular first in the theory of magnetism [Tya67].

Although almost evident, the two-side estimate (3.58) is much more efficient. It is a key for the *Approximating Hamitonian Method* (see [BBZKT81] and [BBZKT84]) in quantum statistical mechanics. The proof of the Bogoliubov convexity inequality (3.58) for the case of a bounded operator B, that is, without the issues with derivatives, is straightforward. Remark 3.24 explains a strategy for the general case. It was realised in [ZBT75], [BBZKT81].

The generalisation of the Peierls-Bogoliubov inequality in Proposition 3.25 as well as Corollary 3.26 for non-exponential convex functions are standard. They are independent of Klein's inequality [Kle31] that has stated at the end in Corollary 3.27. Again the main problem here is dealing with the derivative involved in the Klein inequality (3.65), or (3.67), cf. [Rue69], [Car09].

Chapter 4

Gibbs semigroups

This chapter contains notations and definitions concerning the main subject of the book. Here we introduce the Gibbs semigroups. They are strongly continuous (or degenerate) semigroups with values in the trace-class ideal of bounded operators on a Hilbert space. Although they constitute a subclass of the more general class of compact semigroups, the Gibbs semigroups have many special properties of their own. We present here both *eventually* and *immediately* Gibbs semigroups. The immediately Gibbs semigroups may be in turn *degenerate* or not. The immediately Gibbs semigroups play important rôle in Quantum Statistical Mechanics, see the Notes in Section 4.7. In our discussion of the non-self-adjoint Gibbs semigroups, we define a class of *p-generators*. The perturbation theory is developed first for a restricted class of the $\mathcal{P}_{0+}$-perturbations of *quasi-bounded* semigroups. Then we consider the *holomorphic* Gibbs semigroups, for which this class may be extended to $\mathcal{P}_{b<1}$-perturbations with $b > 0$.

4.1 Gibbs semigroups

The following is a key definition in the theory of the Gibbs semigroup.

Definition 4.1. A strongly continuous semigroup $\{G_t\}_{t\geqslant 0}$ on a Hilbert space $\mathcal{H}$ is called an *(immediately) Gibbs semigroup* if for $t \in \mathbb{R}^+$, $t \mapsto G_t \in \mathcal{C}_1(\mathcal{H})$. We call $\{G_t\}_{t\geqslant 0}$ a *self-adjoint Gibbs semigroup* if $G_t^* = G_t$, for $t \in \mathbb{R}_0^+$.

Although we study both self-adjoint and non-self-adjoint semigroups, the book places a certain emphases on the self-adjoint Gibbs semigroups. The first straightforward corollary of Definition 4.1 and the $\|\cdot\|_p$-continuity of the multiplication for $1 \leqslant p < \infty$, is the following statement.

Proposition 4.2. *Any immediate Gibbs semigroup* $\{G_t\}_{t\geqslant 0}$ *is trace-norm continuous for* $t > 0$.

V. A. Zagrebnov, *Gibbs Semigroups*, Operator Theory: Advances and Applications 273, https://doi.org/10.1007/978-3-030-18877-1_4

Proof. For any $t > 0$, there is a $\delta \in \mathbb{R}$ such that $t/2 + \delta > 0$. From the semigroup property it follows that $G_{t+\delta} = G_{t/2+\delta} G_{t/2}$, where $G_{t/2+\delta} \in \mathcal{L}(\mathcal{H})$ and $G_{t/2} \in \mathcal{C}_1(\mathcal{H})$. Since the semigroup $\{G_t\}_{t\geqslant 0}$ is strongly continuous, s-$\lim_{\delta\to 0} G_{t/2+\delta} = G_{t/2}$. Hence, for $t > 0$

$$\begin{aligned} \|\cdot\|_1\text{-}\lim_{\delta\to 0} G_{t+\delta} &= \|\cdot\|_1\text{-}\lim_{\delta\to 0}(G_{t/2+\delta}\, G_{t/2}) \\ &= (\text{s-}\lim_{\delta\to 0} G_{t/2+\delta})\, G_{t/2} = G_t\ , \end{aligned} \tag{4.1}$$

by the $\|\cdot\|_1$-continuity of multiplication on $\mathcal{C}_1(\mathcal{H})$, Proposition 2.78. □

According to the next statement, the semigroup property of $\{G_t\}_{t\geqslant 0}$ allows one to relax in Definition 4.1 the trace-class condition $G_t \in \mathcal{C}_1(\mathcal{H})$ for any $t > 0$, to the condition $G_t \in \mathcal{C}_p(\mathcal{H})$ for any $t > 0$, and an ideal $\mathcal{C}_p(\mathcal{H}) \supset \mathcal{C}_1(\mathcal{H})$ with $p > 1$.

Proposition 4.3. *A semigroup $\{G_t\}_{t\geqslant 0}$ such that $G_t \in \mathcal{C}_p(\mathcal{H})$ for all $t > 0$ and a number $p > 1$ is an immediate Gibbs semigroup.*

Proof. Let $t > 0$. Then by assumption one has $G_{t/\alpha} \in \mathcal{C}_p(\mathcal{H})$ for any $\alpha > 1$, and in particular for $\alpha = p$. Therefore, by Corollary 3.16,

$$G_t = (G_{t/p})^p \in \mathcal{C}_1(\mathcal{H}), \tag{4.2}$$

that is, $G_t \in \mathcal{C}_1(\mathcal{H})$ for all $t > 0$. □

Remark 4.4. Let $\{e_n\}_{n\geqslant 1}$ be an orthonormal basis in the Hilbert space $\mathcal{H}$. Then the one-parameter family of operators $\{T_t\}_{t\geqslant 0}$ defined for any $u \in \mathcal{H}$ by the mappings

$$u \mapsto T_t u = \sum_{n=1}^{\infty} \mathrm{e}^{-t\ln(n+1)}(u, e_n)e_n, \quad t \geqslant 0, \tag{4.3}$$

is a self-adjoint strongly continuous contraction semigroup on $\mathcal{H}$. The semigroup $\{T_t\}_{t\geqslant 0}$ is obviously compact for $t > 0$ with the singular values $\{s_n(T_t) = (n + 1)^{-t}\}_{n\geqslant 1}$, but it has the Gibbs semigroup property only for $t > t_0 = 1$. Similarly, $T_t \in \mathcal{C}_p(\mathcal{H})$ only for $t > t_0(p) = p^{-1}$.

Definition 4.5. If there exists a threshold $t_0 > 0$ such that $T_{t>t_0} \in \mathcal{C}_1(\mathcal{H})$, then the semigroup $\{T_t\}_{t\geqslant 0}$ is called an *eventually* Gibbs semigroup away from t_0. If $t_0 = 0$ and $\{T_t \in \mathcal{C}_1(\mathcal{H})\}_{t>0}$, then $\{T_t\}_{t\geqslant 0}$ is called an *immediately* Gibbs semigroup, or briefly the Gibbs semigroup, cf. Definition 4.1.

Corollary 4.6. *By Proposition* 4.2 *and Definition* 4.5 *we conclude that an immediately Gibbs semigroup is immediately $\|\cdot\|_1$-continuous for $t > 0$, whereas an eventually Gibbs semigroup (with threshold $t_0 > 0$) is $\|\cdot\|_1$-continuous only for $t > t_0$.*

Since by definition any Gibbs semigroup $\{G_t\}_{t\geqslant 0}$ is strongly continuous, it is quasi-bounded and it has a generator A which satisfies the conditions of Proposition 1.12. Hence, we can write

$$G_t(A) := \mathrm{e}^{-tA}, \quad t \geqslant 0, \tag{4.4}$$

and

$$\|G_t(A)\| \leqslant M\mathrm{e}^{\omega_0 t}, \quad M > 0,\ \omega_0 \geqslant 0,$$

where $(-\infty, -\omega_0) \in \rho(A)$, the resolvent set of the closed linear operator A. Moreover, if $\Re\mathfrak{e}\,\zeta > \omega_0$, the Laplace transform of (4.4) exists and it is given by the *operator-norm* convergent integral

$$\hat{G}_\zeta(A) = \int_0^\infty \mathrm{d}t\, \mathrm{e}^{-\zeta t} G_t(A) = (\zeta\mathbb{1} + A)^{-1}. \tag{4.5}$$

Note that (4.5) is related to the resolvent, $R_\zeta(A) = A_\zeta^{-1}$, of the operator A by

$$R_\zeta(A) = \hat{G}_{-\zeta}(A),$$

see (1.33). We used here the notation $A_\zeta := A - \zeta\mathbb{1}$ introduced in Chapter 1.

Remark 4.7. The inclusion $\mathcal{C}_1(\mathcal{H}) \subset \mathcal{C}_\infty(\mathcal{H})$ implies that every Gibbs semigroup $\{G_t(A)\}_{t\geqslant 0}$ is also a C_0-semigroup of compact operators. Since the integral (4.5) converges in the operator-norm topology and the integrand is a compact operator, the operator $\hat{G}_\zeta(A) \in \mathcal{C}_\infty(\mathcal{H})$. This implies that the resolvent $R_\zeta(A)$ is compact. Hence, the generator A is an unbounded operator, cf. Proposition 1.14.

Similarly to Sections 1.3 and 1.5, the Laplace transform (4.5) is a key formula for understanding the properties of generator of Gibbs semigroup. We start with statements that are valid also in the general case of Banach spaces, cf. Appendix A, Section A.4.

Proposition 4.8. *Let A be a closed linear operator in a Hilbert space $\mathcal{H}$. If $\lambda \in \mathbb{C}$ is such that there exists a sequence $\{x_n\}_{n\geqslant 1}$ with the properties*

$$\liminf_{n\to\infty} \|x_n\| \geqslant c > 0 \quad \text{and} \quad \lim_{n\to\infty} \|A_\lambda x_n\| = 0,$$

then $\lambda \in \sigma(A)$.

Proof. Suppose, by contradiction, that $\lambda \in \mathbb{C}\backslash\sigma(A) = \rho(A)$. Then the resolvent $R_\lambda(A)$ is a bounded operator on $\mathcal{H}$, i.e., there exists a constant $a > 0$ such that $\|R_\lambda(A)u\| \leqslant a\|u\|$ for all $u \in \mathcal{H}$. This means that $\|A_\lambda x\| \geqslant a^{-1}\|x\|$ for any $x \in \operatorname{dom} A_\lambda$, which contradicts the hypothesis of the proposition. □

We use this proposition to study spectral properties of generators of Gibbs semigroups.

Lemma 4.9. *The spectrum of the resolvent $R_\zeta(A)$ of an unbounded, closed, densely defined operator A in $\mathcal{H}$ admits for any $\zeta \notin \sigma(A)$ the representation*

$$\sigma(R_\zeta(A)) = \{0\} \cup \{(\lambda - \zeta)^{-1} : \lambda \in \sigma(A)\}. \tag{4.6}$$

Proof. Since A is not bounded, there is a sequence $\{f_n\}_{n\geqslant 1} \subset \operatorname{dom} A$, with $\|f_n\| = 1$, such that $\limsup_{n\to\infty} \|A_\zeta f_n\| = +\infty$. This implies that there exists a sequence $\{x_n\}_{n\geqslant 1}$ such that $\|x_n\| \geqslant c > 0$ and $\lim_{n\to\infty} \|R_\zeta(A)x_n\| = 0$: for example, one can take $x_n := A_\zeta f_n/\|A_\zeta f_n\|$. Therefore, by virtue of Proposition 4.8, we obtain that $\{0\} \subset \sigma(R_\zeta(A))$.

Next, since $\operatorname{ran} R_\zeta(A) = \operatorname{dom} A$, the operator

$$L := (z - \zeta)A_\zeta R_z(A), \quad z \notin \sigma(A),$$

is bounded, and it commutes with $R_\zeta(A)$ for any $\zeta \in \mathbb{C}$. Moreover, we have

$$L\,[(z - \zeta)^{-1}\mathbb{1} - R_\zeta(A)] = A_\zeta R_z(A) - (z - \zeta)R_z(A) = \mathbb{1}.$$

Hence, $(z - \zeta)^{-1} \notin \sigma(R_\zeta(A))$ for $z \notin \sigma(A)$.

Conversely, suppose that $(\lambda - \zeta)^{-1} \notin \sigma(R_\zeta(A))$ and consider the operator M defined by

$$M := [\mathbb{1} - (\lambda - \zeta)R_\zeta(A)]^{-1}R_\zeta(A) = R_\zeta(A)[\mathbb{1} - (\lambda - \zeta)R_\zeta(A)]^{-1}.$$

If $f \in \mathcal{H}$, then

$$\begin{aligned} A_\lambda M f &= [(\zeta - \lambda)\mathbb{1} + A - \zeta\mathbb{1}]R_\zeta(A)[\mathbb{1} - (\lambda - \zeta)R_\zeta(A)]^{-1}f \\ &= [\mathbb{1} - (\lambda - \zeta)R_\zeta(A)][\mathbb{1} - (\lambda - \zeta)R_\zeta(A)]^{-1}f = f. \end{aligned}$$

On the other hand, if $f \in \operatorname{dom} A$, then

$$\begin{aligned} M A_\lambda f &= [\mathbb{1} - (\lambda - \zeta)R_\zeta(A)]^{-1}R_\zeta(A)[(\zeta - \lambda)\mathbb{1} + A - \zeta\mathbb{1}]f \\ &= [\mathbb{1} - (\lambda - \zeta)R_\zeta(A)]^{-1}[\mathbb{1} - (\lambda - \zeta)R_\zeta(A)]f = f. \end{aligned}$$

Hence, $\lambda \notin \sigma(A)$. □

Corollary 4.10. *If the operator $R_\zeta(A)$ is compact for some $\zeta \notin \sigma(A)$, then*

$$R_z(A) \in \mathcal{C}_\infty(\mathcal{H}), \quad z \notin \sigma(A),$$

and the spectrum $\sigma(A)$ is either empty, or consists of at most countably many eigenvalues $\{\lambda_n\}_{n\geqslant 1}$ of finite multiplicity, such that

$$\lim_{n\to\infty} |\lambda_n| = +\infty.$$

Moreover, we have that the point spectrum

$$\sigma_{\mathrm{p}}(R_\zeta(A)) = \{(\lambda_n - \zeta)^{-1} : \lambda_n \in \sigma(A)\},$$

whereas $\sigma_{\mathrm{ess}}(R_\zeta(A)) = \{0\}$.

Proof. The first part of the statement follows from the resolvent identity

$$R_z(A) = R_\zeta(A) + R_z(A)(z-\zeta)R_\zeta(A),$$

and from Proposition 2.7, which states that $\mathcal{C}_\infty(\mathcal{H})$ is a $*$-ideal in $\mathcal{L}(\mathcal{H})$.

Note that by Lemma 4.9, $\{0\} \subset \sigma(R_\zeta(A))$ for unbounded operator A even if $\sigma(A) = \varnothing$. Then the second part of the statement is a consequence of Lemma 4.9, Proposition 2.18, and Remark 2.23.

In order to prove the third part, recall that $R_\zeta(A) \in \mathcal{C}_\infty(\mathcal{H})$ implies $0 \in \sigma(R_\zeta(A))$, Remark 2.15. By virtue of Remark 2.23, this point may belong to any of the three disjoint components of the spectrum $\sigma(R_\zeta(A))$, Appendix A (Section A.6).

Suppose that $u \in \mathcal{H}$ is an eigenvector of $R_\zeta(A)$ with eigenvalue equals to zero: $R_\zeta(A)u = 0$. Since $R_\zeta(A)u \in \operatorname{dom} A$, we get

$$u = (A - \zeta\mathbb{1})R_\zeta(A)u = 0\,.$$

This means that $\ker R_\zeta(A)$ is trivial, and we infer that the point $0 \notin \sigma_{\mathrm{p}}(R_\zeta(A))$. Next, following Definition 2.14, we consider the case when $\operatorname{ran} R_\zeta(A) = \operatorname{dom} A$. Since A is unbounded, closed, and densely defined, $\operatorname{ran} R_\zeta(A) \neq \mathcal{H}$, but $\overline{\operatorname{ran} R_\zeta(A)} = \mathcal{H}$, i.e., $\{0\} \subseteq \sigma_{\mathrm{ess}}(R_\zeta(A))$. Therefore, $\sigma(R_\zeta(A))\backslash\{0\} = \sigma_{\mathrm{p}}(R_\zeta(A))$ and the equality

$$\sigma_{\mathrm{p}}(R_\zeta(A)) = \{(\lambda_n - \zeta)^{-1} : \lambda_n \in \sigma(A)\}$$

follows from (4.6). □

Now, let A be generator of a strongly continuous semigroup $\{U_t(A)\}_{t\geqslant 0}$ on $\mathcal{H}$, see Chapter 1. In this general setup, the relationship between $\sigma(A)$ and $\sigma(U_t(A))$ is not straightforward, cf. Section A.9.

Proposition 4.11. *Suppose that $\{U_t(A)\}_{t\geqslant 0}$ is a strongly continuous semigroup on $\mathcal{H}$ with generator A. Then*

$$\{\mathrm{e}^{-\lambda t} : \lambda \in \sigma(A)\} \subseteq \sigma(U_t(A)).$$

Proof. Let $\mathcal{A}$ be the maximal Abelian subalgebra of $\mathcal{L}(\mathcal{H})$ containing $U_t(A)$ for all $t \geqslant 0$. This maximal subalgebra $\mathcal{A}$ obviously exists by Zorn's lemma, and it is closed under the strong limit. Representation (4.5) implies that

$$R_\zeta(A) \in \mathcal{A}\,, \quad \zeta \notin \sigma(A).$$

Moreover, if $X \in \mathcal{A}$ is invertible within $\mathcal{L}(\mathcal{H})$, then $X^{-1} \in \mathcal{A}$. Therefore, the spectrum of X as an *operator* coincides with its spectrum as an *element* of the Banach algebra $\mathcal{A}$ and hence, it coincides with $\{\tilde{X}(m) : m \in M\}$, where M is the maximal ideal space of $\mathcal{A}$, and $\tilde{X}$ is the *Gel'fand transform* of X.

Let $\lambda \in \sigma(A)$ and $\zeta \notin \sigma(A)$. Then $(\lambda - \zeta)^{-1} \in \sigma(R_\zeta(A))$, so there exists $m \in M$ with $\tilde{R}_\zeta(m) = (\lambda - \zeta)^{-1} \neq 0$. If $\gamma(t) = \tilde{U}_t(m)$, then $\gamma(0) = 1$ and

$\gamma(s)\gamma(t) = \gamma(s+t)$ for $s,t \geqslant 0$. Since for any $u \in \mathcal{H}$, $R_\zeta(A)u \in \operatorname{dom} A$, the operator-valued function $\mathbb{R}^+ \ni t \mapsto U_t(A)R_\zeta(A) \in \mathcal{L}(\mathcal{H})$ is continuous in the operator norm, cf. Proposition 1.15. Hence, $\gamma(t)(\lambda-\zeta)^{-1}$ depends continuously on $t \in \mathbb{R}^+$. Consequently, $t \mapsto \gamma(t) = \mathrm{e}^{-\beta t}$ for some $\beta \in \mathbb{C}$.

If $\mathfrak{Re}\, z > \omega_0$, the integral

$$\int_0^\infty \mathrm{d}t\, \mathrm{e}^{-zt} U_t(A) R_\zeta(A) = R_{-z}(A) R_\zeta(A)$$

is operator-norm convergent, so its Gel'fand transform is

$$\int_0^\infty \mathrm{d}t\, \mathrm{e}^{-zt} \mathrm{e}^{-\beta t} (\lambda-\zeta)^{-1} = (\lambda-\zeta)^{-1} \tilde{R}_{-z}(m).$$

Therefore, $\tilde{R}_z(m) = (\beta - z)^{-1}$. On the other hand, from the Gel'fand transform of the resolvent identity,

$$\tilde{R}_\zeta(m) - \tilde{R}_z(m) = (\zeta - z)\tilde{R}_\zeta(m)\tilde{R}_z(m),$$

we obtain

$$\tilde{R}_z(m) = (\lambda-\zeta)^{-1}\{1 + (\zeta - z)(\lambda-\zeta)^{-1}\}^{-1} = (\lambda - z)^{-1}.$$

We conclude that $\lambda = \beta$, that is, $\tilde{U}_t(m) = \mathrm{e}^{-\lambda t}$, so $\mathrm{e}^{-\lambda t} \in \sigma(U_t(A))$. □

In general, the converse of Proposition 4.11 is false. However, if some further conditions are imposed on the semigroup $\{U_t\}_{t\geqslant 0}$, see Section 1.4, one gets the following statement.

Proposition 4.12. *Let $\{U_t(A)\}_{t\geqslant\tau}$ be a norm-continuous semigroup on $\mathcal{H}$ for $t \in [\tau, +\infty)$, $\tau > 0$. Then $\mu_t \neq 0$ is a non-zero element of the spectrum, $\mu_t \in \sigma(U_t(A))$, if and only if it can be written as $\mu_t = \mathrm{e}^{-t\lambda}$ for some $\lambda \in \sigma(A)$.*

Proof. For a non-zero element $\mu_t \in \sigma(U_t(A))$, there exists an element $m \in M$, where M is the maximal ideal space, such that the Gel'fand transform $\gamma(t) = \tilde{U}_t(m)$ satisfies $\gamma(t) = \mu_t$. Moreover, $\gamma(s_1 + s_2) = \gamma(s_1)\gamma(s_2)$ for $s_1, s_2 \in (0,\infty)$, and $\gamma(s)$ is continuous for $s \in [\tau, +\infty)$. This implies that $\gamma(s) = \mathrm{e}^{-s\lambda}$ for some λ and $s \in \mathbb{R}^+$.

Since for the operator-norm continuous semigroup $\{U_t(A)\}_{t\geqslant\tau}$ the integral in representation

$$U_\tau(A)(z\mathbb{1} + A)^{-1} = \int_0^\infty \mathrm{d}t\, \mathrm{e}^{-zt} U_{\tau+t}(A)$$

is convergent in the operator norm, the corresponding Gel'fand transform can be written as

$$\begin{aligned}\mathrm{e}^{-\lambda\tau}\{(z\mathbb{1} + A)^{-1}\}^{\sim}(m) &= \int_0^\infty \mathrm{d}t\, \mathrm{e}^{-zt} \mathrm{e}^{-(\tau+t)\lambda} \\ &= \mathrm{e}^{-\lambda\tau}(z+\lambda)^{-1} = \mathrm{e}^{-\lambda\tau}\tilde{R}_{-z}(m).\end{aligned}$$

This means that $(\lambda - z)^{-1} \in \sigma(R_z(A))$ and $\lambda \in \sigma(A)$, by Lemma 4.9. □

If we assume that $\{U_t(A)\}_{t\geqslant 0}$ is a compact operator-valued semigroup for some $\tau > 0$, then even a stronger statement about the spectral properties of $\{U_t(A)\}_{t\geqslant\tau}$ can be proven. We return to this problem in the next section, Proposition 4.21.

4.2 Norm continuity, revisited

In this section we revise the operator-norm continuous semigroups. They were introduced in Section 1.4 and we found (Proposition 1.14) that operator-norm continuity of the semigroup $\{U_t\}_{t\geqslant 0}$ at $t = +0$ implies the boundedness of its generator.

So, first we relax this condition to the usual C_0-continuity at $t = +0$. Then we study the impact on the semigroup *continuity* away from zero of the *topology* of the *image* $\mathrm{Im}(U_{t>0})$ of the map $t \mapsto U_t$. For example, the strongly continuous immediately Gibbs semigroups are defined by $\mathrm{Im}(G_{t>0}) \subseteq \mathcal{C}_1(\mathcal{H})$. We also observed (Corollary 4.6) that away from zero the semigroup continuity *inherits* the trace-norm topology of the image.

Below we elucidate a relation between *compactness* of the image $\mathrm{Im}(U_{t>0}) \subseteq \mathcal{C}_\infty(\mathcal{H})$, and topology of continuity of semigroup $\{U_t\}_{t\geqslant 0}$ away from $t = 0$. It turns out that the semigroup $\{U_t\}_{t>0}$ *inherits* for its continuity the topology of the ideal of compact operators $\mathcal{C}_\infty(\mathcal{H})$, that is, the operator norm topology.

To continue we first introduce few notations and definitions similar to the ones used for the Gibbs semigroups.

Definition 4.13. A strongly continuous semigroup $\{U_t\}_{t\geqslant 0}$ on a Banach space $\mathcal{B}$ is said to be *eventually* norm-continuous if there exists a threshold $t_0 > 0$ such that the mapping $t \mapsto U_t$ is operator-norm continuous for $t \in (t_0, \infty)$. If $t_0 = 0$, then a strongly continuous semigroup $\{U_t\}_{t\geqslant 0}$ is called *immediately* norm-continuous, or briefly a norm-continuous semigroup, cf. Section 1.4.

Next we introduce the *compact* semigroups on a Hilbert space. The related results we shall use for their spectral analysis.

Definition 4.14. A strongly continuous semigroup $\{U_t\}_{t\geqslant 0}$ on a Hilbert space $\mathcal{H}$ is said to be *eventually* compact if there exists $t_0 > 0$ such that

$$t \mapsto U_t \in \mathcal{C}_\infty(\mathcal{H}) \quad \text{for } t > t_0 \,. \tag{4.7}$$

If $\{U_t \in \mathcal{C}_\infty(\mathcal{H})\}_{t>0}$, then the semigroup $\{U_t\}_{t\geqslant 0}$ is called *immediately* compact, or briefly a compact semigroup.

Lemma 4.15. *Any immediately compact C_0-semigroup $\{U_t\}_{t\geqslant 0}$ is norm-continuous for $t > 0$. The same is true for any eventually compact C_0-semigroup $\{U_t\}_{t\geqslant 0}$ with a threshold $t_0 > 0$, when $t > t_0$.*

Proof. We consider only the case of eventually compact C_0-semigroups, since in the case of immediately compact semigroups it is sufficient to set $t_0 = 0$.

Let $0 < t_0 < \tau$. Then for any $\varepsilon > 0$, there is a finite-rank operator $U_\tau^F \in \mathcal{K}(\mathcal{H})$ such that $\|U_\tau - U_\tau^F\| < \varepsilon$, see Proposition 2.11. Hence, for $\tau \leqslant t \leqslant t'$ and $f \in \mathcal{H}$, we have the estimate

$$\begin{aligned}\|U_{t'}f - U_t f\| &= \|(U_{t'-\tau} - U_{t-\tau})U_\tau f\| \\ &\leqslant \|(U_{t'-\tau} - U_{t-\tau})U_\tau^F f\| \\ &\quad + \|U_{t'-\tau} - U_{t-\tau}\|\|U_\tau - U_\tau^F\|\|f\|. \end{aligned} \tag{4.8}$$

Since $U_\tau^F \in \mathcal{K}(\mathcal{H})$ and $\text{s-lim}_{t'\to t} U_{t'-\tau} = U_{t-\tau}$, we get that

$$\lim_{t'\to t} \|U_{t'-\tau} - U_{t-\tau})U_\tau^F\| = 0. \tag{4.9}$$

Note that $\|U_{t'-\tau} - U_{t-\tau}\| \leqslant 2\max\{\|U_{t'-\tau}\|, \|U_{t-\tau}\|\}$. Then the proof follows from (4.8) and (4.9) due to the estimate $\|U_\tau - U_\tau^F\| < \varepsilon$, for any $\varepsilon > 0$. □

Although essentially we are going to deal with semigroups in a Hilbert space, certain results are also relevant in Banach spaces. In fact, Definition 4.14 and Lemma 4.15 hold in the general case of a Banach space $\mathcal{B}$. Indeed, since U_τ maps the unit ball in $\mathcal{B}$ into a precompact set $C \subset \mathcal{B}$, the strong operator convergence $\text{s-lim}_{t'\to t} U_{t'-\tau} = U_{t-\tau}$, implies the operator-norm convergence on C, see (4.8).

An example of strongly continuous, immediately compact, and eventually Gibbs semigroup is given in Remark 4.4. A simple example illustrating that *eventual* does not imply *immediate* comes from *nilpotent* semigroups.

Example 4.16. Define a C_0-semigroup $\{U_t\}_{t\geqslant 0}$ on the Hilbert space $\mathcal{H} = L^2([0,1])$ by

$$(U_t f)(x) = \begin{cases} f(x+t) & \textit{for} \;\; x+t \leqslant 1\,, \\ 0 & \textit{for} \;\; x+t > 1\,. \end{cases}$$

Then $\{U_t\}_{t\geqslant 0}$ is nilpotent since $U_{t>1}f = 0$ for any $f \in \mathcal{H}$. Note that this strongly continuous semigroup is eventually norm-continuous and eventually compact (with threshold $t_0 = 1$), but it is not immediately norm-continuous or compact.

Proposition 4.17. *A quasi-bounded strongly continuous semigroup $\{U_t(A)\}_{t\geqslant 0}$ on $\mathcal{H}$ with generator $A \in Q(M,\omega_0)$ is immediately compact if and only if the map $\mathbb{R}_0^+ \ni t \mapsto U_t(A) \in \mathcal{L}(\mathcal{H})$ is immediately norm-continuous and the resolvent $R_z(A)$ is compact for $z \in \rho(A)$.*

Proof. Let the semigroup $\{U_t(A)\}_{t\geqslant 0}$ be compact. Then by Lemma 4.15, it is norm-continuous for all $t > 0$. Now, since the generator $A \in Q(M,\omega_0)$, it follows from Definition 1.11 for semigroups that $\|\mathrm{e}^{-t\lambda}U_t(A)\| \leqslant M\,\mathrm{e}^{t(\omega_0-\lambda)}$. Hence, for $\lambda > \omega_0$ the resolvent of A can be represented by the Laplace transform (Proposition 1.12):

$$R_{-\lambda}(A) = \int_0^\infty \mathrm{d}t\,\mathrm{e}^{-\lambda t}U_t(A).$$

Note that the norm-continuity of the semigroup $\{U_t(A)\}_{t\geqslant 0}$ away from zero implies that the (Bohner) integral

$$R^{(\varepsilon)}_{-\lambda}(A) = \int_{\varepsilon>0}^{\infty} dt\, e^{-\lambda t} U_t(A), \tag{4.10}$$

for some $\varepsilon > 0$, is the limit of the corresponding operator-norm convergent Darboux-Riemann sums. Since the semigroup $\{U_t(A)\}_{t\geqslant 0}$ is compact away from zero, each of these sums is in turn a compact operator. Note that the set of compact operators is closed in the operator-norm topology (Proposition 2.8), which yields $R^{(\varepsilon)}_{-\lambda}(A) \in \mathcal{C}_\infty(\mathcal{H})$ for any $\varepsilon > 0$ and $\lambda > \omega_0$. Since the semigroup is quasi-bounded and the integrand in (4.10) is norm-continuous,

$$\lim_{\varepsilon\downarrow 0} \| R_{-\lambda}(A) - R^{(\varepsilon)}_{-\lambda}(A)\| = 0, \quad \lambda > \omega_0,$$

and again by Proposition 2.8, we conclude that $R_{-\lambda}(A)$ is compact.

To conclude this part of the proof, we remark that $(-\infty, -\omega_0) \subset \rho(A)$, see Proposition 1.12, and that by the resolvent identity, the fact that $R_\zeta(A) \in \mathcal{C}_\infty(\mathcal{H})$ for some $\zeta \in (-\infty, -\omega_0)$ implies that $R_z(A) \in \mathcal{C}_\infty(\mathcal{H})$ for any $z \in \rho(A)$.

To prove the converse, we assume now that the map $\mathbb{R}^+_0 \ni t \mapsto U_t(A) \in \mathcal{L}(\mathcal{H})$ is immediately norm-continuous and that $R_z(A) \in \mathcal{C}_\infty(\mathcal{H})$ for some $z \in \rho(A)$. Then by the remark just above, the resolvent $R_z(A)$ is compact for every $z \in \rho(A)$. Since $\{U_t(A)\}_{t\geqslant 0}$ is strongly continuous and quasi-bounded, it follows from Proposition 1.12 that the set $(-\infty, -\omega_0) \subset \rho(A)$. We fix a constant $\lambda > \omega_0$ and define a family of operators:

$$C_t := \int_0^t d\tau\, e^{-\lambda\tau} U_\tau(A), \quad t \geqslant 0. \tag{4.11}$$

Then integrating the equation

$$\partial_t (e^{-\lambda t} U_t(A) u) = -e^{-\lambda t} U_t(A)(\lambda \mathbb{1} + A) u, \quad u \in \operatorname{dom} A\,,$$

we obtain

$$(\mathbb{1} - e^{-\lambda t} U_t(A)) u = \int_0^t d\tau\, e^{-\lambda\tau} U_\tau(A)(\lambda\mathbb{1} + A) u,$$

or with $v := (\lambda\mathbb{1} + A)u$, the equation

$$(\mathbb{1} - e^{-\lambda t} U_t(A))(A + \lambda\mathbb{1})^{-1} v = \int_0^t d\tau\, e^{-\lambda\tau} U_\tau(A) v. \tag{4.12}$$

Since $-\lambda \in \rho(A)$, the range $\operatorname{ran}(\lambda\mathbb{1} + A) = \mathcal{H}$. Therefore, equation (4.12) implies that operator (4.11) is compact:

$$C_t = (\mathbb{1} - e^{-\lambda t} U_t(A))(A + \lambda\mathbb{1})^{-1} \in \mathcal{C}_\infty(\mathcal{H}), \tag{4.13}$$

since $R_{-\lambda}(A) \in \mathcal{C}_\infty(\mathcal{H})$. By the assumption of norm-continuity of the semigroup $\{U_t(A)\}_{t>0}$, it follows that for any $t > 0$

$$\begin{aligned} U_t(A) &= \mathrm{e}^{\lambda t} \, \|\cdot\|\text{-}\lim_{\delta\downarrow 0} \frac{1}{\delta} \int_t^{t+\delta} \mathrm{d}\tau \, \mathrm{e}^{-\lambda\tau} U_\tau(A) \\ &= \mathrm{e}^{\lambda t} \, \|\cdot\|\text{-}\lim_{\delta\downarrow 0} \frac{1}{\delta}(C_{t+\delta} - C_t). \end{aligned} \tag{4.14}$$

By virtue of (4.12) and (4.13), the operator-norm limit of compact operators in the right-hand side of (4.14) exists which implies that $U_t(A) \in \mathcal{C}_\infty(\mathcal{H})$ for $t > 0$. □

Now we consider eventually norm-continuous and eventually compact C_0-semigroups. A sufficient condition guaranteeing that the eventual property *does* imply the immediate property for norm-continuity or for compactness of semigroups is their *analyticity*.

Remark 4.18. It is straightforward that if a C_0-semigroup $\{U_t(A)\}_{t\geqslant 0}$ is holomorphic and eventually norm-continuous for threshold $t_0 > 0$, then it is immediately norm-continuous. This follows directly from the construction of $\{U_t(A)\}_{t\geqslant 0}$ in Proposition 1.27. It ensures that a *strongly continuous* holomorphic semigroup (Definition 1.26) is, in fact, *immediately* norm-continuous and even *norm-holomorphic* for t in some open sector $S_\theta \subset \mathbb{C}_+$.

The next statement together with Example 4.16 settle also the problem of existence of immediately, and of elimination of eventually *compact* semigroups.

Proposition 4.19. *Let $\{U_t(A)\}_{t\geqslant 0}$ be a holomorphic semigroup with generator $A \in \mathscr{H}(\theta, \omega_0)$. If $\{U_t(A)\}_{t\geqslant 0}$ is eventually compact with threshold $t_0 > 0$, then it is immediately compact.*

Proof. Let $\widehat{t}_0 > t_0$ such that $\widehat{t}_0 - t_0 < \widehat{t}_0 \sin\theta$. Since by Corollary 1.29 and Proposition 1.33 the semigroup $\{U_t(A)\}_{t\geqslant 0}$ is norm-holomorphic in the open sector $S_\theta \subset \mathbb{C}_+$, one gets the operator-norm convergent power series expansion

$$U_z(A) = \sum_{n=0}^{\infty} \frac{(z - \widehat{t}_0)^n}{n!} \, \partial_t^n U_t(A)|_{t=\widehat{t}_0} \tag{4.15}$$

in the disc $D_{\widehat{t}_0} := \{z \in \mathbb{C}_+ : |z - \widehat{t}_0| < (\mathrm{e}M_1')^{-1}\widehat{t}_0\}$, where $\mathrm{e}M_1' > 1$ and $\theta = \arcsin(\mathrm{e}M_1')^{-1}$, see (1.79). Therefore, $t_0 \in D_{\widehat{t}_0}$.

Note that the analyticity of the map $z \mapsto U_z(A)$ allows us to calculate all derivatives in (4.15). For $|\delta| < \widehat{t}_0 - t_0$ the operators $U_{\widehat{t}_0}(A)$ and $U_{\widehat{t}_0+\delta}(A)$ are compact. Then the derivative

$$\partial_t U_t(A)\big|_{t=\widehat{t}_0} = \|\cdot\|\text{-}\lim_{\delta\to 0} \frac{1}{\delta}[U_{\widehat{t}_0+\delta}(A) - U_{\widehat{t}_0}(A)] \in \mathcal{C}_\infty(\mathcal{H}), \tag{4.16}$$

being a limit in the operator-norm topology, is also a *compact* operator with $\|\partial_t U_t(A)\big|_{t=\widehat{t}_0}\| < M_1'/\widehat{t}_0$, see (1.76).

Since the map $z \mapsto U_z(A)$ is norm-analytic, one can iterate the argument in (4.16) for higher-order derivatives: $\partial_t^n U_t(A)\big|_{t=\hat{t}_0} \in \mathcal{C}_\infty(\mathcal{H})$, for $n \in \mathbb{N}$, with estimates $\|\partial_t^n U_t(A)\big|_{t=\hat{t}_0}\| < (nM_1'/\hat{t}_0)^n$. Then the operator-norm convergent power series (4.15) defines compact operators $\{U_z(A)\}_{z \in D_{\hat{t}_0}}$ in the disc $D_{\hat{t}_0}$ and in particular, $U_{t_0}(A) \in \mathcal{C}_\infty(\mathcal{H})$.

Now we can take $U_{t_0}(A) \in \mathcal{C}_\infty(\mathcal{H})$ and consider the operator-norm convergent in the disc $D_{t_0} := \{z \in \mathbb{C}_+ : |z - t_0| < t_0 \sin\theta\}$ power series expansion (4.15) at $t = t_0$. Then by the same argument as above the operators $\{U_z(A)\}_{z \in D_{t_0}}$ are compact in the disc D_{t_0}, and so $U_{t_1}(A) \in \mathcal{C}_\infty(\mathcal{H})$, for example, at $t_1 = t_0(1 - (\sin\theta)/2)$.

If one considers expansion (4.15) at t_1, then $U_{t_2}(A) \in \mathcal{C}_\infty(\mathcal{H})$ for $t_2 = t_1(1 - (\sin\theta)/2)$, and so on for n steps. Therefore, $U_{t_n}(A) \in \mathcal{C}_\infty(\mathcal{H})$ for $t_n = t_0(1 - (\sin\theta)/2)^n \to 0$, when $n \to \infty$, establishes the assertion. □

Remark 4.20. By Definitions 4.1 and 4.5, the Gibbs semigroups are obviously compact. Remark 4.4 gives an example of a self-adjoint *eventually* Gibbs semigroup. On the other hand, since this semigroup is holomorphic, it is *immediately* compact by Proposition 4.19. The lack of analyticity in Example 4.16 of the nilpotent semigroup leads to an eventually compact and eventually Gibbs semigroup, which has zero trace-norm for $t \geqslant t_0 = 1$, see Appendix A, Section A.7.

Now we are in position to analyse spectra of compact semigroups versus spectra of their generators.

Proposition 4.21. *Let A be generator of a compact semigroup $\{U_t(A)\}_{t \geqslant 0}$. Then the following assertions hold for* $\dim \mathcal{H} = \infty$.

(a) *The spectrum $\sigma(A)$ is either empty, or consists of a finite or countable set of eigenvalues $\{\lambda_n\}_{n \geqslant 1}$ of finite multiplicity, which have no limit point in $\mathbb{C}$.*

(b) *The spectrum of the semigroup has the representation*

$$\sigma(U_t(A)) = \{0\} \cup \{e^{-t\sigma(A)}\}, \quad t > 0, \tag{4.17}$$

where $\sigma(U_t(A))\backslash\{0\} = \sigma_p(U_t(A))\backslash\{0\} = \{e^{-t\sigma(A)}\}$, that is, $A\varphi_n = \lambda_n\varphi_n$ if and only if $U_t(A)\varphi_n = e^{-t\lambda_n}\varphi_n$.

Proof. (a) Since compact semigroups are norm-continuous, A is the generator of a strongly continuous semigroup. Then, by Proposition 1.12, the resolvent set $\rho(A)$ is nonempty, and by virtue of Proposition 4.17, the resolvent $R_z(A)$ is compact for any $z \in \rho(A)$. Therefore, the assertion follows from Corollary 4.10.

(*b*) Note that in contrast to Corollary 4.10, we cannot claim that $0 \notin \sigma_p(U_t(A))$. We assume that the spectrum of A is non-empty, otherwise $\sigma(U_t(A)) = \{0\}$ in (4.17). It then follows from (a) that $\sigma(A) = \{\lambda_n\}_{n \geqslant 1}$. For an eigenvector $\varphi_n \in \operatorname{dom} A$ corresponding to the eigenvalue λ_n, the identity

$$(A - z\mathbb{1})R_z(A)\varphi_n = \varphi_n = R_z(A)(\lambda_n - z)\varphi_n, \quad z \in \rho(A),$$

and the fact that $A \in Q(M, \omega_0)$ imply (see (1.5))

$$\begin{aligned} U_{t,m}(A_{-\omega_0})\varphi_n &= \left(\mathbb{1} + \frac{t}{m}A_{-\omega_0}\right)^{-m}\varphi_n \\ &= \left(1 + \frac{t}{m}(\lambda_n + \omega_0)\right)^{-m}\varphi_n, \end{aligned} \tag{4.18}$$

where $A_{-\omega_0} = A + \omega_0\mathbb{1}$. By virtue of Proposition 1.12 and taking in (4.18) the limit $m \to \infty$, we get

$$U_t(A)\varphi_n = \mathrm{e}^{t\omega_0}U_t(A_{-\omega_0})\varphi_n = \mathrm{e}^{-t\lambda_n}\varphi_n, \tag{4.19}$$

for $t \geqslant 0$.

Conversely, since $U_t(A)$ is compact for $t > 0$, there exists an eigenvector $\psi \in \mathcal{H}$ with an eigenvalue of the form $\mathrm{e}^{-t\mu}$, i.e., $U_t(A)\psi = \mathrm{e}^{-t\mu}\psi$. Calculating the limit

$$\lim_{t\downarrow 0}\frac{1}{t}(\mathbb{1} - U_t(A))\psi = \lim_{t\downarrow 0}\frac{1}{t}(\mathbb{1} - \mathrm{e}^{-t\mu})\psi = \mu\psi,$$

it follows from the definition of the generator A (Definition 1.9) that $\psi \in \operatorname{dom} A$ and $A\psi = \mu\psi$, i.e., $\mu \in \{\lambda_n\}_{n\geqslant 1}$. In the general case, fix $t_0 > 0$ and let $\eta(\neq 0) \in \sigma(U_{t_0}(A))$; since $U_{t_0}(A)$ is compact, this η must be an eigenvalue. We denote the corresponding eigenvector by v. Then for any $t \geqslant 0$, we have

$$U_{t_0}(A)U_t(A)v = U_t(A)U_{t_0}(A)v = \eta\, U_t(A)v\,,$$

which implies that the subspace $\mathcal{H}_\eta := \ker(U_{t_0}(A) - \eta\mathbb{1})$ is invariant under the semigroup. Since $U_{t_0}(A)$ is compact, any proper subspace $\mathcal{H}_\eta$ must be finite-dimensional. Let $\{U_t^{(\eta)} := U_t(A) \restriction \mathcal{H}_\eta\}_{t\geqslant 0}$ be the restriction of the semigroup $\{U_t(A)\}_{t\geqslant 0}$ to the subspace $\mathcal{H}_\eta$. Then $\{U_t^{(\eta)}\}_{t\geqslant 0}$ is a strongly continuous semigroup acting on a finite-dimensional space, and hence, its generator K is a bounded linear operator with $\operatorname{dom} K = \mathcal{H}_\eta$. Since $\{U_t^\eta = \mathrm{e}^{-tK}\}_{t\geqslant 0}$ acts on the finite-dimensional space $\mathcal{H}_\eta$, the spectra $\sigma(U_t^{(\eta)})$ and $\sigma(K)$ are non-empty, discrete, and are related by $\sigma(U_t^{(\eta)}) = \mathrm{e}^{-t\sigma(K)}$. Let $\lambda \in \sigma(K)$. There exists an eigenvector $f_\lambda \in \mathcal{H}_\eta$ such that

$$U_t^{(\eta)} f_\lambda = \mathrm{e}^{-t\lambda} f_\lambda\,, \quad t \geqslant 0. \tag{4.20}$$

Since $U_t^{(\eta)} = U_t(A) \restriction \mathcal{H}_\eta$, (4.20) implies

$$U_t(A) f_\lambda = \mathrm{e}^{-t\lambda} f_\lambda\,, \quad t \geqslant 0. \tag{4.21}$$

Thus, the eigenvalues of $\{U_t(A)\}_{t\geqslant 0}$ indeed have an exponential form, in particular, $\eta = \mathrm{e}^{-t_0\lambda}$. From the arguments developed for this case, we get that $\lambda \in \sigma(A) = \{\lambda_n\}_{n\geqslant 1}$, with f_λ being the corresponding eigenvector. Since when $\dim \mathcal{H} = \infty$, the point $\{0\}$ always belongs to the spectrum of a compact operator, this completes the proof of the representation (4.17). $\square$

Although in general for compact operators one has $\{0\} \subseteq \sigma_{\text{ess}}$, (Appendix A, Section A.6), in the next statement the essential spectrum shrinks to its continuous part.

Proposition 4.22. *Let A be a densely defined generator of a strongly continuous semigroup $\{U_t(A)\}_{t\geqslant 0}$ on a Hilbert space $\mathcal{H}$.*

(a) *$\{U_t(A)\}_{t>0}$ is self-adjoint and compact if and only if the operator A is self-adjoint and has a compact resolvent $R_\zeta(A)$ for some $\zeta \in \rho(A)$.*

(b) *In this case, there is an orthonormal basis $\{\varphi_n\}_{n\geqslant 1}$ in $\mathcal{H}$ consisting of eigenvectors of A with real eigenvalues $\{\lambda_n\}_{n\geqslant 1}$ satisfying $\min_{n\geqslant 1} \lambda_n = -\omega_0$, such that we have the representation*

$$U_t(A)u = \sum_{n\geqslant 1} \mathrm{e}^{-t\lambda_n}(u, \varphi_n)\varphi_n \quad u \in \mathcal{H}, \quad t \geqslant 0. \tag{4.22}$$

(c) *For $\dim \mathcal{H} = \infty$ and $t > 0$, the spectrum of $U_t(A)$ can be written as $\sigma(U_t(A)) = \{0\} \cup \{\mathrm{e}^{-t\sigma(A)}\}$, where $\{0\} = \sigma_{\text{cont}}(U_t(A))$ and $\{\mathrm{e}^{-t\sigma(A)}\} = \sigma_{\text{p}}(U_t(A))$. In this case the self-adjoint operators $U_t(A)$ are compact for $t > 0$.*

Proof. (a) Since the semigroup $\{U_t(A)\}_{t\geqslant 0}$ is strongly continuous, it is quasi-bounded with generator $A \in Q(M, \omega_0)$, see Section 1.3. Suppose that $U_t(A)$ is compact for $t > 0$ and self-adjoint. If $\omega_0 < \lambda$, then by Proposition 4.17, the resolvent

$$R_{-\lambda}(A) = (A + \lambda\mathbb{1})^{-1} = \int_0^\infty \mathrm{d}t\, \mathrm{e}^{-\lambda t} U_t(A), \tag{4.23}$$

is compact and the integral converges in the operator-norm topology. Since $U_t(A)$ is self-adjoint for all $t \geqslant 0$, the latter implies that $R_{-\lambda}(A)$ is also self-adjoint. For real λ we have $R_{-\lambda}(A)^* = R_{-\lambda}(A^*)$, and therefore $A = A^*$. Then the resolvent identity implies that $R_z(A)$ is compact for any $z \in \rho(A)$.

Conversely, assume that $A \geqslant -\omega_0\mathbb{1}$ is self-adjoint and has a compact resolvent. Since the numerical range $\mathrm{Nr}(A + \omega_0\mathbb{1}) \subseteq \mathbb{R}_0^+$, by Proposition 1.33 and Corollary 1.47, the operator $A \in \mathscr{H}(\pi/2, \omega_0)$, i.e., it is the generator of a holomorphic semigroup. Holomorphic semigroups are norm-continuous for $t > 0$, see Corollary 1.29, so Proposition 4.17 implies that $U_t(A)$ is compact for $t > 0$.

(b) Since $R_{-\lambda}(A) \in \mathcal{C}_\infty(\mathcal{H})$ for $\omega_0 < \lambda$, by Corollary 4.10, we have $\ker R_{-\lambda}(A) = \varnothing$, i.e., $R_{-\lambda}(A) \neq 0$ and $\{0\} \subset \sigma_{\text{ess}}(R_{-\lambda}(A))$. Since the operator $R_{-\lambda}(A)$ is non-zero and self-adjoint, the set $\sigma(R_{-\lambda}(A))\backslash\{0\}$ is nonempty, by consequence, also $\sigma(A)$ is nonempty. Furthermore,

$$\sigma(R_{-\lambda}(A))\backslash\{0\} = \sigma_{\text{p}}(R_{-\lambda}(A)) = \{(\lambda_n + \lambda)^{-1} : \lambda_n \in \sigma(A)\}, \tag{4.24}$$

see Corollary 4.10. By Corollary 2.25 and Corollary 2.26, the normalised eigenvectors $\{\varphi_n\}_{n\geqslant 1}$ corresponding to the point spectrum (4.24) of the self-adjoint compact operator $R_{-\lambda}(A)$ form an orthonormal basis in $\mathcal{H}$, i.e.,

$$\begin{aligned} u &= \sum_{n\geqslant 1} (u,\varphi_n)\varphi_n\,, \\ R_{-\lambda}(A)u &= \sum_{n\geqslant 1} (\lambda_n+\lambda)^{-1}(u,\varphi_n)\varphi_n\,, \end{aligned} \tag{4.25}$$

for any $u \in \mathcal{H}$, where $R_{-\lambda}(A)\varphi_n = (\lambda_n+\lambda)^{-1}\varphi_n$. Since $\varphi_n \in \operatorname{dom} A$, the latter implies that $\{\varphi_n\}$ are eigenvectors of the self-adjoint operator A: $A\varphi_n = \lambda_n\varphi_n$, with real eigenvalues $\{\lambda_n\}_{n\geqslant 1} = \sigma(A)$.

By virtue of Proposition 4.21(b), $\{\varphi_n\}$ are also eigenvectors of the elements of the semigroup $\{U_t(A)\}_{t\geqslant 0}$ with eigenvalues $\{\mathrm{e}^{-t\lambda_n}\}_{n\geqslant 1}$. Since $\{\varphi_n\}_{n\geqslant 1}$ is a basis in $\mathcal{H}$, we get the representation (4.22). By Proposition 4.21(a), the spectrum $\sigma(A)$ has no accumulation point in $\mathbb{R}$, and denoting $\lambda_- := \min_{n\geqslant 1}\lambda_n$, expression (4.22) becomes

$$U_t(A)u = \mathrm{e}^{-t\lambda_-}\sum_{n\geqslant 1} \mathrm{e}^{-t(\lambda_n-\lambda_-)}(u,\varphi_n)\varphi_n\,.$$

Hence,

$$\|U_t(A)u\| \leqslant \mathrm{e}^{-t\lambda_-}\|u\|, \quad u \in \mathcal{H}. \tag{4.26}$$

Since $A \in Q(M,\omega_0)$, the estimate (4.26) implies $-\lambda_- \leqslant \omega_0$. On the other hand, $U_t(A)\varphi_n = \mathrm{e}^{-t\lambda_n}\varphi_n$, i.e., $U_t(A)\varphi_- = \mathrm{e}^{-t\lambda_-}\varphi_-$, so that

$$\|U_t(A)\| \geqslant \mathrm{e}^{-t\lambda_-}, \quad \lim_{t\to+\infty}\frac{\ln\|U_t(A)\|}{t} \geqslant -\lambda_-\,,$$

and we infer that, $\lambda_- = -\omega_0$, by virtue of (4.26).

(c) If $\dim\mathcal{H} = \infty$, the point $\{0\}$ belongs to the spectrum of $U_t(A) \in \mathcal{C}_\infty(\mathcal{H})$ for $t > 0$. The structure of $\sigma(U_t(A))$ has already been established in Proposition 4.21(b). It only remains to classify the point $\{0\}$. By virtue of representation (4.22), the relation

$$0 = \|U_t(A)u\|^2 = \sum_{n=1}^{\infty} \mathrm{e}^{-2t\lambda_n}|(u,\varphi_n)|^2, \quad t \geqslant 0, \quad u \in \mathcal{H},$$

implies $u = 0$. Therefore, $\ker U_t(A) = \varnothing$, that is, $0 \notin \sigma_{\mathrm{p}}(U_{t>0}(A))$. Since $U_t(A)$ is self-adjoint, we have

$$\mathcal{H} = \ker U_t(A) \oplus \overline{\operatorname{ran} U_t(A)},$$

and by the remark just above, $\mathcal{H} = \overline{\operatorname{ran} U_t(A)}$. Hence, by Definition 2.14 (cf. Appendix A, Section A.1), we see that $0 \notin \sigma_{\mathrm{res}}(U_t(A))$. Consequently, the non-isolated point zero belongs to the continuous spectrum and we infer that in fact $\{0\} = \sigma_{\mathrm{cont}}(U_t(A))$. □

Since $\mathcal{C}_1(\mathcal{H}) \subset \mathcal{C}_\infty(\mathcal{H})$, our observations concerning compact semigroups, i.e. Propositions 4.17–4.22, are obviously applicable to Gibbs semigroups. More details on the *spectral mapping theorem* for semigroups one finds in Section A.9.

We conclude this section with a version of the *minimax principle* (2.20) for unbounded self-adjoint operators. It is useful for the identification of generators of compact self-adjoint semigroup.

Proposition 4.23. *For an unbounded self-adjoint operator A with a dense domain $\operatorname{dom} A \subset \mathcal{H}$, we define*

$$\mu(\mathcal{M}_n) := \sup_{\substack{u \in \mathcal{M}_n \subset \operatorname{dom} A \\ \|u\|=1}} (Au, u), \tag{4.27}$$

where $\mathcal{M}_n$ is a finite-dimensional subspace with $\dim \mathcal{M}_n = n \geqslant 1$. If we set

$$\mu_n(A) := \inf_{\mathcal{M}_n \subset \operatorname{dom} A} \mu(\mathcal{M}_n), \tag{4.28}$$

then

(a) *$\mu_1(A) > -\infty$ if and only if A is semibounded from below.*

(b) *If $\mu_1(A) > -\infty$ and $\mu_n(A) \to +\infty$, for $n \to \infty$, then $\sigma(A)$ is a point spectrum with finite multiplicities, that is, $\sigma(A) = \sigma_{\mathrm{p}}(A)$.*

(c) *If $\mu_1(A) > -\infty$ and $\sigma(A) = \sigma_{\mathrm{p}}(A)$ with finite multiplicities, then $\mu_n(A) = \lambda_n(A)$, where $\{\lambda_n(A)\}_{n \geqslant 1}$ is the set of eigenvalues in nondecreasing order and repeated according to the multiplicity.*

Proof. (a) If $A \geqslant -\gamma_A \mathbb{1}$, then $\mu_1(A) \geqslant -\gamma_A$, by definitions (4.27) and (4.28). Conversely,

$$\begin{aligned} -\gamma_A \leqslant \mu_1(A) &= \inf_{\mathcal{M}_1 \subset \operatorname{dom} A} \sup_{\substack{u \in \mathcal{M}_1 \subset \operatorname{dom} A \\ \|u\|=1}} (Au, u) \\ &= \inf_{\substack{u \in \operatorname{dom} A \\ \|u\|=1}} (Au, u). \end{aligned}$$

(b) If $\sigma(A) = \sigma_{\mathrm{p}}(A)$ and some $\lambda_s(A) \in \sigma_{\mathrm{p}}(A)$ has infinite multiplicity, the corresponding subspace $\mathcal{H}_s \subset \operatorname{dom} A$ of eigenvectors is infinite-dimensional. Hence, there is a sequence of subspaces $\mathcal{M}_m^{(s)} \subset \mathcal{H}_s$, with $\dim \mathcal{M}_m^{(s)} = m$, such that

$$\mu(\mathcal{M}_m^{(s)}) = \lambda_s(A) < +\infty, \quad \text{for } m \to \infty,$$

which contradicts the hypothesis.

Since for a self-adjoint operator A, the residual spectrum $\sigma_{\mathrm{res}}(A) = \varnothing$, it only remains to show that $\sigma_{\mathrm{cont}}(A) = \varnothing$ (Appendix A). Suppose that there is a $\lambda_0 \in \sigma_{\mathrm{cont}}(A)$. Then λ_0 would be a non-isolated point of *continuous nonconstancy* of the spectral measure $E_A(\lambda)$, that is,

$$E_A(\Delta_0(\varepsilon)) := E_A(\lambda_0 + \varepsilon) - E_A(\lambda_0 - \varepsilon) \neq 0, \quad \Delta_0(\varepsilon) = (\lambda_0 - \varepsilon, \lambda_0 + \varepsilon),$$

such that

$$\dim E_A(\Delta_0(\varepsilon))\mathcal{H} = \infty, \quad \text{for any } \varepsilon > 0. \tag{4.29}$$

Now, considering the subspaces $\mathcal{M}_m \subset E_A(\Delta_0(\varepsilon))\mathcal{H} \subset \operatorname{dom} A$, we get a contradiction, since

$$\mu(\mathcal{M}_m) = \sup_{\substack{u \in \mathcal{M}_m \subset \operatorname{dom} A \\ \|u\|=1}} \int_{\lambda_0-\varepsilon}^{\lambda_0+\varepsilon} (\mathrm{d}E_A(\lambda)u, u), \quad \lambda \leqslant |\lambda_0 + \varepsilon|,$$

is bounded for $m \to \infty$, whereas $\mu_m(A) \to \infty$ (4.28) for $m \to \infty$ by conditions in (b).

(c) Let $\{\lambda_n(A)\}_{n\geqslant 1}$ be the nondecreasing sequence of the eigenvalues of A repeated according to multiplicity, and let $\{\varphi_n\}$ be the corresponding family of orthonormal eigenvectors. We denote by $\mathcal{F}_n := [\varphi_1, \ldots, \varphi_n]$ the space spanned by the first n eigenvectors. From Definitions (4.27) and (4.28), we obviously get

$$\mu_n(A) = \inf_{\mathcal{M}_n \subset \operatorname{dom} A} \mu(\mathcal{M}_n) \leqslant \mu(\mathcal{F}_n) = \lambda_n(A). \tag{4.30}$$

To prove the equality in (4.30), suppose that there is a subspace $\tilde{\mathcal{M}}_n$ such that the strict inequality holds:

$$\mu(\tilde{\mathcal{M}}_n) < \mu(\mathcal{F}_n) = \lambda_n(A). \tag{4.31}$$

Since the subspace $\tilde{\mathcal{M}}_n$ is finite-dimensional, there exists a number N large enough such that $\tilde{\mathcal{M}}_n \subset \mathcal{F}_N$. Counting then the dimensions, we see that $\tilde{\mathcal{M}}_n \cap [\varphi_n, \varphi_{n+1}, \ldots, \varphi_N] \neq \varnothing$, i.e., $\tilde{\mathcal{M}}_n$ and the space spanned by $\{\varphi_j\}_{j=n}^N$ contain a common vector $\tilde{u}$. Therefore,

$$\mu(\tilde{\mathcal{M}}_n) \geqslant (A\tilde{u}, \tilde{u}) = \sum_{j=n}^{N} \lambda_j(A)|(\tilde{u}, \varphi_j)|^2 \geqslant \lambda_n(A),$$

which contradicts the strict inequality (4.31). □

Corollary 4.24. *If A and B are densely defined self-adjoint operators in $\mathcal{H}$ such that $A \leqslant B$, i.e., $(Au, u) \leqslant (Bu, u)$ for $u \in \operatorname{dom} B \subseteq \operatorname{dom} A$, then Proposition 4.23 (see (4.27) and (4.28)) implies that $\lambda_n(A) \leqslant \lambda_n(B)$ for $n \geqslant 1$.*

4.3 Generators

We start by observing that an operator $A \in Q(M, \omega_0)$ with a compact resolvent $R_\zeta(A)$, $\zeta \in \rho(A)$, is not necessarily the generator of a compact semigroup, Proposition 4.17. On the other hand, Proposition 4.22 (a) implies that the compactness of the resolvent $R_\zeta(A)$, $\zeta \in \rho(A)$, of a self-adjoint operator $A \geqslant -\omega_0 \mathbb{1}$ is a sufficient and necessary condition for the compactness of the semigroup $\{U_t(A)\}_{t\geqslant 0}$ generated by A.

Hence, the generators of self-adjoint compact semigroups are exhaustively characterised by the property $R_z(A) \in \mathcal{C}_\infty(\mathcal{H})$ for some $z \in \rho(A)$. Based on Proposition 4.22, we can formulate a sufficient condition on the generator of a self-adjoint strongly continuous semigroup ensuring that this semigroup is a Gibbs semigroup.

Proposition 4.25. *A self-adjoint operator $A \geqslant -\omega_0 \mathbb{1}$ is the generator of a Gibbs semigroup $\{G_t(A)\}_{t\geqslant 0}$ if the resolvent $R_\zeta(A) \in \mathcal{C}_p(\mathcal{H})$ for some $\zeta \in \rho(A)$ and $1 \leqslant p < \infty$.*

Proof. Since A is the generator of a strongly continuous semigroup, that is, $A \in Q(M, \omega_0)$, by hypothesis and by Proposition 4.22, we get the representation (4.22). Hence, $\mathrm{e}^{-t\lambda_n} = s_n(\mathrm{e}^{-tA})$ are the singular values of the compact semigroup $\{\mathrm{e}^{-tA}\}_{t\geqslant 0}$ and, by Lemma 2.36, we have

$$\|\mathrm{e}^{-tA}\|_1 = \sum_{n=1}^{\infty} \mathrm{e}^{-t\lambda_n}, \quad t > 0. \tag{4.32}$$

By expression (4.25), we get that $(\lambda_n + \lambda)^{-1} = s_n(R_{-\lambda}(A))$ are the singular values of the self-adjoint operator $R_{-\lambda}(A) \in \mathcal{C}_p(\mathcal{H})$, $\lambda > \omega_0$. For the resolvent $R_{-\lambda}(A) \in \mathcal{C}_p(\mathcal{H})$, its $\|\cdot\|_p$-norm is well defined, see Definition 2.50, and the series

$$\left[\sum_{n=1}^{\infty} (\lambda_n + \lambda)^{-p}\right]^{1/p} = \|R_{-\lambda}(A)\|_p$$

is convergent. This implies that the sum in (4.32) is convergent for any $t > 0$. □

This observation motivates the introduction of the following class of generators of Gibbs semigroups.

Definition 4.26. A quasi-m-sectorial operator A with vertex $\gamma = -\omega_0$ and with semi-angle α is called a *p-generator* if for some $\zeta_0 \in \rho(A)$, the resolvent $R_{\zeta_0}(A)$ belongs to the von Neumann-Schatten class $\mathcal{C}_p(\mathcal{H})$ for a finite $p \geqslant 1$.

A self-adjoint operator A satisfying the criterion of Proposition 4.25 is a p-generator.

Proposition 4.27. *Any p-generator A is the generator of a holomorphic Gibbs semigroup $\{G_z(A)\}_{z\in S_{\pi/2-\alpha}}$.*

Proof. Proposition 1.46 implies that the operator $A_{-\omega_0} = A + \omega_0 \mathbb{1}$ is the generator of a holomorphic semigroup, that is, $A_{-\omega_0} \in \mathscr{H}(\pi/2 - \alpha, 0)$ and $\{U_z(A_{-\omega_0})\}_{z\in S_{\pi/2-\alpha}}$ is a contraction holomorphic semigroup in the open sector $S_{\pi/2-\alpha} = \{z \in \mathbb{C} : |\arg z| < \pi/2 - \alpha\}$. Hence, for this semigroup we can use representation (1.67)

$$U_z(A_{-\omega_0}) = \frac{1}{2\pi i} \int_\Gamma \mathrm{d}\zeta\, \mathrm{e}^{-z\zeta} (\zeta \mathbb{1} - A_{-\omega_0})^{-1}, \tag{4.33}$$

where Γ is a positively oriented contour enclosing the sector $S_\alpha \supset \mathrm{Nr}(A_{-\omega_0})$, i.e., $\Gamma \subset \rho(A)$. Since the resolvent identity implies that $R_\zeta(A_{-\omega_0}) \in \mathcal{C}_p(\mathcal{H})$ for any $\zeta \in \rho(A)$, whenever $R_{\zeta_0}(A_{-\omega_0}) \in \mathcal{C}_p(\mathcal{H})$, we assume that $\zeta_0 \in \Gamma$. Then formula (4.33) yields

$$\begin{aligned} \|U_z(A_{-\omega_0})\|_p \leqslant \frac{1}{2\pi} \int_\Gamma |\mathrm{d}\zeta|\, \mathrm{e}^{-\mathfrak{Re}(z\zeta)} \|R_{\zeta_0}(A_{-\omega_0})\|_p \\ \times \left(1 + |\zeta_0 - \zeta| \|R_\zeta(A_{-\omega_0})\|\right), \quad z \in S_{\pi/2-\alpha}. \end{aligned} \tag{4.34}$$

By estimate (1.83),

$$\|R_\zeta(A_{-\omega_0})\| \leqslant \frac{M_\varepsilon}{|\zeta|}, \tag{4.35}$$

for $\zeta \in \mathbb{C}\backslash S_{\alpha+\varepsilon}$ and $\varepsilon > 0$. Therefore, we can choose the contour $\Gamma \subset \mathbb{C}\backslash S_{\alpha+\varepsilon}$ in such a way that $\|R_{\zeta\in\Gamma}(A_{-\omega_0})\| \leqslant L$, and there is $\delta > 0$ ensuring that

$$\mathfrak{Re}(z\zeta) \geqslant \delta|z||\zeta|, \quad \text{for } z \in S_{\pi/2-\alpha},\ \zeta \in \Gamma \cap \mathbb{C}_+. \tag{4.36}$$

The estimates (4.34)–(4.36) yield

$$\|U_z(A_{-\omega_0})\|_p \leqslant \tilde{L}\ \|R_{\zeta_0}(A_{-\omega_0})\|_p, \quad z \in S_{\pi/2-\alpha}.$$

By Proposition 4.3, this estimate implies that $U_z(A_{-\omega_0}) =: G_z(A_{-\omega_0}) \in \mathcal{C}_1(\mathcal{H})$ is a Gibbs semigroup for $z \in S_{\pi/2-\alpha}$.

Since by Corollary 1.28, the holomorphic semigroup $U_z(A_{-\omega_0})$ is $\|\cdot\|$-differentiable in the sector $S_{\pi/2-\alpha}$, and since $\mathcal{C}_1(\mathcal{H})$ is the $*$-ideal in $\mathcal{L}(\mathcal{H})$, for $z, z+\Delta z \in S_{\pi/2-\alpha}$ the derivative

$$\begin{aligned} \partial_z G_z(A_{-\omega_0}) &= \lim_{\Delta z\to 0} \frac{1}{\Delta z}\left(G_{z+\Delta z}(A_{-\omega_0}) - G_z(A_{-\omega_0})\right) \\ &= \lim_{\Delta z\to 0} \frac{1}{\Delta z}\left(G_{z/2+\Delta z}(A_{-\omega_0}) - G_{z/2}(A_{-\omega_0})\right) G_{z/2}(A_{-\omega_0}) \\ &= -A\, G_z(A) \in \mathcal{C}_1(\mathcal{H}), \end{aligned} \tag{4.37}$$

exists in the trace-norm by the $\|\cdot\|_1$-continuity of multiplication, Proposition 2.78. Hence, the Gibbs semigroup $\{G_z(A_{-\omega_0})\}_{z\in S_{\pi/2-\alpha}}$ is $\|\cdot\|_1$-holomorphic in the sector $S_{\pi/2-\alpha}$. This implies that the same is true for the Gibbs semigroup with generator A,

$$\{G_z(A)\}_{z\in S_{\pi/2-\alpha}} := \{\mathrm{e}^{\omega_0 z} G_z(A_{-\omega_0})\}_{z\in S_{\pi/2-\alpha}},$$

and completes the proof. □

Remark 4.28. The example in Remark 4.4 shows that the conditions of Proposition 4.27 are suitable for strongly continuous and immediate Gibbs semigroups. Indeed, the self-adjoint operator A with $\sigma(A) = \{\ln(n+1)\}_{n\geq 1}$ is *not* a p-generator. The C_0-semigroup $\{T_t\}_{t\geqslant 0}$ defined by (4.3) is only an eventually Gibbs semigroup with the threshold $t_0 = 1$. On the other hand, the semigroup $\{G_t(A)\}_{t\geqslant 0}$ defined as:

$$\{G_{t=0}(A) = \mathbb{1}\} \vee \{G_t(A) := T_{t+1}\}_{t>0},$$

is an immediate, but *degenerate* Gibbs semigroup since $\lim_{t\downarrow 0} G_t(A) \neq \mathbb{1}$.

To continue we mention a contact with construction of generators by sesquiliniar forms. We develop this approach below (see, e.g., Section 5.5) in a framework of perturbations and in analytic continuations of forms and the associated operators.

Remark 4.29. Let A be an m-sectorial operator with vertex $\gamma = 0$ and semi-angle $\alpha \in [0, \pi/2)$, see Definition 1.41. We recall that there is a one-to-one correspondence between the set of all m-sectorial operators and the set of all densely defined, closed sesquilinear sectorial forms. We denote the form corresponding to the operator A by

$$a[u,v] = \mathfrak{Re}\, a[u,v] + i\,\mathfrak{Im}\, a[u,v]\,, \quad u,v \in \operatorname{dom} a\,.$$

Here $\mathfrak{Re}\, a := \frac{1}{2}(a+a^*)$ and $\mathfrak{Im}\, a := \frac{1}{2i}(a-a^*)$ are the real and the imaginary parts of a. The adjoint form $a^*[u,v] := \overline{a[v,u]}$, $\operatorname{dom} a^* = \operatorname{dom} a$, which corresponds to the operator A^*, is in turn densely defined, closed and sectorial. The form a is called symmetric if $a[u,v] = a^*[u,v]$. Since by definition the symmetric form $\mathfrak{Re}\, a[u,v]$ is closed and $\gamma = 0$, the *representation theorem* for non-negative forms shows that it defines a non-negative self-adjoint operator $A_R := \mathfrak{Re}\, A \geqslant 0$ with $\operatorname{dom} A_R \subset \operatorname{dom} a$, which is called the *real* part of A. Recall that $A_R = \frac{1}{2}(A+A^*)$, if $A \in \mathcal{L}(\mathcal{H})$, but this is *not* true in general for unbounded operators. Since the sum $(a+b)[u,v]$ of the closed forms corresponding to two m-sectorial operators A and B is also closed, one calls the m-sectorial operator $(A \dot{+} B)$ associated via representation theorem with $(a+b)$, the *form-sum* of the operators A and B. This definition yields for unbounded case a non-negative self-adjoint operator $\mathfrak{Re}\, A := \frac{1}{2}(A \dot{+} A^*)$, with $\operatorname{dom} \sqrt{\mathfrak{Re}\, A} = \operatorname{dom} a$.

We refer to Section 5.5 for more details about sesquilinear sectorial forms and their properties that we use below in this chapter.

By Proposition 1.46, any m-sectorial operator A with vertex $\gamma = 0$ and semi-angle $\alpha \in [0, \pi/2)$ is the generator of a contraction holomorphic semigroup: $A \in \mathscr{H}(\pi/2 - \alpha, 0)$. Then one gets a relation between generators A and $\mathfrak{Re}\, A$ in the case of the Gibbs semigroups.

Proposition 4.30. *Let $A \in \mathscr{H}(\pi/2 - \alpha, 0)$ be an m-sectorial operator with real part $\mathfrak{Re}\, A \geqslant 0$. If $\mathrm{e}^{-t\,\mathfrak{Re}\, A} \in \mathcal{C}_1(\mathcal{H})$ for $t > 0$, then $\mathrm{e}^{-tA} \in \mathcal{C}_1(\mathcal{H})$ for $t \in S_\theta$, $\theta = \pi/2 - \alpha$, and*

$$\begin{aligned} \|\mathrm{e}^{-tA}\|_1 &\leqslant \cdots \leqslant \|\mathrm{e}^{-tA/2^p}\|_{2^p}^{2^p} \leqslant \|\mathrm{e}^{-tA/2^{p+1}}\|_{2^{p+1}}^{2^{p+1}} \\ &\leqslant \cdots \leqslant \lim_{p\to\infty} \|\mathrm{e}^{-tA/2^p}\|_{2^p}^{2^p} = \|\mathrm{e}^{-t\,\mathfrak{Re}\, A}\|_1 \,. \end{aligned} \tag{4.38}$$

Proof. First we note that since $\mathfrak{Re}\, A \geqslant 0$ and $\mathrm{e}^{-t\,\mathfrak{Re}\, A} \in \mathcal{C}_1(\mathcal{H})$,

$$(\mathbb{1} + \mathfrak{Re}\, A)^{-1} = \int_0^\infty \mathrm{d}t\, \mathrm{e}^{-t(\mathbb{1}+\mathfrak{Re}\, A)} \in \mathcal{C}_\infty(\mathcal{H}), \tag{4.39}$$

because the right-hand side operator can be arbitrary well approximated in the operator norm by finite-rank operators. By Remark 5.65, this implies that m-sectorial generators A and A^* also have compact resolvents. Since $\overline{\operatorname{dom} a} = \mathcal{H}$ for the sesquilinear form a corresponding to A, this ensures that the Trotter product formula [1]

$$\|\cdot\|\text{-}\lim_{n\to\infty}(\mathrm{e}^{-tA^*/2n}\mathrm{e}^{-tA/2n})^n = \mathrm{e}^{-t(A^*\dot{+}A)/2}, \quad t \in S_\theta , \tag{4.40}$$

converges in the operator-norm topology to the semigroup generated by half of the form-sum of A and A^*. By virtue of Remark 4.29, it coincides with non-negative self-adjoint generator $\mathfrak{Re}\, A = (A \dot{+} A^*)/2 \geqslant 0$.

Secondly, by Lemma 2.56, the operator-norm convergence of the Trotter product formula (4.40) implies the convergence of all singular values of the compact operators involved in expression (4.40), i.e., for $j = 1, 2, 3, \ldots$, we have

$$s_j((\mathrm{e}^{-tA^*/2n}\mathrm{e}^{-tA/2n})^n) \xrightarrow[n\to\infty]{} s_j(\mathrm{e}^{-t\,\mathfrak{Re}\,A}) . \tag{4.41}$$

Furthermore, by the inequalities (3.26) in Corollary 3.8,

$$S_k(p) := \sum_{j=1}^{k} s_j(\mathrm{e}^{-tA/2^p})^{2^p} \leqslant \sum_{j=1}^{k} s_j(\mathrm{e}^{-tA/2^{p+1}})^{2^{p+1}} =: S_k(p+1) . \tag{4.42}$$

The limit (4.41) yields the convergence of any finite sum in (4.42),

$$S_k := \lim_{p\to\infty} S_k(p) = \sum_{j=1}^{k} s_j(\mathrm{e}^{-t\,\mathfrak{Re}\,A}) , \tag{4.43}$$

thus, (4.42) and (4.43) give

$$S_k(p) \leqslant S_k(p+1) \leqslant \cdots \leqslant S_k , \quad k = 1, 2, \ldots .$$

Hence, the series $\{S_k(p)\}_{k\geqslant 1, p\geqslant 0}$ is monotonically increasing in k and p. Since $\mathrm{e}^{-t\,\mathfrak{Re}\,A} \in \mathcal{C}_1(\mathcal{H})$ for $t > 0$, this series is bounded by $\sup_{k\geqslant 1} S_k = \|\mathrm{e}^{-t\,\mathfrak{Re}\,A}\|_1$, see (4.43). In particular,

$$\lim_{k\to\infty} S_k(0) = \|\mathrm{e}^{-tA}\|_1 \tag{4.44}$$

and

$$\lim_{k\to\infty} S_k(p) = \|\mathrm{e}^{-tA/2^p}\|_{2^p}^{2^p} . \tag{4.45}$$

Therefore, taking the limit $k \to \infty$ in the inequalities

$$S_k(0) \leqslant \cdots \leqslant S_k(p) \leqslant S_k(p+1) \leqslant \cdots \leqslant S_k \leqslant \|\mathrm{e}^{-t\,\mathfrak{Re}\,A}\|_1 ,$$

we get in view of the limits (4.44) and (4.45) the announced inequalities (4.38) and

$$\lim_{p\to\infty} \|\mathrm{e}^{-tA/2^p}\|_{2^p}^{2^p} \leqslant \|\mathrm{e}^{-t\,\mathfrak{Re}\,A}\|_1 . \tag{4.46}$$

[1] **See Chapter 5 (Section 5.5, Propositon 5.75).**

On the other hand, by (4.42), we also have

$$\|e^{-tA/2^p}\|_{2^p}^{2^p} \geqslant S_k(p),$$

or, see (4.43),

$$\lim_{p\to\infty} \|e^{-tA/2^p}\|_{2^p}^{2^p} \geqslant \sup_{k\geqslant 1} S_k = \|e^{-t\,\mathfrak{Re}\,A}\|_1\,,$$

which together with the *opposite* inequality (4.46) imply the equality in (4.38). □

Corollary 4.31. *If the generator A satisfying the conditions of Proposition 4.30 is normal, all the inequalities in (4.38) become equalities.*

Proof. If the generator A is normal, then the operator e^{-tA} is also normal and thus $s_j(e^{-tA}) = |\lambda_j(e^{-tA})|$. Since the corresponding spectral representations take the form

$$A = \sum_{j=1}^{\infty} \lambda_j(A)P_j\,, \quad e^{-tA} = \sum_{j=1}^{\infty} e^{-t\lambda_j(A)}P_j\,, \quad \mathfrak{Re}\,A = \sum_{j=1}^{\infty} \mathfrak{Re}\,\lambda_j(A)P_j\,,$$

where P_j are the orthogonal spectral projectors associated with the eigenvalues $\lambda_j(A)$, we obtain:

$$\begin{aligned} s_j(e^{-tA}) &= |\lambda_j(e^{-tA})| = |e^{-t\lambda_j(A)}| \\ &= e^{-t\,\mathfrak{Re}\,\lambda_j(A)} = e^{-t\lambda_j(\mathfrak{Re}\,A)} = s_j(e^{-t\,\mathfrak{Re}\,A}), \end{aligned}$$

which implies $\|e^{-tA}\|_1 = \|e^{-t\,\mathfrak{Re}\,A}\|_1$ in (4.38). □

Corollary 4.32. *An m-sectorial operator A with positive real part $\mathfrak{Re}\,A \geqslant 0$ is the generator of a contraction holomorphic Gibbs semigroup $\{G_z(A)\}_{z\in S_{\pi/2-\alpha}}$ if $e^{-t\,\mathfrak{Re}\,A} \in \mathcal{C}_1(\mathcal{H})$, for $t > 0$.*

Proof. By Proposition 1.46, the operator $A \in \mathscr{H}(\pi/2-\alpha, 0)$, that is, the semigroup $\{e^{-zA}\}_{z\in S_{\pi/2-\alpha}}$ is holomorphic, and it is a contraction. By Proposition 4.30, we have $e^{-tA} \in \mathcal{C}_1(\mathcal{H})$, $t > 0$. Since $\mathcal{C}_1(\mathcal{H})$ is a $*$-ideal in $\mathcal{L}(\mathcal{H})$, in view of the semigroup property, we get $e^{-zA} \in \mathcal{C}_1(\mathcal{H})$ for $z \in S_{\pi/2-\alpha}$. By the same reason, the $\|\cdot\|_1$-derivative of e^{-zA} exists in this sector. Hence, $\{e^{-zA} = G_z(A)\}_{z\in S_{\pi/2-\alpha}}$ is a contraction $\|\cdot\|_1$-holomorphic Gibbs semigroup. □

Corollary 4.33. *Let A be an m-sectorial operator with real part $\mathfrak{Re}\,A \geqslant 0$. If $e^{-t\,\mathfrak{Re}\,A} \in \mathcal{C}_1(\mathcal{H})$ for $t > 0$, then the Trotter product formula (4.40) converges in the trace-norm topology:*

$$\|\cdot\|_1\text{-}\lim_{p\to\infty} (e^{-tA^*/2^p} e^{-tA/2^p})^{2^{p-1}} = e^{-t\,\mathfrak{Re}\,A}\,, \quad t \in S_\theta\,. \tag{4.47}$$

Proof. Let $t > 0$. By the definition of singular values (2.11) and Definition 2.50 one gets

$$\|\mathrm{e}^{-tA/2^p}\|_{2^p}^{2^p} = \|(\mathrm{e}^{-tA^*/2^p}\mathrm{e}^{-tA/2^p})^{2^{p-1}}\|_1 .$$

Since by (4.38) the left-hand side converges, we obtain

$$\lim_{p\to\infty} \|(\mathrm{e}^{-tA^*/2^p}\mathrm{e}^{-tA/2^p})^{2^{p-1}}\|_1 = \|\mathrm{e}^{-t\,\mathfrak{Re}\,A}\|_1 . \tag{4.48}$$

On the other hand, (4.40) yields the operator-norm convergence

$$\lim_{p\to\infty} \|(\mathrm{e}^{-tA^*/2^p}\mathrm{e}^{-tA/2^p})^{2^{p-1}} - \mathrm{e}^{-t\,\mathfrak{Re}\,A}\| = 0. \tag{4.49}$$

Then by virtue of Proposition 2.69 (or Corollary 2.76) the limits (4.48) and (4.49) imply assertion (4.47). □

This result is also a corollary of a more general Proposition 5.78 about Trotter product formula for Gibbs semigroups.

4.4 $\mathcal{P}$-perturbations of generators

In this section we study a class of *infinitesimally* small *unbounded* perturbations, called $\mathcal{P}$-perturbations.

Definition 4.34. Recall that a *closed* operator B belongs to the class of $\mathcal{P}$-perturbations of the C_0-semigroup generator A if

$$\text{(i)} \qquad \operatorname{dom} B \supseteq \bigcup_{t>0} U_t(A)\mathcal{H}\,, \tag{4.50}$$

$$\text{(ii)} \qquad \int_0^1 \mathrm{d}t\,\|B\,U_t(A)\| < \infty\,. \tag{4.51}$$

We denote this property as $B \in \mathcal{P}(A)$, or abbreviate to $B \in \mathcal{P}$, if there is no danger of confusion.

Remark 4.35. For a closed B the condition (i) implies that the product $BU_t(A) \in \mathcal{L}(\mathcal{H})$, for all $t > 0$ by the *closed graph* theorem. Note that then the function $t \mapsto B\,U_t(A)$ is strongly continuous away from zero. This implies that the function $t \mapsto \|B\,U_t(A)\|$ is measurable, locally bounded, and, hence, strongly integrable on any finite interval $I \subset \mathbb{R}^+$. The integral in (ii) is understood in the sense $\epsilon \to 0^+$ for the lower limit. Finally, since I is a separable set, the range of any strongly continuous operator-valued function is a *separable* subset of $\mathcal{L}(\mathcal{H})$. Hence, the corresponding integral exists in the sense of Bochner.

Remark 4.36. In general, for a quasi-bounded semigroup $\{U_t(A)\}_{t\geqslant 0}$ the image $U_t(A)\mathcal{H}$ is not included in $\operatorname{dom} A$, for $t \geqslant 0$, see discussion in Chapter 1. However, it is true that $U_t(A) : \operatorname{dom} A \to \operatorname{dom} A$, for $t > 0$ by Propositions 1.1 and 1.5. If instead of (4.50), we consider the domain

$$\mathcal{D} := \bigcup_{t>0} U_t(A)\operatorname{dom} A\,, \tag{4.52}$$

then $\mathcal{D} \supseteq \operatorname{dom} A$, and for $B \in \mathcal{P}(A)$ one also has $\operatorname{dom} B \supseteq \mathcal{D}$. Since the domain $\mathcal{D}$ is by construction invariant under the semigroup $\{U_t(A)\}_{t\geqslant 0}$, for any $u \in \operatorname{dom} A$ there exists a sequence $\{u_n\}_{n\geqslant 1} \subset \mathcal{D}$ such that $\lim_{n\to\infty} \|u - u_n\| = 0$, i.e., $\mathcal{D}$ is a core of A, see Proposition 1.13. Therefore, $\operatorname{dom} B$ contains a core $A = \mathcal{D}$. The next assertion says more.

Lemma 4.37. *Let $B \in \mathcal{P}(A)$ for generator $A \in \mathcal{Q}(M, \omega_0)$ of C_0-continuous quasi-bounded semigroup $\{U_t(A)\}_{t\geqslant 0}$. Then*

$$\operatorname{dom} B \supseteq \operatorname{dom} A, \tag{4.53}$$

and for any $\varepsilon > 0$ there exists $\lambda_\varepsilon > \omega_0$ such that

$$\|BR_{-\lambda}(A)\| < \varepsilon\,, \tag{4.54}$$

for all $\lambda > \lambda_\varepsilon$.

Proof. By virtue of Proposition 1.12 and condition (4.50) of Definition 4.34, we have:

$$\|BU_t(A)\| \leqslant \|BU_{t=1}(A)\,\|\,M\,\mathrm{e}^{\omega_0(t-1)}\,, \quad t > 1.$$

Hence, by condition (4.51) of Definition 4.34,

$$\int_0^\infty \mathrm{d}t\,\|BU_t(A)\mathrm{e}^{-\lambda t}\| < \infty,$$

for all $\lambda > \omega_0$. Moreover, for any $\varepsilon > 0$, there is a constant $\lambda_\varepsilon > \omega_0$ such that

$$\left\|\int_0^\infty \mathrm{d}t\,BU_t(A)\mathrm{e}^{-\lambda t}u\right\| < \varepsilon\,\|u\|, \quad u \in \mathcal{H}, \tag{4.55}$$

for all $\lambda > \lambda_\varepsilon$. Since

$$\int_0^\infty \mathrm{d}t\,U_t(A)\mathrm{e}^{-\lambda t}u = (A + \lambda\mathbb{1})^{-1}u, \quad u \in \mathcal{H},$$

cf. (1.46), the closedness of B and the bound (4.55) yield $R_{-\lambda}(A)u \in \operatorname{dom} B$. Thus, they imply (4.53) and (4.54). □

Corollary 4.38. *By virtue of* (4.54)

$$\|Bu\| \leqslant \varepsilon\lambda_\varepsilon\|u\| + \varepsilon\|Au\|, \; u \in \operatorname{dom} A \subset \operatorname{dom} B,$$

for any $\varepsilon > 0$. *This means that* $B \in \mathcal{P}(A)$ *is an infinitesimally Kato-small perturbation with the relative* A*-bound equals to* $b = 0^+$, *Section* 1.7. *By Remark* 1.49 *and Definition* 1.50 *we infer that* $B \in \mathcal{P}_{0^+}(A)$, *or that the class of unbounded perturbations* $\mathcal{P}(A) \subset \mathcal{P}_{0^+}(A)$. *Taking into account* (1.102) *one gets the hierarchy*

$$\mathcal{P}_0 \subset \mathcal{P}(A) \subset \mathcal{P}_{0^+}(A) \subset \mathcal{P}_b(A)\,. \tag{4.56}$$

Here $\mathcal{P}_0$ *denotes the class of bounded perturbations* $\mathcal{P}_{b=0}(A)$.

If $B \in \mathcal{P}(A)$, *then the operator* $H := A + B$ *has* $\operatorname{dom} H = \operatorname{dom} A$ *and it is closed, by the stability of closeness under relatively small perturbations* $\mathcal{P}_{b<1}(A)$.

In Section 5 we consider other classes of *infinitesimally* small unbounded perturbations of generators such that the relative bound $b = 0^+$. One of this classes consists of the *relatively compact* perturbations (Remark 5.31), while another one consists of perturbations verifying certain *fractional* power conditions, Remark 5.48.

Before proceeding with the construction of C_0-semigroups for perturbations of class $\mathcal{P}$ we need a preparatory lemma.

Lemma 4.39. *Suppose the densely defined operator* A *is the generator of a* C_0*-semigroup* $\{U_t(A)\}_{t\geqslant 0}$ *on a Hilbert space* $\mathcal{H}$ *and* $H = A + B$ *is the generator of the* C_0*-semigroup* $\{U_t(A+B)\}_{t\geqslant 0}$ *with a bounded perturbation* $B \in \mathcal{P}_0$. *Then this semigroup satisfies the following two integral equations*:

$$U_t(A+B)u = U_t(A)u - \int_0^t \mathrm{d}s\, U_{t-s}(A+B)\, B\, U_s(A)\, u, \tag{4.57}$$

$$U_t(A+B)u = U_t(A)u - \int_0^t \mathrm{d}s\, U_{t-s}(A)\, B\, U_s(A+B)\, u, \tag{4.58}$$

for every $t \geqslant 0$ *and* $u \in \mathcal{H}$. *Taking advantage of the uniform boundedness of integrands the integrals are understood in the operator-norm sense.*

Proof. Since $\operatorname{dom} A = \operatorname{dom}(A+B)$ is invariant under the perturbed $\{U_t(A+B)\}_{t\geqslant 0}$ and nonperturbed $\{U_t(A)\}_{t\geqslant 0}$ semigroups, the function

$$[0,t] \ni s \mapsto W_s^{(1)} v := U_{t-s}(A+B)\, U_s(A) v\ , \tag{4.59}$$

is continuously differentiable for any $v \in \operatorname{dom} A$. By the fundamental semigroup equation (1.30) one obtains

$$\partial_s W_s^{(1)} v := U_{t-s}(A+B)\, B\, U_s(A) v\,. \tag{4.60}$$

Then after integration for any $v \in \operatorname{dom} A$, definitions (4.59) and (4.60) yield

$$U_t(A+B)v - U_t(A)v = -\int_0^t \mathrm{d}s\, U_{t-s}(A+B)\, B\, U_s(A)\, v\,. \tag{4.61}$$

Since dom A is dense in $\mathcal{H}$, the boundedness of operators in the left- and right-hand sides of (4.61) ensures that the integral equation (4.57) holds on $\mathcal{H}$ in the operator-norm sense.

The same arguments, applied to the family

$$[0,t] \ni s \mapsto W_s^{(2)} v := U_{t-s}(A)\, U_s(A+B) v \in \mathcal{H}, \tag{4.62}$$

yield the integral equation (4.58). □

The integral formulae (4.57), (4.58) are a source of the *semigroup-based* method for the construction of perturbed semigroups. In contrast to the *resolvent-based* method, it involves the *Dyson-Phillips series*, see Notes in Section 1.8 and in Section 4.7.

In the next Proposition 4.41 we use the *semigroup-based* method to treat the $\mathcal{P}$-perturbations of quasi-bounded semigroups. According to Lemma 4.39, in order to construct the perturbed semigroup $\{U_t(A+B)\}_{t\geqslant 0}$ one can use the solutions of either of the integral equations (4.57), or (4.58), and then check that they verify all the properties of C_0-semigroups.

Remark 4.40. To proceed, we rewrite (4.57) in the operator-valued function space

$$\mathfrak{F}_t := C([0,t], \mathcal{L}(\mathcal{H})), \tag{4.63}$$

such that for each element $\Phi \in \mathfrak{F}_t$ and any $u \in \mathcal{H}$ the function $[0,t] \ni s \mapsto \Phi_s u$ is continuous in the topology of $\mathcal{H}$. This space endowed with the norm topology

$$\|\Phi\|_{\infty,t} := \sup_{s\in[0,t]} \|\Phi_s\|, \quad \Phi \in \mathfrak{F}_t\,,$$

is a Banach space. Motivated by (4.57) we define on (4.63) the operator $\mathbb{V}^{(1)}$: $\Phi \mapsto \mathbb{V}^{(1)}\Phi$ such that for each $r \in [0,t]$ we have

$$(\mathbb{V}^{(1)}\Phi)(r) := \int_0^r \mathrm{d}s\, \Phi_{r-s}\, B\, U_s(A). \tag{4.64}$$

By virtue of condition (4.51), the *Volterra-type* operator (4.64) on the space $\mathfrak{F}_{t\leqslant 1}$ is a bounded operator:

$$\|\mathbb{V}^{(1)}\|_{\mathfrak{F}_t} = \sup_{\{\Phi\in\mathfrak{F}_t : \|\Phi\|_{\infty,t}=1\}} \|\mathbb{V}^{(1)}\Phi\|_{\infty,t} \leqslant \int_0^t \mathrm{d}s\, \|BU_s(A)\|. \tag{4.65}$$

Moreover, there exists $\tau > 0$ such that in the space $\mathfrak{F}_{\tau<1}$ the norm (4.65) obeys the estimate

$$\|\mathbb{V}^{(1)}\|_{\mathfrak{F}_\tau} \leqslant \int_0^\tau \mathrm{d}s\, \|BU_s(A)\| =: \xi < 1. \tag{4.66}$$

Therefore, $(\mathbb{1} + \mathbb{V}^{(1)}) \in \mathcal{L}(\mathfrak{F}_\tau)$ is invertible and the solution of the integral equation (4.57) for $t \in [0,\tau]$ can be defined in the space $\mathfrak{F}_\tau$ as

$$U_{(\cdot)}(A+B) = (\mathbb{1} + \mathbb{V}^{(1)})^{-1}\, U_{(\cdot)}(A). \tag{4.67}$$

Similarly, we define on $\mathfrak{F}_t$ the operator $\mathbb{V}^{(2)} : \Phi \mapsto \mathbb{V}^{(2)}\Phi$ such that for each $r \in [0,t]$

$$(\mathbb{V}^{(2)}\Phi)(r) := \int_0^r \mathrm{d}s\, U_{r-s}(A)\, B\, \Phi_s\,. \tag{4.68}$$

Then by condition (4.51), the operator (4.68) in the space $\mathfrak{F}_{t\leqslant 1}$ is bounded

$$\|\mathbb{V}^{(2)}\|_{\mathfrak{F}_t} = \sup_{\{\Phi\in\mathfrak{F}_t:\|\Phi\|_{\infty,t}=1\}} \|\mathbb{V}^{(2)}\Phi\|_{\infty,t} \leqslant \int_0^t \mathrm{d}s\, \|U_s(A)B\|, \tag{4.69}$$

and there exists $\tau > 0$ such that in the space $\mathfrak{F}_{\tau<1}$ the norm (4.69) obeys the estimate

$$\|\mathbb{V}^{(2)}\|_{\mathfrak{F}_\tau} \leqslant \int_0^\tau \mathrm{d}s\, \|BU_s(A)\| =: \eta < 1. \tag{4.70}$$

Consequently, $(\mathbb{1} + \mathbb{V}^{(2)}) \in \mathcal{L}(\mathfrak{F}_\tau)$ is invertible and we can define the solution of the integral equation (4.58) in the space $\mathfrak{F}_\tau$ by

$$U_{(\cdot)}(A+B) = (\mathbb{1} + \mathbb{V}^{(2)})^{-1}\, U_{(\cdot)}(A)\,. \tag{4.71}$$

Proposition 4.41. *Let* $B \in \mathcal{P}(A)$ *for a generator* $A \in Q(M,\omega_0)$*. Then* $\operatorname{dom} A \subset \operatorname{dom} B$ *and the operator sum* $H := A + B$*, with* $\operatorname{dom} H = \operatorname{dom} A$*, is the generator of a strongly continuous quasi-bounded semigroup on* $\mathcal{H}$*.*

Proof. Let $t \in [0,\tau]$, where τ is defined by condition (4.66). Then by definition (4.64) the right-hand side of representation (4.67) is the convergent in $\mathfrak{F}_\tau$ infinite series

$$((\mathbb{1} + \mathbb{V}^{(1)})^{-1}\, U_{(\cdot)}(A))(t) = \sum_{n=0}^{\infty} S_n(t), \tag{4.72}$$

where the terms in the sum (4.72) are defined by the recurrence relation

$$\begin{aligned} S_0(t) &= U_t(A), \\ S_n(t) &= -\int_0^t \mathrm{d}s\, S_{n-1}(t-s)BU_s(A), \quad n \geqslant 1. \end{aligned} \tag{4.73}$$

Indeed, since by (4.73) the operators $S_{n\geqslant 1}(t)$ are the n-fold operator-norm convergent Bochner integrals

$$\begin{aligned} S_n(t) = \int_0^t \mathrm{d}s_1 \int_0^{s_1} \mathrm{d}s_2 \ldots \int_0^{s_{n-1}} \mathrm{d}s_n \\ U_{t-s_1}(A)(-B)U_{s_1-s_2}(A)\cdots U_{s_{n-1}-s_n}(A)(-B)U_{s_n}(A), \end{aligned} \tag{4.74}$$

the estimate (4.66) implies that for $0 \leqslant t \leqslant \tau$

$$\|S_n(t)\| \leqslant M e^{\omega_0 t} \xi^n, \quad n \geqslant 1. \tag{4.75}$$

Therefore, (4.72) converges in the $\mathcal{L}(\mathcal{H})$ operator norm, and

$$\|F_t\| \leqslant M e^{\omega_0 t}(1-\xi)^{-1}, \quad 0 \leqslant t \leqslant \tau\,, \tag{4.76}$$

where we denote the series in (4.72) by

$$F_t := \sum_{n=0}^{\infty} S_n(t)\,. \tag{4.77}$$

Now we proceed with the verification of the (semigroup) properties of the solution $\{F_t\}_{t\leqslant\tau}$, which is the operator-norm convergent Dyson-Phillips series (4.77).
(1) For any $u \in \mathcal{H}$ one gets the immediate estimate

$$\|(F_t - \mathbb{1})u\| \leqslant \|(U_t(A) - \mathbb{1})u\| + \|\sum_{n=1}^{\infty} S_n(t)\|\,\|u\|.$$

For the first term of this estimate we obtain $\lim_{t\to+0}\|(U_t(A)-\mathbb{1})u\| = 0$. To make evident the time dependence of the sum, we use (4.74) to refine the upper bound of (4.75) as follows:

$$\|\sum_{n=1}^{\infty} S_n(t)\| \leqslant \sum_{n=1}^{\infty} M e^{\omega_0 \tau}\left[\int_0^t \mathrm{d}s\|B\,U_s(A)\|\right]^n. \tag{4.78}$$

Then by Definition (4.34)(ii) and a comment in Remark 4.35, we have

$$\lim_{t\to+0}\int_0^t \mathrm{d}s\|B\,U_s(A)\| = 0\,.$$

So, the solution (4.77) is strongly continuous at $t = +0$:

$$\lim_{t\downarrow 0}\|(F_t - \mathbb{1})u\| = 0, \quad u \in \mathcal{H}. \tag{4.79}$$

(2) Let $0 \leqslant t \leqslant \tau$. By virtue of (4.73), for any $u \in \mathcal{H}$ the sum (4.77) can be rearranged as follows:

$$\begin{aligned} F_t u &= U_t(A)u - \int_0^t \mathrm{d}s \sum_{n=1}^{\infty} S_{n-1}(t-s)\,B\,U_s(A)u \\ &= U_t(A)u - \int_0^t \mathrm{d}s\, F_{t-s}\,B\,U_s(A)u\,, \quad 0 \leqslant t \leqslant \tau\,. \end{aligned} \tag{4.80}$$

Thus, the family $\{F_t\}_{t\leqslant\tau}$ satisfies the integral equation (4.57).

Similarly, with the help of (4.74) for any $u \in \mathcal{H}$ the sum (4.77) can be rearranged differently:

$$\begin{aligned} F_t u &= U_t(A)u - \int_0^t \mathrm{d}s_1\, U_{t-s_1}(A)\,B \sum_{n=1}^{\infty} S_{n-1}(s_1)u \\ &= U_t(A)u - \int_0^t \mathrm{d}s\, U_{t-s}(A)\,B\,F_s u\,, \quad 0 \leqslant t \leqslant \tau\,. \end{aligned} \tag{4.81}$$

Therefore, the family $\{F_t\}_{t\leqslant\tau}$ satisfies also the integral equation (4.58).
(3) Both equations (4.80) and (4.81) can be extended to any $t \geqslant 0$. Suppose for example that (4.80) is valid for $0 \leqslant t \leqslant n\tau$. Let $n\tau < t' \leqslant (n+1)\tau$, then

$$\begin{aligned} F_{t'} &= F_\tau F_{t'-\tau} \\ &= F_\tau\Big(U_{t'-\tau}(A) - \int_0^{t'-\tau} \mathrm{d}s\, F_{t'-\tau-s}BU_s(A)\Big) \\ &= U_\tau\, U_{t'-\tau} - \int_0^\tau \mathrm{d}s\, F_{\tau-s}BU_s(A)U_{t'-\tau}(A) - \int_0^{t'-\tau} \mathrm{d}s\, F_{t'-s}BU_s(A) \\ &= U_{t'} - \int_0^{t'} \mathrm{d}s\, F_{t'-s}BU_s(A), \end{aligned}$$

and by induction, (4.80) holds for all $t \geqslant 0$. The same is true for (4.81).
(4) Although the above arguments bring forward the semigroup functional equation (Definition 1.2(b)) for the family $\{F_t\}_{t\geqslant 0}$, it is necessary to check this property for *any* choice of $t_1, t_2 \geqslant 0$. Taking into account (4.74) and the representation (4.77), one can verify by a direct multiplication of the series F_{t_1} and F_{t_2} for $t_1, t_2 \geqslant 0$ with $t_1 + t_2 \leqslant \tau$, that

$$F_{t_1}F_{t_2} = F_{t_2}F_{t_1} = F_{t_1+t_2}\,. \tag{4.82}$$

Then due to (3) this property can be extended to $t \geqslant \tau$ by putting

$$F_t = (F_\tau)^n F_{t-n\tau}\,, \quad n\tau < t \leqslant (n+1)\tau\,. \tag{4.83}$$

Summarising (1)–(4), or (4.76)–(4.83), we conclude that the family $\{F_t\}_{t\geqslant 0}$ is in fact a quasi-bounded C_0-semigroup with $M' = M(1-\xi)^{-1}$ and $\omega_0' = \omega_0$.
(5) Finally, with the help of equation (4.80) we shall determine the generator of the semigroup $\{F_t\}_{t\geqslant 0}$. To this end, we consider the domain $\mathcal{D} = \bigcup_{t>0} U_t(A)\mathrm{dom}\, A$, (4.52). If $v \in \mathcal{D}$, there exist a vector $u \in \mathrm{dom}\, A$ and a constant $\tau > 0$ such that $v = U_\tau(A)u$. By virtue of (4.80) and by the definition of the semigroup generator, we get for $B \in \mathcal{P}_{0+}$

$$\begin{aligned} \lim_{t\downarrow 0} \frac{1}{t}(\mathbb{1} - F_t)v &= \lim_{t\downarrow 0} \frac{1}{t}(\mathbb{1} - U_t(A))v \\ &\quad + \lim_{t\downarrow 0} \frac{1}{t}\int_0^t \mathrm{d}s\, F_{t-s}(B\,U_\tau(A))U_s(A)u \\ &= Av + BU_\tau(A)u = (A+B)v\,, \quad v \in \mathcal{D}. \end{aligned} \tag{4.84}$$

This means that the generator H of the semigroup $\{F_t\}_{t\geqslant 0}$ satisfies $\mathrm{dom}\, H \supseteq \mathcal{D}$, and that $H = A + B$ on $\mathcal{D}$. By Remark 4.36, we get that $\mathrm{dom}\, A \subseteq \mathrm{dom}\, H$ and

$$Hu = (A+B)u, \quad u \in \mathrm{dom}\, A. \tag{4.85}$$

To finish the proof, we have to verify that dom H is not larger than dom A. Note that dom $A \subseteq$ dom B and that the bound (4.54) holds by Lemma 4.37. Then the Laplace transform of (4.80) for $\lambda > \omega_0$ and λ large enough (such that $\varepsilon < 1$ in (4.54)), gives

$$\begin{aligned}(H + \lambda\mathbb{1})^{-1} &= (A + \lambda\mathbb{1})^{-1} - (H + \lambda\mathbb{1})^{-1}B(A + \lambda\mathbb{1})^{-1} \\ &= (A + \lambda\mathbb{1})^{-1}\{\mathbb{1} + B(A + \lambda\mathbb{1})^{-1}\}^{-1}. \end{aligned} \tag{4.86}$$

Since $\|B(A+\lambda\mathbb{1})^{-1}\| < \varepsilon < 1$, equation (4.86) implies that dom H = ran $R_{-\lambda}(H) \subseteq$ ran $R_{-\lambda}(A)$ = dom A. Together with expression (4.85), this implies dom H = dom A. □

Corollary 4.42. *The sum of a bounded operator $B \in \mathcal{L}(\mathcal{H})$ and a generator $A \in Q(M,\omega_0)$, $H = A + B$, is the generator of a strongly continuous quasi-bounded semigroup $\{U_t(H)\}_{t\geqslant 0}$, that is, $H \in Q(M',\omega_0')$.*

Note also that in this case the operator (4.64) *is indeed an abstract Volterra operator. Then iterations of* (4.64), (4.65) *yield the estimate*

$$\|\mathbb{V}^n\|_{\mathfrak{F}_t} = \sup_{\{\Phi\in\mathfrak{F}_t:\|\Phi\|_{\infty,t}=1\}} \|\mathbb{V}^n\Phi\|_{\infty,t} \leqslant \frac{1}{n!}(t\,\|B\|Me^{\omega_0 t})^n ,$$

which shows that the spectral radius of this operator is zero, see Section A.7. *Then the proof of Proposition* 4.41 *is straightforward.*

In Proposition 4.41, the perturbed semigroup $\{U_t(H)\}_{t\geqslant 0}$ with perturbation $B \in \mathcal{P}_{0^+}(A)$ is constructed via an operator-norm convergent Dyson-Phillips series (4.77). The extension of this perturbation theory to Gibbs semigroups needs the following preparatory lemma.

Lemma 4.43. *Let the m-sectorial operator A be such that $e^{-t\,\mathfrak{Re}\,A} \in \mathcal{C}_1(\mathcal{H})$ for $t > 0$, and let $V_1, V_2, \ldots, V_n$ be bounded operators on $\mathcal{H}$. For any set of positive numbers $t_1, t_2, \ldots, t_n$,*

$$\Big\|\prod_{j=1}^{n} V_j e^{-t_j A}\Big\|_1 \leqslant \prod_{j=1}^{n} \|V_j\| \|e^{-(t_1+t_2+\ldots+t_n)\,\mathfrak{Re}\,A/4}\|_1. \tag{4.87}$$

Proof. Firstly, let $V_j \in \mathcal{C}_\infty(\mathcal{H})$ for $j = 1,2,\ldots,n$. We set $t_m := \min\{t_j\}_{j=1}^n > 0$ and $T := \sum_{j=1}^n t_j > 0$. For any $1 \leqslant j \leqslant n$, we define an integer $\ell_j \in \mathbb{N}$ by

$$2^{\ell_j} t_m \leqslant t_j \leqslant 2^{\ell_j+1} t_m.$$

We then set $\sum_{j=1}^n 2^{\ell_j} t_m > T/2$ and

$$\prod_{j=1}^{n} V_j e^{-t_j A} = \prod_{j=1}^{n} V_j e^{-(t_j-2^{\ell_j}t_m)A}(e^{-t_m A})^{2^{\ell_j}}. \tag{4.88}$$

By the definition of the $\|\cdot\|_1$-norm and by the inequality (3.25) for singular values, see Corollary 3.7, we get

$$\begin{aligned}\Big\|\prod_{j=1}^{n} V_j \mathrm{e}^{-t_j A}\Big\|_1 &= \sum_{k=1}^{\infty} s_k\Big(\prod_{j=1}^{n} V_j \mathrm{e}^{-(t_j-2^{\ell_j}t_m)A}(\mathrm{e}^{-t_m A})^{2^{\ell_j}}\Big) \\ &\leqslant \sum_{k=1}^{\infty}\prod_{j=1}^{n} s_k\left(\mathrm{e}^{-(t_j-2^{\ell_j}t_m)A}\right)\left[s_k(\mathrm{e}^{-t_m A})\right]^{2^{\ell_j}} s_k(V_j) \\ &\leqslant \sum_{k=1}^{\infty} s_k(\mathrm{e}^{-t_m A})^{\sum_{j=1}^{n} 2^{\ell_j}} \prod_{j=1}^{n}\|V_j\|. \end{aligned} \tag{4.89}$$

Here we used that $s_k(\mathrm{e}^{-(t_j-2^{\ell_j}t_m)A}) \leqslant \|\mathrm{e}^{-(t_j-2^{\ell_j}t_m)A}\| \leqslant 1$ and that $s_k(V_j) \leqslant \|V_j\|$, see Section 4.2. Let $N := \sum_{j=1}^{n} 2^{\ell_j}$ and $T_m := Nt_m > T/2$; then inequality (4.89) yields

$$\Big\|\prod_{j=1}^{n} V_j \mathrm{e}^{-t_j A}\Big\| \leqslant \left(\big\|\mathrm{e}^{-T_m A/N}\big\|_{q=N}\right)^N \prod_{j=1}^{n}\|V_j\|. \tag{4.90}$$

In order to apply Proposition 4.30, we consider an integer $p \in \mathbb{N}$ such that $2^p \leqslant N < 2^{p+1}$. It then follows that $T/4 < T_m/2 < 2^p T_m/N$, and hence we obtain

$$\begin{aligned}\left(\big\|\mathrm{e}^{-T_m A/N}\big\|_{q=N}\right)^N &= \sum_{k=1}^{\infty} s_k^N(\mathrm{e}^{-T_m A/N}) \\ &\leqslant \sum_{k=1}^{\infty} s_k^{2^p}(\mathrm{e}^{-2^p T_m A/2^p N}) \\ &\leqslant \sum_{k=1}^{\infty} s_k^{2^p}(\mathrm{e}^{-TA/2^{p+2}}), \end{aligned} \tag{4.91}$$

where we used that $s_k(\mathrm{e}^{-T_m A/N}) = s_k(\mathrm{e}^{-2^p T_m A/2^p N}) \leqslant \|\mathrm{e}^{-T_m A/N}\| \leqslant 1$, and that $s_k(\mathrm{e}^{-(t+\tau)A}) \leqslant \|\mathrm{e}^{-tA}\| s_k(\mathrm{e}^{-\tau A}) \leqslant s_k(\mathrm{e}^{-\tau A})$ for any $t, \tau > 0$. Therefore, the estimates (4.90), (4.91) and inequalities (4.38) give the bound (4.87), see (2.32).

Secondly, let $V_j \in \mathcal{L}(\mathcal{H})$, $j = 1, 2, \ldots, n$, and set $\tilde{V}_j := V_j \mathrm{e}^{-\varepsilon A}$ for $0 < \varepsilon < t_m$. Hence, $\tilde{V}_j \in \mathcal{C}_1(\mathcal{H})$ and $s_k(\tilde{V}_j) \leqslant \|\tilde{V}_j\| \leqslant \|V_j\|$. If we set $\tilde{t}_j := t_j - \varepsilon$, then

$$\Big\|\prod_{j=1}^{n} V_j \mathrm{e}^{-t_j A}\Big\|_1 \leqslant \prod_{j=1}^{n}\|V_j\| \|\mathrm{e}^{-(\tilde{t}_1+\tilde{t}_2+\cdots+\tilde{t}_n)\,\Re\mathfrak{e}\, A/4}\|_1. \tag{4.92}$$

Since the semigroup $\{\mathrm{e}^{-t\,\Re\mathfrak{e}\,A}\}_{t\geqslant 0}$ is $\|\cdot\|_1$-continuous for $t > 0$, we can now take in (4.92) the limit $\varepsilon \downarrow 0$, which gives the result (4.87) in the general case. □

Proposition 4.44. *Let the m-sectorial operator A be such that $\mathrm{e}^{-t\,\Re\mathfrak{e}\,A} \in \mathcal{C}_1(\mathcal{H})$ for $t > 0$, and let the perturbation $B \in \mathcal{P}(A)$ define the generator $H(\kappa) := A + \kappa B$ of a Gibbs semigroup $\{G_t(H(\kappa))\}_{t\geqslant 0}$, for $\kappa \in \mathbb{C}$. Then the function $\mathbb{C} \ni \kappa \mapsto G_t(H(\kappa)) \in \mathcal{C}_1(\mathcal{H})$ is $\|\cdot\|_1$-holomorphic for any fixed $t > 0$.*

Proof. By virtue of Definition 1.41, Proposition 1.43 and Proposition 1.46, $A \in Q(M = 1, \omega_0 = 0)$. Hence, by Proposition 4.41, we can define a perturbation of the Gibbs semigroup $\{G_t(A)\}_{t\geqslant 0}$, see Proposition 4.30, by the norm-convergent Dyson-Phillips series (4.77)

$$F_t(\kappa) := \sum_{n=0}^{\infty} S_n(t,\kappa). \tag{4.93}$$

Here, as, e.g., in (4.81), we define

$$\begin{aligned} S_0(t,\kappa) &= G_t(A), \\ S_n(t,\kappa) &= \int_0^t \mathrm{d}s\, G_{t-s}(A)(-\kappa B)S_{n-1}(s,\kappa), \quad n \geqslant 1. \end{aligned} \tag{4.94}$$

Since $B \in \mathcal{P}_{0^+}(A)$ implies that the operator $BG_t(A) \in \mathcal{L}(\mathcal{H})$ for $t > 0$, we get that $G_{t-s}(A)(-\kappa B)S_{n-1}(s,\kappa) \in \mathcal{C}_1(\mathcal{H})$ for $s > 0$ and $S_n(t,\kappa)$ is an n-fold $\|\cdot\|_1$-convergent Bochner integral, which can be estimated as

$$\|S_n(t,\kappa)\|_1 \leqslant \int_0^t \mathrm{d}\tau_0 \ldots \int_0^t \mathrm{d}\tau_n\, \chi_n^t(\tau_0,\ldots,\tau_n)\|G_{\tau_0}(A)\kappa B G_{\tau_1}(A)\cdots\kappa B G_{\tau_n}(A)\|_1. \tag{4.95}$$

Here, $\chi_n^t(\tau_0,\ldots,\tau_n)$ is the characteristic function of the set

$$\Big\{\tau_i \geqslant 0, i = 0,1,\ldots,n : \sum_{i=0}^{n} \tau_i = t\Big\} \subset \bigtimes_{i=0}^{n} \mathbb{R}_0^+ .$$

Now, we can use inequality (4.87) to estimate the integrand in (4.95). To this end, we set $V_0 := G_{\tau_0/2}(A)$ and $V_j := BG_{\tau_j/2}(A)$, $j = 1,\ldots,n$, and introduce the functions $t \mapsto q(t) := \|G_t(A)\| \leqslant 1$ and $t \mapsto p(t) := \|BG_t(A)\|$. By Lemma 4.43, see (4.87), we then get from (4.95) that

$$\|S_{n\geqslant 1}(t,\kappa)\|_1 \leqslant |\kappa|^n 2^n (q * \underbrace{p * \cdots * p}_{n})(t/2)\,\mathrm{Tr}\, G_{t/8}(A), \tag{4.96}$$

where

$$(q * \underbrace{p * \cdots * p}_{n})(t/2) = \int_0^{t/2} \mathrm{d}\tau_0 \ldots \int_0^{t/2} \mathrm{d}\tau_n\, \chi_n^{t/2}(\tau_0,\ldots,\tau_n) q(\tau_0)p(\tau_1)\cdots p(\tau_n). \tag{4.97}$$

Since $B \in \mathcal{P}(A)$, (4.51) implies that for any $R > 0$ there is an $\varepsilon > 0$ small enough such that

$$\int_0^{\varepsilon} \mathrm{d}\tau\, p(\tau) := \gamma_\varepsilon < (2R)^{-1}. \tag{4.98}$$

The estimates (4.96)–(4.98) then yield

$$\|S_n(t,\kappa)\|_1 \leqslant |\kappa|^n 2^n \gamma_\varepsilon^n \operatorname{Tr} G_{t/8}(A), \tag{4.99}$$

for $0 < t \leqslant 2\varepsilon$. Therefore, the series (4.93) converges uniformly in the $\|\cdot\|_1$-topology in the disk $D_R = \{\kappa \in \mathbb{C} : |\kappa| < R\}$ and for t in any compact $K \subset (0, 2\varepsilon]$. Therefore, $F_t(\kappa) \in \mathcal{C}_1(\mathcal{H})$.

Finally, similarly to Proposition 4.41, we can use the semigroup integral equation (4.80), or (4.81), to extend this statement to any compact $K \subset \mathbb{R}^+$. Proposition 4.41, (4.84), enables us to identify the generator of the Gibbs semigroup (4.93) with $H(\kappa) = A + \kappa B$, that is, $\{F_t(\kappa)\}_{t\geqslant 0} = \{G_t(H(\kappa))\}_{t\geqslant 0}$. □

4.5 Holomorphic Gibbs semigroups

In this section, we enlarge the class of perturbations discussed in Section 4.4 to the class $\mathcal{P}_b$ with relative bound $b > 0$. First, we consider a rather restricted set of self-adjoint generators A and study perturbations of the self-adjoint Gibbs semigroups. Then we pass to the theory of holomorphic non-self-adjoint Gibbs semigroups.

Note that, by Proposition 4.27 any p-generator A is the generator of a holomorphic Gibbs semigroup. Therefore, below we focus, in particular, on developing of the corresponding perturbation theory.

Proposition 4.45. *The generator $A \geqslant -\omega_0 \mathbb{1}$ of a self-adjoint Gibbs semigroup $\{G_t(A)\}_{t\geqslant 0}$ on $\mathcal{H}$ and a symmetric operator $B \in \mathcal{P}_{b<1}(A)$, define the operator $H = A + B$, with $\operatorname{dom} H = \operatorname{dom} A$, as the generator of a quasi-bounded self-adjoint Gibbs semigroup $\{G_t(H)\}_{t\geqslant 0}$.*

Proof. Recall that $B \in \mathcal{P}_{b<1}(A)$ means that the operator B is Kato-small with respect to A with the relative bound $b < 1$, see Definition 1.50. Since B is symmetric, by the Kato-Rellich theorem, the operator sum $H = A + B$ with $\operatorname{dom} H = \operatorname{dom} A$, see (1.101), is a self-adjoint operator and $H \geqslant -\omega_0' \mathbb{1}$, with $\omega_0' := \omega_0 + \max\{a/(1-b), a + b|\omega_0|\}$. Hence, H is an m-sectorial operator with vertex $\gamma = -\omega_0'$ and semi-angle $\alpha = 0$, i.e., $H \in \mathscr{H}(\pi/2, \omega_0')$.

Since $G_t(A) \in \mathcal{C}_1(\mathcal{H})$ for $t > 0$ and $A = A^*$, by Proposition 4.21, the spectrum $\sigma(A) = \sigma_{\mathrm{p}}(A)$ is pure point with real eigenvalues $\lambda_n(A) \to +\infty$, for $n \to \infty$. Applying the minimax principle to the perturbed operator $H = A + B$, or more precisely to its resolvent $R_{\gamma<-\omega_0'}(H)$, see Section 2.3, we then get that $\sigma(H) = \sigma_{\mathrm{p}}(H)$ and $\lambda_n(H) \geqslant a + (1-b)\lambda_n(A) > 0$ for all n greater than some $n_0 > 1$. This implies that

$$\|\mathrm{e}^{-tH}\|_1 = \sum_{n=1}^{\infty} \mathrm{e}^{-t\lambda_n(H)} \leqslant c_1 \|\mathrm{e}^{-c_2 tA}\|_1$$

for $t > 0$ and some $c_1, c_2 > 0$. Hence, H is the generator of a quasi-bounded self-adjoint Gibbs semigroup. □

Remark 4.46. Any self-adjoint Gibbs semigroup $\{G_t(A)\}_{t\geqslant 0}$ can be extended to a $\|\cdot\|_1$-holomorphic Gibbs semigroup $\{G_z(A)\}_{z\in S_{\pi/2}}$. Indeed, since the operator $A = A^* \geqslant -\omega_0\mathbb{1}$, it is quasi-$m$-sectorial with vertex $\gamma = -\omega_0$ and semi-angle $\alpha = 0$. Thus, $A \in \mathscr{H}(\theta = \pi/2, \omega_0)$ is the generator of the quasi-bounded holomorphic semigroup $\{U_z(A)\}_{z\in S_{\pi/2}}$, see Corollary 1.47. By the semigroup property we get that

$$G_z(A) := U_{z=t+i\tau}(A) = U_t(A)U_{i\tau}(A) \in \mathcal{C}_1(\mathcal{H}), \quad z \in S_{\pi/2},$$

since $U_t(A) = G_t(A) \in \mathcal{C}_1(\mathcal{H})$ for $t > 0$ and $U_{i\tau}(A) \in \mathcal{L}(\mathcal{H})$.

Recall that by Proposition 1.27 and Corollary 1.29, the semigroup $\{G_z(A)\}_{z\in S_{\pi/2}}$ is $\|\cdot\|$-holomorphic in the open sector $S_{\pi/2}$. Therefore, it is $\|\cdot\|_1$-differentiable, cf. (4.37):

$$\begin{aligned}\partial_z G_z(A) &= \|\cdot\|_1\text{-}\lim_{\Delta z\to 0} \frac{1}{\Delta z}\left(G_{z/2+\Delta z}(A) - G_{z/2}(A)\right) G_{z/2}(A)\\ &= -AG_z(A) \in \mathcal{C}_1(\mathcal{H}),\end{aligned} \tag{4.100}$$

for $z \in S_{\pi/2}$ by the $\|\cdot\|_1$-continuity of multiplication, Proposition 2.78. Equation (4.100) now implies that the complex operator-valued function of $z \in S_{\pi/2}$, $z \mapsto G_z(A) \in \mathcal{C}_1(\mathcal{H})$ is $\|\cdot\|_1$-holomorphic.

Remark 4.47. (a) The condition that the symmetric perturbation is Kato small, $B \in \mathcal{P}_{b<1}(A)$, can be relaxed if one assumes that $B \geqslant 0$ and $D := \operatorname{dom} A \cap \operatorname{dom} B$ is dense in $\mathcal{H}$. Let us suppose for simplicity that the generator A of a self-adjoint semigroup $\{G_t(A)\}_{t\geqslant 0}$ satisfies $A = A^* \geqslant \alpha\mathbb{1}$ with $\alpha > 0$. Since the densely defined symmetric operator $T := A + B$, $\operatorname{dom} T = D$, is semi-bounded from below by $\alpha\mathbb{1}$, it has a self-adjoint extension $\tilde{T} \geqslant \alpha\mathbb{1}$, and $\tilde{T} \geqslant A$ by $B \geqslant 0$.

Since $\alpha > 0$, the inverse operators $\tilde{T}^{-1}$ and A^{-1} exist and $0 \leqslant \tilde{T}^{-1} \leqslant A^{-1}$, where $A^{-1} \in \mathcal{C}_\infty(\mathcal{H})$. Then $\tilde{T}^{-1} \in \mathcal{C}_\infty(\mathcal{H})$ and $\lambda_n(\tilde{T}) \geqslant \lambda_n(A)$ for eigenvalues, $n \geqslant 1$ (cf. Proposition 2.63). Hence,

$$\|\mathrm{e}^{-t\tilde{T}}\|_1 \leqslant \|\mathrm{e}^{-tA}\|_1.$$

Therefore $\{G_t(\tilde{T})\}_{t\geqslant 0}$ is a self-adjoint Gibbs semigroup, which can be extended to $z \in S_{\pi/2}$ by Remark 4.46.

(b) Let $A \geqslant 0$ be generator of a self-adjoint Gibbs semigroup $\{G_t(A) = \mathrm{e}^{-tA}\}_{t\geqslant 0}$ and $B \geqslant 0$ be generator of a self-adjoint contraction semigroup. It may happen that domain $\operatorname{dom}(A) \cap \operatorname{dom}(B)$ is not dense in $\mathcal{H}$, or even is trivial, that is, reduces to $\{0\}$. Then the construction of semigroup corresponding to the pair A, B via a perturbation theory is difficult/impossible. On the other hand, in Section 5.4 (Proposition 5.53) we shall show that a construction is possible via the Trotter-Kato *product formulae* approximation. We call it the *product formula-based* construction of semigroups.

(c) For example, let $\operatorname{dom} A \cap \operatorname{dom} B = \{0\}$, whereas $\operatorname{dom} A^{1/2} \cap \operatorname{dom} B^{1/2}$ is nontrivial. Then the closure $\overline{\operatorname{dom} A^{1/2} \cap \operatorname{dom} B^{1/2}} =: \mathcal{H}_0$ is a subspace $\mathcal{H}_0 \subseteq \mathcal{H}$.

The non-negative self-adjoint *form-sum* operator $H := A \dot{+} B$ is densely defined in the Hilbert subspace $\mathcal{H}_0$, see Remark 4.29. The exponential Trotter-Kato product formula converges in the *trace-norm* topology *away from zero* for any self-adjoint operator $B \geqslant 0$ to a *degenerate* Gibbs semigroup:

$$\|\cdot\|_1\text{-}\lim_{n\to\infty}\left(\mathrm{e}^{-tA/n}\mathrm{e}^{-tB/n}\right)^n = \mathrm{e}^{-tH}P_0\,,\quad H = A \dot{+} B\,.$$

Here $t > 0$ and $P_0 : \mathcal{H} \to \mathcal{H}_0$ is the orthogonal projection.

In order to generalise this observation to any $A \geqslant -\gamma_A \mathbb{1}$ and $B \geqslant -\gamma_B \mathbb{1}$, $\gamma_A, \gamma_B \in \mathbb{R}$, one can consider the operators $\tilde{A} = A + (\gamma_A + \varepsilon)\mathbb{1} > 0$ and $\tilde{B} = B + \gamma_B \mathbb{1} \geqslant 0$.

Remark 4.48. According to Remark 4.46, the perturbed self-adjoint semigroup defined in Remark 4.47 can be extended to a holomorphic Gibbs semigroup $\{G_z(A + B)\}_{z\in S_{\pi/2}}$. Similarly to Section 1.7 and Section 4.4, we would like to study here the stability of generators with respect to perturbations, or the analytic properties of the perturbed semigroups with respect to the parameter κ of the family of perturbations $\{\kappa B\}_{\kappa\in\mathbb{C}}$. However, in general, even for $\kappa < 0$ this is not possible if there is no *subordination* between A and B, cf. Proposition 1.52.

Note that the construction of perturbed semigroups in Proposition 4.45 and in Remarks 4.46, 4.47 is not semigroup- or resolvent-based, see notes in Section 1.8 and in Section 4.7. In fact, only the operator values of the one-parametric family were enough to identify that the corresponding semigroups are Gibbs. A particular case one finds in Remark 4.47(c). There the construction of a perturbed semigroup is entirely based on the product formula, which is an example of the *product formula-based* construction of semigroups.

Below we consider the case, when A is a p-generator, Definition 4.26, and $\kappa B \in \mathcal{P}_{b<1}(A)$ for $\kappa \in \mathbb{C}$ and $b \geqslant 0$. Recall that, by Proposition 4.27, any p-generator A is the generator of a holomorphic Gibbs semigroup. For the construction of the perturbed semigroup we use the resolvent-based method to *lift* the result of Proposition 1.54 for the operator-norm topology to the trace-norm topology.

Proposition 4.49. *Let the positive self-adjoint operator $A \geqslant 0$ in $\mathcal{H}$ be a p-generator and let the perturbation operator $B \in \mathcal{P}_b(A)$ for $b < 1$. Then the operator-valued map into the trace-class:*

$$\mathcal{D} \ni (z, \kappa) \mapsto \mathrm{e}^{-z(A+\kappa B)} \in \mathcal{C}_1(\mathcal{H}), \tag{4.101}$$

is holomorphic in the trace-norm topology on the domain

$$\begin{aligned}\mathcal{D} &:= \{z \in S_{\theta(|\kappa|,b)} \subset \mathbb{C}\} \times \{\kappa \in D_{r<1/b} \subset \mathbb{C}\} \\ &= \{(z,\kappa) \in \mathbb{C}^2 : |\arg(z)| < \mathrm{arctg}(\sqrt{1-|\kappa|^2 b^2}/|\kappa| b) \ \wedge\ |\kappa| < 1/b\}.\end{aligned} \tag{4.102}$$

Proof. By Proposition 1.54, the quasi-sectorial for each semi-angle $\alpha(\kappa, b) = \pi/2 - \theta(|\kappa|, b)$ operators $\{H(\kappa) := A + \kappa B\}_{\kappa \in D_{1/b}}$ in the unit disc $D_{1/b}$ are generators of the operator-norm holomorphic family $\{U_z(H(\kappa)) = e^{-zH(\kappa)}\}_{\kappa \in D_{1/b}}$ of holomorphic semigroups $\{U_z(H(\kappa))\}_{z \in S_{\theta(|\kappa|,b)}}$ in sectors with semi-angles $\theta(|\kappa|, b) = \operatorname{arctg}(\sqrt{1 - |\kappa|^2 b^2}/|\kappa| b)$. Then for $B \in \mathcal{P}_b(A)$ the $\|\cdot\|$-convergent Neumann series and the Riesz-Dunford formula (1.118) give the representation

$$U_z(H(\kappa)) = \frac{1}{2\pi i} \int_\Gamma d\zeta \, e^{z\zeta} \, R_{-\zeta}(A) \sum_{n=0}^{\infty} (\kappa B R_{-\zeta}(A))^n \, . \tag{4.103}$$

Here the $\|\cdot\|$-convergent integral is taken along the contour $\Gamma \subset (-\mathcal{M}_{\kappa,\, b})$, see (1.115). Since $(-\mathcal{M}_{\kappa,\, b}) \subset \rho(A)$ and since A is a p-generator, we obtain that

$$R_{-\zeta}(A) = R_{\zeta_0}(A)[\mathbb{1} - (\zeta_0 + \zeta) R_{-\zeta}(A)] \in \mathcal{C}_p(\mathcal{H}) \, ,$$

for any $\zeta \in \Gamma$. Hence, the integral is in fact $\|\cdot\|_p$-convergent and the representation (4.103) leads to the estimate

$$\begin{aligned} \|U_z(H(\kappa))\|_p &\leqslant \frac{1}{2\pi} \int_\Gamma |d\zeta| \, e^{Re(z\zeta)} \|R_\zeta(A)\|_p \{1 - |\kappa| \|BR_{-\zeta}(A)\|\}^{-1} \\ &\leqslant \frac{1}{2\pi} \int_\Gamma |d\zeta| \, e^{Re(z\zeta)} \|R_{\zeta_0}(A)\|_p \frac{1 + |\zeta_0 + \zeta| \|R_{-\zeta}(A)\|}{1 - |\kappa| \|BR_{-\zeta}(A)\|} , \end{aligned} \tag{4.104}$$

where the last integral converges for $z \in S_{\theta(|\kappa|,b)}$ and $\kappa \in D_{1/b}$.

Therefore, the estimate (4.104) implies $U_z(H(\kappa)) \in \mathcal{C}_p(\mathcal{H})$, and by Proposition 4.3 we obtain that $\{U_z(H(\kappa))\}_{z \in S_{\theta(|\kappa|,b)}}$ is a Gibbs semigroup: $U_z(H(\kappa)) =: G_z(H(\kappa)) \in \mathcal{C}_1(\mathcal{H})$. Since $\{U_z(H(\kappa))\}_{z \in S_{\theta(|\kappa|,b)}}$ is $\|\cdot\|$-holomorphic in the open sector $S_{\theta(|\kappa|,b)}$, the map $z \mapsto G_z(H(\kappa))$ is $\|\cdot\|_1$-holomorphic in the same sector by (4.100). We note also that for a fixed $r < b^{-1}$ the family of Gibbs semigroups $\{G_z(H(\kappa))\}_{\kappa \in D_r}$ is $\|\cdot\|_1$-uniformly bounded:

$$\|G_z(H(\kappa))\|_1 \leqslant M_r(z), \tag{4.105}$$

for any $z \in S_{\theta(r,b)}$.

Now we recall that, by Proposition 1.54 (ii), the series in (4.103) is a uniformly $\|\cdot\|$-convergent Taylor series for $|\kappa| \leqslant r < b^{-1}$. Therefore, we can rewrite (4.103) as

$$G_z(H(\kappa)) = \sum_{n=0}^{\infty} \frac{\kappa^n}{n!} \, \partial_\kappa^n G_z(H(0)), \quad z \in S_{\theta(r,b)}. \tag{4.106}$$

Here, the $\|\cdot\|$-derivatives have the representation

$$\partial_\kappa^n G_z(H(0)) = \frac{n!}{2\pi i} \int_C d\kappa \, \kappa^{-(n+1)} G_z(H(\kappa)), \tag{4.107}$$

where the contour $C \subset D_{1/b}$ encircles the point $\kappa = 0$. If the contour $C_{r'}$ is the circle of radius $r < r' < b^{-1}$, then by (4.107) and by the $\|\cdot\|_1$-estimate (4.105) in the disc D_1 one gets the bounds

$$\begin{aligned}\|\partial_\kappa^n G_z(H(0))\|_1 &\leqslant \frac{n!}{2\pi}\int_0^{2\pi} d\varphi (r')^{-n} \sup_{|\kappa|\leqslant r'} \|G_z(H(\kappa))\|_1 \\ &= n!\,(r')^{-n}\, M_{r'}(z), \end{aligned} \tag{4.108}$$

for any $z \in S_{\theta(r,b)}$. Consequently, (4.106) and (4.108) yield for any $z \in S_{\theta(r',b)}$ the estimate

$$\left\|G_z(H(\kappa)) - \sum_{n=0}^{N} \frac{\kappa^n}{n!}\partial_\kappa^n G_z(H(\kappa = 0))\right\|_1 \leqslant \sum_{n=N+1}^{\infty} (|\kappa|/r')^n\, M_{r'}(z),$$

which implies the $\|\cdot\|_1$-convergence of the series (4.106), uniformly in the closed disc $|\kappa| \leqslant r$. Therefore, the function $\kappa \mapsto G_z(H(\kappa))$ is $\|\cdot\|_1$-holomorphic in disc $D_{1/b}$. □

Corollary 4.50. *Let the closed operators $\{H(\kappa)\}_{\kappa\in D_{1/c}\subset\mathbb{C}}$ form in a Hilbert space $\mathcal{H}$, a holomorphic family of type* (A), *Definition* 5.67, *and let $D_{1/c} \ni 0$. If $H(0) \in \mathscr{H}(\theta,\omega)$ is the generator of a quasi-bounded holomorphic semigroup, then for any $u \in \operatorname{dom} H(0)$ by the criterion in Proposition* 5.68 *one can estimate the difference*

$$\|(H(\kappa) - H(0))u\| \leqslant \frac{|\kappa|}{1-|\kappa|c}(a\|u\| + b\|H(0)u\|). \tag{4.109}$$

For $|\kappa| < (b+c)^{-1}$ the difference in (4.109) *is relatively $H(0)$-small. Thus by Propositions* 1.51 *the operator $H(\kappa) \in \mathscr{H}(\theta',\omega')$ generates a holomorphic semigroup and one can use the Riesz-Dunford representation*

$$U_z(H(\kappa)) = \frac{1}{2\pi i}\int_\Gamma d\zeta\, e^{z\zeta}(\zeta\mathbb{1} + H(\kappa))^{-1}, \tag{4.110}$$

for $z \in S_{\theta'}$, $\kappa \in D_{1/(b+c)}$ and with the operator-norm convergent Bochner integral taken along an appropriate contour Γ.

Note that due to the resolvent identity we get the representation

$$\begin{aligned} &U_z(H(\kappa)) - U_z(H(0)) \\ &= \frac{1}{2\pi i}\int_\Gamma d\zeta\, e^{z\zeta}\,(\zeta\mathbb{1} + H(\kappa))^{-1}((H(0) - H(\kappa)))(\zeta\mathbb{1} + H(0))^{-1}\end{aligned}$$

for $\kappa \in D_{1/(b+c)}$. Then the smallness (4.109) ensures the existence of the operator-norm derivative at $\kappa = 0$.

$$\begin{aligned}\partial_\kappa U_z(H(0)) &= \frac{1}{2\pi i}\int_\Gamma d\zeta\ e^{z\zeta}\ \partial_\kappa(\zeta\mathbb{1} + H(0))^{-1} \\ &= \frac{1}{2\pi i}\int_\Gamma d\zeta\ e^{z\zeta}\,\frac{1}{2\pi i}\int_C d\kappa'\,(\kappa')^{-2}\,(\zeta\mathbb{1} + H(\kappa'))^{-1}.\end{aligned}$$

In the last line we use that the type (A) analyticity of $H(\kappa)$ *implies* the operator-norm κ-analyticity of the resolvent $R_\zeta(H(\kappa))$. So, we calculate the above derivative as the integral for a small circle $C \subset D_{1/(b+c)}$ around zero in the κ-plane. Then for derivatives of any order at non-zero points κ one gets the expression

$$\partial_\kappa^n (\zeta\mathbb{1} + H(\kappa))^{-1} = \frac{n!}{2\pi i}\int_C d\kappa' \, (\kappa' - \kappa)^{-n-1} (\zeta\mathbb{1} + H(\kappa'))^{-1}, \tag{4.111}$$

for a small circle C around κ. Due to fundamental inequality (1.69), which is uniform in a small disc $D_{1/I}$ with the centre at κ, (4.111) yields the estimate

$$\|\partial_\kappa^n (\zeta\mathbb{1} + H(\kappa))^{-1}\| \leqslant \frac{M_\varepsilon \, I^n \, n!}{|\zeta|} \ , \ |\arg\zeta| \leqslant \frac{\pi}{2} + \theta - \varepsilon \ . \tag{4.112}$$

The estimate (4.112) ensures for small $|\kappa|$ the uniform in $\zeta \in \rho(H(\kappa))$ operator-norm convergence of the Taylor series

$$(\zeta\mathbb{1} + H(\kappa))^{-1} = \sum_{n=0}^{\infty} \frac{\kappa^n}{n!} \, \partial_\kappa^n (\zeta\mathbb{1} + H(\kappa))^{-1}\Big|_{\kappa=0} . \tag{4.113}$$

Integration of expansion (4.113) in (4.110) gives the $\|\cdot\|$-convergent Taylor series for holomorphic semigroup $\{U_z(H(\kappa))\}_{z\in S_{\theta(|\kappa|,b)}}$, cf. (4.106), and also proves that the family $\{U_z(H(\kappa))\}_{\kappa\in D}$ is norm holomorphic in a small disc D.

We note that most of the above arguments can be used verbatim in the proof for the case of the $\|\cdot\|_p$-norm topology.

Corollary 4.51. *Let closed operators* $\{H(\kappa)\}_{\kappa\in D\subset\mathbb{C}}$ *form a holomorphic family of type* (A). *If* $H(0) \in \mathscr{H}(\theta,\omega)$ *is a p-generator, then the Riesz-Dunford formula* (4.110) *allows to proceed with the same arguments as above using systematically the* $\|\cdot\|_p$*-topology. This also includes the integration.*

Finally, the semigroup property and Proposition 4.3 *allow to lift the holomorphic properties of the map*

$$\mathcal{D} \ni (z,\kappa) \mapsto e^{-z\,H(\kappa)} \in \mathcal{C}_p(\mathcal{H}) \ ,$$

to the trace-norm topology for $p = 1$.

4.6 $\mathcal{P}_b$-perturbations of Gibbs semigroups

In this section we return to the $\mathcal{P}_b$-perturbations of Gibbs semigroups constructed in the previous section using the Riesz-Dunford formula. Here we use for analysis the *semigroup-based method.* In contrast to the resolvent-based, used in the previous section, it makes the perturbation series (cf. Section 4.4) more explicit.

The semigroup-based method is motivated by the abstract Volterra-type integral equations (Lemma 4.39) and their iterative solutions by the convergent series, Propositions 4.41 and 4.44. But within this *direct* method we forfeit the full power of advantages deriving from the analyticity which is essential for the resolvent-based method.

The Proposition 4.49 and Corollaries 4.50, 4.51, show the power of the resolvent-based method for Gibbs semigroups. We learned a similar arguments before in Sections 1.6, 1.7 for the case of the strongly continuous holomorphic semigroups.

A discrepancy between these methods gets even more sound, when one moves from $\mathcal{P}_{0+}$-perturbations to the $\mathcal{P}_b$-perturbations, or to stronger topologies of continuity. To illustrate the limitations of the semigroup-based method for $\mathcal{P}_b$-perturbations we consider here a concrete example of perturbations from the holomorphic family of type (A), $\{H(\kappa) := A + \kappa B\}_{\kappa \in D \subset \mathbb{C}}$, which is a linear function.

If conditions of Proposition 4.49 are satisfied, then the function

$$[0,1] \ni \alpha \mapsto V_\alpha^{(1)} := G_{(1-\alpha)z}(H(\kappa_0 + \kappa))\, G_{\alpha z}(H(\kappa_0)) \in \mathcal{C}_1(\mathcal{H}) \tag{4.114}$$

in the sector $\mathfrak{S}_{\{|\kappa_0|,|\kappa_0+\kappa|\}} := S_{\theta(|\kappa_0|,b)} \cap S_{\theta(|\kappa_0+\kappa|,b)}$, is continuously $\|\cdot\|_1$-differentiable for $s \in (0,1)$ and for the fixed parameters $\kappa_0, (\kappa_0 + \kappa) \in D_1$, with a given $z \in \mathbb{C}$. The derivative of (4.114) has the form

$$\partial_\alpha V_\alpha^{(1)} = z\, G_{(1-\alpha)z}(H(\kappa_0 + \kappa))\, \kappa\, B\, G_{\alpha z}(H(\kappa_0)) \in \mathcal{C}_1(\mathcal{H}). \tag{4.115}$$

Applying to (4.115) the $\|\cdot\|_1$-convergent Bochner integral one gets the $\|\cdot\|_1$-valued relation

$$\begin{aligned} &G_z(H(\kappa_0 + \kappa)) - G_z(H(\kappa_0)) \\ &= -z \int_0^1 d\alpha\, G_{(1-\alpha)z}(H(\kappa_0 + \kappa))\, \kappa\, B\, G_{\alpha z}(H(\kappa_0)). \end{aligned} \tag{4.116}$$

For $\kappa_0 = 0$ the relation (4.116) is the *integral equation*

$$G_z(H(\kappa)) = G_z(A) - \int_{[0,z]} d\zeta\, G_{z-\zeta}(H(\kappa))\, \kappa\, B\, G_\zeta(A), \tag{4.117}$$

for the construction of a perturbed holomorphic Gibbs semigroup

$$\{G_z(H(\kappa))\}_{z \in S_{\theta(|\kappa|,b)}}$$

by the semigroup-based method. Here the integral is taken along the ray $[0,z]$ in the sector $S_{\theta(|\kappa|,b)} \cup \{0\}$.

After a change of variables the equations (4.116) and (4.117) are reduced to the *standard* integral equations with integration along the ray $[0,t] \subset \mathbb{R}_0^+$, Section 4.4 (4.61).

Proposition 4.52. *Let the self-adjoint operator $A \geqslant 0$ be a p-generator. Let the linear holomorphic family of type (A): $\{H(z) := A + zB\}_{z\in D_1}$, for unit disc $D_1 \subset \mathbb{C}$, be generated by $B \in \mathcal{P}_{b<1}(A)$. Suppose also that the adjoint operator $B^* \in \mathcal{P}_{b_* <1}(A)$. Then the family of Gibbs semigroups $\{G_t(H(\kappa))\}_{\{\kappa:|\kappa|<1\}}$, where $t \in \mathbb{R}^+$, is $\|\cdot\|_1$-differentiable with respect to $\kappa \in (-1, 1)$.*

Proof. Since the conditions for Proposition 4.49 are satisfied, $\{G_t(H(\kappa))\}_{t\geqslant 0}$ is a Gibbs semigroup for any $|\kappa| \leqslant 1$. Using the equation (4.116) with the integral along the ray $[0, t] \subset \mathbb{R}_0^+$, we obtain

$$G_t(H(\kappa + \kappa')) - G_t(H(\kappa)) = -\kappa' \int_0^t dsG_{t-s}(H(\kappa + \kappa'))\, B\, G_s(H(\kappa)), \tag{4.118}$$

where $\kappa \in [-1, 1]$ and $(\kappa + \kappa') \in [-1, 1]$. Then to prove the existence of the $\|\cdot\|_1$-derivative $\partial_\kappa G_t(H(\kappa))$ one has to check (at least) that the $\|\cdot\|_1$-limit of the right-hand side of (4.118) exists and is equal to zero, when $\kappa' \to 0$.

For this purpose we split the integral in (4.118) into two parts:

$$\begin{aligned}&\int_0^{t/2} ds\, G_{t-s}(H(\kappa + \kappa'))\, B\, G_s(H(\kappa)) \\ &\quad + \int_{t/2}^{t} ds\, G_{t-s}(H(\kappa + \kappa'))\, B\, G_s(H(\kappa)),\end{aligned} \tag{4.119}$$

to estimate the integrand in the $\|\cdot\|_1$-norm on two intervals.

First we consider the integrand for $s \in [t/2, t]$. Then

$$\begin{aligned}&\|G_{t-s}(H(\kappa + \kappa'))\, B\, G_s(H(\kappa))\|_1 \\ &\leqslant \|G_{t-s}(H(\kappa + \kappa'))\|\|BG_{s/2}(H(\kappa))\|\|G_{s/2}(H(\kappa))\|_1.\end{aligned} \tag{4.120}$$

Since $\{G_z(H(\kappa))\}_{z\in S_{\theta(|\kappa|,b)}}$ is a quasi-bounded holomorphic semigroup (Proposition 1.54), we have $\|G_{t-s}(H(\kappa + \kappa'))\| \leqslant Me^{(s-t)\,\gamma(1,a,b)}$, where its type $\omega_0 \leqslant \omega := -\gamma(1, a, b)$. The last factor $\|G_{s/2}(H(\kappa))\|_1$ is bounded, since $\{G_t(H(\kappa))\}_{t\geqslant 0}$ is a Gibbs semigroup and since $s \geqslant t/2 > 0$.

Let us set $H_\omega(\kappa) := H(\kappa) + \omega \mathbb{1} \geqslant 0$. Then for the factor in the middle we obtain for $\mu > 0$ that

$$\|BG_{s/2}(H(\kappa))\| \leqslant \|B(H_\omega(\kappa) + \mu\mathbb{1})^{-1}\|\|(H_\omega(\kappa) + \mu\mathbb{1})G_{s/2}(H(\kappa))\|. \tag{4.121}$$

By the Definition 1.50 of the Kato-small perturbations $\mathcal{P}_{b<1}(A)$ (1.101) one gets for $u \in \mathcal{H}$ that

$$\begin{aligned}&\|B(H_\omega(\kappa) + \mu\mathbb{1})^{-1}u\| \\ &\leqslant a\|(H_\omega(\kappa) + \mu\mathbb{1})^{-1}u\| + b\|A(H_\omega(\kappa) + \mu\mathbb{1})^{-1}u\| \\ &\leqslant b\|u\| + (a + b(\omega + \mu))\|(H_\omega(\kappa) + \mu\mathbb{1})^{-1}u\| + b|\kappa|\|B(H_\omega(\kappa) + \mu\mathbb{1})^{-1}u\|.\end{aligned} \tag{4.122}$$

Since $H_\omega(\kappa) \geqslant 0$, these inequalities imply the estimate

$$\|B(H_\omega(\kappa) + \mu\mathbb{1})^{-1}\| \leqslant \frac{1}{1 - b|\kappa|}\,(2b + (a + \omega b)/\mu)\,. \tag{4.123}$$

For the bound of the norm of the last factor in (4.121) we use that operator $H_\omega(\kappa) \in \mathscr{H}(\theta(|\kappa|, b), 0)$ is the generator of a holomorphic contraction semigroup. By the same reason as in (1.76) we obtain the upper bound

$$\|(H_\omega(\kappa) + \mu\mathbb{1})G_{s/2}(H(\kappa))\| \leqslant \frac{2\,M_1'}{s} + (\omega + \mu)M\mathrm{e}^{s\,\omega/2}\,. \tag{4.124}$$

Combining the estimates (4.121)–(4.124) proves the $\|\cdot\|_1$-norm uniform boundedness of (4.120):

$$\|G_{t-s}(H(\kappa + \kappa'))\,B\,G_s(H(\kappa))\|_1 \leqslant M_{[t/2,t]}\,, \tag{4.125}$$

and consequently the boundedness of the integral (4.119) on the interval $[t/2, t]$.

To estimate the integral (4.119) on the interval $s \in [0, t/2]$ we use for $\mu > 0$ the inequality:

$$\begin{aligned}
&\|G_{t-s}(H(\kappa + \kappa'))\,B\,G_s(H(\kappa))\|_1 \\
&\leqslant \|G_{(t-s)/2}(H(\kappa + \kappa'))\|_1 \|\overline{G_{(t-s)/2}(H(\kappa + \kappa'))(H_\omega(\kappa + \kappa') + \mu\mathbb{1})}\| \\
&\quad\times \|\overline{(H_\omega(\kappa + \kappa') + \mu\mathbb{1})^{-1}B}\| \|G_s(H(\kappa))\|\,.
\end{aligned} \tag{4.126}$$

Here the overline stands for the *closure* of the indicated operators.

Note that the both of these closed operators are *bounded*. We prove this only for the less evident case of the second operator since the line of reasoning for the first one is similar. To this aim we introduce the auxiliary operator

$$\mathcal{O} := B^*(H_\omega^*(\kappa + \kappa') + \mu\mathbb{1})^{-1}. \tag{4.127}$$

Since $B^* \in \mathcal{P}_{b_*<1}(A)$ and $(\kappa + \kappa') \in [-1, 1]$, we have $H^*(\kappa + \kappa') = A + (\kappa + \kappa')B^*$. Then by the same arguments as for B (see (4.122), (4.123)) we deduce that the operator (4.127) is bounded and its norm is estimated as

$$\|\mathcal{O}\| \leqslant \frac{1}{1 - b_*|\kappa + \kappa'|}\,(2b_* + (a + \omega b_*)/\mu).$$

Consequently, for all $u \in \mathcal{H}$ and any $v \in \operatorname{dom} B$ the scalar product

$$(\mathcal{O}u, v) = (u, \mathcal{O}^* v) = (u, (H_\omega(\kappa + \kappa') + \mu\mathbb{1})^{-1}B\,v), \tag{4.128}$$

is bounded. Since the adjoint of a bounded operator is also bounded, we have $\mathcal{O}^* \in \mathcal{L}(\mathcal{H})$ with $\|\mathcal{O}^*\| = \|\mathcal{O}\|$. Moreover, by (4.128) we obtain

$$(H_\omega(\kappa + \kappa') + \mu\mathbb{1})^{-1}B \subseteq \mathcal{O}^*.$$

This means that the operator $(H_\omega(\kappa+\kappa')+\mu\mathbb{1})^{-1}B$ with a dense domain dom B is closable and its closure is the bounded operator

$$\overline{(H_\omega(\kappa+\kappa')+\mu\mathbb{1})^{-1}B} = ((H_\omega(\kappa+\kappa')+\mu\mathbb{1})^{-1}B)^{**} = \mathcal{O}^*, \tag{4.129}$$

with the norm

$$\|\overline{(H_\omega(\kappa+\kappa')+\mu\mathbb{1})^{-1}B}\| = \|(B^*(H_\omega^*(\kappa+\kappa')+\mu\mathbb{1})^{-1})^*\| = \|\mathcal{O}\|.$$

Since the semigroup $\{G_z(H(\kappa))\}_{z\in S_{\theta(|\kappa|,b)}}$ is holomorphic, the operator $(H_\omega(\kappa+\kappa')+\mu\mathbb{1})G_{(t-s)/2}(H(\kappa+\kappa'))$ is bounded. Then by similar arguments as above, the operator $\overline{G_{(t-s)/2}(H(\kappa+\kappa'))(H_\omega(\kappa+\kappa')+\mu\mathbb{1})}$ is also bounded. This yields the uniform $\|\cdot\|_1$-boundedness of (4.126):

$$\|G_{t-s}(H(\kappa+\kappa'))\,B\,G_s(H(\kappa))\|_1 \leqslant M_{[0,t/2]}\,, \tag{4.130}$$

and, hence of the integral (4.119) on the interval $s\in[0,t/2]$. Together with the estimate on the interval $s\in[t/2,t]$ our arguments yield the $\|\cdot\|_1$-norm boundedness of the $\|\cdot\|_1$-integral in (4.118).

Thanks to the uniform boundedness of integral in the right-hand side of (4.118), one first gets the $\|\cdot\|_1$-continuity of the family $\{G_t(H(\kappa))\}_{\{\kappa:|\kappa|<1\}}$:

$$\lim_{\kappa'\to 0} G_t(H(\kappa+\kappa')) = G_t(H(\kappa)). \tag{4.131}$$

Using the limit (4.131) in the integrand and the uniform $\|\cdot\|_1$-convergence of the integral in (4.118) we obtain for the $\|\cdot\|_1$-derivative the explicit formula

$$\partial_\kappa G_t(H(\kappa)) = -\int_0^t ds\, G_{t-s}(H(\kappa))\,B\,G_s(H(\kappa)). \tag{4.132}$$

This proves the assertion about the $\|\cdot\|_1$-differentiability of the family of Gibbs semigroups $\{G_t(H(\kappa))\}_{\{\kappa:|\kappa|<1\}}$, for $t>0$. □

Corollary 4.53. *Using the estimates* (4.125) *and* (4.130) *one can iterate* (4.132) *to calculation high-order derivatives. This leads to the Dyson-Phillips representation* (4.106), *in correspondence with the semigroup-based method. A difference is that the resolvent-based method gives simultaneously a control of convergence the series* (4.106), *as in* (4.108).

Corollary 4.54. *The same arguments are applicable to the case of integration along the ray* $\zeta\in[0,z]$ (4.117) *in the sector* $S_{\theta(|\kappa|,b)}\cup\{0\}$. *This gives for the* $\|\cdot\|_1$-*derivative the formula*

$$\partial_\kappa G_z(H(\kappa)) = -\int_0^z d\zeta\, G_{z-\zeta}(H(\kappa))\,B\,G_\zeta(H(\kappa))\,, \tag{4.133}$$

for Gibbs semigroups, when $z\in S_{\theta(|\kappa|,b)}$.

4.7 Notes

Notes to Section 4.1. Compact semigroups are well-known for already a long time. They appear naturally, for example, in the study of parabolic partial differential equations in bounded spatial domains [RR93]. The extension to semigroups with values in von Neumann-Schatten $\mathcal{C}_p(\mathcal{H})$-ideals is conceptually straightforward, see for example, [Bal76], [Dav07].

However, it seems that it was D. A. Uhlenbrock [Uhl71] who proposed for the first time the notion of *Gibbs semigroups* (Definition 4.1), motivated by Quantum Statistical Mechanics. Then this concept was supported in [ANB75]. Indeed, the canonical *quantum Gibbs state* of a finite system is defined by the density matrix operator $Z_\Lambda(\beta)^{-1}\exp(-\beta H_\Lambda)$, where $\beta^{-1} \geqslant 0$ is the temperature of the system, H_Λ is its Hamiltonian, and the canonical partition function $Z_\Lambda(\beta) = \operatorname{Tr}\exp(-\beta H_\Lambda)$, see e.g., [BR96].

Remark 4.4 says that, according to our Definition 4.1, the semigroup corresponding to the Quantum Statistical Mechanics is *immediately* Gibbs [Zag80], [Zag89]. In contrast, the semigroup defined by expression (4.3) is an example of an *eventually* Gibbs semigroup, i.e., $\{T_t\}_{t\geqslant 0}$ becomes a Gibbs semigroup only after some threshold $t_0 > 0$. Then, by Corollary 4.6, the semigroup $\{T_t\}_{t\geqslant 0}$ is also *eventually* $\|\cdot\|_1$-continuous, that is, continuous in the trace-norm topology for $t > t_0$. The terminology *immediately/eventually*, is the same as for compact semigroups, see for example, Section 4.2 and [EN00].

We resume this section by discussing the relation between spectra $\sigma(A)$ and $\sigma(U_t(A))$. For details concerning the application of the Gel'fand transform, see [Dav80], Chapter 2.2, and [BR79], Section 2.3.5.

Notes to Section 4.2. For *strongly* continuous semigroups $\{U_t\}_{t\geqslant 0}$ with values in $\mathcal{C}_{1\leqslant p\leqslant\infty}(\mathcal{H})$-ideals for t away from zero there is a curious relation between their *topology of continuity* and *topology of image* $\mathcal{C}_p(\mathcal{H})$ as the Banach space. Here we continue to discuss this observation (started in Section 4.1 for the Gibbs semigroups), but now for the case of compact semigroups, see Proposition 4.17.

Note that in Chapter 5 we shall show that this relation also concerns the *topology of continuity* of semigroups away from zero and the *topology of convergence* of the Trotter-Kato product formulae. The first result was due to V. A. Zagrebnov [Zag88]. In [NZ90a, NZ90b] the existence of such relation was formulated for abstract Gibbs semigroups. Then it was confirmed for symmetrically-normed ideals in [NZ99d]. For the case of compact semigroups this relation with the operator-norm convergence of the Trotter-Kato product formulae was established in [NZ99b].

Definitions 4.13 and 4.14 of *immediately* and *eventually* compact C_0-semigroups as well as Proposition 4.19 for holomorphic semigroups are standard, see [EN00]. Since (in contrast to the Gibbs semigroups, see Remark 4.4) they are *immediately* norm-continuous for $t > 0$ (Lemma 4.15), the relation between the spectral properties of these semigroups and their generators, Proposition 4.21 and Proposition 4.22, is more explicit than in the general case, see Proposition 4.11.

Different formulations of the minimax principle (Proposition 4.23), including formulations for unbounded operators, as well as its history can be found in, e.g., [RS78] and [BEH94].

Notes to Section 4.3. The notion of p-generator was introduced for the particular case of self-adjoint semigroups by V. A. Zagrebnov [Zag89]. Proposition 4.27 is a generalisation of Theorem 4.1 from [Zag89].

Proposition 4.30 and Corollaries 4.31 and 4.32 are due to V. Cachia and V. A. Zagrebnov [CZ01c]. Note that Proposition 4.30 is an extension of the result [BG72] for self-adjoint generators to the case of m-sectorial generators.

Notes to Section 4.4. In [Uhl71] D. A. Uhlenbrock considered the perturbation theory of Gibbs semigroups generators for bounded perturbations from the class $\mathcal{P}_0$.

The class of $\mathcal{P}$-perturbations was introduced in [HP57], see also [Dav07], Chapter 11.4. The perturbation theory of quasi-bounded strongly continuous semigroups $Q(M, \omega_0)$ for the case of $\mathcal{P}$-perturbations (Proposition 4.41) is standard, [Dav80], Chapter 2 and [Dav07], Chapter 11.

For *self-adjoint* Gibbs semigroups, the theory concerned with $\mathcal{P}$-perturbations was developed by N. Angelescu, G. Nenciu, and M. Bundaru in [ANB75]. Both results, [Uhl71] and [ANB75], for Gibbs semigroups are based on the Ginibre-Gruber inequality [GG69].

A generalisation of this inequality to m-sectorial generators (Lemma 4.43) is due to V. Cachia and V. A. Zagrebnov [CZ01c]. This generalisation allows to prove the trace-norm convergence for the $\mathcal{P}$-perturbation series of Gibbs semigroups generated by m-sectorial operators, see Proposition 4.44.

Notes to Section 4.5. Proposition 4.45 on $\mathcal{P}_{b<1}$-perturbations for *self-adjoint* Gibbs semigroups is due to V. A. Zagrebnov [Zag89].

The analytic theory for the case of $\mathcal{P}_{b<1}$-perturbations of Gibbs semigroups with p-generators was developed for the first time by D. Maison [Mai71]. In our exposition, Proposition 4.49 and Corollary 4.50, we follow essentially [Mai71] and [Zag89].

Notes to Section 4.6. The names *resolvent-based* and *semigroup-based* methods are due to [Dav07], Chapters 11.4 and 11.5. They indicate clearly the difference between two ways for construction of semigroups. The *product formula-based* method mentioned in Remark 4.47(b),(c), was advocated in [Che74].

The illustrative example and Proposition 4.52 are motivated by [Mai71] and [Zag89]. Condition $B^* \in \mathcal{P}_{b_*<1}(A)$ does not follow from $B \in \mathcal{P}_{b<1}(A)$. It is indispensable for estimating products (like (4.126)) including unbounded operators. The line of reasoning is standard, see [NZ98] Section 1, or Remark 5.14.

The question whether there exists a modification of Definition 4.34 such that it gives *non*-infinitesimally small perturbations, was positively solved in [Voi77]. Modifying the integral condition (4.51) he proposed a new (Miyadera-Voigt) class of unbounded perturbations $\mathcal{P}_{MV} \subset \mathcal{P}_{b>0}$, such that $\mathcal{P}_{0^+} \subset \mathcal{P}_{MV}$, cf. (4.56). For details see, e.g., [EN00], Chapter III, Section 3c. Similarly to perturbations of class $\mathcal{P}_b$ (Proposition 1.51), the holomorphic semigroups are stable with respect to Miyadera-Voigt perturbations, [EN00], 3.17 Exercises.

Chapter 5

Product formulae for Gibbs semigroups

A wealth of results in the literature deal with the Lie-Trotter and Trotter-Kato product approximations for *strongly* continuous semigroups in the *strong* operator topology. However, it has been known since a long time that for the Gibbs semigroups the Trotter-Kato product formulae converges also in the *trace-norm* topology, see Notes in Section 5.6 and comments in Appendix D.4.

On the other hand, the *operator-norm* convergence of the Trotter-Kato product formulae with *error bound* estimates is a relatively recent result for *self-adjoint* strongly continuous semigroups. We give a detailed proof of this fact in Section 5.2.

In Section 5.3 some general conditions ensuring the operator-norm Trotter-Kato product formulae convergence of the *degenerate* self-adjoint semigroup *without* error bound estimates are considered. We also present there some results about the *non-self-adjoint* strongly continuous semigroups.

In Section 5.4, it is shown how to *lift* the error bound estimates from Section 5.2 to establish the *trace-norm* convergence of the Trotter-Kato product formulae for the *self-adjoint* Gibbs semigroups. This lifting method allows also to prove that the estimates of rate of convergence for the operator-norm and the trace-norm cases coincide.

We complete this chapter by Section 5.5, where we prove the convergence of the Trotter-Kato product formulae in the trace norm for *non-self-adjoint* Gibbs semigroups. First, we develop for our purpose the *analytic continuation* method for associated with generators sesquilinear forms. It allows us to establish the convergence, but it does not yield results on the error bound estimates. Then similarly to Section 5.4 we show how to *lift* the error bound estimates obtained for the operator-norm convergence of the Trotter-Kato product formulae for strongly continuous *non-self-adjoint* semigroups, to trace-norm estimates for *non-self-adjoint*

V. A. Zagrebnov, *Gibbs Semigroups*, Operator Theory: Advances and Applications 273, https://doi.org/10.1007/978-3-030-18877-1_5

Gibbs semigroups.

In the next Section 5.1 we start by recalling the first known result about the product formulae, which is due to Sophus Lie (1875). We also present there a general programme that one follows developing the analysis of the product formulae convergence in various topologies.

5.1 The Lie-Trotter product formula

Since the observation by S. Lie for finite matrices, it is known that the semigroup $\{U_t(A+B)\}_{t\geqslant 0}$ generated by $A+B$, the sum of two bounded operators A and B on a Banach space $\mathcal{B}$, can be approximated in terms of the semigroups $\{U_t(A)\}_{t\geqslant 0}$ and $\{U_t(B)\}_{t\geqslant 0}$.

Proposition 5.1 (Lie product formula). *Let $A, B \in \mathcal{L}(\mathcal{B})$. Then for any $t \geqslant 0$ and $n \in \mathbb{N}$, we have*

(a) *the product approximation*

$$\left\|\left(U_{t/n}(A)U_{t/n}(B)\right)^n - U_t(A+B)\right\| \leqslant \frac{c}{n}, \tag{5.1}$$

where $c = c(t, \|A\|, \|B\|) > 0$;

(b) *the symmetrised product approximation*

$$\left\|\left(U_{t/2n}(A)U_{t/n}(B)U_{t/2n}(A)\right)^n - U_t(A+B)\right\| \leqslant \frac{c_{\mathrm{sym}}}{n^2}, \tag{5.2}$$

where $c_{\mathrm{sym}} = c_{\mathrm{sym}}(t, \|A\|, \|B\|) > 0$.

Proof. (a) Since A, B and $A+B$ are bounded, we can use the norm convergent exponential series (1.1). Let $P(t) := U_t(A)U_t(B)$. Then the difference

$$P(t/n)^n - U^n_{t/n}(A+B) = \sum_{m=0}^{n-1} P(t/n)^{n-m-1}\left(P(t/n) - U_{t/n}(A+B)\right)U^m_{t/n}(A+B) \tag{5.3}$$

can be estimated as follows

$$\begin{aligned} &\left\|P(t/n)^n - U^n_{t/n}(A+B)\right\| \\ &\leqslant n\big(\max\{\|P(t/n)\|, \|U_{t/n}(A+B)\|\}\big)^n \|P(t/n) - U_{t/n}(A+B)\|. \end{aligned} \tag{5.4}$$

Using exponential series (1.1) one gets the estimates

$$\|P(t/n)\| \leqslant \mathrm{e}^{t(\|A\|+\|B\|)/n} \quad \text{and} \quad \|U_{t/n}(A+B)\| \leqslant \mathrm{e}^{t(\|A\|+\|B\|)/n}.$$

Then (5.4) yields

$$\Big\|P(t/n)^n - U_{t/n}^n(A+B)\Big\| \leqslant n\,\mathrm{e}^{t(\|A\|+\|B\|)}\|P(t/n) - U_{t/n}(A+B)\|. \qquad (5.5)$$

Since

$$\begin{aligned}
&\Big\|P(t/n) - U_{t/n}(A+B)\Big\| \\
&= \Big\|\sum_{k=0}^{\infty}\frac{1}{k!}\Big(-\frac{t}{n}A\Big)^k \sum_{\ell=0}^{\infty}\frac{1}{\ell!}\Big(-\frac{t}{n}B\Big)^\ell - \sum_{s=0}^{\infty}\frac{1}{s!}\Big[-\frac{t}{n}(A+B)\Big]^s\Big\| \\
&\leqslant \sum_{k=2}^{\infty}\frac{1}{k!}\Big(\frac{t}{n}\|A\|\Big)^k \sum_{\ell=2}^{\infty}\frac{1}{\ell!}\Big(\frac{t}{n}\|B\|\Big)^\ell + \sum_{k=3}^{\infty}\frac{1}{k!}\Big(\frac{t}{n}\|A\|\Big)^k \\
&\quad + \sum_{\ell=3}^{\infty}\frac{1}{\ell!}\Big(\frac{t}{n}\|B\|\Big)^\ell + \frac{t}{n}\|A\|\sum_{\ell=2}^{\infty}\frac{1}{\ell!}\Big(\frac{t}{n}\|B\|\Big)^\ell \\
&\quad + \frac{t}{n}\|B\|\sum_{k=2}^{\infty}\frac{1}{k!}\Big(\frac{t}{n}\|A\|\Big)^k + \sum_{s=3}^{\infty}\frac{1}{s!}\Big[\frac{t}{n}(\|A\|+\|B\|)\Big]^s \\
&\quad + \frac{t^2}{2n^2}\|AB - BA\| =: \frac{1}{n^2}M_n(t,\|A\|,\|B\|),
\end{aligned} \qquad (5.6)$$

the inequality (5.5) yields the estimate (5.1), where c is defined as the maximum over $n \geqslant 1$ of the right-hand side of (5.6) multiplied by $n^2\,\mathrm{e}^{t(\|A\|+\|B\|)}$. Then $c = \mathrm{e}^{t(\|A\|+\|B\|)}M(t,\|A\|,\|B\|)$, where $M(t,\|A\|,\|B\|) := \max_{n\geqslant 1} M_n(t,\|A\|,\|B\|)$.

(b) Let $T(t) := U_t^{1/2}(A)U_t(B)U_t^{1/2}(A)$ be a *symmetrised* product approximation. Then arguing as in (a) we obtain

$$\Big\|T(t/n)^n - U_{t/n}^n(A+B)\Big\| \leqslant n\,\mathrm{e}^{t(\|A\|+\|B\|)}\Big\|T(t/n) - U_{t/n}(A+B)\Big\|. \qquad (5.7)$$

Since the symmetrisation eliminates the last term in (5.6), we get

$$\begin{aligned}
&\Big\|T(t/n) - U_{t/n}(A+B)\Big\| \\
&= \Big\|\sum_{k_1=0}^{\infty}\frac{1}{k_1!}\Big(-\frac{t}{2n}A\Big)^{k_1}\sum_{\ell=0}^{\infty}\frac{1}{\ell!}\Big(-\frac{t}{n}B\Big)^\ell\sum_{k_2=0}^{\infty}\frac{1}{k_2!}\Big(-\frac{t}{2n}A\Big)^{k_2} \\
&\quad - \sum_{s=0}^{\infty}\frac{1}{s!}\Big[-\frac{t}{n}(A+B)\Big]^s\Big\| \\
&\leqslant \frac{1}{n^3}L(t,\|A\|,\|B\|).
\end{aligned} \qquad (5.8)$$

where L is determined in a similar way as the right-hand side of (5.6). The estimate (5.8) gives a better than (5.1) *rate* of convergence (5.2) for $c_{\mathrm{sym}} = \mathrm{e}^{t(\|A\|+\|B\|)}L(t,\|A\|,\|B\|)$. □

Remark 5.2. (1) In Section 5.6 and Appendix D.4 we present a brief history of the product formulae since the work of S. Lie. Here we would like to notice only that in Hilbert and Banach spaces the (*exponential*) Trotter product formula

$$\gamma\text{-}\lim_{n\to\infty}\left(\mathrm{e}^{-tA/n}\mathrm{e}^{-tB/n}\right)^n = \mathrm{e}^{-tC}, \tag{5.9}$$

has permeated throughout operator and probability theory, for *various* topologies of convergence γ. The challenge is to find general hypotheses under which the formula holds in infinite-dimensional settings, including the *strongest* topology of convergence γ.

(2) A realisation of this *programme* requires at least the following steps:

(a) to give a description of the set of pairs of operators $\{A, B\}$ for which the limit (5.9) exists (in some sense) and yields a *semigroup*;

(b) to find the corresponding *optimal* (or the strongest) topology γ, for which this limit exists;

(c) to reconstruct the *generator* C from the pair of operators $\{A, B\}$;

(d) to generalise, if possible, the *exponential* functions in (5.9) to *Borel measurable functions* $f, g : \mathbb{R}_0^+ \to [0,1]$, with the properties: $f(0) = 1,\ f'(+0) = -1$ and $g(0) = 1,\ g'(+0) = -1$;

(e) to consider product formulae for *different* product *approximants*, for example, nonsymmetrised $\{P(t/n)^n\}_{n\geqslant 1}$ or symmetrised $\{T(t/n)^n\}_{n\geqslant 1}$, including various order of factors, to optimise the rate of convergence.

5.2 Trotter-Kato product formulae: operator-norm convergence, error bounds

For a long time, the Lie-Trotter product formula (5.9) for strongly continuous contraction semigroups was known only for *exponential* functions f, g and *solely* in the *strong* operator topology $\gamma = s$, with a limit that in turn is a contraction C_0-semigroup.

In the present section we show that the arguments used in the proof of Proposition 5.1 can be refined to yield a remarkable and rather nontrivial statement about convergence in the *operator-norm* topology $\gamma = \|\cdot\|$, of *various* Trotter-Kato product formulae corresponding to *different* types of *non-exponential* approximants generated by a set of Borel measurable functions f, g, Proposition 5.8.

We note at this point the following important fact, which explains why we undertake the study below:

- It is the *operator-norm* continuity away from zero, which holds for the *strongly* continuous *self-adjoint* semigroups generated by A and B, that yields $\gamma = \|\cdot\|$.

- For strongly continuous Gibbs semigroups, which are *trace-norm* continuous away from zero, the Trotter-Kato product formulae were established in the trace-norm topology $\gamma = \|\cdot\|_1$.
- These observations confirm a conventional wisdom that a *natural* topology γ for the convergence of the product formulae *inherits* the (strongest) topology in which the semigroup is continuous *away* from zero.
- The γ-limit of the product formulae *approximants* gives one-parameter family for $t \in \mathbb{R}_0^+$, which is either a C_0-semigroup, or a degenerate semigroup in the sense of Definition 1.24, and is in turn γ-continuous *away* from zero.

Before proceeding, we first specify in subsection 5.2.1 (*Preliminaries and Proposition* 5.8) the conditions on the pair of generators $\{A, B\}$, including a *hierarchy* between the operators A and B. Below this is a *Kato-smallness* condition of one operator with respect to the other. These conditions allow us to obtain in Proposition 5.8 an operator-norm error bound estimate for the convergence *rate* of the Trotter-Kato product formulae for a certain subclass of generic Kato functions $f, g \in \mathcal{K}$. For details about these functions, see Appendix C.

Subsection 5.2.2, *Auxiliary lemmata*, is collection of statements indispensable for subsection 5.2.3, *Proof of Proposition* 5.8. The *exponential* Kato functions $f(x) = g(x) = e^{-x}$ trivially satisfy conditions formulated in these subsections.

We conclude Section 5.2 by a discussion of a quite subtle *optimality* property of the convergence rate. In subsection 5.2.4, *Optimal rate of the operator-norm convergence*, we study sufficient conditions on the pairs of $\{A, B\}$ and on the corresponding admissible functions $\{f, g\}$ that ensure in Proposition 5.19 and in Proposition 5.25 the optimality of the error bound estimate for the convergence rate.

Note that a choice of the *suitable* Kato functions from $\mathcal{K}$ is decisive for the convergence of Trotter-Kato product formulae as well as for the error bound optimality. In subsection 5.2.5, *Optimal rate: fractional conditions*, we show how subtle is the tuning of the class of admissible Kato functions $\{f, g\}$, when the operators $\{A, B\}$ are related by some *fractional smallness* conditions, see Proposition 5.27.

5.2.1 Preliminaries and Proposition 5.8

(i) Let A and B be densely defined, semibounded from below, *self-adjoint* operators in a Hilbert space $\mathcal{H}$. Without loss of generality we can assume that

$$A \geqslant \mathbb{1}, \quad B \geqslant \mathbb{1}. \tag{5.10}$$

(ii) Let $B \in \mathcal{P}_b(A)$. Again for simplicity, we assume that $a = 0$ in (1.101), that is, $\operatorname{dom} A \subset \operatorname{dom} B$ and

$$\|Bu\| \leqslant b\|Au\|, \quad u \in \operatorname{dom} A, \tag{5.11}$$

for $0 \leqslant b < 1$.

Hence, the operator $H = A + B$, with $\operatorname{dom} H = \operatorname{dom} A$, is self-adjoint thanks to the $b < 1$ A-smallness of B (5.11). Note that by (5.10) and Corollary 1.46 the operators A, B and H are generators of holomorphic contraction semigroups on $\mathcal{H}$.

Remark 5.3. We note that there are others hierarchy of conditions on the pair of generators $\{A, B\}$ that ensure convergence in the topology $\gamma = \|\cdot\|$, *without* as well as *with* error bound estimate. We consider some of them in Section 5.3, see Propositions 5.36, 5.45 for the operator-norm convergence without estimate and Propositions 5.47, 5.49 for proofs with error bound estimates. In Section 5.4 we show that for the Gibbs semigroups the results of Section 5.3 can be *lifted* to convergence of the Trotter-Kato product formulae in the trace-norm topology, correspondingly without or with the error bound estimates, *inherited* from the operator-norm estimates.

Definition 5.4. A pair of the Borel measurable functions f, g, which is defined on $\mathbb{R}_0^+ = [0, \infty)$ and satisfies conditions

$$0 \leqslant f(x) \leqslant 1, \quad f(0) = 1, \quad f'(+0) = -1 \tag{5.12}$$

$$0 \leqslant g(x) \leqslant 1, \quad g(0) = 1, \quad g'(+0) = -1. \tag{5.13}$$

is said to belong to the class of *generic Kato functions*. We denote the algebra of monomials generated by f, g, including their fractional powers, by $\mathcal{K}$.

(iii) For the Trotter-Kato product formulae we consider the functions f, g which belong to generic Kato functions from the class $\mathcal{K}$.

Note that for the operators (5.10) and for any $\alpha \geqslant 0$, conditions (5.12) and (5.13) yield

$$\operatorname*{s-lim}_{t \to +0} f(tA)^\alpha = \mathbb{1} \quad \text{and} \quad \operatorname*{s-lim}_{t \to +0} g(tB)^\alpha = \mathbb{1}\,,$$

in the strong operator topology on $\mathcal{L}(\mathcal{H})$.

Next, we formulate conditions on the *pair* of Kato functions from $\mathcal{K}$ under which one can lift the Trotter product formula (5.9) to the operator-norm convergent Trotter-Kato product formula:

$$\|\cdot\|\text{-}\lim_{n \to \infty} \left(f(tA/n) g(tB/n)\right)^n = U_t(H) := \mathrm{e}^{-tH}\,. \tag{5.14}$$

(iv) To this aim we assume more about the admissible Kato functions:

$$C_{1/2} := \operatorname*{ess\,sup}_{x>0} \frac{x\sqrt{f(x)}}{1 - f(x)} < \infty, \tag{5.15}$$

$$C_1 := \operatorname*{ess\,sup}_{x>0} \frac{1 - f(x)}{x} < \infty, \tag{5.16}$$

$$C_2 := \operatorname*{ess\,sup}_{x>0} \left| \left(f(x) - \frac{1}{1+x} \right) \frac{1}{x^2} \right| < \infty, \tag{5.17}$$

$$S_1 := \operatorname*{ess\,sup}_{x>0} \frac{1-g(x)}{x} < \infty, \tag{5.18}$$

$$S_2 := \operatorname*{ess\,sup}_{x>0} \left| \left(g(x) - \frac{1}{1+x} \right) \frac{1}{x^2} \right| < \infty. \tag{5.19}$$

(v) Our last condition reads as

$$b\, C_{1/2} S_1 < 1. \tag{5.20}$$

Remark 5.5. Besides the standard (exponential) Kato functions $f(x) = g(x) = \mathrm{e}^{-x}$, the class of Borel functions satisfying conditions (5.12)–(5.19) contains, for example, $f(x) = (1 + x/2)^{-2}$ with $C_{1/2} = 2$, but *not* $f(x) = (1 + x)^{-1}$, since in this case $C_{1/2} = \infty$. On the other hand $g(x) = (1 + x)^{-1}$ belongs to the class described by (iii) and (iv), since there is no condition on $g(x)$ similar to (5.15). This *asymmetry* in conditions is related to the hierarchy (ii) of operators $\{A, B\}$.

Remark 5.6. Note that by (5.12) and (5.13), we always have $C_{1/2} \geqslant 1$ and $S_1 \geqslant 1$. Therefore, condition (v) implies

$$0 \leqslant b < (C_{1/2} S_1)^{-1} \leqslant 1. \tag{5.21}$$

This means that for $B \in \mathcal{P}_{b<1}(A)$, the condition (v) is *not* superfluous.

Remark 5.7. If $f(x) = g(x) = \mathrm{e}^{-x}$, then

$$C_{1/2} = C_1 = C_2 = S_1 = S_2 = 1. \tag{5.22}$$

This yields $0 \leqslant b < 1$ by condition (v), which is already ensured by (ii). We note that the case $b = 1$ (see subsection 5.2.4) demands another approach to prove the operator-norm convergence in (5.14).

To formulate the main statement of this section, we define for $t \in \mathbb{R}_0^+$ the bounded operator-valued functions: $t \mapsto P(t)$ and $t \mapsto T(t)$, by

$$P(t) := f(tA)g(tB) \quad \text{and} \quad T(t) := f^{1/2}(tA)g(tB)f^{1/2}(tA). \tag{5.23}$$

They produce two types of approximants: nonsymmetrised $\{P(t/n)^n\}_{n\geqslant 1}$ and symmetrised (self-adjoint) $\{T(t/n)^n\}_{n\geqslant 1}$, for two different Trotter-Kato product formulae.

Proposition 5.8. *Suppose the operators* A , B *satisfy conditions* (i), (ii), *and the functions* f, g *satisfy conditions* (iii), (iv). *If also the condition* (v) *is satisfied, then there are positive constants* L_P *and* L_T *such that*

$$\left\| P(t/n)^n - \mathrm{e}^{-tH} \right\| \leqslant \frac{L_P}{(1 - C_{1/2} S_1 b)^2 (1-b)} \frac{\ln(n)}{n} \tag{5.24}$$

and

$$\left\| T(t/n)^n - \mathrm{e}^{-tH} \right\| \leqslant \frac{L_T}{(1 - C_{1/2} S_1 b)(1-b)} \frac{\ln(n)}{n}, \tag{5.25}$$

for $n = 3, 4, \ldots$, *uniformly in* $t \geqslant 0$. *Here* $\{e^{-tH}\}_{t\geqslant 0}$ *is the self-adjoint contraction* C_0*-semigroup with generator* $H = A + B$, $\operatorname{dom} H = \operatorname{dom} A$.

Remark 5.9. We see that under conditions (i)–(v), Proposition 5.8 provides a solution of the problems (a)–(e) of Remark 5.2 concerning the product formulae for unbounded generators. This solution includes the *identification* of the topology γ, which is the same as in Proposition 5.1. It inherits the topology $\gamma = \|\cdot\|$ in which the semigroup $\{e^{-tH}\}_{t\geqslant 0}$ is continuous away from zero.

The differences with Lie-Trotter product formulae are the following:

- before it was known only for exponential functions f, g and only in the strong operator topology $\gamma = s$ (Trotter product formula (5.9));
- the error bound estimates of the rate of convergence (5.1) and (5.2) are *optimal*, i.e. the best possible, whereas those in (5.24) and (5.25) are not, see Section 5.4.

5.2.2 Auxiliary lemmata

We prove Proposition 5.8 with the help of the following series of lemmata.

Lemma 5.10. *Let* $X \in \mathcal{L}(\mathcal{H})$ *be a self-adjoint operator such that* $0 \leqslant X \leqslant \mathbb{1}$. *Then*

$$\|X^k(\mathbb{1} - X)\| \leqslant (1+k)^{-1}, \quad k = 0, 1, 2, \ldots \tag{5.26}$$

Proof. This estimate follows from the inequality

$$\lambda^k(1-\lambda) \leqslant (1+k)^{-1}, \quad k = 0, 1, 2, \ldots \quad (0 \leqslant \lambda \leqslant 1),$$

and the spectral representation

$$X^k(\mathbb{1} - X) = \int_0^1 dE_X(\lambda)\, \lambda^k(1-\lambda),$$

which yields (5.26). □

Lemma 5.11. *For any self-adjoint operator* $X \geqslant 0$ *we have the estimate*

$$\|Xe^{-\alpha X}\| \leqslant (\alpha e)^{-1}, \quad \alpha > 0. \tag{5.27}$$

Proof. Since $\lambda e^{-\alpha\lambda} \leqslant (\alpha e)^{-1}$ for $\lambda \geqslant 0$, (5.27) is again a consequence of the spectral representation for self-adjoint operators. □

Lemma 5.12. *Let* $A \geqslant \mathbb{1}$ *be a self-adjoint operator on* $\mathcal{H}$. *If the Borel function* f *obeys* (5.12) *and* (5.15), *then the operator* $\mathbb{1} - f(\tau A)$ *is boundedly invertible, and*

$$\|(\mathbb{1} - f(\tau A))^{-1}u\| \leqslant \|u\| + \frac{C_{1/2}}{\tau}\|A^{-1}u\|, \quad u \in \mathcal{H}, \tag{5.28}$$

for each $\tau > 0$.

Proof. By virtue of $f(x) \leqslant 1$ and (5.15),

$$0 \leqslant \frac{xf(x)}{1-f(x)} \leqslant C_{1/2}\,, \quad x \geqslant 0.$$

or

$$1 - f(x) \geqslant \frac{x}{(C_{1/2} + x)}\,. \tag{5.29}$$

Since $A \geqslant \mathbb{1}$, the spectral theorem yields

$$\mathbb{1} - f(\tau A) \geqslant \int_1^\infty \mathrm{d}E_A(\lambda)\, \frac{\tau}{C_{1/2}\lambda^{-1} + \tau} \geqslant \frac{\tau}{C_{1/2} + \tau}\mathbb{1}\,.$$

Therefore, the operator $\mathbb{1} - f(\tau A)$ is invertible for $\tau > 0$. On the other hand, since (5.29) implies

$$(1 - f(x))^{-1} \leqslant 1 + \frac{C_{1/2}}{x}\,, \quad x > 0,$$

the spectral representation gives the estimate

$$\|(\mathbb{1} - f(\tau A))^{-1}u\| \leqslant \left\| \int_1^\infty \mathrm{d}E_A(\lambda) \left(1 + \frac{C_{1/2}}{\tau\lambda}\right) u \right\|, \quad u \in \mathcal{H},$$

which proves the assertion (5.28) for $\tau > 0$. □

Lemma 5.13. *Let A and B be self-adjoint operators satisfying conditions* (i), (ii). *If the Borel functions f and g satisfy* (iii), (iv), *and condition* (v) *is fulfilled, then the operator $\mathbb{1} - T(\tau)$ is boundedly invertible for each $\tau > 0$, and*

$$\|(\mathbb{1} - T(\tau))^{-1}u\| \leqslant \frac{1}{1 - C_{1/2}S_1 b}\left(\|u\| + \frac{C_{1/2}}{\tau}\|A^{-1}u\|\right), \quad u \in \mathcal{H}. \tag{5.30}$$

Proof. By definition (5.23), we get the representation

$$\mathbb{1} - T(\tau) = \mathbb{1} - f(\tau A) + f^{1/2}(\tau A)(\mathbb{1} - g(\tau B))f^{1/2}(\tau A). \tag{5.31}$$

Since by Lemma 5.12 the operator $\mathbb{1} - f(\tau A)$, $\tau > 0$, is invertible, identity (5.31) gives

$$\mathbb{1} - T(\tau) = (\mathbb{1} - f(\tau A))[\mathbb{1} + f^{1/2}(\tau A)(\mathbb{1} - f(\tau A))^{-1}(\mathbb{1} - g(\tau B))f^{1/2}(\tau A)]. \tag{5.32}$$

Note that for $\tau > 0$, we get

$$f^{1/2}(\tau A)(\mathbb{1} - f(\tau A))^{-1}(\mathbb{1} - g(\tau B))f^{1/2}(\tau A) \tag{5.33}$$

$$= \tau A f^{1/2}(\tau A)(\mathbb{1} - f(\tau A))^{-1}A^{-1}B\frac{1}{\tau}(\mathbb{1} - g(\tau B))B^{-1}f^{1/2}(\tau A). \tag{5.34}$$

Remark 5.14. Recall that since by (5.11) we have $BA^{-1} \in \mathcal{L}(\mathcal{H})$, and since

$$(BA^{-1}u, v) = (u, A^{-1}Bv), \quad u \in \mathcal{H}, \ v \in \operatorname{dom} B,$$

the bounded operator $(BA^{-1})^* \supseteq A^{-1}B$ is an extension of $A^{-1}B$. This implies that the operator $A^{-1}B$ is closable and the closure $\overline{A^{-1}B} = (BA^{-1})^*$. Since $\|(BA^{-1})^*\| = \|BA^{-1}\|$, the norm

$$\|\overline{A^{-1}B}\| = \|BA^{-1}\| \leqslant b. \tag{5.35}$$

Now, we use (5.35) to estimate (5.34) as follows:

$$\begin{aligned}
&\left\|f^{1/2}(\tau A)(\mathbb{1} - f(\tau A))^{-1}(\mathbb{1} - g(\tau B))f^{1/2}(\tau A)\right\| \\
&\leqslant \frac{1}{\tau}\|BA^{-1}\|\|(\mathbb{1} - g(\tau B))B^{-1}\|\|f^{1/2}(\tau A)\| \\
&\quad \times \left\|\tau A f^{1/2}(\tau A)(\mathbb{1} - f(\tau A))^{-1}\right\|.
\end{aligned} \tag{5.36}$$

Then by virtue of (5.10), (5.11), (5.15), (5.18) and (5.20) we get

$$\left\|f^{1/2}(\tau A)(\mathbb{1} - f(\tau A))^{-1}(\mathbb{1} - g(\tau B))f^{1/2}(\tau A)\right\| \leqslant C_{1/2}S_1 b < 1.$$

Hence, the operator $\mathbb{1} + f^{1/2}(\tau A)(\mathbb{1} - f(\tau A))^{-1}(\mathbb{1} - g(\tau B))f^{1/2}(\tau A)$ is invertible with norm

$$\left\|\left[\mathbb{1} + f^{1/2}(\tau A)(\mathbb{1} - f(\tau A))^{-1}(\mathbb{1} - g(\tau B))f^{1/2}(\tau A)\right]^{-1}\right\| \leqslant (1 - C_{1/2}S_1 b)^{-1}.$$

The last estimate and Lemma 5.12 imply that the operator (5.31) is invertible, with estimate

$$\begin{aligned}
\left\|(\mathbb{1} - T(\tau))^{-1}u\right\| &\leqslant (1 - C_{1/2}S_1 b)^{-1}\left\|(\mathbb{1} - f(\tau A))^{-1}u\right\| \\
&\leqslant (1 - C_{1/2}S_1 b)^{-1}\left[\|u\| + \frac{C_{1/2}}{\tau}\|A^{-1}u\|\right], \quad u \in \mathcal{H},
\end{aligned}$$

for any $\tau > 0$. □

Lemma 5.15. *Let conditions* (i)–(v) *be satisfied. Then there exists a constant* $L_1 > 0$ *such that*

$$\frac{1}{\tau}\left\|A^{-1}(T(\tau) - U_\tau(H))\right\| \leqslant L_1 \tag{5.37}$$

for any $\tau > 0$.

Proof. By the definition (5.23) we get

$$\begin{aligned}
T(\tau) - U_\tau(H) = &- f^{1/2}(\tau A)(\mathbb{1} - g(\tau B))f^{1/2}(\tau A) - (\mathbb{1} - f(\tau A)) \\
&+ (\mathbb{1} - U_\tau(H)),
\end{aligned} \tag{5.38}$$

which yields

$$A^{-1}(T(\tau) - U_\tau(H)) = -f^{1/2}(\tau A)A^{-1}B(\mathbb{1} - g(\tau B))B^{-1}f^{1/2}(\tau A) \tag{5.39}$$
$$- (\mathbb{1} - f(\tau A))A^{-1} + A^{-1}H(\mathbb{1} - U_\tau(H))H^{-1}. \tag{5.40}$$

Then, using the same argument as in (5.35), we obtain the estimate

$$\begin{aligned}\frac{1}{\tau}\|A^{-1}(T(\tau) - U_\tau(H))\| &\leqslant \|BA^{-1}\|\frac{1}{\tau}\|(\mathbb{1} - g(\tau B))B^{-1}\| \\ &+ \frac{1}{\tau}\|(\mathbb{1} - f(\tau A))A^{-1}\| + \|HA^{-1}\|\frac{1}{\tau}\|(\mathbb{1} - U_\tau(H))H^{-1}\|.\end{aligned} \tag{5.41}$$

Therefore, by (5.16), (5.18) and the spectral theorem we get

$$\frac{1}{\tau}\|A^{-1}(T(\tau) - U_\tau(H))\| \leqslant \|BA^{-1}\|S_1 + C_1 + \|HA^{-1}\|. \tag{5.42}$$

Since by (5.11) we have $\|HA^{-1}\| \leqslant \|\mathbb{1} + BA^{-1}\| \leqslant 1 + \|BA^{-1}\| \leqslant 1 + b$, we obtain (5.37) from (5.42) by setting $L_1 := bS_1 + C_1 + 1 + b$. □

Lemma 5.16. *Let conditions* (i)–(v) *be satisfied. Then there exists a constant* $L_2 > 0$ *such that*

$$\frac{1}{\tau^2}\|A^{-1}(T(\tau) - U_\tau(H))A^{-1}\| \leqslant L_2 \tag{5.43}$$

for any $\tau > 0$.

Proof. From (5.38) we obtain the representation

$$\begin{aligned}&T(\tau) - U_\tau(H) \\ &= (\mathbb{1} - f^{1/2}(\tau A))(\mathbb{1} - g(\tau B))f^{1/2}(\tau A) \\ &\quad + (\mathbb{1} - g(\tau B))(\mathbb{1} - f^{1/2}(\tau A)) + (f(\tau A) - (\mathbb{1} + \tau A)^{-1}) \\ &\quad + (g(\tau B) - (\mathbb{1} + \tau B)^{-1}) + ((\mathbb{1} + \tau H)^{-1} - U_\tau(H)) \\ &\quad + \tau H(\mathbb{1} + \tau H)^{-1} - \tau A(\mathbb{1} + \tau A)^{-1} - \tau B(\mathbb{1} + \tau B)^{-1}.\end{aligned} \tag{5.44}$$

Since

$$\begin{aligned}&A^{-1}\{\tau H(\mathbb{1} + \tau H)^{-1} - \tau A(\mathbb{1} + \tau A)^{-1} - \tau B(\mathbb{1} + \tau B)^{-1}\}A^{-1} \\ &= \tau^2\{(\mathbb{1} + \tau A)^{-1} + A^{-1}B(\mathbb{1} + \tau B)^{-1}BA^{-1} \\ &\quad - A^{-1}H(\mathbb{1} + \tau H)^{-1}HA^{-1}\},\end{aligned} \tag{5.45}$$

we get from (5.44) the identity

$$\begin{aligned}
&A^{-1}\{T(\tau) - U_\tau(H)\}A^{-1} \\
&= (\mathbb{1} - f^{1/2}(\tau A))A^{-1}(\mathbb{1} - g(\tau B))B^{-1}BA^{-1}f^{1/2}(\tau A) \\
&\quad + A^{-1}B(\mathbb{1} - g(\tau B))B^{-1}(\mathbb{1} - f^{1/2}(\tau A))A^{-1} \\
&\quad + (f(\tau A) - (\mathbb{1} + \tau A)^{-1})A^{-2} \\
&\quad + A^{-1}B(g(\tau B) - (\mathbb{1} + \tau B)^{-1})B^{-2}BA^{-1} \\
&\quad + A^{-1}H((\mathbb{1} + \tau H)^{-1} - U_\tau(H))H^{-2}HA^{-1} + \tau^2(\mathbb{1} + \tau A)^{-1} \\
&\quad + \tau^2 A^{-1}B(\mathbb{1} + \tau B)^{-1}BA^{-1} - \tau^2 A^{-1}H(\mathbb{1} + \tau H)^{-1}HA^{-1},
\end{aligned}$$

which yields the estimate

$$\begin{aligned}
&\frac{1}{\tau^2}\left\|A^{-1}\{T(\tau) - U_\tau(H)\}A^{-1}\right\| \\
&\leqslant \frac{1}{\tau}\left\|(\mathbb{1} - f^{1/2}(\tau A))A^{-1}\right\| \frac{1}{\tau}\left\|(\mathbb{1} - g(\tau B))B^{-1}\right\| \left\|BA^{-1}\right\| \\
&\quad + \|BA^{-1}\| \frac{1}{\tau}\left\|\mathbb{1} - g(\tau B)B^{-1}\right\| \frac{1}{\tau}\left\|(\mathbb{1} - f^{1/2}(\tau A))A^{-1}\right\| \\
&\quad + \frac{1}{\tau^2}\left\|\{f(\tau A) - (\mathbb{1} + \tau A)^{-1}\}A^{-2}\right\| \\
&\quad + \frac{1}{\tau^2}\left\|\{g(\tau B) - (\mathbb{1} + \tau B)^{-1}\}B^{-2}\right\| \|BA^{-1}\|^2 \\
&\quad + \left\|\{(\mathbb{1} + \tau H)^{-1} - U_\tau(H)\}H^{-2}\right\| \|HA^{-1}\|^2 + 1 + \|BA^{-1}\|^2 \\
&\quad + \|HA^{-1}\|^2. \qquad (5.46)
\end{aligned}$$

Note that by (iii) and (iv) we obtain the inequalities

$$\frac{1 - f^{1/2}(x)}{x} \leqslant \frac{1 - f(x)}{x} \leqslant C_1,$$

which by the spectral representation for A give the estimate

$$\frac{1}{\tau}\left\|(\mathbb{1} - f^{1/2}(\tau A))A^{-1}\right\| \leqslant C_1\,. \qquad (5.47)$$

The spectral representation gives also

$$\frac{1}{\tau^2}\left\|\left\{(\mathbb{1} + \tau H)^{-1} - U_\tau(H)\right\}H^{-2}\right\| \leqslant \frac{1}{2}, \qquad (5.48)$$

where $1/2 = \sup_{x>0}[(1+x)^{-1} - \mathrm{e}^{-x}]x^{-2}$. Now, using estimates (5.47) and (5.48), we obtain from (5.46) that

$$\begin{aligned}
\frac{1}{\tau^2}\|A^{-1}\{T(\tau) - U_\tau(H)\}A^{-1}\| &\leqslant 2C_1S_1\|BA^{-1}\| + C_2 + S_2\|BA^{-1}\|^2 \\
&\quad + \frac{1}{2}\|HA^{-1}\|^2 + 1 + \|BA^{-1}\|^2 + \|HA^{-1}\|^2. \qquad (5.49)
\end{aligned}$$

Then since $\|BA^{-1}\| \leqslant b$ and $\|HA^{-1}\| \leqslant 1+b$, by setting

$$L_2 := 2 + C_2 + \frac{1}{2}(3 + 2C_1S_1)b + (5/2 + S_2)b^2 \,,$$

we get from (5.49) the estimate (5.43). □

5.2.3 Proof of Proposition 5.8

Proof of Proposition 5.8. First we consider the symmetric case (5.25), which is simpler to treat since the operator $T(\tau)$ is self-adjoint. To this end we set $\tau = t/n$, $U_\tau = U_\tau(H)$, and we use the identity (5.3)

$$T^n(\tau) - U_\tau^n = T(\tau)^{n-1}(T(\tau) - U_\tau) + \sum_{n=1}^{n-1} T(\tau)^{n-m-1}(T(\tau) - U_\tau)U_\tau^m.$$

Since by Lemma 5.13 the operator $(\mathbb{1} - T(\tau))^{-1} \in \mathcal{L}(\mathcal{H})$ for $\tau > 0$, we find that

$$\begin{aligned} T^n(\tau) - U_\tau^n &= T(\tau)^{n-1}(T(\tau) - U_\tau) \\ &\quad + \sum_{m=1}^{n-1} T(\tau)^{n-m-1}(\mathbb{1} - T(\tau))\{(\mathbb{1} - T(\tau))^{-1}(T(\tau) - U_\tau)U_\tau^m\}, \end{aligned}$$

which in turn leads to the estimate

$$\begin{aligned} \|T(\tau)^n - U_\tau^n\| &\leqslant \|T(\tau)^{n-1}(T(\tau) - U_\tau)\| \\ &\quad + \sum_{m=1}^{n-1} \|T(\tau)^{n-m-1}(\mathbb{1} - T(\tau))\| \|(\mathbb{1} - T(\tau))^{-1}(T(\tau) - U_\tau)U_\tau^m\|. \end{aligned} \tag{5.50}$$

Taking into account (5.30) we get for the last factor in the sum (5.50) the estimate

$$\|(\mathbb{1} - T(\tau))^{-1}(T(\tau) - U_\tau)U_\tau^m\| \leqslant (1 - C_{1/2}S_1 b)^{-1}\{\|(T(\tau) - U_\tau)U_\tau^m\| \tag{5.51}$$

$$+ \frac{C_{1/2}}{\tau}\|A^{-1}(T(\tau) - U_\tau)U_\tau^m\|\}. \tag{5.52}$$

Note that the identity

$$(T(\tau) - U_\tau)U_\tau^m = (T(\tau) - U_\tau)A^{-1}AH^{-1}HU_\tau^m$$

implies the estimate (cf. (5.35)):

$$\|(T(\tau) - U_\tau)U_\tau^m\| \leqslant \|(T(\tau) - U_\tau)A^{-1}\|\|AH^{-1}\|\|HU_\tau^m\|. \tag{5.53}$$

By virtue of Lemma 5.11 we have $\|HU_\tau^m\| \leqslant (e\,\tau m)^{-1}$. Then (5.53) yields

$$\|(T(\tau) - U_\tau)U_\tau^m\| \leqslant \|A^{-1}(T(\tau) - U_\tau)\|(e(1-b)\tau m)^{-1}, \tag{5.54}$$

where we used that $\|AH^{-1}\| = \|(\mathbb{1} + BA^{-1})^{-1}\| \leqslant (1-b)^{-1}$. Similarly,

$$\|A^{-1}(T(\tau) - U_\tau)U_\tau^m\| \leqslant \|A^{-1}(T(\tau) - U_\tau)A^{-1}\|(e(1-b)\tau m)^{-1}. \tag{5.55}$$

Inserting (5.54) and (5.55) into (5.52), we obtain

$$\begin{aligned} &\|(\mathbb{1} - T(\tau))^{-1}(T(\tau) - U_\tau)U_\tau^m\| \\ &\leqslant \left[\frac{1}{\tau}\|A^{-1}(T(\tau) - U_\tau)\| + \frac{C_{1/2}}{\tau^2}\|A^{-1}(T(\tau) - U_\tau)A^{-1}\|\right] \\ &\quad \times (e(1 - C_{1/2}S_1 b)(1-b)m)^{-1}. \end{aligned} \tag{5.56}$$

Now taking into account Lemmata 5.15 and 5.16 we deduce from (5.56) the following estimate of the last factor in in the sum (5.50)

$$\|(\mathbb{1} - T(\tau))^{-1}(T(\tau) - U_\tau)U_\tau^m\| \leqslant \frac{L_3}{(1 - C_{1/2}S_1 b)(1-b)}\frac{1}{m}, \tag{5.57}$$

where we set $L_3 := (L_1 + C_{1/2}L_2)e^{-1}$.

To estimate in the sum (5.50) the factor entirely dependent on the self-adjoint operator $T(\tau)$ it is sufficient to apply Lemma 5.10:

$$\|T(\tau)^{n-m-1}(\mathbb{1} - T(\tau))\| \leqslant \frac{1}{n-m}, \tag{5.58}$$

where $n = 2, 3, \ldots$ and $m = 0, 1, \ldots, n-1$. Inserting the estimates (5.57) and (5.58) in (5.50) we get

$$\|T(\tau)^n - U_\tau^n\| \tag{5.59}$$

$$\leqslant \|T(\tau)^{n-1}(T(\tau) - U_\tau)\| + \frac{L_3}{(1 - C_{1/2}S_1 b)(1-b)}\sum_{m=1}^{n-1}\frac{1}{(n-m)m}. \tag{5.60}$$

Since

$$\sum_{m=1}^{n-1}\frac{1}{(n-m)m} = \frac{2}{n}\sum_{m=1}^{n-1}\frac{1}{m} \leqslant \frac{2}{n}[1 + \ln(n-1)] \leqslant 4\frac{\ln(n)}{n}$$

for $n = 3, 4, \ldots$, we get for (5.60) the inequality

$$\|T(\tau)^n - U_\tau^n\| \leqslant \|T(\tau)^{n-1}(T(\tau) - U_\tau)\| + \frac{4L_3}{(1 - C_{1/2}S_1 b)(1-b)}\frac{\ln(n)}{n}. \tag{5.61}$$

It remains to estimate $\|T(\tau)^{n-1}(T(\tau) - U_\tau)\|$. Again we refer to the existence of $(\mathbb{1} - T(\tau))^{-1}$, see Lemma 5.13, to get the estimate

$$\begin{aligned} &\|(\mathbb{1} - T(\tau))^{-1}(T(\tau) - U_\tau)\| \\ &\leqslant \frac{1}{1 - C_{1/2}S_1 b}\left(\|T(\tau) - U_\tau\| + \frac{C_{1/2}}{\tau}\|A^{-1}(T(\tau) - U_\tau)\|\right) \\ &\leqslant \frac{1}{1 - C_{1/2}S_1 b}\left(2 + \frac{C_{1/2}}{\tau}\|A^{-1}(T(\tau) - U_\tau)\|\right). \end{aligned} \tag{5.62}$$

Then by Lemma 5.15 we obtain from (5.62) that

$$\|(\mathbb{1} - T(\tau))^{-1}(T(\tau) - U_\tau)\| \leqslant \frac{1}{1 - C_{1/2}S_1 b}(2 + C_{1/2}L_1). \tag{5.63}$$

Inserting (5.63) in the inequality

$$\|T(\tau)^{n-1}(T(\tau) - U_\tau)\| \leqslant \|T(\tau)^{n-1}(\mathbb{1} - T(\tau))\| \|(\mathbb{1} - T(\tau))^{-1}(T(\tau) - U_\tau)\|$$

we get

$$\|T(\tau)^{n-1}(T(\tau) - U_\tau)\| \leqslant \frac{2 + C_{1/2}L_1}{1 - C_{1/2}S_1 b}\|T(\tau)^{n-1}(\mathbb{1} - T(\tau))\|. \tag{5.64}$$

Since $T(\tau)$ is self-adjoint, we can use Lemma 5.10 to estimate the right-hand side of (5.64), which gives

$$\|T(\tau)^{n-1}(T(\tau) - U_\tau)\| \leqslant \frac{2 + C_{1/2}L_1}{1 - C_{1/2}S_1 b}\frac{1}{n}. \tag{5.65}$$

Inserting (5.65) in (5.61) we get assertion (5.25):

$$\|T(\tau)^n - U_\tau^n\| \leqslant \frac{L_T}{(1 - C_{1/2}S_1 b)(1 - b)}\frac{\ln(n)}{n}, \quad n = 3, 4, \ldots, \tag{5.66}$$

for $L_T := (2 + C_{1/2}L_1)(1 - b) + 4L_3$.

Now we deal with the estimate (5.24). To this aim we use the identity

$$P(\tau)^n - T(\tau)^n = \sum_{m=0}^{n-1} P(\tau)^{n-m-1}(P(\tau) - T(\tau))T(\tau). \tag{5.67}$$

Note that

$$P(\tau) - T(\tau) = f^{1/2}(\tau A)[f^{1/2}(\tau A)g(\tau B) - g(\tau B)f^{1/2}(\tau A)].$$

Then (5.67) can be rewritten as

$$\begin{aligned} &P(\tau)^n - T(\tau)^n \\ &= f^{1/2}(\tau A)\sum_{m=0}^{n-1} T(\tau)^{n-m-1}[f^{1/2}(\tau A)g(\tau B) - g(\tau B)f^{1/2}(\tau A)]T(\tau)^m. \end{aligned}$$

Since by Lemma 5.13 we have

$$\mathbb{1} = (\mathbb{1} - T(\tau))(\mathbb{1} - T(\tau))^{-1} = (\mathbb{1} - T(\tau))^{-1}(\mathbb{1} - T(\tau)),$$

the last identity may be recast as

$$\begin{aligned}
&P(\tau)^n - T(\tau)^n \\
&= f^{1/2}(\tau A) \sum_{m=0}^{n-1} T(\tau)^{n-m-1}(\mathbb{1} - T(\tau))(\mathbb{1} - T(\tau))^{-1} \\
&\quad \times \left[f^{1/2}(\tau A)g(\tau B) - g(\tau B)f^{1/2}(\tau A) \right] \\
&\quad \times (\mathbb{1} - T(\tau))^{-1}(\mathbb{1} - T(\tau))T(\tau)^m.
\end{aligned}$$

This yields the estimate

$$\begin{aligned}
&\|P(\tau)^n - T(\tau)^n\| \\
&\leqslant \sum_{m=0}^{n-1} \|T(\tau)^{n-m-1}(\mathbb{1} - T(\tau))\| \\
&\times \left\|(\mathbb{1} - T(\tau))^{-1}[f^{1/2}(\tau A)g(\tau B) - g(\tau B)f^{1/2}(\tau A)](\mathbb{1} - T(\tau))^{-1}\right\| \\
&\times \|(\mathbb{1} - T(\tau))T(\tau)^m\|.
\end{aligned} \tag{5.68}$$

Again by Lemma 5.13, see (5.30), we find that

$$\begin{aligned}
&\left\|(\mathbb{1} - T(\tau))^{-1}[f^{1/2}(\tau A)g(\tau B) - g(\tau B)f^{1/2}(\tau A)](\mathbb{1} - T(\tau))^{-1}\right\| \\
&\leqslant \frac{1}{(1 - C_{1/2}S_1 b)^2} \left[\left\| f^{1/2}(\tau A)g(\tau B) - g(\tau B)f^{1/2}(\tau A) \right\| \right. \\
&\quad + \frac{2C_{1/2}}{\tau} \left\| A^{-1}\left(f^{1/2}(\tau A)g(\tau B) - g(\tau B)f^{1/2}(\tau A)\right) \right\| \\
&\quad \left. + \frac{C_{1/2}^2}{\tau^2} \left\| A^{-1}\left(f^{1/2}(\tau A)g(\tau B) - g(\tau B)f^{1/2}(\tau A)\right) A^{-1} \right\| \right].
\end{aligned} \tag{5.69}$$

Since

$$f^{1/2}(\tau A)g(\tau B) - g(\tau B)f^{1/2}(\tau A) \tag{5.70}$$
$$= (\mathbb{1} - g(\tau B))f^{1/2}(\tau A) - f^{1/2}(\tau A)(\mathbb{1} - g(\tau B)), \tag{5.71}$$

we get

$$A^{-1}[f^{1/2}(\tau A)g(\tau B) - g(\tau B)f^{1/2}(\tau A)] \tag{5.72}$$
$$= A^{-1}B(\mathbb{1} - g(\tau B))B^{-1}f^{1/2}(\tau A) - f^{1/2}(\tau A)A^{-1}B(\mathbb{1} - g(\tau B))B^{-1}. \tag{5.73}$$

On the other hand, since (5.71) can also be written as

$$\begin{aligned}
&f^{1/2}(\tau A)g(\tau B) - g(\tau B)f^{1/2}(\tau A) \\
&= (\mathbb{1} - f^{1/2}(\tau A))(\mathbb{1} - g(\tau B)) - (\mathbb{1} - g(\tau B))(\mathbb{1} - f^{1/2}(\tau A)),
\end{aligned}$$

we get

$$\begin{aligned}&A^{-1}[f^{1/2}(\tau A)g(\tau B)-g(\tau B)f^{1/2}(\tau A)]A^{-1}\\&=(\mathbb{1}-f^{1/2}(\tau A))A^{-1}(\mathbb{1}-g(\tau B))B^{-1}BA^{-1}-A^{-1}B(\mathbb{1}-g(\tau B))\\&\times B^{-1}(\mathbb{1}-f^{1/2}(\tau A))A^{-1}.\end{aligned}\tag{5.74}$$

Now we can estimate the right-hand side of (5.69). For the first term, by condition (iii), we simply get

$$\left\|f^{1/2}(\tau A)g(\tau B)-g(\tau B)f^{1/2}(\tau A)\right\|\leqslant 2.\tag{5.75}$$

The estimate of the second term needs (5.11), (5.18) and (5.73), and we obtain

$$\begin{aligned}&\frac{1}{\tau}\left\|A^{-1}[f^{1/2}(\tau A)g(\tau B)-g(\tau B)f^{1/2}(\tau A)]\right\|\\&\leqslant\|BA^{-1}\|\frac{1}{\tau}\|(\mathbb{1}-g(\tau B))B^{-1}\|+\|BA^{-1}\|\frac{1}{\tau}\|(\mathbb{1}-g(\tau B))B^{-1}\|\\&\leqslant 2bS_1.\end{aligned}\tag{5.76}$$

To estimate the third term in (5.69) we use (5.74) and (5.11), (5.16), (5.18), and (5.47), which yields

$$\begin{aligned}&\frac{1}{\tau^2}\left\|A^{-1}[f^{1/2}(\tau A)g(\tau B)-g(\tau B)f^{1/2}(\tau A)]A^{-1}\right\|\\&\leqslant 2\|BA^{-1}\|\frac{1}{\tau}\|(\mathbb{1}-f^{1/2}(\tau A))A^{-1}\|\frac{1}{\tau}\|(\mathbb{1}-g(\tau B))B^{-1}\|\\&\leqslant 2bC_1S_1.\end{aligned}\tag{5.77}$$

Inserting (5.75)–(5.77) in (5.69), we obtain

$$\left\|(\mathbb{1}-T(\tau))^{-1}[f^{1/2}(\tau A)g(\tau B)-g(\tau B)f^{1/2}(\tau A)](\mathbb{1}-T(\tau))^{-1}\right\|\leqslant\frac{L_4}{(1-C_{1/2}S_1b)^2},\tag{5.78}$$

where $L_4:=2+2b(2C_{1/2}S_1+C_{1/2}^2C_1S_1)$. By virtue of (5.68) and (5.78) we get

$$\begin{aligned}&\|P(\tau)^n-T(\tau)^n\|\\&\leqslant\frac{L_4}{(1-C_{1/2}S_1b)^2}\sum_{m=0}^{n-1}\|T(\tau)^{n-m-1}(\mathbb{1}-T(\tau))\|\|(\mathbb{1}-T(\tau))T(\tau)^m\|.\end{aligned}\tag{5.79}$$

Applying Lemma 5.10 to the estimate (5.79) we find that

$$\|P(\tau)^n-T^n(\tau)\|\leqslant\frac{L_4}{(1-C_{1/2}S_1b)^2}\sum_{m=0}^{n-1}\frac{1}{(n-m)(m+1)}.\tag{5.80}$$

Since

$$\sum_{m=0}^{n-1} \frac{1}{(n-m)(m+1)} = \frac{2}{n+1} \sum_{m=1}^{n} \frac{1}{m} \leqslant 4\frac{\ln(n)}{n}, \quad n = 3, 4, \ldots,$$

estimate (5.80) implies that

$$\|P(\tau)^n - T(\tau)^n\| \leqslant \frac{4L_4}{(1 - C_{1/2}S_1 b)^2} \frac{\ln(n)}{n}, \quad n = 3, 4, \ldots \tag{5.81}$$

By (5.66), (5.81), and the inequality

$$\|P(\tau)^n - U_\tau^n\| \leqslant \|P(\tau)^n - T(\tau)^n\| + \|T(\tau)^n - U_\tau^n\|$$

we finally obtain the assertion (5.24):

$$\|P(\tau)^n - U_\tau^n\| \leqslant \frac{L_P}{(1 - C_{1/2}S_1 b)^2(1-b)} \frac{\ln(n)}{n}, \tag{5.82}$$

where $L_P := L_T(1 - C_{1/2}S_1 b) + 4L_4(1-b)$. □

Remark 5.17. In contrast to the Lie product formulae (5.1) and (5.2), the error bound estimates in Proposition 5.8 are (up to coefficients defined by the Kato functions $\mathcal{K}$ and the powers of $(1 - C_{1/2}S_1 b)$) the same for symmetrised and nonsymmetrised versions of them.

Remark 5.18. Note that condition (5.10) is not essential. If A and B are semi-bounded from below by, respectively, $-\gamma_A \mathbb{1}$ and $-\gamma_B \mathbb{1}$, then the operators $\tilde{A} = A + (\gamma_A + 1)\mathbb{1}$ and $\tilde{B} = B + (\gamma_B + 1)\mathbb{1}$ satisfy (5.10). The proof goes through verbatim and gives (5.24) and (5.25) uniformly on any interval $[0, t_0]$ (that is, *locally* uniformly), with $L_P(t_0) = L_P e^{(\gamma_A+1)t_0} e^{(\gamma_B+1)t_0}$ and $L_T(t_0) = L_T e^{(\gamma_A+1)t_0} e^{(\gamma_B+1)t_0}$. Similarly, if $a \neq 0$ in (1.101), one can eliminate a by modifying b, i.e., by taking $b' > b$ and shifting A to obtain (5.11).

5.2.4 Optimal rate of convergence in the operator norm

We conclude this section by a discussion of the *optimality* of the rate of convergence rate for the self-adjoint Trotter-Kato product formulae, since the rate $O(\ln(n)/n)$ in the error bound estimates (5.24) and (5.25) from Proposition 5.8 is not the best possible.

Proposition 5.19. *Let A and B be non-negative self-adjoint operators such that the operator sum $C := A + B$ on* $\operatorname{dom} C = \operatorname{dom} A \cap \operatorname{dom} B$ *is also self-adjoint. Then:*

(a) *the exponential Trotter-Kato product formula*

$$\|(e^{-tA/n} e^{-tB/n})^n - e^{-tC}\| = O(n^{-1}), \quad n \to \infty, \tag{5.83}$$

converges with the same rate $O(n^{-1})$ as its symmetrised version

$$\|(e^{-tB/2n}e^{-tA/n}e^{-tB/2n})^n - e^{-tC}\| = O(n^{-1}), \quad n \to \infty, \tag{5.84}$$

uniformly on each compact t-interval in $[0,\infty)$. If the operator C is strictly positive, then this holds uniformly on $[0,\infty)$;

(b) *the condition of self-adjointness and the error bound $O(n^{-1})$ are ultimate optimal.*

The proof of this proposition requires a method that is more refined than that in Proposition 5.8. This aspect of the Trotter-Kato product formulae theory is out of the scope of the present book, see Notes in Section 5.6. In a few remarks that follow we comment only the *optimality* claimed in Proposition 5.19 (b).

Remark 5.20. One *cannot* relax the self-adjointness assumption on the operator sum $C := A + B$ to *essential* self-adjointness (even enhanced by the additional condition of Kato A-form smallness of B), since there is an example of a couple of non-negative self-adjoint operators A and B such that C on $\operatorname{dom} A \cap \operatorname{dom} B$ is essentially self-adjoint and that

$$\liminf_{n\to\infty} \|(e^{-tA/n}e^{-tB/n})^n - e^{-t\overline{C}}\| \geqslant D(t),$$

where $D(t) > 0$ is continuous for $t > 0$. Note that in this case we still have the well-known *strong* convergent Trotter product formula

$$\lim_{n\to\infty} (e^{-tA/n}e^{-tB/n})^n u = e^{-t\overline{C}}u,$$

for any $u \in \mathcal{H}$. Here self-adjoint operator $\overline{C}$ is the *closure* of the operator sum $C := A + B$.

Remark 5.21. In general, for noncommuting operators A and B the error bound (5.83) cannot be improved, i.e., to be made better than for matrices in the Lie product formula (5.1). Hence, the estimate $O(n^{-1})$ is ultimate optimal. This is much less evident for symmetrised Trotter product formula, since for matrices or *bounded* generators one obtains the rate of convergence $O(n^{-2})$ (5.2).

Proposition 5.22. *There exists an example of a couple of unbounded operators A and B verifying conditions of Proposition 5.19, such that*

$$\|(e^{-tB/2n}e^{-tA/n}e^{-tB/2n})^n - e^{-tC}\| \geqslant L(t)n^{-1},$$

where $L(t) > 0$ is a continuous function for $t > 0$.

This proposition proves optimality of the error bound $O(n^{-1})$ for the rate of convergence of *symmetric* Trotter product formula (5.84).

Remark 5.23. We advise against a conclusion that for *any* couple of noncommuting unbounded operators A and B verifying conditions of Proposition 5.19 the rate of convergence $O(n^{-1})$ is the best possible for the symmetric Trotter product formula. There are examples of Schrödinger C_0-semigroups that manifest for symmetric Trotter product formula a better rate of convergence than $O(n^{-1})$. Whereas for the *nonsymmetric* case the error bound (5.83) is optimal, see Notes in Section 5.6 and Appendix D.

Now we consider a sub-class of generic Kato functions $\mathcal{K}$, which we denote by $\widehat{\mathcal{K}}_\beta$, Appendix C (Section C.1).

Definition 5.24. We say that $h \in \widehat{\mathcal{K}}_\beta$ if:

(i) It is a Borel measurable function $h : [0, \infty) \to [0, 1]$ such that $h(0) = 1$, and $h'(+0) = -1$.

(ii) There exist $\varepsilon > 0$ and $\delta(\varepsilon) < 1$, such that $h(s) \leqslant 1 - \delta(\varepsilon)$ for $s \geqslant \varepsilon$, and

$$[h]_\beta := \sup_{s>0} \frac{|h(s) - 1 + s|}{s^\beta} < \infty, \quad \text{for } 1 < \beta \leqslant 2.$$

The standard examples are $f(x) = e^{-x}$ and $f(x) = (1 + a^{-1}x)^{-a}$, for $a > 0$, see the next subsection 5.2.5 and Appendix C to compare $\widehat{\mathcal{K}}_\beta$ with $\mathcal{K}_\alpha$.

Then the following assertion extends Proposition 5.19 to the Trotter-Kato product formulae.

Proposition 5.25. *Let $f, g \in \widehat{\mathcal{K}}_\beta$ with $\beta = 2$, and let A, B be positive self-adjoint operators in $\mathcal{H}$ such that the operator sum $C := A+B$ is self-adjoint on the domain* $\operatorname{dom} C := \operatorname{dom} A \cap \operatorname{dom} B$. *Then the Trotter-Kato product formulae converge in the operator norm:*

$$\|[g(tB/2n)f(tA/n)g(tB/2n)]^n - e^{-tC}\| = O(n^{-1}) \ ,$$
$$\|[f(tA/n)g(tB/n)]^n - e^{-tC}\| = O(n^{-1}), \quad n \to \infty.$$

and the error bound $O(n^{-1})$ is optimal.

We note that in fact the Propositions 5.19 and 5.25 (when properly extended) include the *majority* of known related results. In all these cases, the operator sums of the two self-adjoint operators under study are self-adjoint, see Proposition 5.8 and Notes in Section 5.6. Nevertheless, these propositions do not cover all known (optimal) results even in the case of strongly continuous self-adjoint semigroups. For example, this is the case when the sum of operators A and B (Remark 5.20) exists only in the sense of quadratic forms, see results in Sections 5.3 and 5.4.

In the next subsection we study optimality for the case when the operators A and B are related by a fractional smallness conditions.

5.2.5 Optimal rate: fractional conditions

The *fractional smallness* conditions for an operator B with respect to an operator A that we consider below are the following:

Let the self-adjoint operators $A \geqslant \mathbb{1}$ and $B \geqslant 0$ in a Hilbert space $\mathcal{H}$ be such that for $\alpha \in (1/2, 1]$ and $b \in (0, 1)$ they satisfy

(i) $\operatorname{dom} A^{\alpha} \subseteq \operatorname{dom} B^{\alpha}$ and $\|B^{\alpha}u\| \leqslant b\|A^{\alpha}u\|$, $u \in \operatorname{dom} A^{\alpha}$,

(ii) $\operatorname{dom} H^{\alpha} \subseteq \operatorname{dom} A^{\alpha}$, $H := A \dot{+} B$.

Definition 5.26. For each $\alpha \in (1/2, 1]$ we denote by $\mathcal{K}_{\alpha}$ the class of Kato functions $f \in \mathcal{K}$ defined by the following conditions:

$$C_{1/2\alpha} := \sup_{x>0} \frac{x f(x)^{1/2\alpha}}{1 - f(x)} < +\infty, \tag{5.85}$$

$$C_1 := \sup_{x>0} \left| (1 - f(x)) \frac{1}{x} \right| < +\infty, \tag{5.86}$$

$$C_2 := \sup_{x>0} \left| \left(f(x) - \frac{1}{1+x} \right) \frac{1}{x^2} \right| < +\infty, \tag{5.87}$$

$$S_1 := \sup_{x>0} \left| (1 - g(x)) \frac{1}{x} \right| < +\infty, \tag{5.88}$$

$$S_2 := \sup_{x>0} \left| \left(g(x) - \frac{1}{1+x} \right) \frac{1}{x^2} \right| < +\infty . \tag{5.89}$$

To link the fractional smallness with properties of Kato functions $f \in \mathcal{K}_{\alpha}$ we assume also the condition

$$b^{1/\alpha} C_{1/2\alpha}\, S_1 < 1 . \tag{5.90}$$

Note that the class $\mathcal{K}_{\alpha}$ and condition (5.90) are *tuned* according to the fractional conditions (i), (ii), and that for $\alpha = 1$ they coincide with the sub-class of Kato functions: (5.15)–(5.19), and the *smallness* conditions (5.11), (5.20), for Proposition 5.8.

Proposition 5.27. *Let the self-adjoint operators $A \geqslant \mathbb{1}$ and $B \geqslant 0$ in a Hilbert space $\mathcal{H}$ be such that for $\alpha \in (1/2, 1)$ and $b \in (0, 1)$the fractional smallness conditions* (i) *and* (ii) *are satisfied. Then the semibounded from below operator $H := A \dot{+} B$ is densely defined self-adjoint form-sum of A and B and for Kato functions $f, g \in \mathcal{K}_{\alpha}$ that satisfy* (5.90) *one gets for some $c > 0$ the operator-norm estimate:*

$$\|(f(tA/n)g(tB/n))^n - e^{-tH}\| \leqslant \frac{c}{n^{2\alpha - 1}}, \quad n \geqslant 2, \tag{5.91}$$

uniformly for $t \geqslant 0$.

The same estimate for asymptotic $O(n^{1-2\alpha})$ is also true for symmetrised Trotter-Kato *approximants*, cf. (5.23):

$$\begin{aligned}(P(t/2n)P^*(t/2n))^n &= (f(tA/2n)g(tB/n)f(tA/2n))^n,\\ T(t/n)^n &= (f^{1/2}(tA/n)g(tB/n)f^{1/2}(tA/n))^n,\\ F(t/n)^n &= (g^{1/2}(tB/n)f(tA/n)g^{1/2}(tB/n))^n.\end{aligned}$$

Proof. Recall that if positive self-adjoint operators C, D verify conditions: $\operatorname{dom} C \subseteq \operatorname{dom} D$ and $\|Du\| \leqslant b\|Cu\|$, for $b \in (0,1)$, and $u \in \operatorname{dom} C$, then the *Heinz-Kato* inequality yields: $\operatorname{dom} C^\beta \subseteq \operatorname{dom} D^\beta$ and $\|D^\beta u\| \leqslant b^\beta \|C^\beta u\|$, for any $u \in \operatorname{dom} C^\beta$ and $\beta \in (0,1)$.

Applying this statement to the operators $C = A^\alpha$ and $D = B^\alpha$ verifying condition (i), we obtain for $\beta = \theta/\alpha$ and $u \in \operatorname{dom} A^\theta$ that

$$\operatorname{dom} A^\theta \subseteq \operatorname{dom} B^\theta, \quad \|B^\theta u\| \leqslant b^{\theta/\alpha}\|A^\theta u\|, \quad \text{for } \theta \in (0,\alpha]. \tag{5.92}$$

If one takes $\theta = 1/2$, then (5.92) implies

$$\operatorname{dom} A^{1/2} \cap \operatorname{dom} B^{1/2} = \operatorname{dom} A^{1/2} \subseteq \operatorname{dom} B^{1/2}\,. \tag{5.93}$$

Since A and consequently $A^{1/2}$ are densely defined, by (5.93) we obtain that $\operatorname{dom} A^{1/2} \cap \operatorname{dom} B^{1/2} = \operatorname{dom} H^{1/2}$ if the operator H with dense domain is defined as the form-sum of A and B. By the construction of the form-sum, the operator $H = A \dotplus B$ is self-adjoint and $H \geqslant \mathbb{1}$.

Due to condition (ii) the arguments in the proof of Propositions 5.8 can be refined. This yields for the operator-norm convergence of the Trotter-Kato product formulae the error bound estimate (5.91) uniformly for $t \geqslant 0$. The corresponding argument is quite involved, see Notes in Section 5.6.

We comment that if one keeps only condition (i) and *skips* (ii), then arguing as in the proof of Propositions 5.8 one obtains the error bound estimate, which (similar to (5.24)) has the $\ln(n)$ correction:

$$\|(f(tA/n)g(tB/n))^n - e^{-tH}\| \leqslant c' \, \frac{\ln(n)}{n^{2\alpha-1}}\,, \quad n \geqslant 2\,, \tag{5.94}$$

for some $c' > 0$ and uniformly for $t \geqslant 0$. □

On the other hand, for *exponential* Kato function, which evidently belongs to $\mathcal{K}_\alpha$, the following assertion is valid.

Proposition 5.28. *For each* $\alpha \in (0,1)$ *and* $b \in (0,1)$, *there exist operators* A *and* B *in a Hilbert space* $\mathcal{H}$ *such that they satisfy the conditions of Propositions* 5.27 *but also the estimates:*

$$\left\| e^{-t(A\dotplus B)} - \left(e^{-tA/n}e^{-tB/n}\right)^n \right\| \geqslant \frac{c_t}{n^{2\alpha-1}}, \quad \text{for } \alpha \in (1/2,1)\,,$$

and for large $n \in \mathbb{N}$. Moreover, there are A and B such that

$$\liminf_{n\to\infty} \left\| e^{-t(A\dot{+}B)} - \left(e^{-tA/n}e^{-tB/n}\right)^n \right\| \geqslant d_t, \quad \text{for } \alpha \in (0,1/2],$$

where c_t and d_t are positive and continuous for $t > 0$. These estimates from below are also valid for self-adjoint symmetrised Trotter approximants, for example:

$$F(t/n)^n = \left(e^{-tB/2n}e^{-tA/n}e^{-tB/2n}\right)^n, \quad n \in \mathbb{N}.$$

Remark 5.29. First, we note that Proposition 5.28 implies that for $\alpha \in (1/2, 1)$ the error bound (5.91) in the Proposition 5.27 is *optimal.* This means that it is the *best* possible in the abstract setting for asymptotics in the power of $n \in \mathbb{N}$. Note that it could be refined for symmetrised approximates since symmetrisation improves the convergence in the Lie product formula (5.2).

Second, for $\alpha = 1$ the condition (i) yields $\operatorname{dom} H = \operatorname{dom} A$. This case is the subject of Propositions 5.8 and further improvements in subsection 5.2.4. See Proposition 5.25 about optimality of the rate $O(n^{-1})$ for Kato functions of class $\widehat{\mathcal{K}}_\beta$.

Third, the case $\alpha \in (0, 1/2)$ is an example for which the Trotter product formula does not hold in the operator-norm convergence sense, but it holds in the strong operator topology by the standard Trotter theorem. This means that the operator-norm convergence of the product formula cannot be extended to the cases $\alpha < 1/2$ in the abstract setting.

As (5.90) and Propositions 5.27, 5.28 make clear, the case $\alpha = 1/2$ needs a special analysis, see Notes in Section 5.6.

Proposition 5.30. *Let self-adjoint operators $A \geqslant \mathbb{1}$ and $B \geqslant 0$ in a Hilbert space $\mathcal{H}$ be such that*

$$\operatorname{dom} A^{1/2} \subseteq \operatorname{dom} B^{1/2}. \tag{5.95}$$

Let the Kato functions $f, g \in \mathcal{K}$.

If in addition to (5.95) the operator $B^{1/2}$ is relatively $A^{1/2}$-compact, that is,

$$B^{1/2}A^{-1/2} \in \mathcal{C}_\infty(\mathcal{H}), \tag{5.96}$$

and the Kato function f satisfies the condition

$$C := \sup_{x>0} \frac{xf(x)}{1-f(x)} < +\infty, \tag{5.97}$$

then the Trotter-Kato product formulae converge in the operator norm locally uniformly away from zero to the C_0-semigroup with self-adjoint generator $H = A\dot{+}B$.

Note that condition (5.97) coincides with (5.85), for $\alpha = 1/2$, and with (5.146). The corresponding sub-class of the Kato functions includes for example, $f(x) = \mathrm{e}^{-x}$ and $f(x) = (1+x)^{-1}$, Appendix C (Section C.3).

Remark 5.31. Suppose the densely defined in a Hilbert space $\mathcal{H}$ operators K and L with $\operatorname{dom} K \subseteq \operatorname{dom} L$ are such that $L(K - z\mathbb{1})^{-1} \in \mathcal{C}_\infty(\mathcal{H})$ for some (and then all) $z \in \rho(K)$, where $\rho(K)$ is the resolvent set of K. Then we say that L is *relatively compact* with respect to K, or simply that it is *K-compact.*

We note that, if L is K-compact and closable, then $L \in \mathcal{P}_{0^+}(K)$. This means that L is K-bounded with the relative bound $b = 0^+$, Definition 1.50. Another class of $\mathcal{P}_{0^+}$-perturbations will be considered in Remark 5.48. For other classes of infinitesimally small *unbounded* perturbations, see Section 4.4.

Since $B^{1/2}$ is closed, one gets that under the condition of relative compactness (5.96) the operator $B^{1/2}$ is relatively $A^{1/2}$-bounded with relative bound $b = 0^+$, that is, infinitesimally small. Then there is no difference between the case $\alpha = 1/2$ and $\alpha \in (0, 1/2)$ with conditions (i),(ii).

We conclude this section by a statement, that extends Proposition 5.27 from $\mathcal{K}_\alpha$ to a larger class of Kato functions $\widehat{\mathcal{K}}_\beta$, Definition 5.24. Moreover, it relaxes the smallness condition in (i). For this purpose we modify the fractional conditions in (i) and (ii), which still ensure the optimal rate of convergence of the Trotter-Kato product formulae.

Proposition 5.32. *Let the self-adjoint operators $A \geqslant \mathbb{1}$ and $B \geqslant 0$ in a Hilbert space $\mathcal{H}$ be such that non-negative operator $H := A\dot{+}B$ is the densely defined self-adjoint form-sum of A and B.*

Let us assume that for some $\alpha \in (1/2, 1)$:

(i) $\operatorname{dom} A^{1/2} \subseteq \operatorname{dom} B^{1/2}$,

(ii) $\operatorname{dom} H^\alpha \subseteq \operatorname{dom} A^\alpha \cap \operatorname{dom} B^\alpha$,

(iii) *the Kato functions $f, g \in \widehat{\mathcal{K}}_\beta$, where $\beta = 2\alpha$,*

(iv) $\sup_{y \geqslant x} f(y) < 1$ *for $x > 0$.*

Then for some $c > 0$ one gets the operator-norm estimate

$$\|(f(tA/n)g(tB/n))^n - e^{-tH}\| \leqslant \frac{c}{n^{2\alpha-1}}, \quad n \geqslant 2, \tag{5.98}$$

uniformly for $t \geqslant 0$.

Remark 5.33. (a) The same example as the one that is the source of Proposition 5.28 shows that the error bound estimate (5.98) cannot be improved under the conditions of the proposition, i.e., it is ultimate optimal.

(b) The application of the Heinz-Kato inequality to condition (i) of Proposition 5.27 yields (5.92), and finally the inclusion (5.93), which coincides with (i) Proposition 5.32. Hence, the smallness condition is relaxed.

(c) Note that by Definition 5.26 of the class $\mathcal{K}_\alpha$, the Kato functions $f, g \in \mathcal{K}_\alpha$ verify (5.87) and (5.89) for any $\alpha \in (1/2, 1]$. This corresponds to the conditions: $[f]_{\beta=2} < \infty$ and $[g]_{\beta=2} < \infty$, which are stronger than condition (iii) in Proposition 5.32. Moreover, besides $f(x) < 1$ the condition (5.85) demands that f decays at infinity: $f(x) = O(1/x^{2\alpha})$. By contrast, condition (iv) of Proposition 5.32 requires only the first.

For further details and references see Notes in Section 5.6.

5.3 Operator-norm convergence: self-adjoint and non-self-adjoint semigroups

The conditions imposed in the previous Section 5.2 on the pair of self-adjoint generators $\{A, B\}$, namely that operator $A + B$ is self-adjoint, and on the couple of Kato functions, that $f, g \in \widehat{\mathcal{K}}_\beta$, seem to be too restrictive. However, they ensure in particular the *optimal* estimate of the rate of convergence in the operator norm for the Kato-Trotter product formulae. If one does not care about the rate of convergence, then for strongly continuous *self-adjoint* semigroups (or for non-exponential Kato functions f, g from $\mathcal{K}$) there is an alternative approach to establish the operator-norm convergence of the Trotter-Kato product formulae.

This approach covers a fairly large family of pairs of self-adjoint generators A and B since self-adjointness of the algebraic sum $A+B$ is *not* required, but it gives *no* estimate for the rate of convergence, see comments in Remark 5.3. We present the main statement in subsection 5.3.1 *Preliminaries and Proposition* 5.36.

The proof of the Proposition 5.36 is quite involved and so it needs certain preparations. First in subsection 5.3.2 *Operator-norm approximation theorem à la Chernoff*, we prove the *lifting* to the operator-norm convergence of some results that are known in the strong operator topology. This allows us to prove in subsection 5.3.3 *Key preliminary statement*, a central for our approach Proposition 5.40, which in the case of self-adjoint generators A and B is applicable for the operator-norm topology. The conditions on A and B are mild and demand only existence of the (perhaps not densely) defined in $\mathcal{H}$ form-sum $A \dot{+} B$.

Then for arbitrary Kato-functions $f, g \in \mathcal{K}$ we prove in subsection 5.3.4 *Proof of Proposition* 5.36 a result about the operator-norm convergence of Trotter-Kato product formulae to a degenerate semigroup without estimates of the rate of convergence. It is curious that Proposition 5.36 allows in some sense a *converse* statement: see Proposition 5.45 for the case when f is a regular Kato-function (Appendix C).

In subsection 5.3.5 *Rate of convergence*: *non-self-adjoint semigroups*, we consider extensions of the convergence in the operator norm of the Trotter product formula (i.e., f, g are exponentials) to non-self-adjoint C_0-semigroups. These extensions include the error bound estimates for the rates of convergence. The estimates are possible due to a hierarchy condition: a relative *smallness* of B with respect to A and of B^* with respect to A^*. In Proposition 5.47 the generator A is supposed to be an m-sectorial operator and the small B to be m-accretive. To relax the smallness condition and to improve the rate of convergence the condition on the generator A in Proposition 5.49 is stricter: A is assumed to be a non-negative self-adjoint operator.

We lift these results to the *trace-norm* convergence of the product formulae for non-self-adjoint Gibbs semigroups in Section 5.5. Two lifting methods that we use there allow to prove the trace-norm convergence *without* and *with* error bound estimates only for exponential function, that is, for the Trotter product formula.

5.3.1 Preliminaries and Proposition 5.36

Let $A \geqslant 0$ and $B \geqslant 0$ be non-negative self-adjoint operators in a Hilbert space $\mathcal{H}$, which are not hierarchically ordered. In contrast to the conditions of Propositions 5.19, here it may happen that $\operatorname{dom} A \cap \operatorname{dom} B = \{0\}$ whereas $\operatorname{dom} A^{1/2} \cap \operatorname{dom} B^{1/2}$ is nontrivial. If one denotes the closure $\overline{\operatorname{dom} A^{1/2} \cap \operatorname{dom} B^{1/2}}$ by $\mathcal{H}_0 \subseteq \mathcal{H}$, then the non-negative self-adjoint *form-sum* operator $H := A \dotplus B$ is densely defined in the Hilbert subspace $\mathcal{H}_0$, see Remark 4.29.

Recall that under these conditions the Trotter product formula converges in the *strong* operator topology:

$$\operatorname*{s-lim}_{n\to\infty} \left(e^{-tA/n}e^{-tB/n}\right)^n = e^{-tH}P_0\,, \quad t>0, \tag{5.99}$$

to the *degenerate semigroup* $\{e^{-tH}P_0\}_{t>0}$, see Definition 1.24. Here P_0 denotes the orthogonal projection of $\mathcal{H}$ onto $\mathcal{H}_0$.

Moreover, the product formula (5.99) is valid not only for strongly continuous self-adjoint semigroups, that is, for exponentials corresponding to generators $A \geqslant 0$ and $B \geqslant 0$, but also for Kato functions $f, g \in \mathcal{K}$ (or $\in \mathcal{K}_\beta$):

$$\operatorname*{s-lim}_{n\to\infty} \left(f(tA/n)g(tB/n)\right)^n = e^{-tH}P_0\,, \quad t>0, \tag{5.100}$$

that is, for Trotter-Kato product formula.

Similar to the programme outlined at the end of Section 5.1, in Remark 5.2 a natural problem is to find conditions such that the strong convergence of the Trotter (5.99) and the Trotter-Kato (5.100) product formulae can be lifted to the limit in the *operator-norm* topology.

The aim of the Section 5.3 is to collect results concerning *sufficient* (and *necessary*) conditions, which guarantee the operator-norm convergence of the Trotter-Kato product formulae for a given pair of non-negative self-adjoint operators A, B and for admissible Kato functions f, g. However, in contrast to Section 5.2 these results do not yield estimates for the corresponding error bounds.

Although the result in Proposition 5.36 is valid for arbitrary Kato functions $f, g \in \mathcal{K}$, to *formulate* the conditions for this assertion one needs certain functions associated with them. Note that a specific subclass of $\mathcal{K}$ selected by *regularity* will be needed only for a converse statement in Proposition 5.45.

Definition 5.34. With any pair of Kato functions $f, g \in \mathcal{K}$ we associate Borel functions f_0, g_0 on $\mathbb{R}_0^+$ constructed as follows:

(1) First we define two auxiliary functions

$$\varphi(x) := x^{-1}\left(\frac{1}{f(x)} - 1\right) \text{ and } \psi(x) := x^{-1}(1 - g(x)) . \tag{5.101}$$

Here it is supposed that $f(x) > 0$ for (5.101) to be correct.

(2) Using (5.101) we define a couple of new auxiliary functions

$$0 \leqslant \varphi_0(x) := \inf_{0<s\leqslant x} s^{-1}\left(\frac{1}{f(s)} - 1\right), \tag{5.102}$$

$$0 \leqslant \psi_0(x) := \inf_{0<s\leqslant x} s^{-1}(1 - g(s)) . \tag{5.103}$$

Here $f(x) = 0$ is allowed and we make the convention that $0^{-1} := +\infty$.

(3) Finally for any $x \in \mathbb{R}_0^+$ we set

$$f_0(x) := \begin{cases} 1, & \text{if } x = 0, \\ (1 + x\varphi_0(x))^{-1}, & \text{if } x > 0, \end{cases} \tag{5.104}$$

and

$$g_0(x) := \begin{cases} 1, & \text{if } x = 0, \\ (1 - x\psi_0(x)), & \text{if } x > 0. \end{cases} \tag{5.105}$$

We note that $0 \leqslant f(x) \leqslant f_0(x) \leqslant 1$ and $0 \leqslant g(x) \leqslant g_0(x) \leqslant 1$, $x \in \mathbb{R}_0^+$, and that in turn the functions f_0, g_0 belong to $\mathcal{K}$. More details and proofs of properties concerning these functions one can found in Appendix C (Section C.2).

For example, if $f(x) = e^{-x}$, $x \geqslant 0$, then $\varphi_0(x) = 1$, which yields $f_0(x) = (1 + x)^{-1}$, whereas if $f(x) = 1/(1 + x)$, $x \geqslant 0$, then $\varphi_0(x) = 1$, and one gets again $f_0(x) = (1 + x)^{-1}$.

To formulate Proposition 5.36 and similar results about convergence of the Trotter-Kato product formulae we use the following terminology, cf. Proposition 5.30.

Definition 5.35. We say that a sequence of families of bounded operators: $\{\Phi_n(t)\}_{t\geqslant 0, n\in\mathbb{N}}$, converges to the family $\{\Phi(t)\}_{t\geqslant 0}$ in the norm topology $\|\cdot\|_\tau$ *locally uniformly away from* $t_0 \geqslant 0$ if

$$\lim_{n\to\infty} \sup_{[a,b]\subset(t_0,\infty)} \|\Phi_n(t) - \Phi(t)\|_\tau = 0 ,$$

for any compact $[a, b] \subset (t_0, \infty)$.

Proposition 5.36. *Let A and B be non-negative densely defined self-adjoint operators in $\mathcal{H}$ and $H = A \dotplus B$ be the form-sum self-adjoint operator defined in the subspace $\mathcal{H}_0$. Let f and g be arbitrary Kato functions from $\mathcal{K}$. If the condition*

$$f_0(t_0 A) \in \mathcal{C}_\infty(\mathcal{H}) \tag{5.106}$$

is satisfied for some $t_0 > 0$, *then the Trotter-Kato product formula* (5.100) *and all other formulae generated by* f *and* g *converge in the operator-norm topology locally uniformly away from zero* ($t_0 = 0$) *to degenerate semigroup* $\{e^{-tH}P_0\}_{t>0}$, *and*

$$(\mathbb{1}_0 + H)^{-1} \in \mathcal{C}_\infty(\mathcal{H}_0), \quad \mathbb{1}_0 = \mathbb{1} \restriction \mathcal{H}_0 . \tag{5.107}$$

Here $\mathbb{1}_0 := \mathbb{1} \restriction \mathcal{H}_0$ *denotes restriction of the unit operator to the subspace* $\mathcal{H}_0 = P_0(\mathcal{H})$.

Remark 5.37. Note that the statements below in subsections 5.3.3 and 5.3.4 are valid for general pairs $f, g \in \mathcal{K}$. We shall let the reader know when our arguments are restricted in subsection 5.3.4 to *regular* Kato functions, or in subsection 5.3.5 to *exponential* Kato function.

5.3.2 Operator-norm approximation theorem à la Chernoff

As the first step toward the proof of Proposition 5.36 we lift to the operator-norm convergence some results that are well-known in the strong operator topology.

Let $\{X(s)\}_{s>0}$ be a family of self-adjoint operators in $\mathcal{H}$ and let X_0 be a self-adjoint operator in the subspace $\mathcal{H}_0 \subseteq \mathcal{H}$. Recall that $\{X(s)\}_{s>0}$ converges in the *uniform resolvent* sense (or in the *operator-norm resolvent* sense) to X_0 as $s \to +0$ if

$$\lim_{s \to +0} \|(X(s) - z\mathbb{1})^{-1} - (X_0 - z\mathbb{1}_0)^{-1} P_0\| = 0 , \tag{5.108}$$

for one (and hence for all) nonreal $z \in \mathbb{C}$. Then the following *operator-norm* analogue of the Trotter-Kato strong convergence theorem holds for semigroups.

Lemma 5.38. *Let* $\{X(s)\}_{s>0}$ *be a family of non-negative self-adjoint operators in a Hilbert space* $\mathcal{H}$ *and let* X_0 *be a non-negative self-adjoint operator in the closed subspace* $\mathcal{H}_0 \subseteq \mathcal{H}$. *Then the following conditions are equivalent:*

(a) $\lim_{s\to+0} \|(\lambda\mathbb{1} + X(s))^{-1} - (\lambda\mathbb{1}_0 + X_0)^{-1}P_0\| = 0$, for all $\lambda > 0$,

(b) $\lim_{r\to+\infty} \sup_{t\in[a,b]} \|e^{-tX(t/r)} - e^{-tX_0}P_0\| = 0$, for all $[a, b] \subset (0, \infty)$.

Proof. (a) $\Rightarrow$ (b). We note that condition (a) implies the uniform resolvent convergence of $\{X(s)\}_{s>0}$ to X_0 as $s \to +0$. Let $\phi(\cdot)$ be a real continuous function defined on $\mathbb{R}$ which tends to zero as $x \to \pm\infty$. Then by a standard argument the uniform resolvent convergence yields

$$\lim_{s\to+0} \|\phi(X(s)) - \phi(X_0)\| = 0 ,$$

provided X_0 is defined on $\mathcal{H}$, i.e., $\mathcal{H} = \mathcal{H}_0$. However, if $\mathcal{H}_0 \subset \mathcal{H}$, then one sees that the proof still remains valid. So we find that

$$\lim_{s\to+0} \|\phi(X(s)) - \phi(X_0)P_0\| = 0 . \tag{5.109}$$

In particular, if $\phi(x) = e^{-t|x|}$, $x \in \mathbb{R}$, $t > 0$, then

$$\lim_{s \to +0} \|e^{-tX(s)} - e^{-tX_0} P_0\| = 0\,. \tag{5.110}$$

The set of elements $\{e^{-t|x|}\}_{t \in [a,b]}$, $[a,b] \subseteq (0,\infty)$, is compact in $C_0(\mathbb{R})$, which is the Banach space of all real continuous functions such that: $\lim_{x \to \pm\infty} \phi(x) = 0$, endowed with the supremum norm. Therefore,

$$\lim_{s \to +0} \sup_{t \in [a,b]} \|e^{-tX(s)} - e^{-tX_0} P_0\| = 0\,, \tag{5.111}$$

for all intervals $[a,b] \subseteq (0,\infty)$. Since for $s = t/r$ one has $s \leqslant b/r$, from (5.111) we immediately obtain (b).

(b) $\Rightarrow$ (a). From (b) it follows that

$$\lim_{s \to +0} \|e^{-X(s)} - e^{-X_0} P_0\| = 0. \tag{5.112}$$

Setting $Y(s) = e^{-X(s)}$ and $Y_0 = e^{-X_0}$ one gets from (5.112) that the sequence $\{Y(s)\}_{s>0}$ converges in the uniform resolvent sense to Y_0 as $s \to +0$. Let $\phi(x) = (\lambda + |\ln(|x|)|)^{-1}$, $x \in \mathbb{R}$, $\lambda > 0$. Then again by a standard argument, applied now to $\{Y(s)\}_{s>0}$ and Y_0 , we obtain (a). □

Lemma 5.38 enables us to prove the *operator-norm* generalisation of the Chernoff *approximation theorem.*

Proposition 5.39. *Let $\{\Phi(s)\}_{s\geqslant 0}$ be a family of self-adjoint non-negative contractions on a Hilbert space $\mathcal{H}$ and let X_0 be a self-adjoint operator on the subspace $\mathcal{H}_0 \subseteq \mathcal{H}$. Define $X(s) := s^{-1}(\mathbb{1} - \Phi(s))$, $s > 0$. Then the family $\{X(s)\}_{s>0}$ converges in the uniform resolvent sense to X_0 as $s \to +0$ if and only if the sequence $\{\Phi(t/r)^r\}_{r\geqslant 1}$, $t > 0$, converges in the operator norm to $e^{-tX_0}P_0$ as $r \to +\infty$, uniformly in t on any compact interval in $\mathbb{R}^+$.*

Proof. Since $\sup_{x\in[0,1]} |x^r - e^{-r(1-x)}| \leqslant 1/r$, for $r \geqslant 1$, the spectral theorem yields

$$\|\Phi(t/r)^r - e^{-tX(t/r)}\| \leqslant \frac{1}{r}, \quad r \geqslant 1, \quad t \geqslant 0. \tag{5.113}$$

Assume that $\{X(s)\}_{s>0}$ converges in the uniform resolvent sense to X_0 for $s \to +0$. Then the representation

$$\Phi(t/r)^r - e^{-tX_0} P_0 = \Phi(t/r)^r - e^{-tX(t/r)} + e^{-tX(t/r)} - e^{-tX_0} P_0\,,$$

yields the estimate

$$\|\Phi(t/r)^r - e^{-tX_0} P_0\| \leqslant \|\Phi(t/r)^r - e^{-tX(t/r)}\| + \|e^{-tX(t/r)} - e^{-tX_0} P_0\|.$$

Using (5.113) one obtains that

$$\sup_{t\in[a,b]} \|\Phi(t/r)^r - e^{-tX_0} P_0\| \leqslant \frac{1}{r} + \sup_{t\in[a,b]} \|e^{-tX(t/r)} - e^{-tX_0} P_0\|\,, \tag{5.114}$$

for any compact interval $[a,b] \subseteq \mathbb{R}^+$. Applying now Lemma 5.38 to (5.114) we complete this part of the proof.

Conversely, assume that $\{\Phi(t/r)^r\}_{r\geqslant 1}$ converges in the operator norm to $e^{-tX_0}P_0$, uniformly for any compact t-interval of $\mathbb{R}^+$. From

$$e^{-tX(t/r)} - e^{-tX_0}P_0 = e^{-tX(t/r)} - \Phi(t/r)^r + \Phi(t/r)^r - e^{-tX_0}P_0 \ ,$$

one gets the inequality

$$\|e^{-tX(t/r)} - e^{-tX_0}P_0\| \leqslant \|e^{-tX(t/r)} - \Phi(t/r)^r\| + \|\Phi(t/r)^r - e^{-tX_0}P_0\|.$$

Now using again (5.113) we obtain the estimate

$$\sup_{t\in[a,b]} \|e^{-tX(t/r)} - e^{-tX_0}P_0\| \leqslant \frac{1}{r} + \sup_{t\in[a,b]} \|\Phi(t/r)^r - e^{-tX_0}P_0\| \, ,$$

for any compact interval $[a,b] \subseteq \mathbb{R}^+$. Therefore, the condition (b) of Lemma 5.38 is satisfied, so, applying this lemma we establish the uniform resolvent convergence of $\{X(s)\}_{s>0}$ to X_0 when $s \to +0$. □

5.3.3 Key preliminary statement

Now let us introduce the operator-valued families

$$T(t) := f(tA)^{1/2}g(tB)f(tA)^{1/2}, \qquad t \geqslant 0, \tag{5.115}$$

$$R(t) := (\mathbb{1} - T(t))\,t^{-1}, \qquad t > 0, \tag{5.116}$$

$$F(t) := g(tB)^{1/2}f(tA)g(tB)^{1/2}, \qquad t \geqslant 0, \tag{5.117}$$

$$S(t) := (\mathbb{1} - F(t))\,t^{-1}, \qquad t > 0. \tag{5.118}$$

We also define the family of subspaces $\mathcal{H}(t) := \overline{\operatorname{ran}(f(tA))} \subseteq \mathcal{H}$, for $t \geqslant 0$. Let $\{Q(t)\}_{t\geqslant 0}$, denote the corresponding family of orthogonal projections from $\mathcal{H}$ onto the subspaces $\{\mathcal{H}(t)\}_{t\geqslant 0}$. By the continuity of Kato functions at $+0$ one gets

$$\operatorname*{s-lim}_{t\to+0} Q(t) = \mathbb{1}. \tag{5.119}$$

Note that for each $t \geqslant 0$ the projection $Q(t)$ commutes with A. We set $\tilde{f}(tA) := f(tA) \restriction \mathcal{H}(t) : \mathcal{H}(t) \to \mathcal{H}(t)$, $t \geqslant 0$. Then by definition one obtains that $\ker(\tilde{f}(tA)) = \{0\}$ and that

$$0 \leqslant \tilde{f}(tA) \leqslant \mathbb{1}(t), \quad t \geqslant 0, \tag{5.120}$$

where *restrictions* $\mathbb{1}(t){:=}\mathbb{1}\restriction\mathcal{H}(t)$.

All this motivates the introduction of the family of operators $\{M(t)\}_{t>0}$, defined as mappings $M(t) : \mathcal{H}(t) \to \mathcal{H}(t)$, by

$$M(t) := \frac{1}{t}[\tilde{f}(tA)^{-1} - \mathbb{1}(t)] + \frac{1}{t}Q(t)[\mathbb{1} - g(tB)]Q(t), \tag{5.121}$$

on the t-dependent domain

$$\operatorname{dom} M(t) = \operatorname{ran}(\tilde{f}(tA)) \subseteq \mathcal{H}(t). \tag{5.122}$$

Finally, we recall that $\mathbb{1}_0 = \mathbb{1} \upharpoonright \mathcal{H}_0$ is the restriction of the unit operator on the subspace $\mathcal{H}_0 = \overline{\operatorname{dom} A^{1/2} \cap \operatorname{dom} B^{1/2}}$. Note that if for the Kato function f satisfies $f(x) > 0$, $x \in \mathbb{R}$, then one readily gets that

$$\mathcal{H}(t) = \mathcal{H} \quad \text{and} \quad Q(t) = \mathbb{1}, \quad t \geqslant 0.$$

Now we are in position to formulate and to prove the key preliminary statement necessary for the proof of Proposition 5.36.

Proposition 5.40. *Let A and B be non-negative self-adjoint operators in a Hilbert space $\mathcal{H}$ and let $f(\cdot)$ and $g(\cdot)$ be Kato functions of class $\mathcal{K}$. Then the following conditions are equivalent:*

(i) $\lim_{r\to+\infty} \sup_{t\in[a,b]} \|T(t/r)^r - P_0 e^{-tH} P_0\| = 0, \quad [a,b] \subseteq \mathbb{R}^+,$

(ii) $\lim_{t\to+0} \|(\lambda\mathbb{1} + R(t))^{-1} - P_0(\lambda\mathbb{1}_0 + H)^{-1} P_0\| = 0, \quad \lambda > 0,$

(iii) $\lim_{t\to+0} \|Q(t)(\lambda\mathbb{1}(t) + M(t))^{-1} Q(t) - P_0(\lambda\mathbb{1}_0 + H)^{-1} P_0\| = 0, \quad \lambda > 0,$

(iv) $\lim_{t\to+0} \|(\lambda\mathbb{1} + S(t))^{-1} - P_0(\lambda\mathbb{1}_0 + H)^{-1} P_0\| = 0, \quad \lambda > 0,$

(v) $\lim_{r\to+\infty} \sup_{t\in[a,b]} \|F(t/r)^r - P_0 e^{-tH} P_0\| = 0, \quad [a,b] \subseteq \mathbb{R}^+.$

Proof. We prove that (i) ⇒ (ii) ⇒ (iii) ⇒ (iv) ⇒ (v) ⇒ (i).

(i) ⇒ (ii) To prove this implication one simply has to set $\Phi(t) := T(t)$ for $t \geqslant 0$, $X(t) := R(t)$ for $t > 0$, and then apply Proposition 5.39.

(ii) ⇒ (iii) Since by definitions (5.115), (5.116) and (5.121) one gets

$$R(t) = \frac{1}{t}(\mathbb{1} - Q(t)) + f(tA)^{1/2} M(t) f(tA)^{1/2},$$

we obtain the representation

$$f(tA)^{1/2}(\mathbb{1} + R(t))^{-1} f(tA)^{1/2} = Q(t)\{\tilde{f}(tA)^{-1} + M(t)\}^{-1} Q(t).$$

Next, we note that

$$\tilde{f}(tA)^{-1} + M(t) = (1+t)\{M(t) + \frac{1}{1+t} Q(t) g(tB) Q(t)\} \geqslant \mathbb{1}(t),$$

which finally yields the representation

$$\begin{aligned} &f(tA)^{1/2}(\mathbb{1} + R(t))^{-1} f(tA)^{1/2} \\ &= \frac{1}{1+t} Q(t)\{M(t) + \frac{1}{1+t} Q(t) g(tB) Q(t)\}^{-1} Q(t). \end{aligned} \tag{5.123}$$

Now we can prove that

$$\lim_{t\to+0}\|f(tA)^{1/2}(\mathbb{1}+R(t))^{-1}f(tA)^{1/2}-P_0(\mathbb{1}_0+H)^{-1}P_0\|=0. \tag{5.124}$$

To this end we use the representation

$$\begin{aligned}
&f(tA)^{1/2}(\mathbb{1}+R(t))^{-1}f(tA)^{1/2}-P_0(\mathbb{1}_0+H)^{-1}P_0\\
&=f(tA)^{1/2}\{(\mathbb{1}+R(t))^{-1}-P_0(\mathbb{1}_0+H)^{-1}P_0\}f(tA)^{1/2}\\
&\quad+[f(tA)^{1/2}-\mathbb{1}]P_0(\mathbb{1}_0+H)^{-1}P_0f(tA)^{1/2}\\
&\quad+P_0(\mathbb{1}_0+H)^{-1}P_0[f(tA)^{1/2}-\mathbb{1}],
\end{aligned}$$

which gives the estimate

$$\begin{aligned}
&\|f(tA)^{1/2}(\mathbb{1}+R(t))^{-1}f(tA)^{1/2}-P_0(\mathbb{1}_0+H)^{-1}P_0\|\\
&\leqslant\|(\mathbb{1}+R(t))^{-1}-P_0(\mathbb{1}_0+H)^{-1}P_0\|+2\|[\mathbb{1}-f(tA)^{1/2}]\\
&\quad\times P_0(\mathbb{1}_0+H)^{-1}P_0\|.
\end{aligned} \tag{5.125}$$

Since by the construction of the operator $H = A \dot{+} B$ in the sense of form-sum one has $\operatorname{dom} H^{1/2} \subseteq \operatorname{dom} A^{1/2}$, we see that

$$\begin{aligned}
&[\mathbb{1}-f(tA)^{1/2}](\mathbb{1}_0+H)^{-1}h\\
&=t^{1/2}[\mathbb{1}-f(tA)^{1/2}](tA)^{-1/2}A^{1/2}(\mathbb{1}_0+H)^{-1/2}(\mathbb{1}_0+H)^{1/2}\\
&\quad\times(\mathbb{1}_0+H)^{-1}h,
\end{aligned}$$

for $h \in \mathcal{H}_0$, which in turn yields the estimate $\|[\mathbb{1}-f(tA)^{1/2}](\mathbb{1}_0+H)^{-1}h\| \leqslant t^{1/2}C_{1/2}\|h\|$, $h \in \mathcal{H}$, where we used that the Kato function $f(\cdot)$ has the property (5.16): $C_{1/2} := \sup_{x>0}(1-f(x)^{1/2})/x^{1/2} < +\infty$. This implies:

$$\|[\mathbb{1}-f(tA)^{1/2}](\mathbb{1}_0+H)^{-1}\|\leqslant C_{1/2}t^{1/2}. \tag{5.126}$$

Then by (ii), (5.125) and (5.126) we obtain (5.124).

Define the operator $\mathcal{M}(t)$ on the domain $\mathcal{H}(t) = \operatorname{dom} M(t) = \operatorname{dom}(Q(t)g(tB)Q(t))$ by

$$\mathcal{M}(t) := M(t) + Q(t)g(tB)Q(t)/(1+t).$$

Then taking into account (5.123) and (5.124) one finds that

$$\lim_{t\to+0}\|Q(t)\mathcal{M}(t)^{-1}Q(t)-P_0(\mathbb{1}_0+H)^{-1}P_0\|=0. \tag{5.127}$$

Hence, to show that (ii) $\Rightarrow$ (iii) (for $\lambda = 1$) we have to verify that

$$\lim_{t\to+0}\|\mathcal{M}(t)^{-1}-\{\mathbb{1}(t)+M(t)\}^{-1}\|=0. \tag{5.128}$$

To this aim we first use the identity

$$\mathcal{M}(t)^{-1} - \{\mathbb{1}(t) + M(t)\}^{-1}$$
$$= \{\mathbb{1}(t) + M(t)\}^{-1} \left\{\mathbb{1}(t) - \frac{1}{1+t} Q(t)g(tB)Q(t)\right\} \mathcal{M}(t)^{-1},$$

to establish that

$$\mathcal{M}(t)^{-1} - \{\mathbb{1}(t) + M(t)\}^{-1} = \frac{t}{1+t}\{\mathbb{1}(t) + M(t)\}^{-1}\mathcal{M}(t)^{-1} + \frac{1}{1+t}\{\mathbb{1}(t) + M(t)\}^{-1}Q(t)[\mathbb{1} - g(tB)]Q(t)\mathcal{M}(t)^{-1}.$$

Therefore, the (5.128) holds if

$$\lim_{t\to+0} \left\|[\mathbb{1} - g(tB)]Q(t)\mathcal{M}(t)^{-1}\right\| = 0. \tag{5.129}$$

To prove (5.129) we note that, by the construction of the operator $H = A \dot{+} B$, one has $\operatorname{dom} H^{1/2} \subseteq \operatorname{dom} B^{1/2}$. This allows to obtain the representation

$$\begin{aligned}&[\mathbb{1} - g(tB)]Q(t)\mathcal{M}(t)^{-1} \\ &= [\mathbb{1} - g(tB)]\left\{Q(t)\mathcal{M}(t)^{-1}Q(t) - P_0(\mathbb{1}_0 + H)^{-1}P_0\right\} \\ &\quad + t^{1/2}[\mathbb{1} + g(tB)^{1/2}][\mathbb{1} - g(tB)^{1/2}](tB)^{-1/2}B^{1/2} \\ &\quad \times (\mathbb{1}_0 + H)^{-1/2}(\mathbb{1}_0 + H)^{-1/2}P_0,\end{aligned} \tag{5.130}$$

and the estimate

$$\|[\mathbb{1} - g(tB)^{1/2}]P_0(\mathbb{1}_0 + H)^{-1}P_0\| \leqslant S_{1/2}t^{1/2}. \tag{5.131}$$

Here we used the property (5.18): $S_{1/2} := \sup_{x>0}(1 - g(x)^{1/2})/x^{1/2} < +\infty$ of the Kato function g.

From (5.130) and (5.131) we obtain the estimate

$$\begin{aligned}&\|[\mathbb{1} - g(tB)]Q(t)\mathcal{M}(t)^{-1}\| \\ &\leqslant \left\|Q(t)\mathcal{M}(t)^{-1}Q(t) - P_0(\mathbb{1}_0 + H)^{-1}P_0\right\| + 2S_{1/2}t^{1/2}.\end{aligned} \tag{5.132}$$

Applying (5.127) to (5.132) we obtain (5.129), which yields (5.128). Then from (5.127) and (5.128) we obtain (iii) for $\lambda = 1$, and consequently for any $\lambda > 0$.

(iii) $\Rightarrow$ (iv). The identity

$$\begin{aligned}g(tB)^{1/2}Q(t)(\lambda\mathbb{1}(t) + M(t))^{-1}Q(t)g(tB)^{1/2} &= (\lambda\mathbb{1} + S(t))^{-1}F(t) \\ + \lambda(\lambda\mathbb{1} + S(t))^{-1}g(tB)^{1/2}Q(t)[\mathbb{1} - f(tA)]Q(t)&(\lambda\mathbb{1}(t) + M(t))^{-1} \\ &\times Q(t)g(tB)^{1/2}\end{aligned} \tag{5.133}$$

shows that

$$\begin{aligned}(\lambda\mathbb{1} + S(t))^{-1} &= tS(t)(\lambda\mathbb{1} + S(t))^{-1} \\ &\quad + g(tB)^{1/2}Q(t)(\lambda\mathbb{1}(t) + M(t))^{-1}Q(t)g(tB)^{1/2} \\ &\quad + \lambda(\lambda\mathbb{1} + S(t))^{-1}g(tB)^{1/2}[f(tA) - \mathbb{1}]Q(t)(\lambda\mathbb{1}(t) + M(t))^{-1}Q(t)g(tB)^{1/2}.\end{aligned}$$

This allows us to rewrite (iv) as

$$\begin{aligned}&(\lambda\mathbb{1} + S(t))^{-1} - P_0(\lambda\mathbb{1}_0 + H)^{-1}P_0 \\ &= tS(t)(\lambda\mathbb{1} + S(t))^{-1} \\ &\quad + g(tB)^{1/2}\left\{Q(t)(\lambda\mathbb{1}(t) + M(t))^{-1}Q(t) - P_0(\lambda\mathbb{1}_0 + H)^{-1}P_0\right\}g(tB)^{1/2} \\ &\quad + [g(tB)^{1/2} - \mathbb{1}]P_0(\lambda\mathbb{1}_0 + H)^{-1}P_0 g(tB)^{1/2} \\ &\quad + P_0(\lambda\mathbb{1}_0 + H)^{-1}P_0[g(tB)^{1/2} - \mathbb{1}] + \lambda(\lambda\mathbb{1} + S(t))^{-1}g(tB)^{1/2}[f(tA) - \mathbb{1}] \\ &\quad \times \left\{Q(t)(\lambda\mathbb{1}(t) + M(t))^{-1}Q(t) - P_0(\lambda\mathbb{1}_0 + H)^{-1}P_0\right\}g(tB)^{1/2} \\ &\quad + \lambda(\lambda\mathbb{1} + S(t))^{-1}g(tB)^{1/2}[f(tA) - \mathbb{1}]P_0(\lambda\mathbb{1}_0 + H)^{-1}P_0 g(tB)^{1/2}.\end{aligned}$$

Then one gets the estimate

$$\begin{aligned}&\|(\lambda\mathbb{1} + S(t))^{-1} - P_0(\lambda\mathbb{1}_0 + H)^{-1}P_0\| \\ &\leqslant \frac{t}{\lambda} + 2\left\|Q(t)(\lambda\mathbb{1}(t) + M(t))^{-1}Q(t) - P_0(\lambda\mathbb{1}_0 + H)^{-1}P_0\right\| \\ &\quad + 2\|[\mathbb{1} - g(tB)^{1/2}]P_0(\lambda\mathbb{1}_0 + H)^{-1}P_0\| + \|[\mathbb{1} - f(tA)]P_0(\lambda\mathbb{1}_0 + H)^{-1}P_0\|.\end{aligned}$$

Combining (iii), (5.126) and (5.131) we obtain (iv).

(iv) $\Rightarrow$ (v). To prove this part we set $\Phi(t) := F(t)$ for $t \geqslant 0$, and $X(t) := S(t)$ for $t > 0$. Then one can apply Proposition 5.39.

(v) $\Rightarrow$ (i). Using (5.115), (5.117), we start with the identity

$$F(t)g(tB)^{1/2}f(tA)^{1/2} = g(tB)^{1/2}f(tA)^{1/2}T(t),$$

for $t \geqslant 0$. Then

$$F(t)^r g(tB)^{1/2}f(tA)^{1/2} = g(tB)^{1/2}f(tA)^{1/2}T(t)^r,$$

for $t \geqslant 0$ and $r \geqslant 1$, or

$$f(tA/r)^{1/2}g(tB/r)^{1/2}F(t/r)^r g(tB/r)^{1/2}f(tA/r)^{1/2} = T(t/r)^{r+1}.$$

It follows that

$$\begin{aligned}T(t/r)^r &= T(t/r)^r(\mathbb{1} - T(t/r)) \\ &\quad + f(tA/r)^{1/2}g(tB/r)^{1/2}F(t/r)^r g(tB/r)^{1/2}f(tA/r)^{1/2},\end{aligned}$$

which leads to the representation

$$\begin{aligned}
&T(t/r)^r - P_0 e^{-tH} P_0 \\
&= T(t/r)^r(\mathbb{1} - T(t/r)) \\
&\quad + f(tA/r)^{1/2} g(tB/r)^{1/2}\{F(t/r)^r - P_0 e^{-tH} P_0\} g(tB/r)^{1/2} f(tA/r)^{1/2} \\
&\quad + f(tA/r)^{1/2}[g(tB/r)^{1/2} - \mathbb{1}] P_0 e^{-tH} P_0 g(tB/r)^{1/2} f(tA/r)^{1/2} \\
&\quad + [f(tA/r)^{1/2} - \mathbb{1}] P_0 e^{-tH} P_0 g(tB/r)^{1/2} f(tA/r)^{1/2} \\
&\quad + P_0 e^{-tH} P_0 [g(tB/r)^{1/2} - \mathbb{1}] f(tA/r)^{1/2} + P_0 e^{-tH} P_0 [f(tA/r)^{1/2} - \mathbb{1}].
\end{aligned}$$

This readily implies the inequality

$$\begin{aligned}
&\|T(t/r)^r - P_0 e^{-tH} P_0\| \\
&\leqslant \|T(t/r)^r(\mathbb{1} - T(t/r))\| \\
&\quad + \|F(t/r)^r - P_0 e^{-tH} P_0\| + 2\|[\mathbb{1} - g(tB/r)^{1/2}] P_0 e^{-tH} P_0\| \\
&\quad + 2\|[\mathbb{1} - f(tA/r)^{1/2}] P_0 e^{-tH} P_0\|.
\end{aligned} \tag{5.134}$$

By the spectral theorem the first term in the right-hand side of (5.134) is estimated as

$$\|T(t/r)^r(\mathbb{1} - T(t/r))\| \leqslant \frac{1}{1+r}, \quad r \geqslant 1. \tag{5.135}$$

Using for $h \in \mathcal{H}_0$ the expression

$$\begin{aligned}
[\mathbb{1} - f(tA/r)^{1/2}] e^{-tH} h = r^{-1/2}[\mathbb{1} - f(tA/r)^{1/2}](tA/r)^{-1/2}(tA)^{1/2}(\mathbb{1}_0 + tH)^{-1/2} \\
\times (\mathbb{1}_0 + tH)^{1/2} e^{-tH} h,
\end{aligned}$$

and the fact that by condition the Kato function f satisfies $C_{1/2} := \sup_{x>0}(1 - f(x)^{1/2})/x^{1/2} < +\infty$, we obtain

$$\begin{aligned}
&\|[\mathbb{1} - f(tA/r)^{1/2}] e^{-tH} h\| \\
&\leqslant r^{-1/2} C_{1/2} \|(\mathbb{1}_0 + tH)^{1/2} e^{-tH}\| \, \|h\| \leqslant r^{-1/2} C_{1/2} \|h\|, \quad h \in \mathcal{H}_0.
\end{aligned}$$

Since $0 \leqslant (1+x)^{1/2} e^{-x} \leqslant 1$ for $x \geqslant 0$, this yields

$$\|[\mathbb{1} - f(tA/r)^{1/2}] P_0 e^{-tH} P_0\| \leqslant r^{-1/2} C_{1/2}, \tag{5.136}$$

for $t \geqslant 0$ and $r \geqslant 1$. Similarly we prove for $t \geqslant 0$ and $r \geqslant 1$ the corresponding estimate for the Kato function g:

$$\|[\mathbb{1} - g(tB/r)^{1/2}] P_0 e^{-tH} P_0\| \leqslant r^{-1/2} S_{1/2}. \tag{5.137}$$

Therefore, the estimates (5.134), (5.135) and (5.136), (5.137) imply

$$\begin{aligned}
&\sup_{t \in [a,b]} \|T(t/r)^r - P_0 e^{-tH} P_0\| \\
&\leqslant \sup_{t \in [a,b]} \|F(t/r)^r - P_0 e^{-tH} P_0\| + \frac{1}{1+r} + \frac{2}{r^{1/2}}(C_{1/2} + S_{1/2}),
\end{aligned}$$

and so (v) yields (i). □

Corollary 5.41. *Under assumptions of Proposition* 5.40 *the conditions* (i) – (v) *are equivalent to*

(vi) $\lim_{n\to\infty}\sup_{t\in[a,b]}\|(f(tA/n)g(tB/n))^n - P_0e^{-tH}P_0\| = 0,\ \ [a,b]\subseteq(0,\infty)$,

(vii) $\lim_{n\to\infty}\sup_{t\in[a,b]}\|(g(tB/n)f(tA/n))^n - P_0e^{-tH}P_0\| = 0,\ \ [a,b]\subseteq(0,\infty)$.

5.3.4 Proof of Proposition 5.36

Proof. The above conditions on A, B imply that these operators are generators of C_0-semigroups. Then together with properties of the Kato functions $f,g\in\mathcal{K}$ this yields for $\lambda>0$ in $\mathcal{H}$ that

$$\operatorname*{s-lim}_{t\to+0}(\lambda\mathbb{1}+S(t))^{-1} = P_0(\lambda\mathbb{1}_0+H)^{-1}P_0, \quad \operatorname{dom}H\subset\mathcal{H}_0. \tag{5.138}$$

Now, from (5.133) we obtain, for $t>0$ and $\lambda>0$, the identity

$$\begin{aligned} g(tB)^{1/2}Q(t)(\lambda\mathbb{1}(t)+M(t))^{-1}Q(t)g(tB)^{1/2} &= (\lambda\mathbb{1}+S(t))^{-1}F(t) \\ &+\lambda g(tB)^{1/2}Q(t)(\lambda\mathbb{1}(t)+M(t))^{-1}Q(t)[\mathbb{1}-f(tA)]g(tB)^{1/2}(\lambda\mathbb{1}+S(t))^{-1}. \end{aligned}$$

Then using the relations

$$\operatorname*{s-lim}_{t\to+0} g(tB)^{1/2} = \operatorname*{s-lim}_{t\to+0} f(tA) = \operatorname*{s-lim}_{t\to+0} F(t) = \operatorname*{s-lim}_{t\to+0} Q(t) = \mathbb{1}$$

and (5.138), we find for each $\lambda>0$ and for $\operatorname{dom}H\subset\mathcal{H}_0$ the limit in $\mathcal{H}$

$$\operatorname*{s-lim}_{t\to+0} Q(t)(\lambda\mathbb{1}(t)+M(t))^{-1}Q(t) = P_0(\lambda\mathbb{1}_0+H)^{-1}P_0\,. \tag{5.139}$$

Note that by definitions (5.102) and (5.104) one has

$$\varphi_0(x)\geqslant\varphi_0(y), \quad \text{for } 0<x\leqslant y, \tag{5.140}$$

and

$$0\leqslant f(x)\leqslant f_0(x)\leqslant 1, \quad \text{for } x\in\mathbb{R}_0^+. \tag{5.141}$$

By virtue of (5.141) we get $f(tA)\leqslant f_0(tA)$, $t\geqslant 0$, which yields (see (5.119) and (5.120)) $\tilde{f}(tA)\leqslant f_0(tA)Q(t)$, $t>0$. Then by the invertibility of the mappings $\tilde{f}(tA)>0$ and $f_0(tA)>0$, one deduces that $f_0(tA)^{-1}Q(t)\leqslant\tilde{f}(tA)^{-1}$. Consequently, in the subspaces $\{\mathcal{H}(t)=Q(t)\mathcal{H}\}_{t\geqslant0}$ we obtain the inequality

$$\frac{1}{t}[f_0(tA)^{-1}-\mathbb{1}(t)]Q(t)\leqslant\frac{1}{t}[\tilde{f}(tA)^{-1}-\mathbb{1}(t)], \quad t>0.$$

This inequality, the definition of $M(t)$ (5.121), and the fact that $(\mathbb{1}-g(tB))\geqslant0$, yield the estimate $[f_0(tA)^{-1}-\mathbb{1}(t)]Q(t)/t\leqslant M(t)$, $t>0$, or equivalently (see (5.104)) $A\varphi_0(tA)Q(t)\leqslant M(t)$. Consequently,

$$\begin{aligned} &Q(t)(\lambda\mathbb{1}(t)+M(t))^{-1}Q(t) \\ &\leqslant(\lambda\mathbb{1}+A\varphi_0(tA))^{-1}Q(t)\leqslant(\lambda\mathbb{1}+A\varphi_0(tA))^{-1}, \quad t>0, \end{aligned} \tag{5.142}$$

where we used that $Q(t)$ commutes with A.

Recall that φ_0 is a non-increasing function. Then (5.140) implies $\varphi_0(tA)) \geqslant \varphi_0(t_0 A))$ for $0 < t \leqslant t_0$ and hence

$$(\lambda \mathbb{1} + A\varphi_0(t_0 A))^{-1} \geqslant (\lambda \mathbb{1} + A\varphi_0(tA))^{-1} .$$

This inequality and (5.142) yield for $0 < t \leqslant t_0$ the estimate

$$Q(t)(\lambda \mathbb{1}(t) + M(t))Q(t) \leqslant (\lambda \mathbb{1} + A\varphi_0(t_0 A))^{-1} . \tag{5.143}$$

Note that since by definition (5.104)

$$(\lambda \mathbb{1} + A\varphi_0(t_0 A))^{-1} = t_0(\mathbb{1} + (\lambda t_0 - 1) f_0(t_0 A))^{-1} f_0(t_0 A) ,$$

condition (5.106) of the proposition implies

$$(\lambda \mathbb{1} + A\varphi_0(t_0 A))^{-1} \in \mathcal{C}_\infty(\mathcal{H}), \quad \text{for } \lambda t_0 > 1. \tag{5.144}$$

Now taking into account (5.143) and (5.144) we infer that convergence (5.139) gives the proof of condition (iii) in Proposition 5.40. Therefore, by Proposition 5.40 and Corollary 5.41, we obtain that:

(a) The Trotter-Kato product formula (5.100) (vi) and all other versions (i),(v), (vii) converge in the operator-norm topology locally uniformly away from zero to the degenerate semigroup $\{e^{-tH}P_0\}_{t>0}$.

(b) For $\lambda > 0$ the resolvent $(\lambda \mathbb{1}_0 + H)^{-1} \in \mathcal{C}_\infty(\mathcal{H}_0)$, where $\operatorname{dom} H \subset \mathcal{H}_0 \subseteq \mathcal{H}$, see (5.107). □

We comment that condition (5.106) of Proposition 5.36 is quite implicitly related to the properties of the non-negative self-adjoint operator A. The following remark makes this more explicit.

Remark 5.42. Recall that, by Definition 5.34 (5.102), $0 \leqslant \varphi_0(x) \leqslant 1$, which implies $1 + x\varphi_0(x) \leqslant 1 + x$, or by the definition of f_0 (5.104), $f_0(x) \geqslant (1+x)^{-1}$. Since by (5.106) $f_0(t_0 A) \in \mathcal{C}_\infty(\mathcal{H})$, this means that

$$f_0(t_0 A) \geqslant (\mathbb{1} + t_0 A)^{-1} \in \mathcal{C}_\infty(\mathcal{H}), \quad t_0 > 0. \tag{5.145}$$

Thus, (5.106) implies that the resolvent of the operator A is compact.

Although it is plausible, we caution the reader that the converse is certainly not true for generic Kato function, $f \in \mathcal{K}$. To provide a converse we consider a class of *regular* Kato functions $\mathcal{K}_r \subset \mathcal{K}$.

Definition 5.43. Let f be a Kato function from $\mathcal{K}$. We set $b(x) := \sup_{0 \leqslant s \leqslant x} sf(s)$ and $r(x) := \sup_{s \in [x,\infty)} f(s)$, for $x \in \mathbb{R}^+$. The Kato function f is called *regular* if $\lim_{x \to +\infty} b(x)/x = 0$ and $0 \leqslant r(x) < 1$.

For instance, $f(x) = e^{-x}$ and $f(x) = 1/(1+x)$ are regular Kato functions, whereas $f(x) = 1 - |\sin(x)|$ is *not*, see Appendix C (Section C.3).

Lemma 5.44. *Let $f \in \mathcal{K}$ and let A be a non-negative self-adjoint operator in a (infinite-dimensional) Hilbert space $\mathcal{H}$. Let $(\mathbb{1} + A)^{-1} \in \mathcal{C}_\infty(\mathcal{H})$, then $f_0(A) \in \mathcal{C}_\infty(\mathcal{H})$ if and only if the Kato function f is regular.*

An obvious consequence of Proposition 5.36 and Lemma 5.44 is that the Trotter-Kato product formulae converge in the operator-norm topology locally uniformly away from zero for any regular $f \in \mathcal{K}_r$, *any* $g \in \mathcal{K}$, and *any* self-adjoint operator $B \geqslant 0$ if the resolvent of the operator A is *compact.* It turns out that these assumptions are not only *sufficient* but also *necessary.*

Proposition 5.45. *Let $A \geqslant 0$ and $B \geqslant 0$ be self-adjoint operators in a Hilbert space $\mathcal{H}$ such that $H = A \dotplus B$ is self-adjoint in $\operatorname{dom} H \subset \mathcal{H}_0 \subseteq \mathcal{H}$. Then the Trotter-Kato product formulae converge in the operator-norm topology to the degenerate semigroup $\{e^{-tH}P_0\}_{t\geqslant 0}$ locally uniformly away from zero for any regular Kato function $f \in \mathcal{K}_r$ and for arbitrary Kato function g if and only if $(\mathbb{1}+A)^{-1} \in \mathcal{C}_\infty(\mathcal{H})$.*

Proof. If $(\mathbb{1} + A)^{-1} \in \mathcal{C}_\infty(\mathcal{H})$, then by Lemma 5.44 one gets that $f_0(A) \in \mathcal{C}_\infty(\mathcal{H})$. Now applying Proposition 5.36 we verify the sufficient part of proposition.

The converse part needs more preparation. The interested reader can find the corresponding reference in Notes in Section 5.6 □

Comments about some other results when the *operator compactness* condition lifts the strong convergence of the Trotter-Kato product formulae to the convergence in the operator-norm topology one can also find in the Notes in Section 5.6.

For completeness, we quote here a result when instead of *regularity* constrains on the class of admissible Kato functions $\mathcal{K}$ one imposes rather explicit conditions on the operators A and B in order to ensure the convergence of the Trotter-Kato product formulae in the operator-norm topology. Note that in the next statement the operators A, B and the pairs f, g of Kato functions are treated *equally.*

Proposition 5.46. *Let $A \geqslant 0$ and $B \geqslant 0$ be self-adjoint operators in a Hilbert space $\mathcal{H}$ and let f and g be Kato functions, from the class $\mathcal{K}'$, obeying the conditions*

$$C := \sup_{x>0} \frac{xf(x)}{1-f(x)} < +\infty \quad \text{and} \quad S := \sup_{x>0} \frac{xg(x)}{1-g(x)} < +\infty. \tag{5.146}$$

Then the Trotter-Kato product formulae converge to the degenerate semigroup $\{e^{-tH}P_0\}_{t\geqslant 0}$ in the operator-norm topology locally uniformly away from zero and one also has $(\mathbb{1}_0 + H)^{-1} \in \mathcal{C}_\infty(\mathcal{H}_0)$, if

$$(\mathbb{1} + A)^{-1}(\mathbb{1} + B)^{-1} \in \mathcal{C}_\infty(\mathcal{H}). \tag{5.147}$$

Here $\operatorname{dom} H \subset \mathcal{H}_0 \subseteq \mathcal{H}$, $\mathcal{H}_0 = P_0(\mathcal{H})$ *and* $\mathbb{1}_0 = \mathbb{1} \restriction \mathcal{H}_0$.

For more details concerning the class $\mathcal{K}'$ of functions obeying the conditions (5.146) we refer to Appendix C, Section C.3.

5.3.5 Rate of Convergence: non-self-adjoint semigroups

We conclude Section 5.3 by two results that extend the operator-norm convergence of the *exponential* Trotter-Kato product formula to *non-self-adjoint* C_0-semigroups. In contrast to Propositions 5.36, 5.45 and in particular to Proposition 5.46, these results are based on a definite *hierarchy* between generators A and B, where operator B is considered as a *small* perturbation.

Proposition 5.47. *Let* $\{e^{-tA}\}_{t\geqslant 0}$ *be a holomorphic contraction semigroup on a Hilbert space* $\mathcal{H}$. *Suppose that the m-sectorial generator* A *is invertible and* B *is an m-accretive operator in* $\mathcal{H}$ *such that* $\operatorname{dom} A^{\alpha} \subseteq \operatorname{dom} B$ *for some* $\alpha \in [0,1)$ *and* $\operatorname{dom} A^* \subseteq \operatorname{dom} B^*$. *Then the Trotter product formula converges in the operator-norm topology to the* C_0*-semigroup* $\{e^{-t(A+B)}\}_{t\geqslant 0}$ *with rate estimated by*

$$\begin{aligned} & \left\| \left(e^{-tA/n} e^{-tB/n}\right)^n - e^{-t(A+B)} \right\| \\ & \leqslant O(\ln(n)/n^{1-\alpha})\,(\alpha \neq 0) \vee O((\ln(n))^2/n)\,(\alpha = 0). \end{aligned} \tag{5.148}$$

For operators A and B as above, we recall that B is A-small with relative bound $b \geqslant 0$ and $a \geqslant 0$ if

$$\operatorname{dom} A \subseteq \operatorname{dom} B \quad \text{and} \quad \|Bu\| \leqslant a\|u\| + b\|Au\|, \quad u \in \operatorname{dom} A. \tag{5.149}$$

For $b < 1$ these conditions guarantee the existence of the perturbed semigroup $\{e^{-t(A+B)}\}_{t\geqslant 0}$, see Section 1.7.

Remark 5.48. The condition $\operatorname{dom} A^{\alpha} \subseteq \operatorname{dom} B$ in Proposition 5.47 restricts the smallness of the operator B even further. Indeed, since this condition implies $\|BA^{-\alpha}\| < \infty$, one gets that for any $\eta > 0$ and $\alpha < 1$

$$\|Bu\| \leqslant \|BA^{-\alpha}\| \|A^{\alpha}(A+\eta)^{-1}\| (\|Au\| + \eta\|u\|), \quad u \in \operatorname{dom} A.$$

If A is the generator of a holomorphic semigroup, then for each $\alpha \in [0,1)$ there exists a constant C_α, such that $\|A^{\alpha}(A+\eta)^{-1}\| \leqslant C_\alpha/\eta^{1-\alpha}$ for all $\eta > 0$. Then we obtain the estimate

$$\|Bu\| \leqslant a_\eta\|u\| + b_\eta\,\|Au\|, \quad u \in \operatorname{dom} A,$$

where the relative bound $b_\eta := C_\alpha\|BA^{-\alpha}\|/\eta^{1-\alpha}$ of the operator B can be arbitrarily small for large η, cf. (5.149).

This means that the unbounded perturbation B is *infinitesimally* small with respect to A. We denote this by $B \in \mathcal{P}_{0+}(A)$, Definition 1.50. Other infinitesimally small perturbations from the class $\mathcal{P}_{0+}$ are considered in Section 4.4 and Remark 5.31.

To relax this severe smallness condition and improve the rate of convergence one can demand more restrictions on the C_0-semigroup generator A than it is done in Proposition 5.47. Then the following assertion gets true.

Proposition 5.49. *Let A be a positive invertible self-adjoint operator in $\mathcal{H}$. If B is an m-accretive operator such that* $\operatorname{dom} A \subseteq \operatorname{dom} B$, $\operatorname{dom} A \subseteq \operatorname{dom} B^*$, *and*

$$\begin{aligned} \|Bu\| &\leqslant b\|Au\|, \qquad && u \in \operatorname{dom} A,\ 0 < b < 1, \\ \|B^*u\| &\leqslant b_*\|Au\|, && u \in \operatorname{dom} A,\ 0 < b_* < 1, \end{aligned}$$

then the Trotter product formula converges in the operator-norm topology to the C_0-semigroup $\{e^{-t(A+B)}\}_{t\geqslant 0}$ with the error bound estimate for the rate given by

$$\left\| \left(e^{-tA/n} e^{-tB/n}\right)^n - e^{-t(A+B)} \right\| \leqslant O\left(\frac{\ln(n)}{n}\right). \tag{5.150}$$

Remark 5.50. Note that in Proposition 5.47 and in Proposition 5.49 the *hierarchy* of adjoint operators $\overline{\operatorname{dom} A^*} \subseteq \operatorname{dom} B^*$, is indispensable for ensuring the boundedness of the closure $\overline{(A^{-1}B)} = (B^* A^{*-1})^*$, cf. Remark 5.14.

In Section 5.5 we use these results to prove the convergence of the *exponential* Trotter-Kato product formula for Gibbs semigroups and to obtain estimates of the rate of convergence in the *trace-norm* topology.

5.4 Trotter-Kato product formulae: trace-norm convergence, error bounds

In this section we start by an analysis of the *exponential* Trotter-Kato (or simply Trotter) product formula for Gibbs semigroups. The first result is that for these semigroups the convergence of the Trotter product formula can be strengthened from the operator-norm to the trace-norm convergence. The main tool for obtaining this improvement of convergence is the *lifting lemma*, see subsection 5.4.1, Lemma 5.51. For a certain class of Gibbs semigroups it allows to lift, for the Trotter product formula, the operator-norm convergence to convergence in the trace-norm topology due to a direct error bound estimates.

In subsection 5.4.2, *Trace-norm convergence without rate estimate*, we consider the case of *self-adjoint* Gibbs semigroups and prove the trace-norm convergence of the Trotter product formula *without* error bound estimate. In this case the trace-norm limit may be a *degenerate* Gibbs semigroup, Proposition 5.53. Note that here the generators A and B in the Trotter approximants are not conditioned by any hierarchy.

Next we use the operator-norm error bound from Proposition 5.8 and the lifting Lemma 5.51 to prove in Proposition 5.55 the trace-norm convergence of the Trotter product formula to the Gibbs semigroup with an error bound estimate. Here the self-adjoint generators A and B are subject to a *smallness* condition on B. In Proposition 5.56 we show that if the operator-norm error-bound is optimal, then the trace-norm error bound estimate *inherits* the same property.

In subsection 5.4.3, *Smallness conditions and optimal rate of convergence*, we modify Proposition 5.8 by fractional-power relative *smallness* conditions for the pair of self-adjoint operators A and B. Then they yield an *optimal* operator-norm error bound estimate for the rate of convergence of the Trotter-Kato product formulae. Note that these hierarchical fractional power conditions (cf. Remark 5.3) are less restrictive than the self-adjointness of the algebraic sum $A + B$ in Proposition 5.19. Similarly to Proposition 5.36 and Proposition 5.45 they allow the sum $A \dot{+} B$ in the wider sense of quadratic forms.

Finally in subsection 5.4.4, *Trotter-Kato product formulae: trace-norm convergence*, we study sufficient conditions for the trace-norm convergence of the Trotter-Kato product formulae. This extension from the Trotter product formula requires some nontrivial conditions on the pair of generic Kato functions $f, g \in \mathcal{K}$ (Definition 5.4). We mention here that the lifting Lemma 5.51 is well-suited for the Trotter formula, but not for the Trotter-Kato product formulae.

5.4.1 Lifting lemma

The following key *lifting lemma* allows us to lift the norm-convergence to the trace-norm convergence with the $\|\cdot\|_1$-error bound estimate.

Lemma 5.51. *Let A be an m-sectorial operator with vertex $\gamma = 0$ and such that $\mathrm{e}^{-t\,\mathfrak{Re}\,A} \in \mathcal{C}_1(\mathcal{H})$ for $t > 0$. Let B be generator of a contraction C_0-semigroup such that the family $\{\mathrm{e}^{-tH}\}_{t\geqslant 0}$ is a Gibbs semigroup with $\|\mathrm{e}^{-tH}\| \leqslant 1$. Let*

$$\varepsilon(m,t) := \sup_{(2m/(2m+1))t\leqslant\xi\leqslant(2m/(2m-1))t} \left\| \left(\mathrm{e}^{-\xi A/m}\mathrm{e}^{-\xi B/m}\right)^m - \mathrm{e}^{-\xi H} \right\|, \tag{5.151}$$

for $m = 2, 3, \ldots$, and $t > 0$. Then for each $t_0 > 0$ there are numbers $L_1(t_0)$ and $L_2(t_0)$ such that

$$\begin{aligned} &\left\| \left(\mathrm{e}^{-tA/n}\mathrm{e}^{-tB/n}\right)^n - \mathrm{e}^{-tH} \right\|_1 \\ &\leqslant L_1(t_0)\varepsilon([n/2], t/2) + L_2(t_0)\varepsilon\left([(n+1)/2], t/2\right) \end{aligned} \tag{5.152}$$

for all $n = 2, 3, \ldots$, and for $t \geqslant t_0 > 0$, that is, away from zero. Here $[x]$ denotes the integer part of $x \geqslant 0$.

Proof. For $n > 1$, we define two variables: $k_n := [n/2] \leqslant n/2$ and $m_n := [(n+1)/2] \geqslant n/2$. Then $n = k_n + m_n$, and we have the representation

$$\begin{aligned} &(\mathrm{e}^{-tA/n}\mathrm{e}^{-tB/n})^n - \mathrm{e}^{-tH} \\ &= \left[(\mathrm{e}^{-tA/n}\mathrm{e}^{-tB/n})^{k_n} - \mathrm{e}^{-k_n tH/n}\right](\mathrm{e}^{-tA/n}\mathrm{e}^{-tB/n})^{m_n} \\ &\quad + \mathrm{e}^{-k_n tH/n}\left[(\mathrm{e}^{-tA/n}\mathrm{e}^{-tB/n})^{m_n} - \mathrm{e}^{-m_n tH/n}\right]. \end{aligned} \tag{5.153}$$

By Lemma 4.43, we get the inequality:

$$\left\| \left(\mathrm{e}^{-tA/n}\mathrm{e}^{-tB/n}\right)^{m_n} \right\|_1 \leqslant \left\| \mathrm{e}^{-tB/n} \right\|^{m_n} \left\| \mathrm{e}^{-m_n t\,\mathfrak{Re}\,A/4n} \right\|_1 .$$

The self-adjoint operator $\mathfrak{Re}\, A \geqslant 0$ (Remark 4.29) generates a contraction Gibbs semigroup $\{e^{-t\,\mathfrak{Re}\, A}\}_{t\geqslant 0}$ and $\{e^{-tB}\}_{t\geqslant 0}$ is also a contraction semigroup. Then since $m_n \geqslant n/2$, one obtains the uniform in $n > 1$ estimate

$$\left\|\left(e^{-tA/n}e^{-tB/n}\right)^{m_n}\right\|_1 \leqslant \left\|e^{-t\,\mathfrak{Re}\, A/8}\right\|_1 . \tag{5.154}$$

On the other hand, since by $\|e^{-tH}\| \leqslant 1$ the Gibbs semigroup $\{e^{-tH}\}_{t\geqslant 0}$ is a contraction semigroup and since $(n/2 - \epsilon_n) \leqslant k_n \leqslant n/2$, where $\lim_{n\to\infty} \epsilon_n = 0$, we get (for example) the estimate

$$\left\|e^{-k_n tH/n}\right\|_1 \leqslant \left\|e^{-tH/3}\right\|_1 . \tag{5.155}$$

Then (5.153)–(5.155) yield for $n > 1$ the inequality

$$\begin{aligned} &\left\|\left(e^{-tA/n}e^{-tB/n}\right)^{n} - e^{-tH}\right\|_1 \\ &\leqslant \left\|\left(e^{-tA/n}e^{-tB/n}\right)^{k_n} - e^{-k_n tH/n}\right\| \left\|e^{-t\,\mathfrak{Re}\, A/8}\right\|_1 \\ &\quad + \left\|e^{-tH/3}\right\|_1 \left\|\left(e^{-tA/n}e^{-tB/n}\right)^{m_n} - e^{-m_n tH/n}\right\| . \end{aligned} \tag{5.156}$$

Let $t_n = tk_n/n$ and $s_n = tm_n/n$, i.e., $\lim_{n\to\infty} t_n = \lim_{n\to\infty} s_n = t/2$, and let $t_0 > 0$. Then by definition (5.151) and by the estimate (5.156) one gets

$$\begin{aligned} &\left\|\left(e^{-tA/n}e^{-tB/n}\right)^{n} - e^{-tH}\right\|_1 \\ &\leqslant \left\|\left(e^{-t_n A/k_n}e^{-t_n B/k_n}\right)^{k_n} - e^{-t_n H}\right\| \left\|e^{-t\,\mathfrak{Re}\, A/8}\right\|_1 \\ &\quad + \left\|e^{-tH/3}\right\|_1 \left\|\left(e^{-s_n A/m_n}e^{-s_n B/m_n}\right)^{m_n} - e^{-s_n H}\right\| \\ &\leqslant \varepsilon(k_n, t/2) \left\|e^{-t_0\,\mathfrak{Re}\, A/8}\right\|_1 + \left\|e^{-t_0 H/3}\right\|_1 \varepsilon(m_n, t/2) , \end{aligned} \tag{5.157}$$

for $t \geqslant t_0$ and $n > 1$. Hence, we obtain the assertion (5.152), if we set $L_1(t_0) := \|e^{-t_0\,\mathfrak{Re}\, A/8}\|_1$ and $L_2(t_0) := \|e^{-t_0 H/3}\|_1$, which are finite for each $t_0 > 0$. □

Corollary 5.52. *Let $\{e^{-t\,\mathfrak{Re}\, A}\}_{t\geqslant 0}$ and $\{e^{-tH}\}_{t\geqslant 0}$ be eventually Gibbs semigroups away from $t_0 > 0$. Then (5.157) yields the estimate*

$$\begin{aligned} &\left\|\left(e^{-tA/n}e^{-tB/n}\right)^{n} - e^{-tH}\right\|_1 \\ &\leqslant L_1(t)\varepsilon([n/2], t/2) + L_2(t)\varepsilon\left([(n+1)/2], t/2\right) , \end{aligned} \tag{5.158}$$

for all $n = 2, 3, \ldots$ and for $t > 8t_0$, that is, away from $8t_0$.

We note that the threshold $8t_0$ in (5.158) have as its source estimates (5.154) and (5.155) which are originally not optimal.

5.4.2 Trace-norm convergence without rate estimate

First we prove the convergence of the Trotter product formula *away from zero* in the trace-norm topology *without* estimate of the rate of convergence.

Proposition 5.53. *Let $A \geqslant 0$ be the generator of the self-adjoint Gibbs semigroup $\{G_t(A) = \mathrm{e}^{-tA}\}_{t \geqslant 0}$. Then the exponential Trotter-Kato product formula converges in the trace-norm topology, away from zero, for Kato functions $f(x) = g(x) = \mathrm{e}^{-x}$ and for any self-adjoint operator $B \geqslant 0$, to a degenerate Gibbs semigroup:*

$$\|\cdot\|_1\text{-}\lim_{n\to\infty} \left(\mathrm{e}^{-tA/n}\mathrm{e}^{-tB/n}\right)^n = \mathrm{e}^{-tH} P_0 , \quad H = A \dot{+} B, \tag{5.159}$$

where P_0 is the orthogonal projection $P_0 : \mathcal{H} \to \overline{\mathrm{dom}\, H}$ and $t > 0$.

Proof. Operator A is obviously m-sectorial and B is the generator of a contraction semigroup. Then the form-sum operator $H = A \dot{+} B$ is self-adjoint in the subspace $\mathcal{H}_0 := \overline{\mathrm{dom}\, A^{1/2} \cap \mathrm{dom}\, B^{1/2}}$ and generates a degenerate contraction semigroup $\|e^{-tH} P_0\| \leqslant 1$ on $\mathcal{H}$. Moreover, it is known that the Trotter product formula converges strongly to $e^{-tH} P_0$ away from $t = 0$:

$$\text{s-}\lim_{n\to\infty} \left(e^{-tA/n}e^{-tB/n}\right)^n = e^{-tH} P_0 ,$$

where P_0 is the orthogonal projection $P_0 : \mathcal{H} \to \mathcal{H}_0$.

Now we have to check (see Lemma 5.51) that $e^{-tH} \in \mathcal{C}_1(\mathcal{H}_0)$. To this aim, let $\{B_k \geqslant 0\}_{k \geqslant 1} \subset \mathcal{L}(\mathcal{H})$ be a monotonically increasing to B sequence of positive bounded operators. Then the positivity of generator A implies

$$e^{-t(A\dot{+}B)} P_0 \leqslant e^{-t(A+B_{k+1})} \leqslant e^{-t(A+B_k)} \leqslant e^{-tA}. \tag{5.160}$$

By monotonicity the weak operator limit on $\mathcal{H}$ yields

$$\text{w-}\lim_{k\to\infty} e^{-t(A+B_k)} = e^{-tH} P_0 , \quad H = A \dot{+} B. \tag{5.161}$$

Note that by Proposition 4.45 the perturbed semigroup is positive and Gibbs: $e^{-t(A+B_k)} \in \mathcal{C}_{1,+}(\mathcal{H})$. Then by the lower weak semi-continuity of the trace (see Proposition 2.48(f) and Corollary 2.75) and by monotonicity (5.160) one gets

$$0 \leqslant \mathrm{Tr}\, e^{-tH} P_0 \leqslant \liminf_{k\to\infty} \mathrm{Tr}\, e^{-t(A+B_k)} \leqslant \mathrm{Tr}\, e^{-tA}. \tag{5.162}$$

Consequently, the positivity of the semigroups implies that $\|e^{-tH} P_0\|_1 \leqslant \|e^{-tA}\|_1$ on $\mathcal{H}$, or $e^{-tH} \in \mathcal{C}_1(\mathcal{H}_0)$.

Note that by taking the Laplace transform of the Gibbs semigroup $\{G_t(A)\}_{t \geqslant 0}$ one gets that resolvent

$$(\lambda \mathbb{1} + A)^{-1} = \int_0^\infty dt\; e^{-t\lambda} e^{-tA} \in \mathcal{C}_\infty(\mathcal{H}), \quad \lambda > 0, \tag{5.163}$$

is compact. Then Proposition 5.36 yields the operator-norm convergence of the Trotter product formula locally uniformly away from zero, and consequently

$$\lim_{s\to\infty} \varepsilon(s,t) = \lim_{s\to\infty} \sup_{(2s/(2s+1))t\leqslant\xi\leqslant(2s/(2s-1))t} \left\| \left(\mathrm{e}^{-\xi A/s}\mathrm{e}^{-\xi B/s}\right)^s - \mathrm{e}^{-\xi H}P_0 \right\| = 0, \tag{5.164}$$

for $s = 2, 3, \ldots$, and $t > 0$. Therefore, by Lemma 5.51, (5.152) (for immediately Gibbs semigroups) and by the limit (5.164) we obtain

$$\lim_{n\to\infty} \left\| \left(\mathrm{e}^{-tA/n}\mathrm{e}^{-tB/n}\right)^n - \mathrm{e}^{-tH}P_0 \right\|_1 = 0,$$

for $t \geqslant t_0$ and any $t_0 > 0$, that is away from zero. □

Corollary 5.54. *If $\{G_t(A)\}_{t\geqslant 0}$ is an eventually Gibbs semigroup away from $t_0 > 0$, then by Corollary 5.52 and (5.158) one gets that*

$$\lim_{n\to\infty} \left\| \left(\mathrm{e}^{-tA/n}\mathrm{e}^{-tB/n}\right)^n - \mathrm{e}^{-tH}P_0 \right\|_1 = 0,$$

away from $8t_0$ and that $\{\mathrm{e}^{-tH}P_0\}_{t\geqslant 0}$ is a degenerate eventually Gibbs semigroup away from $8t_0$.

5.4.3 Smallness conditions and optimal rate of convergence

Since Lemma 5.51 is established for exponential Kato function, we shall prove the lifting of the operator-norm to the trace-norm convergence with the *error-bound* estimate only for $f(x) = g(x) = \mathrm{e}^{-x}$. To do that, we borrow the operator-norm error-bound from Proposition 5.8 for the *self-adjoint* A and B to prove under a *smallness* condition on B the following statement:

Proposition 5.55. *Suppose that the self-adjoint operators A and B satisfy conditions (5.10) and (5.11) (see (i), (ii) of Section 5.2) with $b < 1$. If A is generator of the Gibbs semigroup $\{G_t(A)\}_{t\geqslant 0}$, then the Trotter product formula converges in the trace-norm topology to the Gibbs semigroup $\{G_t(A+B)\}_{t\geqslant 0}$, with the error bound estimate for $t \geqslant t_0 > 0$:*

$$\left\| (\mathrm{e}^{-tA/n}\mathrm{e}^{-tB/n})^n - \mathrm{e}^{-t(A+B)} \right\|_1 \leqslant c_1 \frac{\ln(n)}{n}, \tag{5.165}$$

for some t_0-depended number $c_1 > 0$. Here $n = 6, 7, \ldots$.

Proof. Since $B \in \mathcal{P}_{b<1}(A)$, the semibounded by (5.10) operator $H = A + B \geqslant 2\mathbb{1}$ with $\operatorname{dom} H = \operatorname{dom} A$ is self-adjoint, and it is the generator of holomorphic contraction semigroup. Since A is the generator of a Gibbs semigroup and $B = B^* \geqslant \mathbb{1}$, Proposition 4.45 shows that the perturbed semigroup is Gibbs, i.e., the operator H is the generator of the Gibbs semigroup $\{G_t(H)\}_{t\geqslant 0}$ with $\|\mathrm{e}^{-tH}\| \leqslant 1$.

Hence, the operators A, B and H satisfy the conditions of Lemma 5.51 and Proposition 5.8 for $f(x) = g(x) = \mathrm{e}^{-x}$, see (5.10). Then by virtue of (5.24) we get for (5.151) the estimate

$$\varepsilon(s,t) \leqslant \frac{L_F}{(1-b)^3}\frac{\ln(s)}{s}, \tag{5.166}$$

where $s = 3, 4, \ldots$, uniformly for $t \geqslant 0$. Since for $n \geqslant 6$ we get from (5.166) that $\varepsilon([n/2], t/2) \geqslant \varepsilon([(n+1)/2], t/2)$, inserting (5.166) into (5.152) we get assertion (5.165) uniformly away from $t_0 > 0$ (i.e. for $t \geqslant t_0 > 0$) with t_0-dependent constant

$$c_1 := \frac{L_F}{(1-b)^3}\left(\|\mathrm{e}^{-t_0 A/8}\|_1 + \|\mathrm{e}^{-t_0 H/3}\|_1\right) \leqslant \frac{2L_F}{(1-b)^3}\|\mathrm{e}^{-t_0 A/8}\|_1 ,$$

which is bounded for any $t_0 > 0$. Here we used that the perturbation $B \geqslant \mathbb{1}$ yields $H \geqslant A$ and Corollary 4.24 to estimate the $\|\cdot\|_1$-norms. □

The error bound estimate in Proposition 5.55 is evidently *not* optimal. Similar to analysis for operator-norm topology (Proposition 5.19) we can get an *optimal* trace-norm asymptotic rate for convergence of the Trotter product formula in the case of self-adjoint Gibbs semigroups.

Proposition 5.56. *Let A and B be non-negative self-adjoint operators in a Hilbert space $\mathcal{H}$ such that the operator sum $C := A + B$ on $\operatorname{dom} C = \operatorname{dom} A \cap \operatorname{dom} B$ is also self-adjoint. Let A be the generator of the Gibbs semigroup $\{G_t(A)\}_{t\geqslant 0}$. Then the operator $C \geqslant 0$ is the generator of the Gibbs semigroup $\{G_t(C) = \mathrm{e}^{-tC}\}_{t\geqslant 0}$ and:*

(a) *the Trotter product formula*

$$\left\|(e^{-tA/n}e^{-tB/n})^n - e^{-tC}\right\|_1 = O(n^{-1}), \quad n \to \infty, \tag{5.167}$$

converges in the trace-norm topology with the same rate $O(n^{-1})$ as its symmetrised version:

$$\left\|(e^{-tB/2n}e^{-tA/n}e^{-tB/2n})^n - e^{-tC}\right\|_1 = O(n^{-1}), \quad n \to \infty, \tag{5.168}$$

locally uniformly away from $t = 0$ (Definition 5.35);

(b) *the condition of self-adjointness and the error bound $O(n^{-1})$ are ultimate optimal.*

Proof. We remark that the condition $e^{-tA} \in \mathcal{C}_1(\mathcal{H})$, $t > 0$ is symmetric with respect to the choice A or B. To check that the operators A, B and H satisfy the conditions of Lemma 5.51 and Proposition 5.8 for $f(x) = g(x) = \mathrm{e}^{-x}$, one follows the same arguments as in the proof of Proposition 5.55. The extension to the symmetrised case uses exactly the same argument as the proof of Proposition 5.8. By virtue of (5.83) and (5.84), for the operator-norm rate of convergence we obtain for (5.151) the estimate

$$\varepsilon(n,t) = O(n^{-1})\ ,\ t > 0\ ,\quad n \to \infty,$$

uniformly away from zero. Then by the *lifting* Lemma 5.51 we obtain that (5.167) and (5.168) hold uniformly for $t \geqslant t_0 > 0$.

Finally, the optimality of (5.167) and (5.168) is a corollary of the optimality of the error bounds (5.83) and (5.84) of the convergence rate in operator norm. □

Another result, when the trace-norm error bound for the Trotter product formula has an *optimal* estimate of the convergence rate, provides a certain clarification on Proposition 5.8 and Proposition 5.55.

First we recall the *fractional smallness* conditions of subsection 5.2.5 (Proposition 5.27) that ensure the optimal asymptotic for the operator-norm rate convergence of the Trotter product formula.

Proposition 5.57. *Let the self-adjoint operators $A \geqslant \mathbb{1}$ and $B \geqslant 0$ in a Hilbert space $\mathcal{H}$ be such that for some $\alpha \in (1/2, 1)$ and $b \in (0,1)$ the following conditions are satisfied:*

(i) $\operatorname{dom} A^\alpha \subset \operatorname{dom} B^\alpha$ *and* $\|B^\alpha u\| \leqslant b\|A^\alpha u\|$, $u \in \operatorname{dom} A^\alpha$;

(ii) $\operatorname{dom} H^\alpha \subset \operatorname{dom} A^\alpha$.

Then semibounded from below operator $H := A\dot{+}B$ is the densely defined self-adjoint form-sum of A and B and for Kato functions $f, g \in \mathcal{K}_\alpha$ one gets the following operator-norm estimate:

$$\left\|(f(tA/n)g(tB/n))^n - e^{-tH}\right\| \leqslant \frac{c_2}{n^{2\alpha-1}}, \quad n \geqslant 2, \tag{5.169}$$

for some $c_2 > 0$, uniformly for $t \geqslant 0$.

Here the class of Kato functions $\mathcal{K}_\alpha$ is specified in Definition 5.26, cf. Appendix C. The application of the operator-norm error bound (5.169) to estimate the trace-norm rate of convergence of the *Trotter product formula* is again based on our *lifting* strategy. It allows us to prove the following statement.

Proposition 5.58. *Suppose that in addition to the conditions of Propositions* 5.57, (iii) *the operator A is the generator of the immediate self-adjoint Gibbs semigroup $\{e^{-tA}\}_{t\geqslant 0}$. Then $\{e^{-tH}\}_{t\geqslant 0}$ is also an immediate Gibbs semigroup and the Trotter product formula converges in the trace-norm with the rate error bound:*

$$\|(e^{-tA/n}e^{-tB/n})^n - e^{-tH}\|_1 \leqslant \frac{c_3}{n^{2\alpha-1}}, \quad n \geqslant 2, \tag{5.170}$$

for some $c_3 > 0$, uniformly away from zero.

According to the lifting strategy, to prove (5.170) one has to check the conditions of Lemma 5.51. To continue, we note that the properties of operators A, B and H are evident by virtue of the conditions of Propositions 5.57. Moreover, the estimate of $\varepsilon(n,t)$ in (5.151) is ensured by Proposition 5.27 and Proposition 5.57 either due to (i), which gives for the Trotter-Kato product formulae an operator-norm rate of convergence of the order $O((\ln(n))/n^{2\alpha-1})$, or due to (i) and (ii), which yield the *optimal* rate of the order $O(1/n^{2\alpha-1})$.

Hence, it remains only to verify that the additional condition (iii) ensures that $e^{-tH} \in \mathcal{C}_1(\mathcal{H})$ for $t > 0$. Note that, in contrast to the operator *smallness* and the operator-sum for H as in Proposition 5.55, now the self-adjoint H is defined as the *form-sum* in Propositions 5.57.

Lemma 5.59. *Condition* (i) *of Proposition* 5.57 *implies that there exists a bounded operator* $Y \in \mathcal{L}(\mathcal{H})$ *obeying* $\|Y\| < 1$ *such that for* $\alpha \in (1/2, 1)$ *one has the representation*

$$H^{-1} = A^{-(1-\alpha)}(\mathbb{1} + Y)^{-1}A^{-\alpha} = A^{-\alpha}(\mathbb{1} + Y^*)^{-1}A^{-(1-\alpha)}. \tag{5.171}$$

Proof. Since $A \geqslant \mathbb{1}$ and $\alpha \in (1/2, 1)$, we set

$$Y := (B^{\alpha}A^{-\alpha})^*(B^{1-\alpha}A^{-(1-\alpha)}).$$

Then by (i) and by (5.92) with $\theta = 1-\alpha$, one gets $\|B^{(1-\alpha)}v\| \leqslant b^{(1-1/\alpha)}\|A^{(1-\alpha)}v\|$, $v \in \operatorname{dom} A^{(1-\alpha)}$. This implies $\|B^{1-\alpha}A^{-(1-\alpha)}\| \leqslant b^{(1-1/\alpha)}$ as well as $\|B^{\alpha}A^{-\alpha}\| \leqslant b$, which is due to (i). Concequently, $\|Y\| \leqslant b^{2-1/\alpha} < 1$.

Now, by monotonicity: $\operatorname{dom} A^{\alpha_+} \subseteq \operatorname{dom} A^{\alpha_-}$ for $\alpha_- \leqslant \alpha_+$, and by (5.93) we find that $\operatorname{dom} A^{\alpha} \subseteq \operatorname{dom} A^{1/2} = \operatorname{dom} H^{1/2} \subseteq \operatorname{dom} A^{1-\alpha}$ for $1/2 < \alpha < 1$. Therefore,

$$\begin{aligned}(u, Hw) &= (A^{1/2}u, A^{1/2}w) + (B^{1/2}u, B^{1/2}w) = (A^{\alpha}u, A^{(1-\alpha)}w)\\ &\quad + (B^{\alpha}A^{-\alpha}A^{\alpha}u, B^{(1-\alpha)}A^{-(1-\alpha)}A^{(1-\alpha)}w) = (A^{\alpha}u, (\mathbb{1} + Y)A^{(1-\alpha)}w),\end{aligned}$$

for $u \in \operatorname{dom} A^{\alpha}$ and $w \in \operatorname{dom} H$. Hence, by the self-adjointness of A^{α} and if $w \in \operatorname{dom} H$, then $(\mathbb{1} + Y)A^{(1-\alpha)}w \in \operatorname{dom} A^{\alpha}$, and

$$Hw = A^{\alpha}(\mathbb{1} + Y)A^{(1-\alpha)}w\,.$$

Consequently, together with $\|Y\| \leqslant b^{2-1/\alpha} < 1$ this yields the representation (5.171), which in turn gives $\operatorname{dom} H \subset \operatorname{dom} A^{\alpha}$ for $\alpha \in (1/2, 1)$. □

Proof of Proposition 5.58. Since by conditions $A \geqslant \mathbb{1}$ and (iii) the operator A^{-1} is compact, the same is true for $A^{-(1-\alpha)}$ and for $A^{-\alpha}$. Therefore, by (5.171) the operator H^{-1} is also compact. Consequently, generator $H \geqslant \mathbb{1}$ has a discrete spectrum $\sigma(H) = \sigma_{\mathrm{d}}(H)$ with the only accumulation point of positive eigenvalues $\{\lambda_n(H)\}_{n\geqslant 1}$ at infinity, see Appendix A (Section A.6).

Note that since $B \geqslant 0$, the *minimax* principle for $H = A\dot{+}B$ (see Corollary 4.24) yields $\{\lambda_n(H) \geqslant \lambda_n(A)\}_{n\geqslant 1}$, or $\operatorname{Tr} e^{-tH} \leqslant \operatorname{Tr} e^{-tA}$ for $t > 0$. This proves that $e^{-tH} \in \mathcal{C}_1(\mathcal{H})$ for $t > 0$.

Now we have all necessary ingredients of Lemma 5.51 to conclude that the trace-norm convergence of the Trotter product formula holds uniformly away from zero with the optimal rate (5.170), Remark 5.29. □

5.4.4 Trotter-Kato product formulae, trace-norm convergence

The lifting Lemma 5.51 is well suited to prove (including the error bound estimates for the convergence rate) the trace-norm convergence of the Trotter product formula. Moreover, it is still efficient for trace-norm estimates in the case of non-self-adjoint generators A and B in the Trotter approximants, see Section 5.5 (subsection 5.5.3), but it is not very useful for the Trotter-Kato product formulae.

To prove the operator-norm or the trace-norm convergence of the Trotter-Kato product formulae for a certain subclass of *non-exponential* Kato functions $\mathcal{K}$ away from zero, for example

$$\tau\text{-}\lim_{n\to\infty}(f(tA/n)g(tB/n))^n = e^{-tH}P_0, \quad \tau \text{ is } \|\cdot\| \vee \|\cdot\|_1, \tag{5.172}$$

we need to impose *additional* conditions on the *generic* functions $f, g \in \mathcal{K}$ (Definition 5.4 and Appendix C) involved in the Trotter-Kato product formulae approximants. Here τ stands for the topology of convergence, either *operator-*, or the *trace-norm.*

Recall that to prove in Proposition 5.8 the *operator-norm* convergence of Trotter-Kato product formulae under assumption of *smallness* of B we put on the Kato functions $\mathcal{K}$ restrictions (iv),(5.15)-(5.19) and (v),(5.20). On the other hand, to prove the operator-norm convergence under conditions of *fractional smallness* (Proposition 5.27) we assumed that the Kato functions belong to the sub-class $\mathcal{K}_\alpha$, Definition 5.26 and Appendix C. In Proposition 5.45 without a smallness condition the *operator-norm* convergence holds if one of the Kato functions f, g is *regular*, Definition 5.43 and Appendix C.

The aim of this subsection is to determine conditions on the pair $f, g \in \mathcal{K}$ and on the pair A and B that ensure the convergence (5.172) in the trace-norm topology, (that is, τ is $\|\cdot\|_1$), to a *degenerate* Gibbs semigroup $\{e^{-tH}P_0\}_{t\geqslant 0}$, where $H = A\dot{+}B$.

Note that Proposition 6.18 shows that to establish (5.172) in the $\|\cdot\|_1$-norm for any of the Trotter-Kato product formulae it is enough to prove convergence of the Trotter-Kato product formula only for one family of approximants, for example, for the family $\{(f(tA/n)g(tB/n))^n\}_{n\geqslant 1}$.

Remark 5.60. To proceed with selection from $\mathcal{K}$ admissible functions for the next proposition we assume that the generic Kato functions $f, g \in \mathcal{K}$ verify the following *additional* condition (cf. (5.101) in Definition 5.34):

(1)* $f(x)$ is *strictly* positive and the auxiliary functions

$$\varphi(x) := s^{-1}\left(\frac{1}{f(x)} - 1\right) \quad \text{and} \quad \psi(x) := x^{-1}(1 - g(x)),$$

$$\text{are monotone non-increasing for } x \in \mathbb{R}^+. \tag{5.173}$$

Then we say that $f \in \mathcal{K}_*$, see Appendix C, Section C.2 (Comments 6 and 7).

Proposition 5.61. *Let A and B be non-negative densely defined self-adjoint operators in $\mathcal{H}$ and let the form-sum $H = A \dotplus B$ be a self-adjoint operator defined in the subspace $\mathcal{H}_0 \subseteq \mathcal{H}$. Suppose the Kato functions $f, g \in \mathcal{K}_*$. If*

$$f(tA) \in \mathcal{C}_p(\mathcal{H}), \quad t > 0, \quad 1 \leqslant p < +\infty, \tag{5.174}$$

then the Trotter-Kato product formula (5.172) *converge in the norm-topology $\tau = \|\cdot\|_{2p}$, to the degenerate Gibbs semigroup $\{e^{-tH}P_0\}_{t\geqslant 0}$, locally uniformly away from $t = 0$.*

In the Notes to Section 5.4 one finds a reference for the proof and some remarks.

A few comments are in order. Condition $f(tA) \in \mathcal{C}_p(\mathcal{H})$ directly relates the operator A and the Kato function f, but now the additional condition (5.173) excludes some natural choices for f. For example, $f(x) = e^{-x}$ does not satisfy (5.173), whereas $f(x) = (1 + x/\kappa)^{-\kappa}$ for $0 < \kappa \leqslant 1$, does. This demonstrates how sensitive is the Trotter-Kato product formula (5.172) to the choice of admissible Kato functions. Besides the elimination of the exponential Kato function f the strongest topology of convergence is now the Hilbert-Schmidt norm $\|\cdot\|_2$ under the rather severe requirement (5.174) that operator $f(tA)$ must belong to the trace-class for $t > 0$.

To improve the convergence in Proposition 5.61 to the trace-norm topology $\tau = \|\cdot\|_1$, we formulate *another* type of conditions on the functions $f, g \in \mathcal{K}$ and on the operators A and B.

Proposition 5.62. *Let A and B be non-negative densely defined self-adjoint operators in $\mathcal{H}$ and let the form-sum $H = A \dotplus B$ be a self-adjoint operator defined in the subspace $\mathcal{H}_0 \subseteq \mathcal{H}$. Let f, g be arbitrary Kato functions. If*

$$f_0(tA) \in \mathcal{C}_p(\mathcal{H}), \quad t > 0, \quad 1 \leqslant p < +\infty, \tag{5.175}$$

then the Trotter-Kato product formulae (5.172) *converge in the trace-norm topology $\tau = \|\cdot\|_1$ to the degenerate Gibbs semigroup $\{e^{-tH}P_0\}_{t\geqslant 0}$, locally uniformly away from zero.*

Here the function f_0 is given by (5.104) in Definition 5.34. We shall present the proofs of this assertion in Chapter 6 (Section 6.5). Therein we consider the problem of the convergence of the Trotter-Kato product formula in the more general framework of the *symmetrically-normed* ideals $\mathcal{C}_\phi(\mathcal{H})$. Then the trace-norm ideal $\mathcal{C}_1(\mathcal{H}) = \mathcal{C}_{\phi_1}(\mathcal{H})$ is simply a particular case corresponding to the special choice (6.20) of the *symmetric norming* function $\phi = \phi_1$.

In fact (5.175) imposes implicit conditions on the original f and A. However, since for the exponential $f(x) = e^{-x}$ one gets $f_0(x) = (1 + x)^{-1}$, the condition (5.175) coincides with the definition of the p-generator

$$f_0(tA) = (\mathbb{1} + tA)^{-1} \in \mathcal{C}_p(\mathcal{H}), \quad t > 0, \quad 1 \leqslant p < +\infty, \tag{5.176}$$

of a Gibbs semigroup, cf. Definition 4.26. In particular, (5.176) yields $f(tA) = e^{-tA} \in \mathcal{C}_p(\mathcal{H})$, $t > 0$, but not converse.

We conclude these observations by remarking that for the particular case of the *exponential* Kato function $f(x)$ Proposition 5.62 can be improved. To this end one assumes instead of (5.175), that is, instead of (5.176), the weaker condition (5.174). Then using the *Araki inequality* (Appendix B, Section B.1) it is possible to refine the corresponding estimates up to the trace-norm convergence. For details see Section 6.5.

5.5 Product formulae: non-self-adjoint Gibbs semigroups

In this section we extend the trace-norm convergence of the *exponential* Trotter-Kato product formula (i.e., the Trotter product formula) to the non-self-adjoint case. This is a subtle problem since self-adjointness served for important estimates. Our extension of the Trotter product formula to non-self-adjoint Gibbs semigroups covers essentially two cases.

The first one concerns the Trotter product formula when the involved C_0-semigroups have *m-sectorial* generators A and B. The strategy of the proof of the convergence in this case is based on the *lifting* the corresponding results for the self-adjoint generators. To this aim we propose the *analytic extension method* for holomorphic families of sectorial forms (generators) and of the corresponding semigroups. We develop this method in the preliminary subsection 5.5.1, *Holomorphic families of generators and semigroups.*

In subsection 5.5.2 we use this method to establish the trace-norm compactness of the holomorphic families of Trotter *approximants*. Then in Proposition 5.78 we prove (without an estimate of the convergence rate) that the Trotter product formula converges in the trace-norm topology to a degenerate Gibbs semigroup. Note that in this statement the m-sectorial generators A and B are *not* subordinated.

To control the rate of the trace-norm convergence we impose in subsection 5.5.3 more restrictive conditions of *relative smallness* on the pair of m-sectorial generators A and B. In Proposition 5.80 and Proposition 5.81 we establish the convergence of the Trotter product formula to Gibbs semigroups in the trace-norm topology, with estimates of the rate of convergence. Note that instead of the analytic extension method we use here the lifting Lemma 5.51.

Following the idea of *lifting* and similar to Section 5.4 we first begin by establishing the convergence of non-self-adjoint exponential product formula in the operator-norm topology. Then we proceed with improving of this convergence to the trace-norm topology.

5.5.1 Holomorphic families of generators and semigroups

We start with conditions imposed on the pair of generators A and B. Suppose A and B are densely defined m-sectorial operators in $\mathcal{H}$ with corresponding semi-angles α_A and α_B belonging to $[0, \pi/2)$. We denote by a and b the densely defined closed sectorial sesquilinear forms associated to the operators A and B, respectively, see Definition 1.41 and Remark 4.29.

Recall that if A is an m-sectorial operator with vertex $\gamma = 0$ and semi-angle $\alpha \in [0, \pi/2)$, then we can associate with A a densely defined, closed, sectorial sesquilinear form $a[u, v] : \operatorname{dom} a \times \operatorname{dom} a \to \mathbb{C}$ (with the same vertex $\gamma = 0$ and semi-angle $\alpha \in [0, \pi/2)$), such that

$$a[u, v] = (Au, v), \quad u \in \operatorname{dom} A (= \operatorname{core} a), \ v \in \operatorname{dom} a.$$

Then $a[u] := a[u, u]$, for $u \in \operatorname{dom} a$, is the quadratic form associated with the sesquilinear form. Quadratic form determines the sesquilinear form uniquely by the *polarisation* identity

$$a[u, v] = \frac{1}{4}\{a[u + v] - a[u - v] + i\, a[u + i\, v] - i\, a[u - i\, v]\}.$$

Similarly to A, the adjoint operator A^* is also m-sectorial with the same semi-angle and the vertex. Then the associated densely defined, sectorial and closed sesquilinear form $a^*[u, v] = (A^* u, v) = \overline{a[v, u]}$ extended from dom A^* to domain dom a^* = dom a, is the *adjoint* form to a. A form a is symmetric if $a = a^*$, i.e., $a[u, v] = \overline{a[v, u]}$. By polarisation identity the sesquilinear form $a[u, v]$ is symmetric if and only if the quadratic form $a[u]$, $u \in \operatorname{dom} a$, is real-valued.

For the sesquilinear form a we associate two symmetric forms

$$\mathfrak{Re}\, a := \frac{1}{2}(a + a^*) \quad \text{and} \quad \mathfrak{Im}\, a := \frac{1}{2i}(a - a^*),$$

such that one has the representation

$$a[u, v] = \mathfrak{Re}\, a[u, v] + i\, \mathfrak{Im}\, a[u, v], \quad u, v \in \operatorname{dom} a. \tag{5.177}$$

Recall that these two forms are not real-valued and $\mathfrak{Re}\, a[u, v] \neq \mathfrak{Re}(a[u, v])$, $\mathfrak{Im}\, a[u, v] \neq \mathfrak{Im}(a[u, v])$, but they are *real* and *imaginary* parts for the quadratic form $a[u]$:

$$\mathfrak{Re}\, a[u] = \mathfrak{Re}(a[u]) \quad \text{and} \quad \mathfrak{Im}\, a[u] = \mathfrak{Im}(a[u]), \quad u \in \operatorname{dom} a.$$

Note that the vertex γ and the semi-angle α are (not uniquely) determined by the sesquilinear form a via the inequalities

$$\mathfrak{Re}\, a[u] \geqslant \gamma, \quad |\mathfrak{Im}\, a[u]| \leqslant \operatorname{tg} \alpha_A\, \mathfrak{Re}\, a[u], \quad u \in \operatorname{dom} a. \tag{5.178}$$

Proposition 5.63 (Representation Theorem). *Let the sesquilinear form* $t : \mathrm{dom}\, t \times \mathrm{dom}\, t \to \mathbb{C}$ *be densely defined, closed, and sectorial on the domain* $\mathrm{dom}\, t \subset \mathcal{H}$. *Then there exists a unique m-sectorial operator* T *with* $\mathrm{dom}\, T \subset \mathrm{dom}\, t$ *such that* $t[u,v] = (Tu, v)$ *for* $u \in \mathrm{dom}\, T$, $v \in \mathrm{dom}\, t$, *and* $\mathrm{dom}\, T$ *is a core of the form* t.

If, in addition, the form t *is symmetric and non-negative:* $t[u] \geqslant 0$, $u \in \mathrm{dom}\, t$, *then* T *is a non-negative self-adjoint operator such that* $\mathrm{dom}\, T^{1/2} = \mathrm{dom}\, t$ *and* $t[u,v] = (T^{1/2}u, T^{1/2}v)$, $u, v \in \mathrm{dom}\, t$. *Moreover, a subset* $D \subset \mathrm{dom}\, t$ *is a core of* t *if and only if* D *is a core of* $T^{1/2}$.

Corollary 5.64. *The mapping* $t \mapsto T_t (= T)$ *is a one-to-one correspondence between the set of all densely defined, closed sectorial forms and the set of all m-sectorial operators. The form* t *is bounded if and only if* T *is bounded,* $T_{t^*} = T^*$, *and* t *is symmetric it and only if operator* T *is self-adjoint.*

Since there are *no maximal* sectorial forms, there are advantages in using the Representation Theorem to exploit sesquilinear forms for constructing m-sectorial operators. The first is that a densely defined sectorial operator S naturally generates a *closable* sesquilinear form $s[u,v] = (Su, v)$ for $u, v \in \mathrm{dom}\, S$. Let $T = T_t$ be the m-sectorial operator associated with the closed form $t := \widetilde{s}$. Then T is an extension of S such that $\mathrm{dom}\, S = \mathrm{dom}\, s$ is a core of t. Hence, even when the closure $\widetilde{S}$ is not m-sectorial, one can always associate with S a *minimal* m-sectorial extension T with $\mathrm{dom}\, T \subset \mathrm{dom}\, t$.

The second advantage comes from the observation that in contrast to the case of the set of the closed operators $\mathcal{C}(\mathcal{H})$, the set of the closed forms $\mathcal{C}_f(\mathcal{H})$ is a *linear space.* If a and b are densely defined, closed sectorial forms, then by the representation theorem there exists the m-sectorial operator C associated with the closed form $c = a + b$. Since in turn there are m-sectorial operators A and B associated with the forms a and b, the operator $C := A \dotplus B$ is called the *form-sum* of these operators. On the other hand, if a and b are densely defined, closed sectorial forms generated by densely defined sectorial operator A and B, then the corresponding form-sum is an extension of the operator-sum: $A + B \subset A \dotplus B$.

Recall that the form-sum $A \dotplus B$ may exist and be well-defined even if $\mathrm{dom}\, A \cap \mathrm{dom}\, B = \{0\}$. We have also seen another case, when for m-sectorial operators one has $\overline{\mathrm{dom}\, A \cap \mathrm{dom}\, B} \subset \overline{\mathrm{dom}\, a \cap \mathrm{dom}\, b} = \mathcal{H}_0 \subset \mathcal{H}$, that is, the operator sum $A + B$ is *not* densely defined in $\mathcal{H}$. Then the m-sectorial form-sum operator $C = A \dotplus B$ is well-defined in the Hilbert space $\mathcal{H}_0 = P_0(\mathcal{H})$, where P_0 is the orthogonal projection $P_0 : \mathcal{H} \to \mathcal{H}_0 = \overline{\mathrm{dom}\, C}$.

Now we return to the form (5.177). Since a is closed and $\gamma = 0$, the *symmetric* form $\mathfrak{Re}\, a[u,v]$ is also closed and $\mathfrak{Re}\, a[u,u] \geqslant 0$. Then by Proposition 5.63 it defines a non-negative self-adjoint operator $A_R := \mathfrak{Re}\, A \geqslant 0$, which is called the *real* part of the operator A. Similar arguments for the closed non-negative symmetric form $\mathfrak{Re}\, a^*[u,v]$ yield also that $A_R = \mathfrak{Re}\, A^*$.

Note that for a bounded operator $A \in \mathcal{L}(\mathcal{H})$ one readily obtains the identity $A_R = (A + A^*)/2$, although in general it is *not* true for unbounded operators. On the hand, by the definitions of a, a^* and by the representation theorem we obtain that the real part of A is the form-sum $A_R = (A \dotplus A^*)/$, which is a non-negative self-adjoint operator with $\operatorname{dom}\sqrt{A_R} = \operatorname{dom} a$.

We also recall that with help of the real part $A_R = \mathfrak{Re}\, A$ the m-sectorial operator A with vertex $\gamma = 0$ and semi-angle $\alpha \in [0, \pi/2)$ has the following representation

$$A = A_R^{1/2}(\mathbb{1} + i\,L)A_R^{1/2}\,. \tag{5.179}$$

Here symmetric bounded operator $L \in \mathcal{L}(\mathcal{H})$ is such that $\|L\| \leqslant \operatorname{tg}\alpha$.

Remark 5.65. By virtue of representation (5.179) the resolvent of the m-sectorial operator A is compact if and only if the resolvent of its real part A_R is compact.

Step 1. The first step in our *analytic extension method* is the complex extension of sectorial forms, cf. the representation (5.177).

Let a and b be densely defined closed sesquilinear sectorial forms with vertex $\gamma = 0$ and semi-angles $\alpha_A, \alpha_B \in [0, \pi/2)$. We define two families of *auxiliary* closed sesquilinear forms for $z \in \mathbb{C}$ by parametrisation of the imaginary parts

$$a_z := \mathfrak{Re}\, a + z\,\mathfrak{Im}\, a, \quad \operatorname{dom} a_z = \operatorname{dom} a, \tag{5.180}$$

$$b_z := \mathfrak{Re}\, b + z\,\mathfrak{Im}\, b, \quad \operatorname{dom} b_z = \operatorname{dom} b. \tag{5.181}$$

Lemma 5.66. *The forms $z \mapsto a_z$ and $z \mapsto b_z$ are sectorial (with $\gamma = 0$) and holomorphic functions in the strip*

$$D_{AB} := \{z \in \mathbb{C} : \ |\mathfrak{Re}\, z| < \min\{\operatorname{ctg}\alpha_A, \operatorname{ctg}\alpha_B\}\}. \tag{5.182}$$

Proof. Let $x, y \in \mathbb{R}$, $z = x + iy$, and $u \in \operatorname{dom} a$. Since the form a is sectorial, (5.178) yields

$$|x\,\mathfrak{Im}\, a[u]| \leqslant |x| \operatorname{tg}\alpha_A\, \mathfrak{Re}\, a[u]. \tag{5.183}$$

Hence, for $|x| \operatorname{tg}\alpha_A < 1$ one gets $\mathfrak{Re}\, a_z = \mathfrak{Re}\, a + x\,\mathfrak{Im}\, a > 0$, i.e., $\gamma = 0$, see (5.178). Moreover, for these values of x and $u \in \operatorname{dom} a$, we get

$$\mathfrak{Re}\, a_z[u] \geqslant \mathfrak{Re}\, a[u] - |x\,\mathfrak{Im}\, a[u]| \geqslant (1 - |x| \operatorname{tg}\alpha_A)\, \mathfrak{Re}\, a[u]. \tag{5.184}$$

By (5.183) and (5.184), we obtain

$$\begin{aligned} |\,\mathfrak{Im}\, a_z[u]| = |y\,\mathfrak{Im}\, a[u]| &\leqslant |y| \operatorname{tg}\alpha_A\, \mathfrak{Re}\, a[u] \\ &\leqslant \frac{|y| \operatorname{tg}\alpha_A}{1 - |x| \operatorname{tg}\alpha_A}\, \mathfrak{Re}\, a_z[u], \end{aligned} \tag{5.185}$$

and by (5.178) the form a_z is sectorial for $z \in \{z \in \mathbb{C} : \ |x| < \operatorname{ctg}\alpha_A\}$.

Similarly, the form b_z is sectorial with $\gamma = 0$ for $|x| < \operatorname{ctg}\alpha_B$. Therefore, the two forms a_z and b_z are sectorial with $\gamma = 0$ in the strip D_{AB} defined by (5.182).

Note that the families (5.180) and (5.181) are explicit functions of the complex variable $z \in \mathbb{C}$. Their derivatives with respect to this variable exist and have the form:

$$\partial_z a_z[u] = \mathfrak{Im}\, a[u], \quad u \in \operatorname{dom} a,$$

$$\partial_z b_z[u] = \mathfrak{Im}\, b[u], \quad u \in \operatorname{dom} b.$$

For each $u \in \operatorname{dom} a$, respectively $u \in \operatorname{dom} b$, they are linear holomorphic complex functions in the domain D_{AB}. □

Now we recall that a set of sesquilinear forms $\{t_z\}_{z\in D\subset\mathbb{C}}$ is called a *holomorphic family* of *type* (a) if it verifies two conditions:

(1) Each t_z is sectorial and closed with dense domain: $\operatorname{dom} t_z = Q$, which is independent of $z \in D$.

(2) $D \ni z \mapsto t_z[u]$ is holomorphic for each fixed $u \in Q$. By the polarisation identity, this implies that the sesquilinear form $t_z[u,v]$ is holomorphic in $z \in D$ for each fixed pair $u, v \in Q$.

Recall that the concept of *holomorphic family* of operators is based on bounded-holomorphic properties of the corresponding resolvents. To this aim, consider a family of closed operators in a Banach space, $\{A(z) \in \mathcal{C}(\mathcal{B})\}_{z\in D}$, defined in a neighborhood of $z_0 \in D$. If $\zeta \in \mathbb{C}$ belongs to the resolvent set $\rho(A(0))$, then $z \mapsto A(z)$ is said to be *holomorphic* at z_0 if there exists a disc $D_\varepsilon(z_0) := \{z \in \mathbb{C} : |z - z_0| < \varepsilon\}$ such that $\zeta \in \rho(A(z))$ and the resolvent $R_\zeta(A(z)) = (A(z) - \zeta\mathbb{1})^{-1}$ is bounded-holomorphic for $z \in D_\varepsilon(z_0)$, $\varepsilon > 0$.

A special type of holomorphic family of operators that will be used in the text, is the family of *type* (A).

Definition 5.67. A holomorphic family $\{A(z)\}_{z\in D}$ of type (A) is defined by two conditions:

(1) Each $A(z)$ is closed with dense domain $\operatorname{dom} A(z) = \mathfrak{D}$, which is independent of $z \in D$.

(2) The mapping $z \mapsto A(z)u$ is a holomorphic vector-valued function for $z \in D$ and for every $u \in \mathfrak{D}$.

Conditions (1) and (2) imply that $\{A(z)\}_{z\in D}$ is a holomorphic family in the bounded-holomorphic resolvent sense. The following criterion for type (A) is useful.

Proposition 5.68. *Let A be a closable operator with domain $\operatorname{dom} A = \mathfrak{D}$ in a Banach space $\mathcal{B}$. Let $\{A^{(n)}\}_{n\geqslant 1}$ be operators with domains $\operatorname{dom} A^{(n)} \subset \mathfrak{D}$, and let a, b, c be non-negative constants such that*

$$\|A^{(n)}u\| \leqslant c^{n-1}\,(a\|u\| + b\|Au\|), \quad u \in \mathfrak{D},\ n \in \mathbb{N}.$$

Then for $|z| < 1/c$ and $u \in \mathfrak{D}$ the series

$$A(z)u = \sum_{n=0}^{\infty} z^n\, A^{(n)}u, \quad A := A^{(0)},$$

defines an operator $A(z)$ with $\operatorname{dom} A(z) = \mathfrak{D}$. If $|z| < (b+c)^{-1}$, then operators $A(z)$ are closable and the closures $\{\tilde{A}(z)\}_{\{z:|z|<(b+c)^{-1}\}}$ form a holomorphic family of type (A).

Corollary 5.69. (i) *The forms* $z \mapsto a_z$ (5.180) *and* $z \mapsto b_z$ (5.181) *are holomorphic families of type* (a) *in the strip* D_{AB} (5.182). *Since* $a_z^* = a_{\overline{z}}$ *and* $b_z^* = b_{\overline{z}}$, *they are called self-adjoint holomorphic families. By the representation theorem (Proposition* 5.63), *there exist* m*-sectorial operators* $A(z)$ *and* $B(z)$ *with* $\gamma = 0$, *which are associated with the closed sectorial forms* a_z *and* b_z, $z \in D_{AB}$.

(ii) *These operators form resolvent bounded-holomorphic families, called holomorphic families of type* (B). *The operators* $\{A(z)\}_{z \in D_{AB}}$ *and* $\{B(z)\}_{z \in D_{AB}}$ *are locally uniformly* m*-sectorial with vertex* $\gamma = 0$ *and for the particular value* $z = i$ *yield* $A(i) = A$, $B(i) = B$. *Therefore (Proposition* 1.46) *for each* $z \in D_{AB}$ *the operators* $A(z)$ *and* $B(z)$ *are generators of holomorphic contraction semigroups.*

(iii) *Operator holomorphic families of type* (B) *inherit the properties of the form self-adjoint families. This yields* $A^*(z) = A(\overline{z})$ *and* $B^*(z) = B(\overline{z})$.

Remark 5.70. Let the family $\{C(z)\}_{z \in D}$ be holomorphic of type (A) and have non-empty resolvent sets such that $\rho_D = \bigcap_{z \in D} \rho(C(z)) \neq \varnothing$. Then $C(z)$ has a compact resolvent $R_\zeta(C(z))$ either for all $z \in D$, or for no z.
Let $\{C(z)\}_{z \in D}$ be a type (B) holomorphic family of (m-sectorial) operators. If the operator $C(z_0)$ has compact resolvent for $z_0 \in D$, then the resolvent of $C(z)$ is compact for any $z \in D$.

Step 2. The second step is the complex extension of the semigroup generators corresponding to sectorial forms.

Proposition 5.71. *Let* $\{C(z)\}_{z \in D}$ *be a holomorphic family of type* (B) *with vertex* $\gamma = 0$ *and semi-angles* $\alpha(z) < \pi/2$. *Then for each* $z \in D$ *the operator* $C(z)$ *is the generator of a holomorphic contraction semigroup* $\{U_t(C(z))\}_{t \in S_{\theta(z)}}$ *with semi-angle* $\theta(z) \in [0, \pi/2)$. *The function* $z \mapsto U_t(C(z))$ *is holomorphic in* $z \in D$ *for any* t *in the open sector* S_{θ_0}, *with semi-angle* $\theta_0 := \inf_{z \in D}(\pi/2 - \alpha(z))$.

Proof. For each $z \in D$ the m-sectorial operator $C(z) \in \mathscr{H}(\theta(z) = \pi/2 - \alpha(z), 0)$ is the generator of a holomorphic contraction semigroup, see Corollary 5.69(ii). Then by the Riesz-Dunford representation (1.67) for holomorphic semigroups we have

$$U_t(C(z)) = \frac{1}{2\pi i} \int_\Gamma d\zeta \, \frac{e^{-t\zeta}}{\zeta \mathbb{1} - C(z)}. \tag{5.186}$$

Recall that this integral is absolutely $\|\cdot\|$-convergent for $t > 0$ if the contour $\Gamma \subset \bigcap_{z \in D} \rho(C(z)) := \rho_D$, running from infinity with $\arg \zeta = \alpha_{\max} + \varepsilon$, and then back to infinity with $\arg \zeta = -(\alpha_{\max} + \varepsilon)$, where $\alpha_{\max} := \sup_{z \in D} \alpha(z)$ for some $0 < \varepsilon < \pi/2 - \alpha_{\max}$, see Section 1.5. Since $\{C(z)\}_{z \in D}$ is a holomorphic family, the resolvents $\{R_\zeta(C(z))\}_{z \in D}$ form a holomorphic family for any $\zeta \in \rho_D$. By Proposition 1.27 the generator $C(z)$ satisfies the condition (1.69), which yields the estimate

$$\|\partial_z(\zeta\mathbb{1} - C(z))^{-1}\| = \left\|\frac{1}{2\pi i}\int_{\gamma_r} \mathrm{d}z'\,(z'-z)^{-2}(\zeta\mathbb{1} - C(z'))^{-1}\right\| \leqslant \frac{\tilde{M}_\varepsilon}{|\zeta|\,r}, \tag{5.187}$$

for any $\zeta \in \{\zeta \in \mathbb{C} : -(\alpha_{\max} + \varepsilon) < \arg\zeta < \alpha_{\max} + \varepsilon\}$. Here γ_r is a small circle of radius r around z and

$$\tilde{M}_\varepsilon := \sup_{\substack{\zeta\in\mathbb{C}\setminus\bar{S}_{\pi/2-\theta_0+\varepsilon}\\ z'\in\gamma_r}} \|(\zeta\mathbb{1} - C(z'))^{-1}\|\,.$$

The estimate (5.187) shows that the representation (5.186) is operator-norm differentiable under the integral since the integrand is operator-valued holomorphic complex function in D.

Note that the arguments developed above are also valid for $t \in \mathbb{C}$ such that $|\arg(t\,\zeta)| < \pi/2$. Then we get the statement for any t in the open sector S_{θ_0} with semi-angle $\theta_0 = \pi/2 - \alpha_{\max}$, which implies the assertion. □

Corollary 5.72. *Applying Proposition* 5.71 *to the particular case of generators* $\{A(z)\}_{z\in D_{AB}}$ *and* $\{B(z)\}_{z\in D_{AB}}$ *in Corollary* 5.69*, we obtain that the contraction semigroups*

$$\{U_t(A(z))\}_{t\in S_{\theta_0}} \quad \textit{and} \quad \{U_t(B(z))\}_{t\in S_{\theta_0}} \tag{5.188}$$

are holomorphic families for $z \in D^0_{AB}$*, and for any* t *in the sector* S_{θ_0}*. Here, the domain of analyticity in* z *is defined as*

$$\begin{aligned} D^0_{AB} = \{z \in \mathbb{C} : |\,\Re\mathfrak{e}\, z| < \min\{\operatorname{ctg}\alpha_A, \operatorname{ctg}\alpha_B\} - \delta_0 := \Delta_R,\\ |\,\Im\mathfrak{m}\, z| < \Delta_I\}, \end{aligned} \tag{5.189}$$

for a small $\delta_0 > 0$ *and for some* $\Delta_I > 1$*, cf.* (5.182)*. Then by* (5.185)*, the corresponding semi-angle* $\theta_0 > 0$ *is defined by*

$$\operatorname{tg}\left(\frac{\pi}{2} - \theta_0\right) = \frac{\Delta_I \max\{\operatorname{tg}\alpha_A, \operatorname{tg}\alpha_B\}}{1 - \Delta_R \max\{\operatorname{tg}\alpha_A, \operatorname{tg}\alpha_B\}}. \tag{5.190}$$

Step 3 (Vitali's theorem). The last step in our *analytic extension method* is application of the Vitali theorem to *holomorphic families* of Trotter's approximants.

Remark 5.73. Recall that for holomorphic families of bounded operators in $\mathcal{L}(\mathcal{H})$ there is no distinction between *uniform, strong,* or *weak* operator analyticity. Moreover, these holomorphic operator-valued functions inherit some properties known from standard complex analysis. For example, let $\{\Phi_n(z)\}_{n\geqslant 1} \subset \mathcal{L}(\mathcal{H})$ be a sequence of operator-valued complex functions such that

$$\|\Phi_n(z)\| < M,$$

for all $n \geqslant 1$ and $z \in D \subset \mathbb{C}$. If $\{\Phi_n(z)\}_{n\geqslant 1}$ converges (in any of the three topologies) on a subset of D having a *limit point* in D, then by the *Vitali theorem* the limit $\lim_{n\to\infty} \Phi_n(z) = \Phi(z)$ exists for any $z \in D$, the convergence is uniform on any compact $K \subset D$, and the limiting function $\Phi(z)$ is holomorphic in D.

5.5.2 Norm holomorphic families of the Trotter approximants

To continue our analysis of the convergence of the Trotter product formula in the trace-norm we recall the *exponential* case of Proposition 5.36.

Proposition 5.74. *Suppose A and B are non-negative densely defined self-adjoint operators in $\mathcal{H}$ and the form-sum $H = A \dot{+} B$ is defined in the subspace $\mathcal{H}_0 = P_0(\mathcal{H})$, where P_0 is the orthogonal projection $P_0 : \mathcal{H} \to \overline{\mathrm{dom}\, A^{1/2} \cap \mathrm{dom}\, B^{1/2}}$. If the resolvent of A or of B is compact: $(\mathbb{1} + A)^{-1} \vee (\mathbb{1} + B)^{-1} \in \mathcal{C}_\infty(\mathcal{H})$, then the Trotter product formula*

$$\| \cdot \|\text{-}\lim_{n\to\infty} \left(\mathrm{e}^{-tA/n}\mathrm{e}^{-tB/n}\right)^n = \mathrm{e}^{-t(A\dot{+}B)} P_0 \,, \tag{5.191}$$

converges in the operator-norm topology for $t > 0$, locally uniformly away from zero.

Now we can extend Proposition 5.74 to m-sectorial generators.

Proposition 5.75. *If A and B are m-sectorial operators (with vertex $\gamma = 0$) in a Hilbert space $\mathcal{H}$ and if $(\mathbb{1} + A)^{-1} \vee (\mathbb{1} + B)^{-1} \in \mathcal{C}_\infty(\mathcal{H})$, then the Trotter product formula converges in the operator-norm topology:*

$$\| \cdot \|\text{-}\lim_{n\to\infty} \left(\mathrm{e}^{-tA/n}\mathrm{e}^{-tB/n}\right)^n = \mathrm{e}^{-t(A\dot{+}B)} P_0 \,, \tag{5.192}$$

for any $t \in S_\theta$, where $\theta = \pi/2 - \max\{\alpha_A, \alpha_B\}$. The convergence is uniform on the compact subsets of the sector S_θ. The generator $H = A \dot{+} B$ is the m-sectorial form-sum of A and B, and P_0 is the orthogonal projection $P_0 : \mathcal{H} \to \mathcal{H}_0 = \overline{\mathrm{dom}\,(H)}$.

Proof. Following Corollary 5.69, we include the operators A and B into holomorphic type (B) self-adjoint families $\{A(z)\}_{z\in D_{AB}}$ and $\{B(z)\}_{z\in D_{AB}}$ of m-sectorial operators with $\gamma = 0$, which are generators of holomorphic contraction semigroups. Then using Corollary 5.72 we construct from the two families (5.188) a sequence of operator-valued uniformly bounded and holomorphic in the operator-norm topology Trotter product formula *approximants*

$$z \mapsto \Phi_n(t, z) := \left(\mathrm{e}^{-tA(z)/n}\mathrm{e}^{-tB(z)/n}\right)^n, \quad \|\Phi_n(t, z)\| \leqslant 1, \tag{5.193}$$

$n \in \mathbb{N}$, with domain of analyticity D^0_{AB}, which is determined by (5.189) for any t in the sector S_{θ_0} with semi-angle defined by (5.190).

Note that by the assumption of the proposition and by Remark 5.70 for $D = D^0_{AB}$, the $(\mathbb{1} + A(z))^{-1}$ and $(\mathbb{1} + B(z))^{-1}$ are holomorphic families of *compact* operators. In particular, the resolvents $\{(\mathbb{1} + A(x))^{-1}\}_{x\in\Delta_R}$ (or $\{(\mathbb{1} + B(x))^{-1}\}_{x\in\Delta_R}$) are compact.

Since the generators in (5.193) are self-adjoint holomorphic families of m-sectorial operators with $\gamma = 0$, the operators $A(x)$ and $B(x)$ are self-adjoint and non-negative for each $x \in \Delta_R$. Then (5.180) and (5.181) yield that

$\operatorname{dom} a = \operatorname{dom} A(x)^{1/2}$, $\operatorname{dom} b = \operatorname{dom} B(x)^{1/2}$, and that the self-adjoint non-negative form-sum $A(x) \dot{+} B(x)$ exists in domain $\mathcal{D} \subseteq \mathcal{H}_0$, which is dense in $\mathcal{H}_0 = \overline{(\operatorname{dom} a \cap \operatorname{dom} b)}$.

Now we are in the position to apply Proposition 5.74 to the sequence of functions (5.193) for $z = x$ and $t \in S_\theta$. This yields

$$\|\cdot\| \text{-} \lim_{n\to\infty} \Phi_n(t,x) = e^{-t(A(x)\dot{+}B(x))} P_0, \quad x \in \Delta_R. \tag{5.194}$$

Therefore, the operator-norm convergence (5.194) of the operator-valued uniformly bounded and holomorphic family (5.193) on the interval Δ_R, together with the Vitali theorem imply the operator-norm convergence in (5.194) for any $z \in D^0_{AB}$. Since $i \in D^0_{AB}$, we obtain (5.192) as well as the equality $H = A(i) \dot{+} B(i)$. □

The analytic extension method of Proposition 5.75 allows us to *lift* the operator-norm convergence of the Trotter-Kato product formulae for non-self-adjoint semigroups to the trace-norm convergence for non-self-adjoint Gibbs semigroups. The first step is the following assertion.

Proposition 5.76. *If A is an m-sectorial operator with vertex $\gamma = 0$ such that $e^{-t\,\mathfrak{Re}\,A} \in \mathcal{C}_1(\mathcal{H})$ for $t > 0$, then $\{e^{-tA(z)}\}_{t\geqslant 0}$ is a family of holomorphic contraction Gibbs semigroups for $z \in D^0_A$, where*

$$\begin{aligned} D^0_A = \{z \in \mathbb{C} : |\,\mathfrak{Re}\, z| < \Delta_R(A) = \operatorname{ctg}\alpha_A - \delta_0, \\ |\,\mathfrak{Im}\, z| < \Delta_I(A)\}, \end{aligned} \tag{5.195}$$

for some $\delta_0 > 0$ and $\Delta_I(A) > 1$, in the sector $S_{\theta_0(A)}$ with the semi-angle $\theta_0(A)$ defined by the equation

$$\operatorname{tg}\left(\frac{\pi}{2} - \theta_0(A)\right) = \frac{\Delta_I(A)\operatorname{tg}\alpha_A}{1 - \Delta_R(A)\operatorname{tg}\alpha_A}. \tag{5.196}$$

Moreover, the function $z \mapsto e^{-tA(z)}$ is $\|\cdot\|_1$-holomorphic in $z \in D^0_A$ for any $t \in S_{\theta_0(A)}$.

Proof. Proposition 4.30 shows that $e^{-tA} \in \mathcal{C}_1(\mathcal{H})$, for $t > 0$. Then by representation (4.39) the operators $\mathfrak{Re}\,A$ and A have compact resolvents. By Corollary 5.69 and Remark 5.70, $\{A(z)\}_{z\in D^0_A}$ is a holomorphic family of uniformly sectorial operators and $A(z = 0) = \mathfrak{Re}\,A$. Hence, $A(z)$ has a compact resolvent and $A(z) \in \mathscr{H}(\theta_0(A), 0)$ for any $z \in D^0_A$, where $\theta_0(A)$ is defined by (5.196). The same is true for $\mathfrak{Re}\,A(z)$.

Taking into account inequality (5.185) we obtain the estimate

$$\{1 - |\,\mathfrak{Re}\, z|\operatorname{tg}\alpha_A\}\,\mathfrak{Re}\, a[u] \leqslant \mathfrak{Re}\, a_z[u], \quad z \in D^0_A, \quad u \in \operatorname{dom} a. \tag{5.197}$$

Then by the *minimax principle* for the self-adjoint operators $\mathfrak{Re}\,A(z)$ and $\mathfrak{Re}\,A$, see Proposition 4.23 and Corollary 4.24, and by the condition $e^{-t\,\mathfrak{Re}\,A} \in \mathcal{C}_1(\mathcal{H})$ for

$t > 0$, we conclude that $\mathrm{e}^{-t\,\Re\mathrm{e}\,A(z)} \in \mathcal{C}_1(\mathcal{H})$ for $t > 0$ and $z \in D_A^0$. In addition, by Proposition 4.30 we have the estimate $\|\mathrm{e}^{-tA(z)}\|_1 \leqslant \|\mathrm{e}^{-t\,\Re\mathrm{e}\,A(z)}\|_1$.

Finally, by (5.185), (5.195) and (5.196), the family $\{A(z)\}_{z\in D_A^0}$ consists of uniformly sectorial operators with numerical range $\mathrm{Nr}(A(z)) \subset S_{\pi/2-\theta_0(A)}$. By Corollary 4.32 we get that $\{G_t(A(z))\}_{t\in S_{\theta_0(A)}}$ is a holomorphic contraction Gibbs semigroup for any $z \in D_A^0$.

On the other hand, by Proposition 5.71, the family $\{\mathrm{e}^{-tA(z)}\}_{z\in D_A^0}$ is $\|\cdot\|$-holomorphic (cf. Remark 5.73) for any $t \in S_{\theta_0(A)}$. Therefore, we have the representation

$$\mathrm{e}^{-tA(z)} = \frac{1}{2\pi i}\oint\limits_{\gamma_r} dw\,\frac{\mathrm{e}^{-tA(w)}}{w-z}\,, \tag{5.198}$$

where the $\|\cdot\|$-convergent Cauchy integral is taken along a small circle $\gamma_r = \{w \in \mathbb{C} : |w-z| = r\} \subset D_A^0$. Since the trace-norm of the semigroup $\{G_t(A(z))\}_{t\in S_{\theta_0(A)}}$ is bounded, using (5.197) one gets for $w \in \gamma_r$ the estimate

$$\left\|\partial_z \frac{\mathrm{e}^{-tA(w)}}{w-z}\right\|_1 \leqslant \frac{\|\mathrm{e}^{-t\,\Re\mathrm{e}\,A(w)}\|_1}{r^2} \leqslant \|\mathrm{e}^{-\beta t\,\Re\mathrm{e}\,A}\|_1 \frac{1}{r^2}\,,$$

where $\beta := \inf_{w\in\gamma_r}(1 - |\Re\mathrm{e}\,w|\,\mathrm{tg}\,\alpha_A)$. Thus, the Cauchy integral (5.198) is $\|\cdot\|_1$-differentiable, i.e., the function $z \mapsto \mathrm{e}^{-tA(z)}$ is $\|\cdot\|_1$-holomorphic in D_A^0 for any $t \in S_{\theta_0(A)}$. □

Now we can apply the *analytic extension method* to prove the trace-norm convergence of the Trotter-Kato product formulae for holomorphic Gibbs semigroups with non-self-adjoint m-sectorial generators A and B. This approach does not give an error estimate for the rate of convergence but treats the generators on the equal level and even for a trivial common domain: $\mathrm{dom}\,A \cap \mathrm{dom}\,B = \{0\}$.

To this aim we collect in Remark 5.77 the elements that we established for application of the Vitali theorem in the $\|\cdot\|_1$-topology.

Remark 5.77. Let A and B be m-sectorial operators with vertex $\gamma = 0$. Then according to definitions (5.180) and (5.181), and by virtue of Corollary 5.69, we constructed two holomorphic type (B) families of m-sectorial operators $\{A(z)\}_{z\in D_A^0}$ and $\{B(z)\}_{z\in D_B^0}$ with D_A^0 and D_B^0 defined by (5.195), such that:

(a) $A(i) = A$ and $B(i) = B$.

(b) For $x = \Re\mathrm{e}\,z \in [-\Delta_R, \Delta_R]$, where $\Delta_R = \min\{\mathrm{ctg}\,\alpha_A, \mathrm{ctg}\,\alpha_B\} - \delta_0$, the non-negative operators $A(x) = \Re\mathrm{e}\,A(z) \geqslant 0$ and $B(x) = \Re\mathrm{e}\,B(z) \geqslant 0$ are self-adjoint.

(c) If $\mathrm{e}^{-t\,\Re\mathrm{e}\,A}$ and $\mathrm{e}^{-t\,\Re\mathrm{e}\,B}$ are trace-class operators for $t > 0$, then by Proposition 5.76 the m-sectorial operators $\{A(z)\}_{z\in D_A^0}$ and $\{B(z)\}_{z\in D_B^0}$ generate two $\|\cdot\|_1$-holomorphic in D_A^0 and in D_B^0 families of holomorphic for $t \in S_{\theta_0(A)}$, respectively for $t \in S_{\theta_0(B)}$, contraction Gibbs semigroups. Here the sectors $S_{\theta_0(A)}$, respectively $S_{\theta_0(B)}$, are defined by condition (5.196).

Proposition 5.78. *Let A and B be m-sectorial operators with vertex $\gamma = 0$ in a Hilbert space $\mathcal{H}$. If $\mathrm{e}^{-t\,\mathfrak{Re}\,A} \vee \mathrm{e}^{-t\,\mathfrak{Re}\,B}$ is trace-class operator for $t > 0$, then the Trotter product formula for contraction Gibbs semigroups converges in the trace-norm topology:*

$$\|\cdot\|_1\text{-}\lim_{n\to\infty} \left(\mathrm{e}^{-tA/n}\mathrm{e}^{-tB/n}\right)^n = \mathrm{e}^{-t(A\dot{+}B)}P_0\,, \tag{5.199}$$

for any $t \in S_\theta$, where $\theta = \pi/2 - \max\{\alpha_A, \alpha_B\}$. The convergence is uniform on the compact subsets of the open sector S_θ. The generator $H = A \dot{+} B$ is the m-sectorial form-sum of A and B, and P_0 is the orthogonal projection $P_0 : \mathcal{H} \to \mathcal{H}_0 = \overline{\mathrm{dom}\,(H)}$.

Proof. Essentially we argue as in the proof of Proposition 5.75. To this end we extend to the trace-norm topology the Vitali theorem about the sequence of the operator-valued functions identical with product formula approximants (5.193):

$$\Phi_n(t,z) = \left(\mathrm{e}^{-tA(z)/n}\mathrm{e}^{-tB(z)/n}\right)^n, \qquad z \in D^0_{AB}\,,\; t \in S_{\theta_0}\,,\; n \in \mathbb{N}. \tag{5.200}$$

The assumption of the proposition, Corollary 5.72, Proposition 5.76 and the $\|\cdot\|_1$-continuity in z of the product $\mathrm{e}^{-tA(z)}\mathrm{e}^{-tB(z)}$ imply that the functions $\{z \mapsto \mathrm{e}^{-tA(z)}\mathrm{e}^{-tB(z)}\}_{t\in S_{\theta_0}}$ are $\|\cdot\|_1$-holomorphic for $z \in D^0_{AB}$ if either of exponentials (or both) belong to the ideal $\mathcal{C}_1(\mathcal{H})$. Therefore, $\Phi_n(t,z) \in \mathcal{C}_1(\mathcal{H})$ and the function $z \mapsto \Phi_n(t,z)$ is $\|\cdot\|_1$-holomorphic in $z \in D^0_{AB}$ for any $t \in S_{\theta_0}$.

Since the conditions of the proposition are symmetric with respect to A and B, suppose that $\mathrm{e}^{-t\,\mathfrak{Re}\,A} \in \mathcal{C}_1(\mathcal{H})$. Then, by Lemma 4.43, the sequence $\{\Phi_n(t,z)\}_{n\geqslant 1}$ is uniformly bounded for $z \in D^0_{AB}$ and $t \in S_{\theta_0}$ in the trace-norm topology:

$$\|\Phi_n(t,z)\|_1 \leqslant \|\mathrm{e}^{-tB(z)/n}\|^n \|\mathrm{e}^{-t\,\mathfrak{Re}\,A(z)/4}\|_1 \leqslant \|\mathrm{e}^{-\beta t\,\mathfrak{Re}\,A/4}\|_1. \tag{5.201}$$

Here we used that, by Corollary 5.72, for $z \in D^0_{AB}$ the operator $B(z)$ is the generator of a contraction semigroup: $\|\mathrm{e}^{-tB(z)}\| \leqslant 1$, and that by (5.197) and Corollary 4.24, the last inequality in (5.201) is valid for $\beta = \inf_{z\in D^0_A}(1 - |\mathfrak{Re}\,z|\,\mathrm{tg}\,\alpha_A)$.

Note that for any real $x \in [-\Delta_R, \Delta_R]$, $\Delta_R = \min\{\mathrm{ctg}\,\alpha_A, \mathrm{ctg}\,\alpha_B\} - \delta_0$, the operators $A(x) \geqslant 0$ and $B(x) \geqslant 0$ are self-adjoint, see Remark 5.77(b). Then by Proposition 5.53 the sequence $\{\Phi_n(t,x)\}_{n\geqslant 1}$ converges in the $\|\cdot\|_1$-norm to $\mathrm{e}^{-tH(x)}P_0$ for $t \in S_{\theta_0}$ and for $x \in [-\Delta_R, \Delta_R]$. Here the m-sectorial operator $H(z) = A(z)\dot{+}B(z)$, with $z \in D^0_{AB}$, is the form-sum corresponding to the sum of two closed densely defined sectorial forms (5.180) and (5.181).

The interval $[-\Delta_R, \Delta_R]$ is compact in D^0_{AB}. Since the family $\{\Phi_n(t,z)\}_{n\geqslant 1}$ is $\|\cdot\|_1$-holomorphic in $z \in D^0_{AB}$ and $\|\cdot\|_1$-uniformly bounded in this domain by (5.201), the Vitali theorem (Remark 5.73) yields

$$\|\cdot\|_1\text{-}\lim_{n\to\infty} \left(\mathrm{e}^{-tA(z)/n}\mathrm{e}^{-tB(z)/n}\right)^n = \mathrm{e}^{-tH(z)}P_0\,, \quad t \in S_{\theta_0}\,, \tag{5.202}$$

for any $z \in D^0_{AB}$ (5.189). Since $z = i \in D^0_{AB}$, one gets the assertion (5.199) in a smaller sector $\theta_0 < \theta$, where θ_0 is defined by (5.190).

The convergence (5.199) for $\theta_0 = \theta$ follows from (5.202) by shrinking domain D^0_{AB}, Remark 5.77. To this aim one takes in (5.190) the limits $\Delta_R \to 0$ and $\Delta_I \to 1$ that localise, in particular, the point $z = i$. □

Corollary 5.79. *Let A be an m-sectorial operator with semi-angle α_A and vertex $\gamma = 0$. If $e^{-t\,\mathfrak{Re}\,A}$ is a trace-class operator for $t > 0$, then the Trotter product formula for contraction Gibbs semigroups converges in the trace-norm topology*

$$\|\cdot\|_1\text{-}\lim_{n\to\infty}\left(e^{-tA^*/2n}e^{-tA/2n}\right)^n = e^{-t\,\mathfrak{Re}\,A}, \tag{5.203}$$

for any $t \in S_\theta$, where $\theta = \pi/2 - \alpha_A$. The convergence is uniform on the compact subsets of the open sector S_θ.

5.5.3 Trace-norm convergence under smallness conditions

In this subsection we impose additional conditions on the generators A and B to prove the trace-norm convergence of the Trotter product formula with error bound estimates for the rate of convergence. Then we do not need a lifting via the *analytic extension method*, but the one that we used in Section 5.4, where the *lifting* Lemma 5.51 plays the key rôle.

Proposition 5.80. *Let A be an m-sectorial operator in a Hilbert space $\mathcal{H}$. Suppose that $\mathfrak{Re}\,A > 0$ and $e^{-t\,\mathfrak{Re}\,A} \in \mathcal{C}_1(\mathcal{H})$ for $t > 0$. If B is an m-accretive operator in $\mathcal{H}$ such that $\operatorname{dom} A^\alpha \subseteq \operatorname{dom} B$ for some $\alpha \in [0,1)$ and $\operatorname{dom} A^* \subseteq \operatorname{dom} B^*$, then the Trotter formula converges in the trace-norm with convergence rate estimates:*

$$\begin{aligned}&\left\|\left(e^{-tA/n}e^{-tB/n}\right)^n - e^{-t(A+B)}\right\|_1 \\ &\leqslant O(\ln(n)/n^{1-\alpha})\,(\alpha \neq 0) \vee O((\ln(n))^2/n)\,(\alpha = 0),\end{aligned} \tag{5.204}$$

uniformly away from $t = 0$.

Proof. Since $e^{-t\,\mathfrak{Re}\,A} \in \mathcal{C}_1(\mathcal{H})$ for $t > 0$, Proposition 4.30 shows that $G_t(A) = e^{-tA} \in \mathcal{C}_1(\mathcal{H})$ is a holomorphic contraction Gibbs semigroup, which is generated by A.

We express the fractional power $0 < \alpha < 1$ of A by the integral

$$A^\alpha u = \frac{1}{\Gamma(-\alpha)}\int_0^\infty d\tau\,\tau^{-\alpha-1}(G_\tau(A) - \mathbb{1})u, \quad u \in \operatorname{dom} A,$$

where τ^α is chosen to be positive for $\tau > 0$. Notice that for any $u \in \operatorname{dom} A$ this integral is convergent, thus $\operatorname{dom} A \subseteq \operatorname{dom} A^\alpha$ and consequently $\operatorname{dom} A^\alpha G_t(A) \in \mathcal{L}(\mathcal{H})$ for $t > 0$. It follows that

$$A^\alpha G_t(A) = \frac{1}{\Gamma(-\alpha)}\left(\int_0^t + \int_t^\infty\right)d\tau\,\tau^{-\alpha-1}(G_{t+\tau}(A) - G_t(A)).$$

Taking into account the estimate for the derivative of a holomorphic semigroup (Corollary 1.28 and Remark 1.31, for $n = 0$) one gets that $\|G_{t+\tau}(A) - G_t(A)\| \leqslant C_A \tau/t$, and therefore

$$\|A^\alpha G_t(A)\| \leqslant \frac{M_\alpha}{t^\alpha}. \tag{5.205}$$

Recall that a closed operator B belongs to the class of $\mathcal{P}_{0^+}$-perturbations of the C_0-semigroup $U_t(A)$ if

$$\operatorname{dom} B \supseteq \bigcup_{t>0} U_t(A)\mathcal{H} \quad \text{and} \quad \int_0^1 dt \|B\, U_t(A)\| < \infty,$$

see Definition 1.50 and Definition 4.34. Since the condition $\operatorname{dom} A^\alpha \subseteq \operatorname{dom} B$ and (5.205) for the holomorphic semigroup $\{G_t(A)\}_{t\geqslant 0}$ yield the estimate

$$\begin{aligned}\int_0^1 dt\, \|BG_t(A)\| &\leqslant \int_0^1 dt\, \|BA^{-\alpha}\| \|A^\alpha e^{-tA}\| \\ &\leqslant \|BA^{-\alpha}\| \int_0^1 dt\, \frac{M_\alpha}{t^\alpha} < \infty,\end{aligned}$$

the perturbation $B \in \mathcal{P}_{0^+} \subset \mathcal{P}_{b<1}$. Then by Proposition 4.44 the operator $H := A + B$ is the generator of the Gibbs semigroup $\{G_t(A + B)\}_{t\geqslant 0}$.

Since the assumptions of the proposition imply the operator-norm convergence (Proposition 5.47), with the convergence rate estimate, as well as the validity of the *lifting* Lemma 5.51, we use (5.148) to calculate $\varepsilon(m, t)$ in the inequality (5.151). Then the estimate (5.152) yields the rate of the trace-norm convergence (5.204). □

Another case of lifting the operator-norm convergence of the Trotter product formula to the trace-norm topology is related to Proposition 5.49.

Proposition 5.81. *Let A be a non-negative self-adjoint operator in $\mathcal{H}$ such that $(\lambda\mathbb{1} + A)^{-1} \in \mathcal{C}_p(\mathcal{H})$ for some $\lambda \in \mathbb{C}$ and for a finite $p \geqslant 1$. If B is an m-accretive operator such that* $\operatorname{dom} A \subseteq \operatorname{dom} B$ *and* $\operatorname{dom} A \subseteq \operatorname{dom} B^*$*, with*

$$\|Bu\| \leqslant b\|Au\|, \qquad u \in \operatorname{dom} A,\ 0 < b < 1, \tag{5.206}$$

$$\|B^*u\| \leqslant b_*\|Au\|, \qquad u \in \operatorname{dom} A,\ 0 < b_* < 1, \tag{5.207}$$

then the Trotter product formula converges in the trace-norm topology with the error bound for the rate of convergence estimated by

$$\left\| \left(\mathrm{e}^{-tA/n}\mathrm{e}^{-tB/n}\right)^n - \mathrm{e}^{-t(A+B)} \right\|_1 \leqslant O\left(\frac{\ln(n)}{n}\right), \tag{5.208}$$

uniformly away from $t = 0$.

Proof. By the hypotheses (5.206), (5.207) and by Definition 1.50 we have $B \in \mathcal{P}_{b<1}(A)$. The m-accretive operator B generates the contraction semigroup $\{U_t(B)\}_{t\geqslant 0}$.

On the other hand, since $(\lambda\mathbb{1}+A)^{-1} \in \mathcal{C}_p(\mathcal{H})$, Definition 4.26 and Proposition 4.27 show that the operator A is the self-adjoint p-generator for the holomorphic contraction Gibbs semigroup $\{G_t(A)\}_t$ with $t \in S_{\pi/2}$. Then by Proposition 4.49 the operator $A + B \in \mathscr{H}(\pi/2 - \delta, 0)$ generates a holomorphic contraction Gibbs semigroup $\{G_t(A + B)\}_t$ for $t \in S_{\pi/2-\delta}$.

Therefore, the semigroups $G_t(A)$, $U_t(B)$ and $G_t(H = A + B)$, satisfy the corresponding requirements of the *lifting* Lemma 5.51. Since the conditions of the proposition imply the operator-norm convergence of the Trotter product formula (Proposition 5.49), with explicit estimate of the rate of convergence, we use (5.150) to calculate $\varepsilon(m, t)$ in inequality (5.151) of Lemma 5.51. Then the estimate (5.152) yields the rate (5.208) of convergence in the trace-norm. □

5.6 Notes

Notes to Section 5.1. Proposition 5.1 is due to S. Lie (1875), who established it for matrices. The symmetrised version of the Trotter product formula in Proposition 5.1 was a source of many other generalisations for matrices and for the case of bounded generators, as well as for the so-called quantum Monte Carlo simulations, see e.g. [Suz96]. In this case the semigroups are *norm-continuous*, Section 1.4.

The important step forward was done by H. Trotter [Tro59], who extended the Lie product formula to Banach spaces. A similar extension was formulated by Yu. L. Dalestskii [Dal60], [Dal61], in the context of solutions of operator-valued equations.

The programme (a)-(e) outlined at the end of Section 5.1 for *strongly* continuous semigroups was formulated by V. A. Zagrebnov [Zag88], and then refined in [NZ90a]. Therein this programme was realised for the *trace-norm* convergence of the Trotter product formula first in the case of the Gibbs semigroups generated by Schrödinger operators.

In [NZ90a], [NZ90b] H. Neidhardt and V. A. Zagrebnov extended this result to the abstract strongly continuous Gibbs semigroups and to convergence of the *Trotter-Kato* product formulae in the trace-norm topology. This and later the operator-norm convergence confirm the *conventional wisdom* [Zag03b] that a *natural* topology γ for the convergence of the product formulae *inherits* the topology in which the semigroup is continuous *away from zero.* (We discuss the concept of continuity away from zero in Section 4.2)

Although for a long time, the convergence of the Lie-Trotter product formula for *strongly* continuous semigroups was known only in the *strong* operator topology $\gamma = s$ ([Tro59], [RS80]), as a trivial byproduct one obtains sufficient conditions for the operator-norm ($\gamma = \|\cdot\|$) convergence for this formula in [NZ90a, NZ90b].

Moreover, these results were obtained in the more general context of the Trotter-Kato product formulae.

A few years later, Dzh. L. Rogava in a short note [Rog93] announced the $\|\cdot\|$-convergence of the Trotter product formula for (*non-Gibbs*) strongly continuous self-adjoint contraction semigroups. This statement with improved rate of convergence and generalisation to the Trotter-Kato product formulae is due to H. Neidhardt and V. A. Zagrebnov [NZ98]. See also Chapter D.4 for further details concerning this episode.

Notes to Section 5.2. Proposition 5.8 is due to H. Neidhardt and V. A. Zagrebnov [NZ98]. In the proof presented here we essentially follow this paper.

For *self-adjoint* semigroups the proof of the operator-norm convergence (with estimate of the error bound for the rate) was improved in [ITTZ01]. This refinement allows to obtain the *optimal* estimate of the rate of convergence for the Trotter product formula (Proposition 5.19) as well as for a special class $\widehat{\mathcal{K}}_\beta$ of Kato functions for the Trotter-Kato product formulae (Proposition 5.25). Although these statements cover a majority of results for the self-adjoint case, they (in contrast to the method [NZ98]) are too restrictive to be applicable in the case of non-self-adjoint semigroups.

The fractional power conditions and result (Proposition 5.27) are due to H. Neidhardt and V. A. Zagrebnov [NZ99a]. The optimality of the error bound estimate under the assumptions of Proposition 5.27 was proved by Hiroshi Tamura [Tam00], see Proposition 5.28. He also proved (Proposition 5.28) that the operator-norm convergence of the Trotter product formula cannot be extended (as conjectured in [NZ99a]) to the case $\alpha \in (0, 1/2]$ in the abstract setting. It was shown that the case $\alpha = 1/2$ is an example where the Trotter product formula does not hold in the operator-norm topology, but still holds for convergence in the strong operator sense.

The case of the fractional power $\alpha = 1/2$ was studied by H. Neidhardt and V. A. Zagrebnov in [NZ99b]. It is of special interest since one cannot expect to have operator-norm convergence without subordination of the generators A and B. This condition in Proposition 5.30 is the $B^{1/2}$ relative $A^{1/2}$-compactness.

Proposition 5.27 was improved by T. Ichinose, H. Neidhardt and V. A. Zagrebnov in [INZ04], see Proposition 5.32. Their result extends Proposition 5.27 from the class $\mathcal{K}_\alpha$ to a larger class $\widehat{\mathcal{K}}_\beta$ of Kato functions, see Definition 5.24. Moreover, it also relaxes the smallness conditions of Proposition 5.27. To this aim we modified accordingly the fractional conditions in Proposition 5.32. Note that Tamura's optimal theorem (Proposition 5.28) still ensures that under the conditions of Proposition 5.32 the rate of convergence of Trotter-Kato product formulae is optimal for nonsymmetric case.

For further discussion of optimality we refer to [IT09]. The explicit example (one-dimensional harmonic oscillator), in which the symmetric Lie-Trotter product formula converges as fast as for matrices, i.e., with the *sharp*, and thus optimal, rate $O(1/n^2)$, was proposed and studied in details by Y. Azuma and T. Ichinose in [AzIch08].

Notes to Section 5.3. Proposition 5.36 is due to H. Neidhardt and V. A. Zagrebnov [NZ99b]. This paper establishes sufficient conditions that ensure the convergence for a class of the Trotter-Kato product formulae in the operator-norm topology *without* an estimate of the error bound for self-adjoint generators. In the proof presented here we essentially follow [NZ99b].

The result for the non-self-adjoint case with estimates presented in Proposition 5.47 is due to V. Cachia and V. A. Zagrebnov [CZ01a]. The line of reasoning is based on the method of [NZ98] and it heavily uses the analyticity and the *infinitesimal* smallness of m-accretive perturbations. Although the estimates are far from optimal, the method is strong enough to cover even the case of Banach spaces.

Proposition 5.49 is due to V. Cachia, H. Neidhardt and V. A. Zagrebnov [CNZ01]. It drastically relaxes the smallness condition of [CZ01a]. The proof is again motivated by [NZ98].

Notes to Section 5.4. The key Lemma 5.51 for exponential Kato function is due to V. Cachia and V. A. Zagrebnov [CZ01c].

It allows us to prove the trace-norm convergence *without* error bound estimate for the self-adjoint Trotter product formula, see Proposition 5.53. This result is a generalisation of [NZ90a, NZ90b] and of [Hia95].

If one adds a smallness condition, then Lemma 5.51 *lifts* the operator-norm convergence of the self-adjoint Trotter product formula to the trace-norm convergence *with* an error bound estimate that *coincides* with that for the operator-norm convergence, see Proposition 5.55. Here we follow [CZ01c].

Propositions 5.56 and 5.58 show that Lemma 5.51 lifts the *optimal* operator-norm convergence error bound to the *optimal* trace-norm convergence error bound.

The lifting Lemma 5.51 is well suited to prove the trace-norm convergence of the Trotter product formula, but it is not very useful for the Trotter-Kato product formulae. In [NZ90a] (Lemma 3.2) we established the first result in this direction under the conditions of Proposition 5.61. Note that Kato functions of the class $\mathcal{K}_*$ were historically the first non-exponential functions proposed in [Kat74] for product formula approximants converging in the strong operator topology. To this aim he introduced an *auxiliary* functions which define the class $\mathcal{K}_*$.

Under stronger (than (5.174)) condition (5.175) the class of Kato functions admissible for the trace-norm convergence is enlarged in Proposition 5.62 from $\mathcal{K}_*$ to generic class $\mathcal{K}$. This assertion was proven in [NZ90a] (Theorem 3.4). In [NZ99d] we developed further the *auxiliary* function control of $\mathcal{K}$ to produce more sufficient conditions ensuring the trace-norm convergence of Trotter-Kato product formulae, but now in the general framework of the symmetrically-normed ideals, see Chapter 6.

Notes to Section 5.5. The *analytic extension method* is due to V. Cachia and V. A. Zagrebnov [CZ99], [CZ01c]. We proposed it for analysis of the Trotter product formula approximants in case of m-sectorial generators.

The Representation Theorem (Proposition 5.63) is proved in [Kat80], Ch.IX. In the rest of the preliminaries (Lemma 5.66, Corollary 5.69, Proposition 5.71 and Corollary 5.72) we develop the idea of analytic extension from [Kat78] and [CZ99]. Then an application of the *Vitali theorem* to the operator-norm holomorphic families (of *approximants*) allows us to prove Propositions 5.75.

The first step in lifting the operator-norm convergence, established in Propositions 5.75, to convergence in the trace-norm is based on Propositions 5.76, [CZ01c]. Proposition 5.75 allows us also to prove (see Proposition 5.78) the non-self-adjoint Trotter product formula in the case of the trace-norm convergence *without* error bound estimate [CZ01c]. See an important in the context of the present book Corollary 5.79, in particular because of its application to inequality (3.45) in Section 3.3.

We note that idea of the *analytic extension method* was formulated by B. Simon and published by T. Kato as Addendum in [Kat78]. It was realised in [CZ99] and in [CZ01c]. There we constructed extensions of non-self-adjoint holomorphic families of the Trotter product formula approximants (5.193) and then proved their convergence in operator-norm and in trace-norm topologies.

In conclusion, we use the *lifting* Lemma 5.51 to prove the trace-norm convergence of the non-self-adjoint Trotter product formula *with* error bound estimates. In Proposition 5.80 we use the lifting lemma and the error bound estimate obtained in [CNZ01]. Another application of the lifting lemma for the Trotter product formula and non-self-adjoint Gibbs semigroups is Proposition 5.81. Both of these results are from [CZ01c]. See also a well-written review in [Ca10].

Chapter 6

Product formulae in symmetrically-normed ideals

In this chapter we continuing realisation of the *lifting* error bound estimate programme outlined in Chapter 5. The central problem is still the proof and the estimate of the rate of convergence of the Trotter-Kato product formulae, but now in the general setting of symmetrically-normed ideals of compact operators, where a particular case important for the Gibbs semigroups is the trace-class.

Section 6.1 serves for recalling a *three-step* strategy for lifting the convergence of operator-norm product formulae to convergence in the symmetric-norm topology. Since we consider the Trotter-Kato product formulae, the properties of various classes of Kato functions $\mathcal{K}$ are important for establishing the convergence. A class of *self-dominated* Kato functions has a particular importance for our arguments.

Section 6.2 is a brief introduction to *symmetrically-normed* ideals, whereas Section 6.3 contains a recall about convergence in these ideals.

The results presented in Section 6.4 involve the operator-norm convergence of the Trotter-Kato product formulae. Sufficient conditions for lifting them to the case of symmetric-norm topologies are formulated in terms of properties of auxiliary dominating Kato functions.

In Section 6.5 the symmetric-norm convergence of the Trotter-Kato product formulae is proven for self-dominated Kato functions without reference to the operator-norm convergence. This statement (Proposition 6.34) is of special importance for the theory of Gibbs semigroups.

To obtain the symmetric-norm error bound estimates we return in Section 6.6 to the lifting arguments, Proposition 6.38. This allows one to express the symmetric-norm error bound estimates via operator-norm error bound estimates. We find that the convergence rates in the two topologies coincide. Therefore, the optimality of the operator-norm rate remains valid in the case of the symmetric norm.

V. A. Zagrebnov, *Gibbs Semigroups*, Operator Theory: Advances and Applications 273, https://doi.org/10.1007/978-3-030-18877-1_6

6.1 Preliminaries

For non-negative self-adjoint operators A,B and Kato functions $f, g \in \mathcal{K}$ we set

$$F(t) := g(tB)^{1/2} f(tA) g(tB)^{1/2}, \quad t \geqslant 0, \tag{6.1}$$

and

$$T(t) := f(tA)^{1/2} g(tB) f(tA)^{1/2}, \quad t \geqslant 0, \tag{6.2}$$

for symmetrised self-adjoint families $\{F(t)\}_{t\geqslant 0}$ and $\{T(t)\}_{t\geqslant 0}$.

The question of convergence of the Trotter-Kato product formulae for Gibbs semigroups in $\mathcal{C}_1(\mathcal{H})$, formulated in Section 5.4 (subsection 5.4.4), is part of the following *general* problem.

Let ϕ be any *symmetric norming function* and $\mathcal{C}_\phi(\mathcal{H})$ be the corresponding *symmetrically-normed ideal* of compact operators, with the norm denoted by $\|\cdot\|_\phi$, see Section 6.2. Then the main problem solved in the present chapter can be stated as follows:
Find conditions that guarantee the converges of the Trotter-Kato product formulae *locally uniformly* away from $t_0 > 0$ (Definition 5.35) in the $\|\cdot\|_\phi$-topology on the ideal $\mathcal{C}_\phi(\mathcal{H})$, for example, for the Trotter-Kato approximants generated by $\{F(t)\}_{t\geqslant 0}$.

Equivalently, this would mean that for any bounded interval $[\tau_0, \tau] \subseteq (t_0, +\infty)$ there should exist $r_0 \geqslant 1$ (or $n_0 \in \mathbb{N}$) such that

$$\mathrm{e}^{-tH} P_0 \in \mathcal{C}_\phi(\mathcal{H}) \quad \text{and} \quad F(t/r)^r \in \mathcal{C}_\phi(\mathcal{H}), \tag{6.3}$$

uniformly in $t \in [\tau_0, \tau]$ and in $r \geqslant r_0$ (or in $n \geqslant n_0$), and

$$\lim_{r\to\infty} \sup_{t\in[\tau_0,\tau]} \|F(t/r)^r - \mathrm{e}^{-tH} P_0\|_\phi = 0. \tag{6.4}$$

Recall that the generator $H = A \dotplus B$ is the form-sum of A and B, and P_0 is the orthogonal projection $P_0 : \mathcal{H} \to \mathcal{H}_0 = \overline{\mathrm{dom}(H)}$.

Similarly, one can extend this notion of convergence of the Trotter-Kato product formulae to the families $\{T(t)\}_{t\geqslant 0}$, $\{f(tA)g(tB)\}_{t\geqslant 0}$ and $\{g(tB)f(tA)\}_{t\geqslant 0}$, though in the last two cases one has to use the *discrete* parameter $n \in \mathbb{N}$ instead of the *continuous* parameter $r \geqslant 1$. The continuous parameter r is suitable for self-adjoint *approximants* corresponding to the families (6.1) and (6.2).

If the convergence holds for these families, then we say that the Trotter-Kato product formulae converge in the topology of $\mathcal{C}_\phi(\mathcal{H})$ locally uniformly away from t_0 for all families generated by f and g.

It turns out (Proposition 6.18) that if the Trotter-Kato product formula converges locally uniformly away from $t_0 > 0$ in $\mathcal{C}_\phi(\mathcal{H})$ for *one* of the families $\{F(t)\}_{t\geqslant 0}$, $\{T(t)\}_{t\geqslant 0}$, $\{f(tA)g(tB)\}_{t\geqslant 0}$, $\{g(tB)f(tA)\}_{t\geqslant 0}$, then the product formulae converge locally uniformly away from t_0 in $\mathcal{C}_\phi(\mathcal{H})$ for all approximants generated by these families. To solve the problem formulated in (6.3), (6.4) one has to provide answers to the following *three* sub-problems:

First, find the conditions under which that the Trotter-Kato product formulae converge locally uniformly away from some $t_0 > 0$ in the *operator-norm* topology, i.e., for example

$$\lim_{r\to\infty} \sup_{t\in[a,b]} \|F(t/r)^r - \mathrm{e}^{-tH}P_0\| = 0, \tag{6.5}$$

for any bounded interval $[a, b] \subset \mathbb{R}^+$.

Second, one has to guarantee that the sequence $\{F(t/r)^r\}_{r\geqslant 1}$ is locally uniformly *bounded* away from $t_0 > 0$ in the $\|\cdot\|_\phi$-topology. This means that there is $r_0 \geqslant 1$ such that for $r \geqslant r_0$ and $t \in [\tau_0, \tau]$ one has $F(t/r)^r \in \mathcal{C}_\phi(\mathcal{H})$ and

$$M([\tau_0, \tau]) := \sup_{r\geqslant r_0} \sup_{t\in[\tau_0,\tau]} \|F(t/r)^r\|_\phi < +\infty, \tag{6.6}$$

for any bounded interval $[\tau_0, \tau] \subseteq (t_0, +\infty)$.

Third, we have to verify that the limit e^{-tH} belongs to the ideal $\mathcal{C}_\phi(\mathcal{H})$ (or $\mathrm{e}^{-tH}P_0 \in \mathcal{C}_\phi(\mathcal{H}_0)$) for $t \geqslant t_0 > 0$.

In fact these steps are the ingredients of the *lifting method* for establishing the $\|\cdot\|_\phi$-convergence in (6.4). We have already used this approach in Section 5.4 for lifting the operator-norm convergence to the trace-norm convergence for the *exponential* Trotter-Kato product formula, see subsections 5.4.1–5.4.3.

To solve the first sub-problem, we essentially use the results of Sections 5.2 and 5.3. The solution of the second sub-problem relies on Section 5.4 and especially on the remarks concerning the (non-exponential) Trotter-Kato approximants in subsection 5.4.4. To this aim we introduce the notion of *dominated* Kato functions $\mathcal{K}_D$, cf. Appendix C.

Definition 6.1. We say that the generic Kato functions f and g are *dominated* by, respectively, Borel measurable functions $f^D : \mathbb{R}_0^+ \to \mathbb{R}_0^+$ and $g^D : \mathbb{R}_0^+ \to \mathbb{R}_0^+$, if

$$f(qx)^{1/q} \leqslant f^D(x) \quad \text{and} \quad g(qx)^{1/q} \leqslant g^D(x), \quad 0 < q \leqslant 1. \tag{6.7}$$

Therefore, if for some $t_0 > 0$ the condition

$$F^D(t_0) := g^D(t_0B)^{1/2} f^D(t_0A) g^D(t_0B)^{1/2} \in \mathcal{C}_\phi(\mathcal{H}). \tag{6.8}$$

is satisfied, then one can prove that away from t_0 the local uniform boundedness of $\{F(t/r)\}_{r\geqslant 1}$ holds in $\mathcal{C}_\phi(\mathcal{H})$ and, moreover, that the third sub-problem is also solvable: $\mathrm{e}^{-tH}P_0 \in \mathcal{C}_\phi(\mathcal{H})$ for $t > t_0$.

A combination of these results with the operator-norm convergence of the Trotter-Kato product formulae established in Sections 5.2 and 5.3, yields a list of conditions that allow to establish the locally uniform $\|\cdot\|_\phi$-convergence of the Trotter-Kato product formulae away from t_0.

We start this list by the notion of *self-dominated* Kato functions $\mathcal{K}_{s\text{-}d} \subset \mathcal{K}$, see Section C.3.

Definition 6.2. The Kato functions $f, g \in \mathcal{K}$ obeying

$$f(qx)^{1/q} \leqslant f(x) \quad \text{and} \quad g(qx)^{1/q} \leqslant g(x), \quad 0 < q \leqslant 1, \tag{6.9}$$

for $x \geqslant 0$, and such that

$$C := \sup_{x>0} \frac{xf(x)}{1-f(x)} < +\infty \quad \text{and} \quad S := \sup_{x>0} \frac{xg(x)}{1-g(x)} < +\infty \;, \tag{6.10}$$

are called *self-dominated.* In this case we write $f, g \in \mathcal{K}_{s\text{-}d}$.

Restricting now the class of admissible Kato functions to the *self-dominated* functions we obtain the main result. It states that there is a $t_0 > 0$ such that the Trotter-Kato product formulae converge in $\mathcal{C}_\phi(\mathcal{H})$ locally uniformly away from $t_0 > 0$ if and only if for some positive $s_0 \leqslant t_0$ and some integer $p \geqslant 1$ we have

$$F(s_0)^p \in \mathcal{C}_\phi(\mathcal{H}). \tag{6.11}$$

We recall that the trace-class $\mathcal{C}_1(\mathcal{H})$ is a particular symmetrically-normed ideal in the ring of compact operators $\mathcal{C}_\infty(\mathcal{H})$. So this result yields also the proofs of assertions announced in subsection 5.4.4.

Note that in addition we also get the following extension of results from Sections 5.2 and 5.3. For self-dominated functions $\mathcal{K}_{s\text{-}d}$ the Trotter-Kato product formulae converge uniformly away from zero in the *operator-norm* topology if for some $t_0 > 0$ the condition $F(t_0) \in \mathcal{C}_\infty(\mathcal{H})$ is satisfied.

In the next Section 6.2 we recall some basic facts about symmetric norming functions ϕ and the corresponding symmetrically-normed ideals $\mathcal{C}_\phi$. Section 6.3 contains certain indispensable information about convergence in the ideals $\mathcal{C}_\phi(\mathcal{H})$. Section 6.4 concerns the *lifting approach* and related results about the Trotter-Kato product formulae convergence in $\mathcal{C}_\phi(\mathcal{H})$. In Section 6.5 we indicate certain *compactness* conditions, which imply the $\|\cdot\|_\phi$-convergence of the Trotter-Kato product formulae. Section 6.6 is devoted to the lifting approach. We show that under suitable compactness assumptions the operator-norm error bounds can be *lifted* to error bounds in symmetrically-normed ideals. In particular, the operator-norm rate of convergence obtained in Sections 5.4 for the operator *smallness* conditions lifts to the same rate of convergence in symmetrically-normed ideals $\mathcal{C}_\phi(\mathcal{H})$. Section 6.7 is devoted to notes and references.

6.2 Symmetrically-normed ideals

Let $c_0 \subset l^\infty$ be the subspace of bounded sequences $\xi = \{\xi_j\}_{j=1}^\infty \in l^\infty$ of real numbers, that converge to *zero*. We denote by c_f the subspace of c_0 consisting of all sequences with a *finite* number of non-zero terms (*finite sequences*).

Definition 6.3. A real-valued function $\phi : \xi \mapsto \phi(\xi)$ defined on c_f is called a *norming function* if it has the following properties:

$$\phi(\xi) > 0, \quad \forall \xi \in c_f,\ \xi \neq 0, \tag{6.12}$$
$$\phi(\alpha\xi) = |\alpha|\phi(\xi), \quad \forall \xi \in c_f,\ \forall \alpha \in \mathbb{R}, \tag{6.13}$$
$$\phi(\xi + \eta) \leqslant \phi(\xi) + \phi(\eta), \quad \forall \xi, \eta \in c_f, \tag{6.14}$$
$$\phi(1, 0, \ldots) = 1. \tag{6.15}$$

A norming function ϕ is said to be *symmetric* if it has the additional property that

$$\phi(\xi_1, \xi_2, \ldots, \xi_n, 0, 0, \ldots) = \phi(|\xi_{j_1}|, |\xi_{j_2}|, \ldots, |\xi_{j_n}|, 0, 0, \ldots) \tag{6.16}$$

for any $\xi \in c_f$ and any permutation $j_1, j_2, \ldots, j_n$ of integers $1, 2, \ldots, n$.

It turns out that for any *symmetric norming* function ϕ and for any elements ξ, η from the positive *cone* c^+ of *non-negative* and *non-increasing* sequences such that $\xi, \eta \in c_f$ obey $\xi_1 \geqslant \xi_2 \geqslant \ldots \geqslant 0$, $\eta_1 \geqslant \eta_2 \geqslant \ldots \geqslant 0$, and

$$\sum_{j=1}^{n} \xi_j \leqslant \sum_{j=1}^{n} \eta_j, \quad n = 1, 2, \ldots, \tag{6.17}$$

one has the Ky Fan inequality

$$\phi(\xi) \leqslant \phi(\eta). \tag{6.18}$$

Therefore, inequality (6.18) together with the properties (6.12), (6.13) and (6.15) yield inequalities

$$\xi_1 \leqslant \phi(\xi) \leqslant \sum_{j=1}^{\infty} \xi_j, \quad \xi \in c_f^+ := c_f \cap c^+. \tag{6.19}$$

Note that the left- and right-hand sides of (6.19) are the simplest examples of symmetric norming functions on the domain c_f^+. We denote them by

$$\phi_\infty(\xi) := \xi_1 \quad \text{and} \quad \phi_1(\xi) := \sum_{j=1}^{\infty} \xi_j. \tag{6.20}$$

Then by Definition 6.3 the estimates (6.19) and (6.20) yield

$$\phi_\infty(\xi) := \max_{j \geqslant 1} |\xi_j|, \quad \phi_1(\xi) := \sum_{j=1}^{\infty} |\xi_j|, \tag{6.21}$$
$$\phi_\infty(\xi) \leqslant \phi(\xi) \leqslant \phi_1(\xi), \quad \text{for all } \xi \in c_f.$$

Next we denote by $\xi^* := \{\xi_1^*, \xi_2^*, \dots\}$ a *decreasing rearrangement*:

$$\xi_1^* = \sup_{j\geqslant 1} |\xi_j|, \ \xi_1^* + \xi_2^* = \sup_{i\neq j}\{|\xi_i| + |\xi_j|\}, \ \dots ,$$

of the sequence of absolute values $\{|\xi_n|\}_{n\geqslant 1}$, that is, $\xi_1^* \geqslant \xi_2^* \geqslant \dots$. Then $\xi \in c_f$ implies $\xi^* \in c_f$ and also

$$\phi(\xi) = \phi(\xi^*), \quad \xi \in c_f, \tag{6.22}$$

see (6.16). Therefore, any symmetric norming function ϕ is uniquely determined by its values on the positive cone c^+.

Now, let $\xi = \{\xi_1, \xi_2, \dots\} \in c_0$. We define

$$\xi^{(n)} := \{\xi_1, \xi_2, \dots, \xi_n, 0, 0, \dots\} \in c_f . \tag{6.23}$$

If ϕ is a symmetric norming function, we set

$$c_\phi := \{\xi \in c_0 : \sup_{n\geqslant 1} \phi(\xi^{(n)}) < +\infty\}, \tag{6.24}$$

which is a *natural domain* of ϕ. Therefore, one gets

$$c_f \subseteq c_\phi \subseteq c_0. \tag{6.25}$$

Note that by (6.16)-(6.18) and (6.24) one gets

$$\phi(\xi^{(n)}) \leqslant \phi(\xi^{(n+1)}) \leqslant \sup_{n\geqslant 1} \phi(\xi^{(n)}) , \quad \text{for any } \xi \in c_\phi . \tag{6.26}$$

Therefore, by (6.26), the limit

$$\phi(\xi) := \lim_{n\to\infty} \phi(\xi^{(n)}), \quad \xi \in c_\phi , \tag{6.27}$$

on the *natural domain* c_ϕ exists and $\phi(\xi) = \sup_{n\geqslant 1} \phi(\xi^{(n)})$. This means that the symmetric norming function ϕ is a *normal* functional on the linear (over real numbers) set $c_\phi \subset l^\infty$.

By virtue of (6.14) and (6.21), every symmetric norming function is *continuous* on c_f:

$$|\phi(\xi) - \phi(\eta)| \leqslant \phi(\xi - \eta) \leqslant \phi_1(\xi - \eta), \quad \forall \xi, \eta \in c_f . \tag{6.28}$$

Suppose that X is a compact operator on a Hilbert space $\mathcal{H}$, i.e. $X \in \mathcal{C}_\infty(\mathcal{H})$. Then we denote by

$$s(X) := \{s_1(X), s_2(X), \dots\} , \tag{6.29}$$

the sequence of singular values of X, counting multiplicities. We always assume that

$$s_1(X) \geqslant s_2(X) \geqslant \dots \geqslant s_n(X) \geqslant \dots . \tag{6.30}$$

To define *symmetrically-normed* ideals of the ring of compact operators $\mathcal{C}_\infty(\mathcal{H})$ we introduce the notion of the *symmetric norm*.

Definition 6.4. Let $\mathfrak{J}$ be a two-sided ideal of $\mathcal{C}_\infty(\mathcal{H})$. A functional $\|\cdot\|_{\mathrm{sym}} : \mathfrak{J} \to \mathbb{R}_0^+$ is called a *symmetric norm* if besides the usual properties of the operator norm $\|\cdot\|$:

$$\|X\|_{\mathrm{sym}} > 0, \quad \forall X \in \mathfrak{J},\ X \neq 0, \tag{6.31}$$
$$\|\alpha X\|_{\mathrm{sym}} = |\alpha|\|X\|_{\mathrm{sym}}, \quad \forall X \in \mathfrak{J},\ \forall \alpha \in \mathbb{C}, \tag{6.32}$$
$$\|X+Y\|_{\mathrm{sym}} \leqslant \|X\|_{\mathrm{sym}} + \|Y\|_{\mathrm{sym}}, \quad \forall X, Y \in \mathfrak{J}, \tag{6.33}$$

it verifies also the following additional properties:

$$\|AXB\|_{\mathrm{sym}} \leqslant \|A\|\|X\|_{\mathrm{sym}}\|B\|, \quad X \in \mathfrak{J},\ A, B \in \mathcal{L}(\mathcal{H}), \tag{6.34}$$
$$\begin{aligned} \|\alpha X\|_{\mathrm{sym}} &= |\alpha|\|X\| \\ &= |\alpha|\, s_1(X), \quad \text{for any rank-one operator } X \in \mathfrak{J}. \end{aligned} \tag{6.35}$$

If condition (6.34) is replaced by

$$\begin{aligned} &\|UX\|_{\mathrm{sym}} = \|XU\|_{\mathrm{sym}} = \|X\|_{\mathrm{sym}}, \quad X \in \mathfrak{J}, \\ &\text{for any unitary operator } U \text{ on } \mathcal{H}, \end{aligned} \tag{6.36}$$

then, one arrives to the definition of the *invariant* symmetric norm $\|\cdot\|_{\mathrm{inv}}$.

First, we note that the ordinary operator norm $\|\cdot\|$ on any ideal $\mathfrak{J} \subseteq \mathcal{C}_\infty(\mathcal{H})$ is evidently a symmetric norm as well as an invariant norm.

Second, we observe that every symmetric norm is automatically invariant. Indeed, for any unitary operators U and V one gets by (6.34) that

$$\|UXV\|_{\mathrm{sym}} \leqslant \|X\|_{\mathrm{sym}}, \quad X \in \mathfrak{J}. \tag{6.37}$$

Since $X = U^{-1}UXVV^{-1}$, we also get $\|X\|_{\mathrm{sym}} \leqslant \|UXV\|_{\mathrm{sym}}$, which together with (6.37) yields (6.36).

Third, we claim that $\|X\|_{\mathrm{sym}} = \|X^*\|_{\mathrm{sym}}$. For the proof, let $X = U|X|$ be the polar decomposition of the operator $X \in \mathfrak{J}$. Since $U^*X = |X|$, by (6.36) we obtain $\|X\|_{\mathrm{sym}} = \||X|\|_{\mathrm{sym}}$. The same argument applied to the adjoint operator $X^* = |X|U^*$ yields $\|X^*\|_{\mathrm{sym}} = \||X|\|_{\mathrm{sym}}$, which proves the claim.

Now we can apply the concept of symmetric norming function to describe the symmetrically-normed ideals of the unital algebra of bounded operators $\mathcal{L}(\mathcal{H})$, or in general, the symmetrically-normed ideals generated by symmetric norming functions. Recall that any *proper* two-sided ideal $\mathfrak{J}(\mathcal{H})$ of the ring $\mathcal{L}(\mathcal{H})$ is contained in $\mathcal{C}_\infty(\mathcal{H})$ and contains the set $\mathcal{K}(\mathcal{H})$ of finite-rank operators:

$$\mathcal{K}(\mathcal{H}) \subseteq \mathfrak{J}(\mathcal{H}) \subseteq \mathcal{C}_\infty(\mathcal{H}). \tag{6.38}$$

To clarify the relationships between symmetric norming functions and symmetrically-normed ideals we mention that on the cone c^+ there is an obvious one-to-one correspondence between functions ϕ (Definition 6.3) and the symmetric norms $\|\cdot\|_{\mathrm{sym}}$ on $\mathfrak{J}(\mathcal{H})$. To carry on with a general setting we need the following relation.

Definition 6.5. Let c_ϕ be the set of vectors (6.24) generated by a symmetric norming function ϕ. We associate with c_ϕ the following subset of compact operators:

$$\mathcal{C}_\phi(\mathcal{H}) := \{X \in \mathcal{C}_\infty(\mathcal{H}) : s(X) \in c_\phi\}, \tag{6.39}$$

which is a proper two-sided ideal of the algebra $\mathcal{L}(\mathcal{H})$ of all bounded operators on $\mathcal{H}$. It is called a *symmetrically-normed* ideal generated by the symmetric norming function ϕ.

Since the limit (6.27) yields $\phi(\xi) = \sup_{n\geqslant 1}\phi(\xi^{(n)})$, we define on the symmetrically-normed ideal $\mathcal{C}_\phi(\mathcal{H})$ the functional

$$\|X\|_\phi := \phi(s(X)), \quad X \in \mathcal{C}_\phi(\mathcal{H}). \tag{6.40}$$

If one equips on the ideal $\mathfrak{I}(\mathcal{H}) = \mathcal{C}_\phi(\mathcal{H})$ (Definition 6.4) the *symmetric norm* $\|\cdot\|_{\mathrm{sym}} := \|\cdot\|_\phi$, then this symmetrically-normed ideal becomes a Banach space. Hence, in accordance with (6.38) and (6.39) we obtain by (6.21) that

$$\mathcal{K}(\mathcal{H}) \subseteq \mathcal{C}_1(\mathcal{H}) \subseteq \mathcal{C}_\phi(\mathcal{H}) \subseteq \mathcal{C}_\infty(\mathcal{H}). \tag{6.41}$$

Here the *trace-class* operators $\mathcal{C}_1(\mathcal{H}) := \mathcal{C}_{\phi_1}(\mathcal{H})$, where the symmetric norming function ϕ_1 is defined in (6.20), (6.21), and consequently

$$\|X\|_\phi \leqslant \|X\|_1 , \quad X \in \mathcal{C}_1(\mathcal{H}). \tag{6.42}$$

Remark 6.6. By virtue of the Ky Fan inequality (6.18) and the definition of the symmetric norm (6.40), the so-called *dominance* property holds: if $X \in \mathcal{C}_\phi(\mathcal{H})$, $Y \in \mathcal{C}_\infty(\mathcal{H})$ and

$$\sum_{j=1}^{n} s_j(Y) \leqslant \sum_{j=1}^{n} s_j(X), \quad n = 1, 2, \ldots, \tag{6.43}$$

then $Y \in \mathcal{C}_\phi(\mathcal{H})$ and $\|Y\|_\phi \leqslant \|X\|_\phi$.

Remark 6.7. To distinguish in (6.41) *nontrivial* ideals $\mathcal{C}_\phi$ we need some criteria based on the properties of ϕ or of the norm $\|\cdot\|_\phi$. For example, any symmetric norming function (6.22) defined by

$$\phi^{(r)}(\xi) := \sum_{j=1}^{r} \xi_j^* , \quad \xi \in c_f , \tag{6.44}$$

generates for fixed $r \in \mathbb{N}$ a symmetrically-normed ideal, which is trivial in the sense that $\mathcal{C}_{\phi^{(r)}}(\mathcal{H}) = \mathcal{C}_{\phi_\infty}(\mathcal{H}) = \mathcal{C}_\infty(\mathcal{H})$. A criterion for the operator A to belong to a nontrivial ideal $\mathcal{C}_\phi$ is

$$M = \sup_{m\geqslant 1} \|P_m A P_m\|_\phi < \infty, \tag{6.45}$$

where $\{P_m\}_{m\geqslant 1}$ is a monotonically increasing sequence of finite-dimensional orthogonal projectors on $\mathcal{H}$ that converges strongly to the identity operator. Note that for $A \in \mathcal{C}_\infty$ condition (6.45) is trivial.

We consider now a couple of examples to explain the concept of the symmetrically-normed ideals $\mathcal{C}_\phi(\mathcal{H})$ generated by symmetric norming functions ϕ and the rôle of the *trace* functional, or simply trace $\mathrm{Tr}(\cdot)$, on these ideals.

Example 6.8. The von Neumann-Schatten ideals $\mathcal{C}_p(\mathcal{H})$ (Chapter 2). These ideals of $\mathcal{C}_\infty(\mathcal{H})$ are generated by the symmetric norming functions $\phi_p(\xi)$, where

$$\phi_p(\xi) := \Big(\sum_{j=1}^{\infty} |\xi_j|^p\Big)^{1/p}, \quad \xi \in c_f, \tag{6.46}$$

coincides with the standard l^p-norm for $1 \leqslant p < \infty$, or with the standard l^∞-norm,

$$\phi_\infty(\xi) := \sup_{1\leqslant j} |\xi_j|, \quad \xi \in c_f, \tag{6.47}$$

for $p = \infty$.

Indeed, if we set $\{\xi_j^* := s_j(X)\}_{j\geqslant 1}$ for $X \in \mathcal{C}_\infty(\mathcal{H})$, then the symmetric norm $\|X\|_{\mathrm{sym}} = \phi_p(s(X))$ (6.40) coincides with (2.37):

$$\|X\|_p = \Big(\sum_{j=1}^{\infty} s_j(X)^p\Big)^{1/p}. \tag{6.48}$$

The corresponding symmetrically-normed ideal $\mathcal{C}_{\phi_p}(\mathcal{H})$ is identical to the von Neumann-Schatten class $\mathcal{C}_p(\mathcal{H})$, see Definition 2.50.

By Definition 2.41, for $X \in \mathcal{C}_p(\mathcal{H})$ one can consider the functional $|X| \mapsto \mathrm{Tr}\,|X| = \sum_{j\geqslant 1} s_j(X) > 0$. Then the trace-norm $\|X\|_1 := \mathrm{Tr}\,|X| \geqslant \|X\|$, is *finite* on the trace-class operators $\mathcal{C}_1(\mathcal{H})$ and it is *infinite* for $X \in \mathcal{C}_{p>1}(\mathcal{H})\backslash\mathcal{C}_1(\mathcal{H})$. We say that for $p > 1$ the von Neumann-Schatten ideals admit *no* trace, whereas for $p = 1$ the map: $X \mapsto \mathrm{Tr}\,X$, is continuous in the $\|\cdot\|_1$-topology.

Note that since the *trace* is linear, the trace-norm: $\mathcal{C}_{1,+}(\mathcal{H}) \ni X \mapsto \|X\|_1$ is *linear* on the positive cone $\mathcal{C}_{1,+}(\mathcal{H})$ of the trace-class operators.

Example 6.9. The symmetrically-normed ideals $\mathcal{C}_\Pi(\mathcal{H})$ and $\mathcal{C}_\Pi^0(\mathcal{H})$. Let $\Pi = \{\pi_j\}_{j=1}^{\infty} \in c^+$ be a non-increasing sequence of positive numbers with $\pi_1 = 1$. We associate with Π the function

$$\phi_\Pi(\xi) = \sup_{n\geqslant 1}\left\{\frac{1}{\sum_{j=1}^{n}\pi_j}\sum_{j=1}^{n}\xi_j\right\}, \quad \xi \in c_f. \tag{6.49}$$

Definition 6.3 shows that ϕ_Π is a symmetric norming function, that is, $\phi_\Pi(\xi) = \phi_\Pi(\xi^*)$, where ξ^* is decreasing rearrangement of $\xi \in c_f$ (6.22). Then the corresponding to (6.24) set c_{ϕ_Π} is defined by

$$c_{\phi_\Pi} := \left\{\xi \in c_f : \sup_{n\geqslant 1}\frac{1}{\sum_{j=1}^{n}\pi_j}\sum_{j=1}^{n}\xi_j^* < +\infty\right\}. \tag{6.50}$$

Hence, the two-sided symmetrically-normed ideal $\mathcal{C}_\Pi(\mathcal{H}) := \mathcal{C}_{\phi_\Pi}(\mathcal{H})$ of $\mathcal{C}_\infty(\mathcal{H})$ generated by symmetric norming function (6.49) consists of all compact operators X for which

$$\|X\|_{\phi_\Pi} := \sup_{n\geqslant 1} \frac{1}{\sum_{j=1}^n \pi_j} \sum_{j=1}^n s_j(X) < +\infty. \tag{6.51}$$

By Definition 6.4 and (6.40), this formula defines on the ideal $\mathcal{C}_\Pi(\mathcal{H})$ of compact operators a symmetric norm $\|X\|_{\text{sym}} = \|X\|_{\phi_\Pi}$.

If the non-increasing sequence $\Pi = \{\pi_j \geqslant 0\}_{j=1}^\infty$ with $\pi_1 = 1$ satisfies

$$\sum_{j=1}^\infty \pi_j = +\infty \quad \text{and} \quad \lim_{j\to\infty} \pi_j = 0, \tag{6.52}$$

then the ideal $\mathcal{C}_\Pi(\mathcal{H})$ is *nontrivial*: $\mathcal{C}_\Pi(\mathcal{H}) \neq \mathcal{C}_\infty(\mathcal{H})$ and $\mathcal{C}_\Pi(\mathcal{H}) \neq \mathcal{C}_1(\mathcal{H})$, see Remark 6.7. Moreover, by Lemma 3.5, (3.16), $\mathcal{C}_\Pi(\mathcal{H})$ contains as a *proper* subset the (separable) symmetrically-normed ideal $\mathcal{C}_\Pi^0(\mathcal{H})$ given by

$$\mathcal{C}_\Pi^0(\mathcal{H}) := \{X \in \mathcal{C}_\Pi(\mathcal{H}) : \lim_{n\to\infty} \frac{1}{\sum_{j=1}^n \pi_j} \sum_{j=1}^n s_j(X) = 0\}. \tag{6.53}$$

Thus,

$$\mathcal{C}_1(\mathcal{H}) \subset \mathcal{C}_\Pi^0(\mathcal{H}) \subset \mathcal{C}_\Pi(\mathcal{H}) \subset \mathcal{C}_\infty(\mathcal{H}). \tag{6.54}$$

If in addition to (6.52) the sequence $\Pi = \{\pi_j\}_{j=1}^\infty$ is *regular*, that is,

$$\sum_{j=1}^n \pi_j = O(n\pi_n), \quad n \to \infty, \tag{6.55}$$

then $X \in \mathcal{C}_\Pi(\mathcal{H})$ if and only if

$$s_n(X) = O(\pi_n), \quad n \to \infty, \tag{6.56}$$

cf. condition (6.45). On the other hand, the asymptotics

$$s_n(X) = o(\pi_n), \quad n \to \infty, \tag{6.57}$$

implies that $X \in \mathcal{C}_\Pi^0(\mathcal{H}) \subset \mathcal{C}_\Pi(\mathcal{H})$.

Remark 6.10. A natural choice of sequence $\{\pi_j\}_{j=1}^\infty$ that satisfies (6.52) is $\pi_j = j^{-\alpha}$, $0 < \alpha \leqslant 1$. Note that if $0 < \alpha < 1$, then the sequence $\Pi = \{\pi_j\}_{j=1}^\infty$ satisfies also (6.55). Consequently, the two-sided symmetrically-normed ideal $\mathcal{C}_\Pi(\mathcal{H})$ generated by the symmetric norming function (6.49) consists of the compact operators X whose singular values obey (6.56):

$$s_n(X) = O(n^{-\alpha}), \quad 0 < \alpha < 1,\ n \to \infty. \tag{6.58}$$

Definition 6.11. Let $\alpha := 1/p,\ p > 1$. The symmetrically-normed ideal corresponding to (6.58), that is,

$$\mathcal{C}_{p,\infty}(\mathcal{H}) := \{X \in \mathcal{C}_\infty(\mathcal{H}) : s_n(X) = O(n^{-1/p}),\ p > 1\}, \tag{6.59}$$

is known as the *weak-$\mathcal{C}_p$* ideal. By definition (6.51), the symmetric norm on the ideal $\mathcal{C}_{p,\infty}(\mathcal{H})$ is equivalent for the regular case $p > 1$ to

$$\|X\|_{p,\infty} := \sup_{n \geqslant 1} \frac{1}{n^{1-1/p}} \sum_{j=1}^{n} s_j(X). \tag{6.60}$$

Note that if $p' > p$, then definition (6.60) yields

$$\mathcal{C}_1(\mathcal{H}) \subset \mathcal{C}_{p,\infty}(\mathcal{H}) \subset \mathcal{C}_{p',\infty}(\mathcal{H}) \subset \mathcal{C}_\infty(\mathcal{H}). \tag{6.61}$$

On the other hand, (6.48), (6.59) and (6.60), show that

$$\mathcal{C}_1(\mathcal{H}) \subset \mathcal{C}_{p,\infty}(\mathcal{H}) \subset \mathcal{C}_q(\mathcal{H}), \quad \text{for } 1 < p < q\,. \tag{6.62}$$

Remark 6.12. The *weak-$\mathcal{C}_p$* ideal defined for $p = 1$ by

$$\mathcal{C}_{1,\infty}(\mathcal{H}) := \{X \in \mathcal{C}_\infty(\mathcal{H}) : \sum_{j=1}^{n} s_j(X) = O(\ln(n)),\ n \to \infty\}, \tag{6.63}$$

is of a special interest. Note that since $\Pi = \{j^{-1}\}_{j=1}^{\infty}$ does *not* satisfy (6.55), the characterisation $s_n(X) = O(n^{-1})$ is *not* true, see (6.56) and (6.58). In this case the equivalent symmetric norm can be defined on (6.63) as

$$\|X\|_{1,\infty} := \sup_{n \geqslant 1} \frac{1}{1+\ln(n)} \sum_{j=1}^{n} s_j(X). \tag{6.64}$$

By (6.64), one readily gets that $\mathcal{C}_1(\mathcal{H}) \ni X \mapsto \|X\|_1 \geqslant \|X\|_{1,\infty}$. Therefore, similarly to (6.62), we obtain $\mathcal{C}_1(\mathcal{H}) \subset \mathcal{C}_{1,\infty}(\mathcal{H})$.

Example 6.13. With a non-increasing sequence of positive numbers $\Pi = \{\pi_j\}_{j=1}^{\infty}$, $\pi_1 = 1$, and with the decreasing rearrangement ξ^* of $\xi \in c_f$ one can associate the symmetric norming function ϕ_π given by

$$\phi_\pi(\xi) := \sum_{j=1}^{\infty} \pi_j \xi_j^*\,, \quad \xi \in c_f\,. \tag{6.65}$$

We denote the corresponding symmetrically-normed ideal by $\mathcal{C}_\pi(\mathcal{H}) := \mathcal{C}_{\phi_\pi}(\mathcal{H})$. If the sequence Π satisfies (6.52), then the ideal $\mathcal{C}_\pi(\mathcal{H})$ is different from $\mathcal{C}_\infty(\mathcal{H})$ and from $\mathcal{C}_1(\mathcal{H})$. If, in particular, $\pi_j = j^{-\alpha}$, $j = 1, 2, \ldots$, for $0 < \alpha \leqslant 1$, then the corresponding ideal is denoted by $\mathcal{C}_{\infty,p}(\mathcal{H})$, $p = 1/\alpha$. The norm on this ideal is given by

$$\|X\|_{\infty,p} := \sum_{j=1}^{\infty} j^{-1/p} s_j(X), \quad p \geqslant 1. \tag{6.66}$$

The symmetrically-normed ideal $\mathcal{C}_{\infty,1}(\mathcal{H})$ is called the *Macaev ideal.* It turns out that the *dual* of the Macaev ideal is $\mathcal{C}_{\infty,1}(\mathcal{H})^* = \mathcal{C}_{1,\infty}(\mathcal{H})$.

We conclude this section by remarks concerning the problem of existence of a *trace* $\mathrm{Tr}_\omega(\cdot)$ on symmetrically normed ideals $\mathcal{C}_{p,\infty}(\mathcal{H})$, $p \geqslant 1$. It should be a linear, positive, and unitarily invariant functional $X \mapsto \mathrm{Tr}_\omega(X)$, for any $X \in \mathcal{C}_{p,\infty}(\mathcal{H})$. On the other hand, X does not admit the standard trace functional $X \mapsto \mathrm{Tr}(X)$, if $X \in \mathcal{C}_{p,\infty}(\mathcal{H})\backslash\mathcal{C}_1(\mathcal{H})$.

In contrast to the linearity of the trace-norm $\|\cdot\|_1$ on the positive cone $\mathcal{C}_{1,+}(\mathcal{H})$ (Example 6.8), the map $X \mapsto \|X\|_{p,\infty}$ on the positive cone $\mathcal{C}_{p,\infty,+}(\mathcal{H})$ is *not* linear. Although this map is *homogeneous*: $\alpha X \mapsto \alpha\|X\|_{p,\infty}$, $\alpha \geqslant 0$, it turns out that for $X, Y \in \mathcal{C}_{p,\infty,+}(\mathcal{H})$ in general $\|X + Y\|_{p,\infty} \neq \|X\|_{p,\infty} + \|Y\|_{p,\infty}$.

However, on the space l^∞ there exists an appropriate *state* $\omega \in \mathcal{S}(l^\infty)$ such that for $p = 1$ the mapping

$$X \mapsto \mathrm{Tr}_\omega(X) := \omega(\{(1 + \ln(n))^{-1} \sum_{j=1}^{n} s_j(X)\}_{n=1}^{\infty}), \tag{6.67}$$

is *linear* and enjoys for $X \in \mathcal{C}_{1,\infty}(\mathcal{H})$, the properties of the trace. Note that ω is a linear positive normalised functional (*state*) on the Banach space l^∞.

Recall that the set of states $\mathcal{S}(l^\infty)$ belongs to $(l^\infty)^*$, where $(l^\infty)^*$ is the dual of the Banach space l^∞. For a particular choice of the state ω this gives the *Dixmier trace* $\mathrm{Tr}_\omega(\cdot)$ on the ideal $\mathcal{C}_{1,\infty}(\mathcal{H})$, which in turn is known as the *Dixmier ideal.* Since the Dixmier trace is *zero* on the set of finite-rank operators $\mathcal{K}(\mathcal{H})$ (and hence on $\mathcal{C}_1(\mathcal{H})$), it is a *singular* trace. We construct the trace $\mathrm{Tr}_\omega(\cdot)$ in Chapter 7.

The Dixmier trace (6.67) is continuous in the topology defined by the norm (6.64). This property is basic for our discussion in Chapter 7 of the Trotter-Kato product formulae in the $\|\cdot\|_{p,\infty}$-topology, for $p \geqslant 1$.

6.3 Convergence in $\mathcal{C}_\phi(\mathcal{H})$-ideals

Let $X \geqslant 0$ be a compact operator. Let $\lambda(X) = \{\lambda_j(X)\}_{j=1}^{\infty}$ be the sequence of its eigenvalues, counting multiplicities. In this case they coincide with the singular values $\lambda(X) = s(X)$ and we always assume that we leave the decreasing arrangement:

$$\lambda_1(X) \geqslant \lambda_2(X) \geqslant \ldots \geqslant 0. \tag{6.68}$$

Definition 6.14. If $X \geqslant 0$ and $Y \geqslant 0$ are two compact operators, then we say that they are *log-ordered*, and write,

$$Y \prec_{\log} X, \tag{6.69}$$

if

$$\prod_{j=1}^{n} \lambda_j(Y) < \prod_{j=1}^{n} \lambda_j(X), \quad n = 1, 2, \ldots . \tag{6.70}$$

Lemma 6.15. *Let X and Y be non-negative self-adjoint operators and let ϕ be a symmetric norming function. If $X \in \mathcal{C}_\phi(\mathcal{H})$ and $Y \prec_{\log} X$, then $Y \in \mathcal{C}_\phi(\mathcal{H})$ and*

$$\|Y\|_\phi < \|X\|_\phi . \tag{6.71}$$

Proof. By (6.69) and (6.70) we get that

$$\sum_{j=1}^{n} \ln(\lambda_j(Y)) < \sum_{j=1}^{n} \ln(\lambda_j(X)), \qquad n = 1, 2, \ldots . \tag{6.72}$$

Now let $x_+ := \max(x, 0)$ for any $x \in \mathbb{R}$ and denote $\mu_j(\cdot) := \ln(\lambda_j(\cdot))$. Then one gets the representation

$$e^{\mu_j(\cdot)} = \int_{\mathbb{R}} du\, e^u \, (\mu_j(\cdot) - u)_+ . \tag{6.73}$$

Note that

$$\sum_{j=1}^{n} (\mu_j(\cdot) - u)_+ = \begin{cases} \sum_{j=1}^{k} (\mu_j(\cdot) - u), & k = \text{the largest integer such that } \mu_k(\cdot) \geqslant u , \\ 0, & \text{if } \mu_1(\cdot) < u , \end{cases}$$

which together with (6.72) implies

$$\begin{aligned} \sum_{j=1}^{n} (\mu_j(Y) - u)_+ &= \sum_{j=1}^{k} \mu_j(Y) - k\,u \\ &\leqslant \sum_{j=1}^{k} \mu_j(X) - k\,u \leqslant \sum_{j=1}^{n} (\mu_j(X) - u)_+ . \end{aligned} \tag{6.74}$$

Therefore, (6.73) and (6.74) yield

$$\sum_{j=1}^{n} e^{\ln(\lambda_j(Y))} < \sum_{j=1}^{n} e^{\ln(\lambda_j(X))}, \quad n = 1, 2, \ldots ,$$

whence

$$\sum_{j=1}^{n} \lambda_j(Y) < \sum_{j=1}^{n} \lambda_j(X), \quad n = 1, 2, \ldots . \tag{6.75}$$

Then by (6.75) and by the dominance property of $\mathcal{C}_\phi(\mathcal{H})$ (Remark 6.6), we find that $Y \in \mathcal{C}_\phi(\mathcal{H})$ and $\|Y\|_\phi < \|X\|_\phi$. □

Below we shall apply Lemma 6.15 in a slightly modified form.

Lemma 6.16. *Let X, X_0, Y and Y_0 be non-negative self-adjoint operators and let ϕ be a symmetric norming function. If for some $r \geqslant 1$ one has $X^r \leqslant X_0$ and $Y^r \leqslant Y_0$ and if $Y_0^{1/2} X_0 Y_0^{1/2} \in \mathcal{C}_\phi(\mathcal{H})$, then*

$$\left(Y^{1/2} X Y^{1/2}\right)^r \prec_{\log} Y_0^{1/2} X_0 Y_0^{1/2}, \tag{6.76}$$

$\left(Y^{1/2} X Y^{1/2}\right)^r \in \mathcal{C}_\phi(\mathcal{H})$ and

$$\|\left(Y^{1/2} X Y^{1/2}\right)^r\|_\phi \leqslant \|Y_0^{1/2} X_0 Y_0^{1/2}\|_\phi . \tag{6.77}$$

Proof. By the Araki inequality (Section B.1), we have

$$\left(Y^{1/2} X Y^{1/2}\right)^r \prec_{\log} Y^{r/2} X^r Y^{r/2}, \quad r > 1. \tag{6.78}$$

Since $X^r \leqslant X_0$, we get

$$Y^{r/2} X^r Y^{r/2} \leqslant Y^{r/2} X_0 Y^{r/2} . \tag{6.79}$$

Moreover, since $Y^r \leqslant Y_0$, there exists a contraction $\Gamma_0 : \overline{\operatorname{ran} Y_0} \to \mathcal{H}$ such that

$$Y^{r/2} = \Gamma_0 Y_0^{1/2}, \tag{6.80}$$

which leads to

$$Y^{r/2} X^r Y^{r/2} \leqslant \Gamma_0 Y_0^{1/2} X_0 Y_0^{1/2} \Gamma_0^* . \tag{6.81}$$

This in turn implies that

$$\lambda_j\left(Y^{r/2} X^r Y^{r/2}\right) \leqslant \lambda_j\left(\Gamma_0 Y_0^{1/2} X_0 Y_0^{1/2} \Gamma_0^*\right) \leqslant \lambda_j\left(Y_0^{1/2} X_0 Y_0^{1/2}\right), \tag{6.82}$$

for $j = 1, 2, \ldots$, and we finally obtain

$$\left(Y^{1/2} X Y^{r/2}\right)^r \prec_{\log} Y^{r/2} X^r Y^{r/2} \prec_{\log} Y_0^{1/2} X_0 Y_0^{1/2}. \tag{6.83}$$

Now applying Lemma 6.15 we complete the proof. □

The next lemma provides a sufficient condition that allows to *lift* the strong operator convergence to convergence in the $\|\cdot\|_\phi$-topology on the symmetrically-normed ideal $\mathcal{C}_\phi(\mathcal{H})$.

Lemma 6.17. *Let $X = X^* \in \mathcal{C}_\phi(\mathcal{H})$, $Y = Y^* \in \mathcal{C}_p(\mathcal{H})$ and $Z = Z^* \in \mathcal{L}(\mathcal{H})$. If $\{Z(t)\}_{t \geqslant 0}$, $Z(t) = Z(t)^*$ is a family of bounded self-adjoint operators such that*

$$\operatorname*{s-lim}_{t \to +0} Z(t) = Z, \tag{6.84}$$

then

$$\lim_{r \to \infty} \sup_{t \in [0,\tau]} \|(Z(t/r) - Z) Y X\|_\phi = \lim_{r \to \infty} \sup_{t \in [0,\tau]} \|X Y (Z(t/r) - Z)\|_\phi = 0, \tag{6.85}$$

for any $\tau \in (0, \infty)$.

Proof. Note that by (6.84) one has s-$\lim_{r\to\infty} Z(t/r) = Z$ uniformly in $t \in [0,\tau]$. Since $Y \in \mathcal{C}_p(\mathcal{H})$, this implies

$$\lim_{r\to\infty} \sup_{t\in[0,\tau]} \|(Z(t/r) - Z)Y\| = 0. \tag{6.86}$$

By virtue of (6.34) one gets the estimate

$$\|(Z(t/r) - Z)YX\|_\phi \leqslant \|(Z(t/r) - Z)Y\| \|X\|_\phi ,$$

which in conjunction with (6.86) yields (6.85). □

Although straightforward, this line of reasoning for lifting to a stronger topology of convergence will be useful in the following.

6.4 Lifting for Trotter-Kato product formulae

This section contains some basic results about the *lifting approach* motivated by Lemma 6.17. Essentially they are statements about the $\mathcal{C}_\phi(\mathcal{H})$-norm convergence of the Trotter-Kato product formulae *conditioned* by the operator-norm convergence. Note that sufficient conditions for this lifting of the topology of convergence require that the admissible Kato functions $\mathcal{K}$ satisfy *domination* condition, see Definition 6.1.

We start by showing that it is enough to establish the convergence of the Trotter-Kato product formula for only *one* representative of the families of operators $\{F(t)\}_{t\geqslant 0}$, $\{T(t)\}_{t\geqslant 0}$, $\{f(tA)g(tB)\}_{t\geqslant 0}$, or $\{g(tB)f(tA)\}_{t\geqslant 0}$.

Recall that the generator $H = A \dot{+} B$ is the form-sum of A and B, and P_0 is the orthogonal projection $P_0 : \mathcal{H} \to \mathcal{H}_0 = \overline{\operatorname{dom} H}$.

Proposition 6.18. *Let A and B be non-negative self-adjoint operators in a Hilbert space $\mathcal{H}$ and let f, g be Kato functions of class $\mathcal{K}$. If the Trotter-Kato product formula for approximants corresponding to one representative of the families $\{F(t)\}_{t\geqslant 0}$, $\{T(t)\}_{t\geqslant 0}$, $\{f(tA)g(tB)\}_{t\geqslant 0}$, or $\{g(tB)f(tA)\}_{t\geqslant 0}$, converges locally uniformly away from $t_0 > 0$ in $\mathcal{C}_\phi(\mathcal{H})$, then the Trotter-Kato product formulae converge locally uniformly away from t_0 in $\mathcal{C}_\phi(\mathcal{H})$ for all families of approximants generated by f and g.*

Proof. We prove the chain of implications of convergence $\{F(t)\}_{t\geqslant 0} \Longrightarrow \{T(t)\}_{t\geqslant 0} \Longrightarrow \{f(tA)g(tB)\}_{t\geqslant 0} \Longrightarrow \{g(tB)f(tA)\}_{t\geqslant 0} \Longrightarrow \{F(t)\}_{t\geqslant 0}$.

$\{F(t)\}_{t\geqslant 0} \Longrightarrow \{T(t)\}_{t\geqslant 0}$. To this aim we use the representation

$$T(t/r)^r = T(t/r)^p f(tA/r)^{1/2} g(tB/r)^{1/2} F(t/r)^{n-1} g(tB/r)^{1/2} f(tA/r)^{1/2}. \tag{6.87}$$

for $t \geqslant 0$ and $r \geqslant 1$, where $r := n + p$, $n = 1, 2 \ldots$, $0 \leqslant p < 1$. Since

$$F(t/r)^{n-1} = F(t_n/(n-1))^{n-1}, \tag{6.88}$$

where $t_n := t(n-1)/(n+p) \leqslant \tau_0'$, we obtain from (6.87) that $T(t/r)^r \in \mathcal{C}_\phi(\mathcal{H})$ if $t_n \in [\tau_0', \tau] \subset (t_0, \infty)$. Since $t_n \in [\tau_0', \tau] \subset (t_0, \infty)$ is equivalent to $t \in [(1 + \frac{p+1}{n-1})\tau_0', (1 + \frac{p+1}{n-1})\tau]$, we find that for $[\tau_0, \tau] \subseteq (\tau_0', \tau]$ and for sufficiently large r one has $[\tau_0, \tau] \subseteq [(1 + \frac{p+1}{n-1})\tau_0', (1 + \frac{p+1}{n-1})\tau]$ and $T(t/r)^r \in \mathcal{C}_\phi(\mathcal{H})$ for $t \in [\tau_0, \tau]$.

To prove that the Trotter-Kato product formula for $\{T(t)\}_{t\geqslant 0}$ converges locally uniformly away from t_0 in $\mathcal{C}_\phi(\mathcal{H})$ we note that this kind of convergence for $\{F(t)\}_{t\geqslant 0}$ implies

$$\lim_{n\to\infty} \sup_{t\in[\tau_0', \tau]} \|F(t/(n-1))^{n-1} - \mathrm{e}^{-tH} P_0\|_\phi = 0. \tag{6.89}$$

Replacing t by t_n we get

$$\lim_{n\to\infty} \sup_{t_n\in[\tau_0', \tau]} \|F(t_n/(n-1))^{n-1} - \mathrm{e}^{-tH} P_0\|_\phi = 0, \tag{6.90}$$

and since $t_n \in [\tau_0', \tau]$ for $t \in [\tau_0, \tau]$, we obtain

$$\lim_{n\to\infty} \sup_{t\in[\tau_0, \tau]} \|F(t/r)^{n-1} - \mathrm{e}^{-tH} P_0\|_\phi = 0. \tag{6.91}$$

Taking into account the representation

$$\begin{aligned}
&T(t/r)^r - \mathrm{e}^{-tH} P_0 \qquad (6.92)\\
&= T(t/r)^p f(tA/r)^{1/2} g(tB/r)^{1/2}\\
&\quad \times \{F(t/r)^{n-1} - \mathrm{e}^{-tH} P_0\} g(tB/r)^{1/2} f(tA/r)^{1/2}\\
&\quad + T(t/r)^p f(tA/r)^{1/2} g(tB/r)^{1/2} \mathrm{e}^{-tH} P_0 \{g(tB/r)^{1/2} - \mathbb{1}\} f(tA/r)^{1/2}\\
&\quad + T(t/r)^p f(tA/r)^{1/2} g(tB/r)^{1/2} \mathrm{e}^{-tH} P_0 \{f(tA/r)^{1/2} - \mathbb{1}\}\\
&\quad + T(t/r)^p f(tA/r)^{1/2} \{g(tB/r)^{1/2} - \mathbb{1}\} \mathrm{e}^{-tH} P_0\\
&\quad + T(t/r)^p \{f(tA/r)^{1/2} - \mathbb{1}\} \mathrm{e}^{-tH} P_0\\
&\quad + T(t/r)^p \{f(tA/r)^{1/2} - \mathbb{1}\} \mathrm{e}^{-tH} P_0 + \{T(t/r)^p - \mathbb{1}\} \mathrm{e}^{-tH} P_0\,,
\end{aligned}$$

we obtain the estimate

$$\begin{aligned}
&\|T(t/r)^r - \mathrm{e}^{-tH} P_0\|_\phi \qquad (6.93)\\
&\leqslant \|F(t/r)^{n-1} - \mathrm{e}^{-tH} P_0\|_\phi + 2\|\{\mathbb{1} - g(tB/r)^{1/2}\} \mathrm{e}^{-tH} P_0\|_\phi\\
&\quad + 2\|\{\mathbb{1} - f(tA/r)^{1/2}\} \mathrm{e}^{-tH} P_0\|_\phi + \|\{\mathbb{1} - T(t/r)^p\} \mathrm{e}^{-tH} P_0\|_\phi\,.
\end{aligned}$$

Since $\text{s-lim}_{t\to+0} f(tA)^{1/2} = \mathbb{1}$, $\text{s-lim}_{t\to+0} g(tB)^{1/2} = \mathbb{1}$ and $\text{s-lim}_{t\to+0} T(t) = \mathbb{1}$, Lemma 6.17 shows that

$$\begin{aligned}
&\lim_{r\to\infty} \sup_{t\in[\tau_0, \tau]} \|\{\mathbb{1} - g(tB/r)^{1/2}\} \mathrm{e}^{-tH} P_0\|_\phi \qquad (6.94)\\
&= \lim_{r\to\infty} \sup_{t\in[\tau_0, \tau]} \|\{\mathbb{1} - f(tA/r)^{1/2}\} \mathrm{e}^{-tH} P_0\|_\phi = 0\,,
\end{aligned}$$

and

$$\begin{aligned}&\lim_{r\to+\infty}\sup_{t\in[\tau_0,\tau]}\|\{\mathbb{1}-T(t/r)^p\}\mathrm{e}^{-tH}P_0\|_\phi \\ &\leqslant \lim_{r\to+\infty}\sup_{t\in[\tau_0,\tau]}\|\{\mathbb{1}-T(t/r)\}\mathrm{e}^{-tH}P_0\|_\phi = 0\,.\end{aligned} \tag{6.95}$$

Combining (6.91), (6.93), (6.94) and (6.95), one gets the proof of the locally uniform convergence away from t_0 of the Trotter-Kato product formula for $\{T(t)\}_{t\geqslant 0}$.

$\{T(t)\}_{t\geqslant 0} \Longrightarrow \{f(tA)g(tB)\}_{t\geqslant 0}$. Using the representation

$$(f(tA/n)g(tB/n))^n = f(tA/n)^{1/2}T(t/n)^{n-1}f(tA/n)^{1/2}g(tB/n), \tag{6.96}$$

and arguing as in the proof that $\{F(t)\}_{t\geqslant 0} \Longrightarrow \{T(t)\}_{t\geqslant 0}$, one verifies that the Trotter-Kato product formula converges locally uniformly away from $t_0 > 0$ in $\mathcal{C}_\phi(\mathcal{H})$.

$\{f(tA)g(tB)\}_{t\geqslant 0} \Longrightarrow \{g(tB)f(tA)\}_{t\geqslant 0}$. The proof is based on the representation

$$(g(tB/n)f(tA/n))^n = g(tB/n)(f(tA/n)g(tB/n))^{n-1}f(tA/n), \tag{6.97}$$

and the same argument.

$\{g(tB)f(tA)\}_{t\geqslant 0} \Longrightarrow \{F(t)\}_{t\geqslant 0}$. The assertion is a consequence of the representation

$$F(t/r)^r = F(t/r)^p g(tA/r)^{1/2}f(tA/r)(g(tB/r)f(tA/r))^{n-1}g(tB/r)^{1/2}, \tag{6.98}$$

where $r = n + p$, $n = 1, 2, \ldots$, $0 \leqslant p < 1$, arguing again as above. □

Proposition 6.18 shows that it is enough to consider and to prove convergence of the Trotter-Kato product formulae in the $\|\cdot\|_\phi$-norm only for one family, for example, for the family $\{F(t)\}_{t\geqslant 0}$.

Proposition 6.19. *Let A and B be non-negative self-adjoint operators in a Hilbert space $\mathcal{H}$ and let f, g be Kato functions in $\mathcal{K}$. If*

(a) *the Trotter-Kato product formula for $\{F(t)\}_{t\geqslant 0}$ and for given f, g converges locally uniformly away from zero in the operator-norm;*

(b) *the family of the approximants $\{F(t/r)\}^r$ is bounded locally uniformly away from some $t_0 > 0$ in $\mathcal{C}_\phi(\mathcal{H})$;*

(c) *the semigroup $\mathrm{e}^{-tH} \in \mathcal{C}_\phi(\mathcal{H}_0)$, for $t > t_0$,*

then the Trotter-Kato product formulae converge locally uniformly away from t_0 in $\mathcal{C}_\phi(\mathcal{H})$ for all families generated by the Kato functions f and g.

Proof. If $t \in [\tau_0, \tau] \subset (t_0, \infty)$, then there is an $\alpha \in (0,1)$ such that $\tau_0' := \alpha\tau_0 > t_0$ and $\alpha t \in [\tau_0', \tau] \subset (t_0, \infty)$. Since

$$F(t/r)^{\alpha r} = F(\alpha t/\alpha r)^{\alpha r}, \quad t \geqslant 0, \tag{6.99}$$

there is a $r_0 \geqslant 1$ such that $F(t/r_0)^{\alpha r_0} \in \mathcal{C}_\phi(\mathcal{H})$ and

$$\|F(t/r)^{\alpha r}\|_\phi \leqslant M([\tau_0', \tau]), \tag{6.100}$$

for $t \in [\tau_0, \tau]$ and $r \geqslant r_0$, see (b).

Since $\mathrm{e}^{-tH} \in \mathcal{C}_\phi(\mathcal{H})$ for $t > t_0$ (c), the operator e^{-tH} is compact for $t > 0$, which yields that the spectrum of H is discrete with the only accumulation point at infinity. Denoting by $E_H(\cdot)$ the spectral measure of H, we consider the decomposition

$$\begin{aligned} \mathrm{e}^{-tH} &= \mathrm{e}^{-tH} E_H([0,N)) + \mathrm{e}^{-tH} E_H([N,\infty)), \\ N &= 1,2,\ldots\,, \ t \geqslant 0\,. \end{aligned} \tag{6.101}$$

Since the spectrum of H is discrete, the projection $E_H([0,N))$ is a finite-dimensional operator for each $N = 1,2,\ldots$. Hence for any $N = 1,2,\ldots$ one gets

$$\lim_{r\to\infty} \sup_{t\in[0,\tau]} \|E_H([0,N))P_0(F(t/r)^{\alpha r} - \mathrm{e}^{-\alpha t H})\|_\phi = 0. \tag{6.102}$$

Since $HE_H([N,\infty)) \geqslant NE_H([N,\infty))$, we find that

$$\|\mathrm{e}^{-(1-\alpha)tH} E_H([N,\infty))\| \leqslant \mathrm{e}^{-(1-\alpha)\tau_0 N}, \quad N = 1,2,\ldots, \tag{6.103}$$

for $t \in [\tau_0, \tau]$. Taking into account the estimate

$$\begin{aligned} &\|\mathrm{e}^{-(1-\alpha)tH} E_H([0,N))(F(t/r)^{\alpha r} - \mathrm{e}^{-\alpha t H} P_0)\|_\phi \\ &\leqslant \|\mathrm{e}^{-(1-\alpha)tH} E_H([N,\infty))\| \left\{\|F(t/r)^{\alpha r}\|_\phi + \|\mathrm{e}^{-\alpha t H} P_0\|_\phi\right\}, \end{aligned} \tag{6.104}$$

as well as (6.100) and (6.103) we finally obtain

$$\begin{aligned} &\|\mathrm{e}^{-(1-\alpha)tH} E_H([N,\infty))(F(t/r)^{\alpha r} - \mathrm{e}^{-\alpha t H} P_0)\|_\phi \\ &\leqslant \mathrm{e}^{-(1-\alpha)\tau_0 N} \left\{M([\tau_0', \tau]) + \|\mathrm{e}^{-\alpha\tau_0 H}\|_\phi\right\}, \end{aligned} \tag{6.105}$$

for $N = 1,2,\ldots$, $t \in [\tau_0, \tau] \subseteq (\tau_0', \tau]$ and $r \geqslant r_0$.

Using the representation

$$\begin{aligned} F(t/r)^r - \mathrm{e}^{-tH} P_0 = &\left(F(t/r)^{(1-\alpha)r} - \mathrm{e}^{-(1-\alpha)tH} P_0\right) F(t/r)^{\alpha r} \\ &+ \mathrm{e}^{-(1-\alpha)tH} E_H([0,N)) P_0 \left(F(t/r)^{\alpha r} - \mathrm{e}^{-\alpha t H} P_0\right) \\ &+ \mathrm{e}^{-(1-\alpha)tH} E_H([N,\infty)) P_0 \left(F(t/r)^{\alpha r} - \mathrm{e}^{-\alpha t H} P_0\right)\,, \end{aligned} \tag{6.106}$$

we argue as in the proof of Lemma 6.17. From (6.34), (6.100), and (6.105) we obtain the estimate

$$\begin{aligned} &\sup_{t\in[\tau_0,\tau]} \|F(t/r)^r - \mathrm{e}^{-tH}P_0\|_\phi \\ &\leqslant M([\tau_0',\tau]) \sup_{t\in[\tau_0,\tau]} \|F(t/r)^{(1-\alpha)r} - \mathrm{e}^{-(1-\alpha)tH}P_0\| \\ &\quad + \sup_{t\in[\tau_0,\tau]} \|E_H([0,N))P_0(F(t/r)^{\alpha r} - \mathrm{e}^{-\alpha tH}P_0)\|_\phi \\ &\quad + \mathrm{e}^{-(1-\alpha)\tau_0 N}\left\{M([\tau_0',\tau]) + \|\mathrm{e}^{-\alpha\tau_0 H}\|_\phi\right\}, \end{aligned} \tag{6.107}$$

for $N = 1,2,\ldots$, $t \in [\tau_0,\tau]$ and $r \geqslant r_0$. For any $\epsilon > 0$ we find an integer $N = 1,2,\ldots$ such that

$$\mathrm{e}^{-(1-\alpha)\tau_0 N}\left\{M([\tau_0',\tau]) + \|\mathrm{e}^{-\alpha\tau_0 H}\|_\phi\right\} \leqslant \epsilon\,. \tag{6.108}$$

Then we fix this integer N. Since for given f, g the Trotter-Kato product formula converges locally uniformly away from zero in the *operator norm* (a), we obtain for sufficiently large r that

$$M([\tau_0',\tau]) \sup_{t\in[\tau_0,\tau]} \|F(t/r)^{(1-\alpha)r} - \mathrm{e}^{-(1-\alpha)tH}P_0\| \leqslant \epsilon\,, \tag{6.109}$$

and

$$\sup_{t\in[\tau_0,\tau]} \|E_H([0,N))P_0(F(t/r)^{\alpha r} - \mathrm{e}^{-\alpha tH}P_0)\|_\phi \leqslant \epsilon\,. \tag{6.110}$$

Hence, for any $\epsilon > 0$ we get

$$\sup_{t\in[\tau_0,\tau]} \|F(t/r)^r - \mathrm{e}^{-tH}P_0\|_\phi \leqslant 3\epsilon\,, \tag{6.111}$$

for the integer N, which is fixed above, and sufficiently large r. Therefore, for $\{F(t)\}_{t\geqslant 0}$ the Trotter-Kato product formula converges locally uniformly away from $t_0 > 0$ in $\mathcal{C}_\phi(\mathcal{H})$. Applying Proposition 6.18 we complete the proof. □

To verify that the family of approximants $\{F(t/r)^r\}_{r\geqslant 1}$ is locally uniformly bounded away from $t_0 > 0$ in $\mathcal{C}_\phi(\mathcal{H})$ we need the following lemma.

Lemma 6.20. *Let A and B be non-negative self-adjoint operators in a Hilbert space $\mathcal{H}$ and let $f^D : \mathbb{R}_0^+ \to \mathbb{R}_0^+$ and $g^D : \mathbb{R}_0^+ \to \mathbb{R}_0^+$ be bounded Borel measurable functions such that $F^D(t_0)^p \in \mathcal{C}_\phi(\mathcal{H})$ for $t_0 > 0$ and some integer $p \geqslant 1$. If the Kato functions f and g are dominated by, respectively, f^D and g^D, then*

$$F(t/r)^r \in \mathcal{C}_\phi(\mathcal{H}), \tag{6.112}$$

and

$$\|F(t/r)^r\|_\phi \leqslant \|F^D(t_0)^p\|_\phi\,, \tag{6.113}$$

for $pt_0 \leqslant t \leqslant prt_0$ and $r \geqslant p$.

Proof. Since f and g are dominated by f^D and g^D, the inequality (6.7) yields

$$f(tA/r)^r \leqslant f(tA/r)^{rt_0/t} \leqslant f^D(t_0 A), \tag{6.114}$$

and

$$g(tB/r)^r \leqslant g(tB/r)^{rt_0/t} \leqslant g^D(t_0 B), \tag{6.115}$$

for $t_0 \leqslant t \leqslant rt_0$ and $r \geqslant 1$. Setting $X_0 := f^D(t_0 A)$ and $Y_0 := g^D(t_0 B)$ we obtain from Lemma 6.16 that

$$F(t/r)^r \prec_{\log} F^D(t_0), \tag{6.116}$$

for $t_0 \leqslant t \leqslant rt_0$ and $r \geqslant 1$. Thus,

$$F(t/r)^{rp} \prec_{\log} F^D(t_0)^p, \tag{6.117}$$

which yields $F(t/r)^{rp} \in \mathcal{C}_\phi(\mathcal{H})$ and by Lemma 6.15

$$\|F(t/r)^{rp}\|_\phi \leqslant \|F(t_0)^p\|_\phi\,, \tag{6.118}$$

for $t_0 \leqslant t \leqslant rt_0$ and $r \geqslant 1$. Since $F(t/r)^{rp} = F(pt/rp)^{rp}$, we get (6.112) and (6.113) for $pt_0 \leqslant t \leqslant prt_0$ and $r \geqslant p$. □

From Lemma 6.20 we immediately see that for any interval $[\tau_0, \tau] \subseteq (pt_0, +\infty)$ there exist an $r_0 \geqslant 1$ such that for $r \geqslant r_0$ and $t \in [\tau_0, \tau]$ the conditions $F(t/r)^r \in \mathcal{C}_\phi(\mathcal{H})$ and (6.6) are satisfied. This means that the sequence $\{F(t/r)^r\}_{r\geqslant 1}$ is locally uniformly bounded away from pt_0 in $\mathcal{C}_\phi(\mathcal{H})$. Moreover, one gets that $M([\tau_0, \tau]) \leqslant \|F^D(t_0)^p\|_\phi$ for sufficiently large $r \geqslant 1$ and any interval $[\tau_0, \tau] \subseteq (pt_0, +\infty)$.

Next, we show that under the assumptions of Lemma 6.20 one has $\mathrm{e}^{-tH} \in \mathcal{C}_\phi(\mathcal{H}_0)$ for $t > pt_0$. To this end we note that if the Kato function f is dominated by f^D, then, by (6.7),

$$f(x/r)^r \leqslant f^D(x), \quad r \geqslant 1, \tag{6.119}$$

for $x \geqslant 0$. Since $\lim_{r\to\infty} f(x/r)^r = \mathrm{e}^{-x}$ for each $x \geqslant 0$, we find that

$$\mathrm{e}^{-x} \leqslant f^D(x), \tag{6.120}$$

for $x \geqslant 0$. In other words, if f^D dominates the Kato function f, then it also dominates the *exponential* Kato function $\hat{f}(x) = \mathrm{e}^{-x}$, $x \geqslant 0$, cf. Appendix C.

Lemma 6.21. *Let A and B be non-negative self-adjoint operators in a Hilbert space $\mathcal{H}$, and let $f^D : \mathbb{R}_0^+ \to \mathbb{R}_0^+$ and $g^D : \mathbb{R}_0^+ \to \mathbb{R}_0^+$ be bounded Borel measurable functions such that $F^D(t_0)^p \in \mathcal{C}_\phi(\mathcal{H})$ for some $t_0 > 0$ and some integer $p \geqslant 1$. If the Kato functions f and g are dominated by f^D and g^D, respectively, then*

$$\mathrm{e}^{-tH} \in \mathcal{C}_\phi(\mathcal{H}_0), \tag{6.121}$$

and

$$\|\mathrm{e}^{-tH} P_0\|_\phi \leqslant \|F^D(t_0)^p\|_\phi\,, \tag{6.122}$$

for $t \geqslant pt_0$.

Proof. Since $\hat{f}(x) = \hat{g}(x) = \mathrm{e}^{-x}$, $x \geqslant 0$, are dominated by f^D and g^D, respectively, we find by Lemma 6.20 that

$$\hat{F}(t/r)^r \in \mathcal{C}_\phi(\mathcal{H}) \tag{6.123}$$

and

$$\|\hat{F}(t/r)^r\|_\phi \leqslant \|F^D(t_0)^p\|_\phi\,, \quad r \geqslant 1, \tag{6.124}$$

for $pt_0 \leqslant t \leqslant prt_0$ and $r \geqslant p$, where

$$\hat{F}(t) := \mathrm{e}^{-tB/2}\mathrm{e}^{-tA}\mathrm{e}^{-tB/2}, \quad t \geqslant 0. \tag{6.125}$$

Using the Araki inequality (Appendix B) and formula (3.20) one obtains that

$$\mathrm{e}^{-tH}P_0 \prec_{\log} \hat{F}(t/n)^n, \quad n = 1, 2, \dots\,, \tag{6.126}$$

for $t > 0$, where the left-hand side is the operator-norm limit of Proposition 6.19(a). Then by (6.123), (6.124), and Lemma 6.15 we complete the proof of (6.121) and (6.122). □

Now, by virtue of Propositions 6.18, 6.19 and of Lemmata 6.20, 6.21 we obtain the following result:

Proposition 6.22. *Let A and B be non-negative self-adjoint operators in a Hilbert space $\mathcal{H}$ and let $f^D : \mathbb{R}_0^+ \to \mathbb{R}_0^+$ and $g^D : \mathbb{R}_0^+ \to \mathbb{R}_0^+$ be bounded Borel measurable functions such that $F^D(t_0)^p \in \mathcal{C}_\phi(\mathcal{H})$ for some $t_0 > 0$ and for some integer $p \geqslant 1$. Suppose*

(a) *the Kato functions $f, g \in \mathcal{K}$ are dominated by f^D,g^D, respectively;*

(b) *the Trotter-Kato product formulae converge locally uniformly away from zero in the operator norm for the family $\{F(t)\}_{t\geqslant 0}$.*

Then the Trotter-Kato product formulae converge locally uniformly away from pt_0 in $\mathcal{C}_\phi(\mathcal{H})$ for all families of approximants generated by Kato functions f and g.

Proof. By Lemma 6.21, we obtain that $\mathrm{e}^{-tH}P_0 \in \mathcal{C}_\phi(\mathcal{H})$ for $t > pt_0$. Lemma 6.20 implies that for each bounded interval $[\tau_0, \tau] \subseteq (pt_0, \infty)$ there is an $r_0 \geqslant 1$ such that the approximants $F(t/r)^r \in \mathcal{C}_\phi(\mathcal{H})$ for $r \geqslant r_0$ and $t \in [\tau_0, \tau]$. By (6.113) one has the upper bound $\sup_{r\geqslant r_0} \sup_{t\in[\tau_0,\tau]} \|F(t/r)^r\|_\phi \leqslant \|F^D(t_0)\|_\phi < \infty$, which ensures that the family $\{F(t/r)^r\}_{r\geqslant 1}$ is locally uniformly bounded away from pt_0 in $\mathcal{C}_\phi(\mathcal{H})$. Then applying Proposition 6.19 we obtain that the Trotter-Kato product formulae for $\{F(t)\}_{t\geqslant 0}$ converge locally uniformly away from pt_0 in $\mathcal{C}_\phi(\mathcal{H})$. Finally, using Proposition 6.18 we conclude that this holds for all families generated by f and g, which are dominated by f^D and g^D. □

6.5 Product formulae in $\mathcal{C}_\phi(\mathcal{H})$-ideals

In this section we study conditions that ensure the $\mathcal{C}_\phi(\mathcal{H})$-norm convergence of the Trotter-Kato product formulae *without* reference to the operator-norm convergence. As we established in Sections 5.3, 5.4, and 6.4 the properties of the Kato functions are crucial for controlling the convergence. Thus, one has first to specify a set of admissible Kato functions f, g from the generic class $\mathcal{K}$.

In the next statement we impose on the pair f, g the *regularity* condition on f, keeping $g \in \mathcal{K}$ *arbitrary*, see Appendix C, Section C.3.

Proposition 6.23. *Let A and B be non-negative self-adjoint operators in a Hilbert space $\mathcal{H}$ and let $f^D : \mathbb{R}_0^+ \to \mathbb{R}_0^+$ be a Borel measurable function such that $f^D(t_0 A) \in \mathcal{C}_\phi(\mathcal{H})$ for some $t_0 > 0$. If the Kato function f is regular and dominated by f^D, and $g \in \mathcal{K}$ is an arbitrary Kato function, then the Trotter-Kato product formulae converge locally uniformly away from t_0 in $\mathcal{C}_\phi(\mathcal{H})$ for all families of approximants generated by f, g.*

Proof. Since for any dominating function f^D one has $e^{-x} \leqslant f^D(x)$, $x \geqslant 0$ (6.120), the condition $f^D(t_0 A) \in \mathcal{C}_\phi(\mathcal{H})$ yields $e^{-t_0 A} \in \mathcal{C}_\phi(\mathcal{H})$, $t_0 > 0$, which evidently implies $(\mathbb{1} + t_0 A)^{-1} \in \mathcal{C}_\infty(\mathcal{H})$. By Proposition 5.45 (or Proposition 5.36), this last observation together with the regularity of f ensure that the Trotter-Kato product formulae converge locally uniformly away from zero in the *operator-norm* topology.

Note that any Kato function g is evidently dominated by $g^D(x) = 1$. Then condition $f^D(t_0 A) \in \mathcal{C}_\phi(\mathcal{H})$ yields $F^D(t_0)^p \in \mathcal{C}_\phi(\mathcal{H})$ for any $p \geqslant 1$ and $t_0 > 0$. Consequently, by the *lifting* Proposition 6.22, the Trotter-Kato product formulae converge locally uniformly in $\mathcal{C}_\phi(\mathcal{H})$ away from some $t_0 > 0$ for all families of the approximants generated by the Kato functions f, g from the statement of the proposition. □

To continue with the next assertion we first give a useful characterisation of the *self-dominated* Kato functions, see Definition 6.2 and Section C.3.

Lemma 6.24. *A Borel function $f \in \mathcal{K}$ is a self-dominated Kato function if and only if there is a non-increasing function $h : \mathbb{R}_0^+ \to \mathbb{R}_0^+$, satisfying*

$$\lim_{x \to +0} h(x) = 1, \tag{6.127}$$

such that f admits the representation

$$f(x) = e^{-x h(x)}, \qquad x \geqslant 0. \tag{6.128}$$

The function h is unique.

Proof. Let f be a self-dominated Kato function and let $t := qx$, $x \geqslant 0$ for $0 < q \leqslant 1$, which yields $0 \leqslant t \leqslant x$. Then by (6.9) we obtain that

$$f(t) \leqslant f(x)^{t/x}, \quad 0 \leqslant t \leqslant x, \ x > 0, \tag{6.129}$$

whence

$$f(t)^{1/t} \leqslant f(x)^{1/x}, \quad 0 < t \leqslant x. \tag{6.130}$$

If $f(x) = 0$ for some $x > 0$, then one gets $f(t) = 0$ for $0 < t \leqslant x$, which is impossible by (5.12) in Definition 5.4. Therefore, $f(x) \neq 0$ and one can define

$$h(x) := -\frac{1}{x} \ln(f(x)) \geqslant 0, \quad x > 0. \tag{6.131}$$

Taking into account (6.130) we get

$$h(t) \geqslant h(x),$$

for $0 < t \leqslant x$. Thus h is a non-negative and non-increasing function.

Note that $f(x) \neq 1$ in a small vicinity of $x = 0$. Otherwise, there would exist a sequence $\{x_n > 0\}_{n=1}^{\infty}$ such that $\lim_{n\to\infty} x_n = 0$ and $f(x_n) = 1$. This is impossible because of (5.12), since then one would get

$$f'(+0) = \lim_{n\to\infty} \frac{f(x_n) - 1}{x_n} = 0\,,$$

instead of $f'(+0) = -1$. Therefore, the representation

$$h(x) = -\frac{1 - f(x)}{x} \ln\left(f(x)^{1/(1-f(x))}\right), \tag{6.132}$$

makes sense for sufficiently small $x > 0$. Since

$$\lim_{y\to 1-0} y^{1/(1-y)} = e^{-1},$$

by (6.132) and $f'(+0) = -1$ one gets (6.127). Now the representation (6.128) follows from the definition (6.131).

Conversely, let f be given by (6.128), where h is a non-increasing function which satisfies (6.127). Since h is non-increasing, we get for $x \geqslant 0$ and $0 < q \leqslant 1$ that

$$f(qx) = \mathrm{e}^{-qxh(qx)} \leqslant \mathrm{e}^{-qxh(x)} = f(x)^q, \tag{6.133}$$

which verifies (6.9). Furthermore, by (6.127) one gets $f(0) = 1$ and

$$\frac{f(x) - 1}{x} = \sum_{n=1}^{\infty} (-1)^n x^{n-1} h(x)^n \frac{1}{n!}, \quad x > 0, \tag{6.134}$$

which leads to the estimate

$$\left|\frac{1 - f(x)}{x} + h(x)\right| \leqslant \sum_{n=2}^{\infty} x^{n-1} \frac{1}{n!}, \quad x > 0. \tag{6.135}$$

Then by (6.127) we immediately obtain $f'(+0) = -1$. Thus (6.128) defines a self-dominated Kato function. The uniqueness of h is obvious by construction. □

Corollary 6.25. *If f is a self-dominated Kato function, then*

$$f_0(x) \leqslant \frac{1}{1-\ln(f(x))}, \tag{6.136}$$

for $x \geqslant 0$, where f_0 is defined by (5.102) *and* (5.104).

Proof. By definition (5.102) and representation (6.128) one has

$$\varphi_0(x) = \inf_{0<t\leqslant x} \frac{1}{t}\left(\frac{1}{f(t)} - 1\right) = \inf_{0<t\leqslant x} \frac{1}{t}(\mathrm{e}^{th(t)} - 1), \tag{6.137}$$

that yields

$$f_0(x) = \frac{1}{1+x\varphi_0(x)} \leqslant \frac{1}{1+xh(x)} = \frac{1}{1-\ln(f(x))}, \tag{6.138}$$

for $x \geqslant 0$. □

Lemma 6.26. *Let A be a non-negative self-adjoint operator in a Hilbert space $\mathcal{H}$ and let f be a self-dominated Kato function. If $f(t_0A) \in \mathcal{C}_\phi(\mathcal{H})$ for some $t_0 > 0$, then*

$$f_0(t_0A) \in \mathcal{C}_\infty(\mathcal{H}), \tag{6.139}$$

where f_0 is defined by (5.102) *and* (5.104).

Proof. Since $f(t_0A) \in \mathcal{C}_\phi(\mathcal{H})$ implies $f(t_0A) \in \mathcal{C}_\infty(\mathcal{H})$, where $0 \leqslant f(t_0A) \leqslant \mathbb{1}$, we see that the positive operator

$$K := \frac{\mathbb{1}}{\mathbb{1} + \ln(\mathbb{1}/f(t_0A))} \in \mathcal{C}_\infty(\mathcal{H}), \tag{6.140}$$

is also compact. By Corollary 6.25, the positive bounded operator $f_0(t_0A)$ is dominated by the compact operator (6.140):

$$f_0(t_0A) \leqslant (\mathbb{1} - \ln(f(t_0A)))^{-1} = K.$$

Hence, $f_0(t_0A)$ is in turn a compact operator (see Section 2.4), which yields (6.139). □

Besides the representation established in Lemma 6.24, the following property of the Kato functions is indispensable.

Lemma 6.27. *Let the function $f \in \mathcal{K}$ obey the inequality $f(x) \leqslant d(x)$ for a Borel function $d(x)$ and $x \geqslant 0$. If d is a self-dominated Kato function, then in turn the function f:*

(1) *is dominated by d;*

(2) *satisfies the condition*

$$C_0 := \sup_{x>0} \frac{xf(x)}{1-f(x)} < +\infty. \tag{6.141}$$

Proof. (1) Since $f(x) \leqslant d(x)$, by (6.9) for the self-dominated Kato function d one gets

$$f(qx)^{1/q} \leqslant d(qx)^{1/q} \leqslant d(x), \quad 0 < q \leqslant 1, \ x \geqslant 0. \tag{6.142}$$

Hence, the function f is dominated by d in the sense of Definition 6.1.

(2) Since $f(x) \leqslant d(x)$ and the self-dominated Kato function d satisfies (6.10) for some constant $C_{0,d}$, we obtain the estimates:

$$C_0 = \sup_{x>0} \frac{x\, f(x)}{1 - f(x)} \leqslant \sup_{x>0} \frac{x\, d(x)}{1 - d(x)} =: C_{0,d} < \infty. \tag{6.143}$$

This proves (6.141) and the lemma. □

Note that similarly to Proposition 6.23, in the next assertion we consider the case of an *arbitrary* $g \in \mathcal{K}$. However, instead of regularity of f, we study Kato functions that are bounded from above by some self-dominated functions. Although these conditions sound very different from regularity, they lead to the same conclusion as Proposition 6.23.

Proposition 6.28. *Let A and B be non-negative self-adjoint operators in a Hilbert space $\mathcal{H}$ and let d be a self-dominated Kato function such that $d(t_0 A) \in C_\phi(\mathcal{H})$ for some $t_0 > 0$. Then for any Kato function f obeying $f(x) \leqslant d(x)$, $x \geqslant 0$ and for arbitrary $g \in \mathcal{K}$ the Trotter-Kato product formulae converge in $C_\phi(\mathcal{H})$ locally uniformly away from t_0 for all approximants generated by f and g.*

Proof. By Lemma 6.27, the inequality $f(x) \leqslant d(x)$ implies that f is *dominated* by the self-dominated Kato function d: $f(qx)^{1/q} \leqslant d(x) = f^D(x)$, $0 < q \leqslant 1$.

On the other hand, by $f(x) \leqslant d(x)$ and by Definition 5.34, see (5.102), one gets

$$\varphi_0(x) = \inf_{0<t\leqslant x} \frac{1}{t}\left(\frac{1}{f(t)} - 1\right) \geqslant \inf_{0<t\leqslant x} \frac{1}{t}\left(\frac{1}{d(t)} - 1\right) = \varphi_{0,d}(x),$$

which in turn by definition (5.104) implies that

$$f_0(x) = \frac{1}{1 + x\varphi_0(x)} \leqslant \frac{1}{1 + x\varphi_{0,d}(x)} = f_{0,d}(x). \tag{6.144}$$

By Lemma 6.26, for a self-dominated Kato function $d \in C_\phi(\mathcal{H})$ we obtain compactness: $f_{0,d}(t_0 A) \in C_\infty(\mathcal{H})$ and then by (6.144), the inequalities

$$0 \leqslant f_0(t_0 A) \leqslant f_{0,d}(t_0 A) \in C_\infty(\mathcal{H}).$$

Therefore, the self-adjoint operator $f_0(t_0 A)$ is also compact.

Since $f_0(t_0 A) \in C_\infty(\mathcal{H})$, Proposition 5.36 ensures that the Trotter-Kato product formula for the approximantes $\{(f(tA/n)g(tB/n))^n\}_{n\geqslant 1}$ converges in the operator-norm topology locally uniformly away from zero ($t_0 = 0$) to the degenerate semigroup $\{e^{-tH} P_0\}_{t>0}$.

Recall that any Kato function g is dominated by $g^D(x) = 1$, Appendix C. Then condition $d(t_0A) \in \mathcal{C}_\phi(\mathcal{H})$ yields that for $F^D(t) = g^D(tB)^{1/2} f^D(tA) g^D(tB)^{1/2} = d(tA)$ and any $p \geqslant 1$, $t_0 > 0$, one has $F^D(t_0)^p \in \mathcal{C}_\phi(\mathcal{H})$. Therefore, by the *lifting* Proposition 6.22, the Trotter-Kato product formulae converge in the topology of $\mathcal{C}_\phi(\mathcal{H})$ locally uniformly away from some $t_0 > 0$ for all families of the product approximants generated by the Kato functions f, g verifying the conditions of the proposition. □

Until now no *additional* conditions were imposed on the Kato function g. But under certain conditions, there are new formulations implying the convergence of the Trotter-Kato product formulae in the ideal $\mathcal{C}_\phi(\mathcal{H})$.

Proposition 6.29. *Let A and B be non-negative self-adjoint operators in a Hilbert space $\mathcal{H}$ and let $f^D : \mathbb{R}_0^+ \to \mathbb{R}_0^+$ and $g^D : \mathbb{R}_0^+ \to \mathbb{R}_0^+$ be Borel measurable functions such that $F^D(t_0) \in \mathcal{C}_\phi(\mathcal{H})$ for some $t_0 > 0$. If the Kato functions f and g are dominated by f^D and g^D, respectively, and if they satisfy conditions* (6.10)*:*

$$C_0 := \sup_{x>0} \frac{xf(x)}{1-f(x)} < +\infty \quad \textit{and} \quad S_0 := \sup_{x>0} \frac{xg(x)}{1-g(x)} < +\infty,$$

then the Trotter-Kato product formulae converge in $\mathcal{C}_\phi(\mathcal{H})$ locally uniformly away from t_0, for all approximants generated by f and g.

Proof. First we check that under hypothesis of the proposition,

$$(\mathbb{1} + A)^{-1}(\mathbb{1} + B)^{-1} \in \mathcal{C}_\infty(\mathcal{H}). \tag{6.145}$$

Arguing as in the proof of Lemma 6.21 we find (in accordance with (6.123)) that

$$e^{-t_0B/2} e^{-t_0A} e^{-t_0B/2} \in \mathcal{C}_\infty(\mathcal{H}). \tag{6.146}$$

Setting $C := e^{-t_0A/2} e^{-t_0B/2}$ we get $C^*C \in \mathcal{C}_\infty(\mathcal{H})$, which yields

$$e^{-t_0A/2} e^{-t_0B/2} \in \mathcal{C}_\infty(\mathcal{H}) \quad \text{and} \quad e^{-t_0B/2} e^{-t_0A/2} e^{-t_0B/2} \in \mathcal{C}_\infty(\mathcal{H}).$$

Iterating, we obtain

$$e^{-t_0A/2^m} e^{-t_0B/2} \in \mathcal{C}_\infty(\mathcal{H}), \tag{6.147}$$

for $m = 1, 2, \ldots$. This evidently gives

$$e^{-tA} e^{-t_0B/2} \in \mathcal{C}_\infty(\mathcal{H}), \tag{6.148}$$

for any $t > 0$. Similarly, we prove that

$$e^{-tA} e^{-t_0B/2^k} \in \mathcal{C}_\infty(\mathcal{H}), \tag{6.149}$$

for $t > 0$ and $k = 0, 1, 2, \ldots$. Therefore,

$$e^{-tA} e^{-sB} \in \mathcal{C}_\infty(\mathcal{H}), \tag{6.150}$$

for $t > 0$ and $s > 0$. Taking into account the representation

$$(\mathbb{1} + A)^{-1}(\mathbb{1} + B)^{-1} = \int_0^\infty \int_0^\infty dt\, ds\, \mathrm{e}^{-(t+s)}\mathrm{e}^{-tA}\mathrm{e}^{-sB}, \qquad (6.151)$$

we obtain from (6.150) and (6.151) that (6.145) holds.

By Proposition 5.46, the conditions (6.10) and the property (6.145) are indispensable for the proof that the Trotter-Kato product formulae converge to the degenerate semigroup $\{\mathrm{e}^{-tH}P_0\}_{t\geqslant 0}$ in the operator-norm topology, locally uniformly away from zero. Moreover, by Proposition 5.46 we also obtain that $(\mathbb{1}_0 + H)^{-1} \in \mathcal{C}_\infty(\mathcal{H}_0)$. Then Proposition 6.22 completes the proof of the assertion. □

Instead of (6.10) and the compactness (6.145) of the product of resolvent, in the next statement the main condition on admissible functions of class $\mathcal{K}$ that ensures the convergence of the Trotter-Kato product formulae is formulated in terms of *self-dominantness.*

Proposition 6.30. *Let A and B be non-negative self-adjoint operators in a Hilbert space $\mathcal{H}$ and let $f^D : \mathbb{R}_0^+ \to \mathbb{R}_0^+$ and $g^D : \mathbb{R}_0^+ \to \mathbb{R}_0^+$ be self-dominated Kato functions such that condition $F^D(t_0) \in \mathcal{C}_\phi(\mathcal{H})$ is satisfied for some $t_0 > 0$. If the Kato functions f and g obey $f(x) \leqslant f^D(x)$ and $g(x) \leqslant g^D(x)$ for $x \geqslant 0$, then the Trotter-Kato product formulae converge locally uniformly away from t_0 in $\mathcal{C}_\phi(\mathcal{H})$ for all families of approximants generated by f and g.*

Proof. This statement is essentially a corollary of Proposition 6.29.

First, we note that the upper bounds $f(x) \leqslant f^D(x)$ and $g(x) \leqslant g^D(x)$ for $x \geqslant 0$, imply that the functions f and g are dominated by f^D and g^D, respectively, see Lemma 6.27, (6.142).

Second, since f^D and g^D are self-dominated Kato functions, Lemma 6.27, (6.141), proves the conditions (6.10). □

Corollary 6.31. *Let A and B be non-negative self-adjoint operators in a Hilbert space $\mathcal{H}$ such that*

$$(\mathbb{1} + B)^{-1/2}(\mathbb{1} + A)^{-1}(\mathbb{1} + B)^{-1/2} \in \mathcal{C}_\phi(\mathcal{H}). \qquad (6.152)$$

If f and g are Kato functions such that $f(x) \leqslant 1/(1+x)$ and $g(x) \leqslant 1/(1+x)$ for $x \geqslant 0$, then the Trotter-Kato product formulae converge locally uniformly away from zero (sic!) in $\mathcal{C}_\phi(\mathcal{H})$ for all families of approximants generated by f and g.

Proof. First of all we note that condition (6.152) implies

$$(\mathbb{1} + t_0 B)^{-1/2}(\mathbb{1} + t_0 A)^{-1}(\mathbb{1} + t_0 B)^{-1/2} \in \mathcal{C}_\phi(\mathcal{H}), \qquad (6.153)$$

for *any* $t_0 > 0$. Further, we set $f^D(x) := (1+x)^{-1}$ and $g^D(x) := (1+x)^{-1}$ for $x \geqslant 0$. The Kato functions f^D and g^D are obviously self-dominated, i.e., they obey conditions (6.9) and (6.10). Therefore, by Lemma 6.27, the Kato functions f and

g satisfy the condition (6.10) too. Now, applying Proposition 6.30 one finishes the proof. □

Corollary 6.32. *Let A and B be non-negative self-adjoint operators in a Hilbert space $\mathcal{H}$ such that*

$$e^{-t_0 B/2} e^{-t_0 A} e^{-t_0 B/2} \in \mathcal{C}_\phi(\mathcal{H}), \quad t_0 > 0. \tag{6.154}$$

If f and g are Kato functions such that $f(x) \leqslant e^{-x}$ and $g(x) \leqslant e^{-x}$ for $x \geqslant 0$, then the Trotter-Kato product formula converges locally uniformly away from t_0 in $\mathcal{C}_\phi(\mathcal{H})$ for all families generated by f and g.

Proof. We set $f^D(x) = g^D(x) = e^{-x}$, $x \geqslant 0$. The exponential Kato functions f^D and g^D are obviously self-dominated, that is, satisfy conditions (6.9) and (6.10), and by Lemma 6.27 the Kato functions f and g satisfy the condition (6.10) too. Then Proposition 6.30 completes the proof. □

Note that Proposition 6.30 allows us to state, in some sense, a converse.

Proposition 6.33. *Let A and B be non-negative self-adjoint operators in a Hilbert space $\mathcal{H}$ and let f and g be self-dominated Kato functions. Then the Trotter-Kato product formulae converge in $\mathcal{C}_\phi(\mathcal{H})$ locally uniformly away from some $t_0 > 0$ for all families of the product approximants generated by f and g if and only if there exist $s_0 > 0$ and an integer $p \geqslant 1$ such that $F(s_0)^p \in \mathcal{C}_\phi(\mathcal{H})$.*

Proof. If the Trotter-Kato product formulae converge locally uniformly away from t_0 in $\mathcal{C}_\phi(\mathcal{H})$ for all families of approximants generated by f and g, then they converge in particular for the family $\{F(t)\}_{t\geqslant 0}$. Hence, there exists $t > t_0$ such that for sufficiently large r one has $F(t/r)^r \in \mathcal{C}_\phi(\mathcal{H})$. Setting $p = r$ and $s_0 := t/r > 0$, we find that $F(s_0)^p \in \mathcal{C}_\phi(\mathcal{H})$. Thus, the necessity is proven.

To prove the converse, we note that if $F(s_0)^p \in \mathcal{C}_\phi(\mathcal{H})$ for some integer $p \geqslant 1$, then $F(s_0) \in \mathcal{C}_\infty(\mathcal{H})$, cf. (6.146). Now, following the proof of Proposition 6.29 we obtain (6.145). By Proposition 5.46, the self-dominants of the Kato functions f, g and the property (6.145) ensure that the Trotter-Kato product formulae converge in the operator-norm topology locally uniformly away from zero. Finally, setting $t_0 = ps_0$ and taking into account Proposition 6.22, one completes the proof of the sufficiency. □

Proposition 6.33 allows to establish the following result, important for the Gibbs semigroups theory.

Proposition 6.34. *Let A and B be non-negative self-adjoint operators in a Hilbert space $\mathcal{H}$ and let f and g be self-dominated Kato functions. Then there is a $t_0 \geqslant 0$ such that the Trotter-Kato product formulae converge locally uniformly away from t_0 in $\mathcal{C}_1(\mathcal{H})$ for all families of the product approximants generated by f, g if and only if there exists an $s_0 > 0$ and an integer p such that $F(s_0) \in \mathcal{C}_p(\mathcal{H})$.*

Proof. From Proposition 6.33 one gets that the condition $F(s_0)^p \in \mathcal{C}_1(\mathcal{H})$ has to be satisfied for some $s_0 > 0$ and some integer $p \geqslant 1$. However, this is equivalent to $F(s_0) \in \mathcal{C}_p(\mathcal{H})$. □

In particular, let $f(x) = g(x) = 1/(1+x)$ or $f(x) = g(x) = e^{-x}$, $x \geqslant 0$. Then Proposition 6.34 shows that the Trotter-Kato product formulae converge in the trace-class norm locally uniformly away from some t_0 for all families of approximants generated by f, g if and only if for some $s_0 > 0$ one has

$$(\mathbb{1} + s_0 B)^{-1/2}(\mathbb{1} + s_0 A)^{-1}(\mathbb{1} + s_0 B)^{-1/2} \in \mathcal{C}_p(\mathcal{H}), \tag{6.155}$$

or

$$e^{-s_0 B/2} e^{-s_0 A} e^{-s_0 B/2} \in \mathcal{C}_p(\mathcal{H}), \tag{6.156}$$

for some $p \geqslant 1$, respectively.

In the case $f(x) = g(x) = 1/(1+x)$ one gets in addition that the convergence in Proposition 6.34 holds away from *zero*, see Corollary 6.31. It is related to the fact that in this case (6.155) implies that either $\{e^{-tA}\}_{t\geqslant 0}$ or $\{e^{-tB}\}_{t\geqslant 0}$ is an *immediately* Gibbs semigroup, see Definition 4.1.

Moreover, the notion of *self-dominated* Kato functions allows one to generalise our results in Proposition 5.36 and Proposition 5.46 about the *operator-norm* convergence of the Trotter-Kato product formulae locally uniformly away from zero, cf. Section 5.3.

Proposition 6.35. *Let A and B be non-negative self-adjoint operators in a Hilbert space $\mathcal{H}$ and let f and g be self-dominated Kato functions. If there is a $t_0 > 0$ such that $F(t_0) \in \mathcal{C}_\infty(\mathcal{H})$, then the Trotter-Kato product formulae converge locally uniformly away from zero in the operator-norm for all families of the product approximants generated by f, g.*

Proof. Following the proof of Proposition 6.29 one obtains (6.145). Then Proposition 5.46 shows that the Trotter-Kato product formulae converge in the operator-norm locally uniformly away from zero, as needed. □

6.6 Product formulae: error bound estimates

Let $\{\eta : \mathbb{R}^+ \vee \mathbb{N} \to \mathbb{R}_0^+\}$ be a non-negative real function defined on $\mathbb{R}^+$ *or* on $\mathbb{N} = \{1, 2, \ldots\}$, such that

$$\lim_{x\to\infty} \eta(x) = 0. \tag{6.157}$$

Definition 6.36. The function η is called the *operator-norm error bound* of the Trotter-Kato product formulae away from $t_0 > 0$ for the family $\{F(t)\}_{t\geqslant 0}$, if for any interval $[a, b] \subseteq (t_0, +\infty)$ there exist an $x_0 \geqslant 1$, such that

$$\|F(t/x)^x - e^{-tH} P_0\| \leqslant \eta(x), \tag{6.158}$$

for $t \in [a, b]$ and $x \geqslant x_0$, where $x = r \vee n$, is either the continuous parameter $r \geqslant r_0 \geqslant 1$, or the discrete parameter $n \in \mathbb{N}$, $n \geqslant n_0 \geqslant 1$.
Recall that the generator $H = A \dot{+} B$ is a form-sum of A and B, and P_0 is the orthogonal projection $P_0 : \mathcal{H} \to \mathcal{H}_0 = \overline{\text{dom}(H)}$.

We note that in contrast to the real number x_0 in (6.158), the error bound η does not depend on the interval $[a, b]$, that is, η is *locally uniform away from* t_0.

Similarly, the notion of an operator-norm error bound can be extended to the families $\{T(t)\}_{t\geqslant 0}$, $\{f(tA)g(tB)\}_{t\geqslant 0}$ and $\{g(tB)f(tA)\}_{t\geqslant 0}$, yet in the last two cases only the discrete parameter $n \in \mathbb{N}$ makes sense for the approximants.

Definition 6.37. The function η_ϕ is called a *locally uniform* error bound in $\mathcal{C}_\phi(\mathcal{H})$ *away from* $t_0 > 0$ for the Trotter-Kato product formulae for the family $\{F(t)\}_{t\geqslant 0}$ if for any interval $[\tau_0, \tau] \subseteq (t_0, +\infty)$ there exist an $x_\phi \geqslant 1$, such that

$$F(t/x)^x - \mathrm{e}^{-tH} P_0 \in \mathcal{C}_\phi(\mathcal{H}), \tag{6.159}$$

and

$$\|F(t/x)^x - \mathrm{e}^{-tH} P_0\|_\phi \leqslant \eta_\phi(x), \tag{6.160}$$

for $t \in [\tau_0, \tau]$ and $x \geqslant x_\phi$, where $x = r \vee n$. As above, this notion has an extension to approximants generated by the families $\{T(t)\}_{t\geqslant 0}$, $\{f(tA)g(tB)\}_{t\geqslant 0}$ and $\{g(tB)f(tA)\}_{t\geqslant 0}$.

The following *lifting* problem (cf. Section 5.4, Lemma 5.51) arises naturally in this setup:
Assume that we know the *operator-norm* locally uniform error bound η (6.158) away from $t_0 > 0$. Can we find a locally uniform error bound η_ϕ away from t_0 in a symmetrically-normed ideal $\mathcal{C}_\phi(\mathcal{H})$?

For simplicity, below we omit systematically the term *locally uniform* as an evidence.

Next we prove the main lifting statement for the Trotter-Kato product formulae with locally uniform error bounds in the symmetrically-normed ideals $\mathcal{C}_\phi(\mathcal{H})$. Recall the trace-class ideal $\mathcal{C}_1(\mathcal{H}) = \mathcal{C}_{\phi_1}(\mathcal{H})$, see Definition 6.5 and (6.20),(6.21)

Proposition 6.38. *Let A and B be non-negative self-adjoint operators in a Hilbert space $\mathcal{H}$, and let $f^D : \mathbb{R}_0^+ \to \mathbb{R}_0^+$ and $g^D : \mathbb{R}_0^+ \to \mathbb{R}_0^+$ be Borel measurable functions such that $F^D(t_0) \in \mathcal{C}_\phi(\mathcal{H})$ for some $t_0 > 0$. Assume that the Kato functions f and g are dominated by f^D and g^D, respectively. Then:*

If $\eta(r)$, $r \geqslant r_0 = 1$ (continuous case), is an operator-norm error bound away from $t_0 > 0$ for the Trotter-Kato product formula for the self-adjoint family $\{F(t)\}_{t\geqslant 0}$ (respectively for $\{T(t)\}_{t\geqslant 0}$), then the function $\eta_\phi(r) = 2\,\|F^D(t_0)\|_\phi\, \eta(r/2)$, $r \geqslant 2$, is an error bound in $\mathcal{C}_\phi(\mathcal{H})$ away from $2t_0$ for the Trotter-Kato product formula for the family $\{F(t)\}_{t\geqslant 0}$ (respectively, for the family $\{T(t)\}_{t\geqslant 0}$).

If $\eta(n)$, $n \geqslant n_0$ (discrete case), is an operator-norm error bound away from $t_0 > 0$ for the Trotter-Kato product formula for $\{f(tA)g(tB)\}_{t\geqslant 0}$, respectively for

$\{g(tB)f(tA)\}_{t\geqslant 0}$, $\{F(t)\}_{t\geqslant 0}$ or $\{T(t)\}_{t\geqslant 0}$, then the function

$$\eta_\phi(n) = \|F^D(t_0)\|_\phi \{\eta([n/2]) + \eta([(n+1)/2])\}, \quad n \geqslant n_0\,,$$

is error bound in $\mathcal{C}_\phi(\mathcal{H})$ away $2t_0$ for the Trotter-Kato product formula for the family $\{f(tA)g(tB)\}_{t\geqslant 0}$, respectively, for $\{g(tB)f(tA)\}_{t\geqslant 0}$, $\{F(t)\}_{t\geqslant 0}$ or $\{T(t)\}_{t\geqslant 0}$.

Here $[\alpha]$ denotes the integer part of $\alpha \geqslant 0$.

Proof. Continuous case. We use the representation

$$\begin{aligned} &F(t/r)^r - \mathrm{e}^{-tH}P_0 \qquad (6.161)\\ &= \big(F(t/r)^{r/2} - \mathrm{e}^{-tH/2}P_0\big)F(t/r)^{r/2} + \mathrm{e}^{-tH/2}P_0\big(F(t/r)^{r/2} - \mathrm{e}^{-tH/2}P_0\big). \end{aligned}$$

Then Lemma 6.20 yields the estimate

$$\|F(t/r)^{r/2}\|_\phi \leqslant \|F^D(t_0)\|_\phi\,, \qquad (6.162)$$

for $t_0 \leqslant t/2 \leqslant rt/2$ and $r/2 \geqslant 1$, while Lemma 6.21 yields

$$\|\mathrm{e}^{-tH/2}P_0\|_\phi \leqslant \|F^D(t_0)\|_\phi\,, \qquad (6.163)$$

for $t \geqslant 2t_0$. Hence, (6.162) and (6.163) yield for (6.161) the estimate

$$\|F(t/r)^r - \mathrm{e}^{-tH}P_0\|_\phi \leqslant 2\|F^D(t_0)\|_\phi\|F(t/r)^{r/2} - \mathrm{e}^{-tH/2}P_0\|, \qquad (6.164)$$

for $2t_0 \leqslant t \leqslant rt_0$ and $r \geqslant 2$.

Since η is an operator-norm error bound (6.158) away from t_0, the estimate

$$\|F(t/r)^{r/2} - \mathrm{e}^{-tH/2}P_0\| \leqslant \eta(r/2), \qquad (6.165)$$

is valid for any interval $[\tau_0, \tau] \subseteq (2t_0, \infty)$ and $r \geqslant 2r_0$. By condition, $r_0 = 1$. Then inserting (6.165) in (6.164) we find that $\eta_\phi(r) := 2\,\|F^D(t_0)\|_\phi\,\eta(r/2)$, $r \in [2, +\infty)$ is the error bound in $\mathcal{C}_\phi(\mathcal{H})$ away from $2t_0$ for approximants $\{F(t/r)^r\}_{t>0, r\geqslant 2}$.

In order to verify the statement for the family $\{T(t)\}_{t\geqslant 0}$, we have to show only that $F^D(t_0) \in \mathcal{C}_\phi(\mathcal{H})$ implies $T^D(t_0) \in \mathcal{C}_\phi(\mathcal{H})$, where

$$T^D(t) := f^D(tA)^{1/2} g^D(tB) f^D(tA)^{1/2}, \quad t \geqslant 0, \qquad (6.166)$$

by a direct estimate. Let

$$G^D(t) := f^D(tA)^{1/2} g^D(tB)^{1/2}, \quad t \geqslant 0. \qquad (6.167)$$

Using the polar decomposition

$$G^D(t) = V(t)|G^D(t)| = V(t)F^D(t)^{1/2}, \quad t \geqslant 0, \qquad (6.168)$$

where $V(t)$ is a partial isometry acting from $\overline{\operatorname{ran} G^D(t)^*}$ onto $\overline{\operatorname{ran} G^D(t)}$, we find that

$$T^D(t) = G^D(t)G^D(t)^* = V(t)F^D(t)V(t)^* \in \mathcal{C}_\phi(\mathcal{H}), \qquad (6.169)$$

for $t \geqslant 0$. This yields $\|T^D(t_0)\|_\phi = \|F^D(t_0)\|_\phi$.

Discrete case. To prove the assertion for the family $\{f(tA)g(tB)\}_{t\geqslant 0}$ we use the decompositions $n = k + m$, $k \in \mathbb{N} := \{1, 2, \dots\}$, $m = 2, 3, \dots, n \geqslant 3$, and

$$\begin{aligned} &(f(tA/n)g(tB/n))^n - \mathrm{e}^{-tH} P_0 \\ &= \big((f(tA/n)g(tB/n))^k - \mathrm{e}^{-ktH/n} P_0\big)\big(f(tA/n)g(tB/n)\big)^m \\ &\quad + \mathrm{e}^{-ktH/n} P_0\big((f(tA/n)g(tB/n))^m - \mathrm{e}^{-mtH/n} P_0\big). \end{aligned} \tag{6.170}$$

Since

$$(f(tA/n)g(tB/n))^m = f(tA/n)g(tB/n)^{1/2} F(t/n)^{m-1} g(tB)^{1/2}, \tag{6.171}$$

Lemma 6.20 shows that $(f(tA/n)g(tB/n))^m \in \mathcal{C}_\phi(\mathcal{H})$ and

$$\|(f(tA/n)g(tB/n))^m\|_\phi \leqslant \|F^D(t_0)\|_\phi\,, \tag{6.172}$$

for $t_0 \leqslant (m-1)t/n \leqslant (m-1)t_0$ and $m - 1 \geqslant 1$. Note that Lemma 6.21 implies $\mathrm{e}^{-ktH/n} \in \mathcal{C}_\phi(\mathcal{H})$ and

$$\|\mathrm{e}^{-ktH/n} P_0\|_\phi \leqslant \|F^D(t_0)\|_\phi\,, \tag{6.173}$$

for $kt/n \geqslant t_0$. Hence, (6.170) yields the estimate

$$\begin{aligned} &\|(f(tA/n)g(tB/n))^n - \mathrm{e}^{-tH} P_0\|_\phi \\ &\leqslant \|F^D(t_0)\|_\phi \,\|(f(tA/n)g(tB/n))^k - \mathrm{e}^{-ktH/n} P_0\| \\ &\quad + \|F^D(t_0)\|_\phi \,\|(f(tA/n)g(tB/n))^m - \mathrm{e}^{-mtH/k} P_0\|, \end{aligned} \tag{6.174}$$

for $(1 + (k+1)/(m-1))t_0 \leqslant t \leqslant nt_0$, $m \geqslant 2$ and $t \geqslant (1 + m/k)t_0$. Since η is an operator-norm error bound away from t_0, by definition (6.158) for any interval $[a, b] \subseteq (t_0, +\infty)$ there exist an $n_0 \geqslant 1$, such that

$$\|(f(tA/n)g(tB/n))^k - \mathrm{e}^{-ktH/n} P_0\| \leqslant C\eta(k), \tag{6.175}$$

for $kt/n \in [a, b] \Leftrightarrow t \in [(1 + m/k)a, (1 + m/k)b]$, and

$$\|(f(tA/n)g(tB/n))^m - \mathrm{e}^{-mtH/n} P_0\| \leqslant \eta(m), \tag{6.176}$$

for $mt/n \in [a, b] \Leftrightarrow t \in [(1 + k/m)a, (1 + k/m)b]$. Setting $m := [(n + 1)/2]$ and $k = [n/2]$, $n \geqslant 3$, we satisfy the conditions $n = k + m$ and $m \geqslant 2$ as well as $\lim_{n\to\infty} (k + 1)/(m - 1) = 1$, $\lim_{n\to\infty} m/k = 1$ and $\lim_{n\to\infty} k/m = 1$. Hence, for any interval $[\tau_0, \tau] \subseteq (2t_0, +\infty)$ we find that $[\tau_0, \tau] \subseteq [(1+(k + 1)/(m - 1))t_0, nt_0]$ for sufficiently large n. Moreover, choosing $[\tau_0/2, \tau/2] \subseteq (a, b) \subseteq (t_0, +\infty)$ we ensure that $[\tau_0, \tau] \subseteq [(1+m/k)a, (1+m/k)b]$ and $[\tau_0, \tau] \subseteq [(1+k/m)a, (1+k/m)b]$ for sufficiently large n too. Thus for any interval $[\tau_0, \tau] \subseteq (2t_0, +\infty)$ there exist an $n_0 \geqslant 1$, such that (6.174), (6.175) and (6.176) hold for $t \in [\tau_0, \tau]$ and $n \geqslant n_0$. Therefore, from (6.174) we obtain the estimate

$$\begin{aligned} &\|(f(tA/n)g(tB/n))^n - \mathrm{e}^{-tH}P_0\|_\phi \\ &\leqslant \|F^D(t_0)\|_\phi \{\eta([n/2]) + \eta([(n+1)/2])\}, \end{aligned} \tag{6.177}$$

for $t \in [\tau_0, \tau] \subseteq (2t_0, +\infty)$ and $n \geqslant n_0$. Hence,

$$\eta_\phi(n) := \|F^D(t_0)\|_\phi \{\eta([n/2]) + \eta([(n+1)/2])\},$$

is the error bound in $\mathcal{C}_\phi(\mathcal{H})$ for the approximants $\{(f(tA/n)g(tB/n))^n\}_{t>0,n\geqslant 1}$, away from $2t_0$ and for $n \geqslant n_0$.

Similarly one proves the result for the families $\{g(tB)f(tA)\}_{t\geqslant 0}$, $\{F(t)\}_{t\geqslant 0}$ and $\{T(t)\}_{t\geqslant 0}$. □

Corollary 6.39. *Note that if the rate of convergence in the operator-norm estimate η (6.158) is optimal, then so are the rates of convergence in the $\|\cdot\|_\phi$-norm estimates: $\eta_\phi(r)$ and $\eta_\phi(n)$.*

We next consider the application of Proposition 6.38 for the Kato functions of class $\mathcal{K}_\alpha$, see Definition 5.26 and Appendix C. They are suitable for pairs of C_0-semigroup generators A and B related by the fractional smallness conditions, subsection 5.2.5.

Combining Proposition 6.38 with operator-norm estimates in Proposition 5.27, we obtain the following result.

Proposition 6.40. *Let the self-adjoint operators $A \geqslant \mathbb{1}$ and $B \geqslant 0$ in a Hilbert space $\mathcal{H}$ be such that for some $\alpha \in (1/2, 1)$ and $b \in (0,1)$,* $\operatorname{dom} A^\alpha \subset \operatorname{dom} B^\alpha$ *and $\|B^\alpha u\| \leqslant b\|A^\alpha u\|$, $u \in \operatorname{dom} A^\alpha$, $H := A \dot{+} B$.*

Let $f^D : \mathbb{R}_0^+ \to \mathbb{R}_0^+$ and $g^D : \mathbb{R}_0^+ \to \mathbb{R}_0^+$ be Kato functions such that $F^D(t_0) = g^D(t_0B)^{1/2} f^D(t_0A) g^D(t_0B)^{1/2} \in \mathcal{C}_\phi(\mathcal{H})$ for some $t_0 > 0$.

Assume that the Kato functions f and g are dominated by f^D and g^D, respectively.

(i) *If for $\alpha \in (1/2, 1)$ the functions $f, g \in \mathcal{K}_\alpha$ and*

$$b^{1/\alpha} C_{1/2\alpha} S_1 < 1, \tag{6.178}$$

then for some $\Gamma_1 > 0$ the inequality

$$\eta_\phi(n) \leqslant \Gamma_1 \ln(n)/n^{2\alpha-1}, \quad n \geqslant 2, \tag{6.179}$$

gives estimate for the error bound in $\mathcal{C}_\phi(\mathcal{H})$ for the Trotter-Kato product formula for the family $\{f(tA)g(tB)\}_{t\geqslant 0}$ away from $2t_0$.

(ii) *If in addition, for $\alpha \in (1/2, 1)$ and (6.178),* $\operatorname{dom} H^\alpha \subseteq \operatorname{dom} A^\alpha$, *then for some $\Gamma_2 > 0$ the inequality*

$$\eta_\phi(n) \leqslant \Gamma_2/n^{2\alpha-1}, \quad n \geqslant 2, \tag{6.180}$$

gives estimate for the error bound in $\mathcal{C}_\phi(\mathcal{H})$ for the Trotter-Kato product formula for the family $\{f(tA)g(tB)\}_{t\geqslant 0}$ away from $2t_0$.

Proof. (i) Taking into account Proposition 5.27, with condition (i), and (5.94) we find that

$$\eta(n) = c' \ln(n)/n^{2\alpha-1}, \quad n \geqslant 2, \tag{6.181}$$

is an operator-norm error bound of the Trotter-Kato product formula for $\{f(tA)g(tB)\}_{t\geqslant 0}$ away from $t_0 > 0$. Applying now Proposition 6.38 we obtain that $\eta_\phi(n) := \|F^D(t_0)\|_\phi \{\eta([n/2]) + \eta([(n+1)/2])\}$, $n \in \mathbb{N}$, is the error bound in $\mathcal{C}_\phi(\mathcal{H})$ for the family $\{f(tA)g(tB)\}_{t\geqslant 0}$ away from $2t_0$. By a straightforward computation one finds a constant $\Gamma_1 := 2^{2\alpha}c'\|F^D(t_0)\|_\phi$ such that $\eta_\phi(n) \leqslant \Gamma_1 \ln(n)/n^{2\alpha-1}$, for $n \geqslant 2$, as in (6.179).

(ii) Similarly, if in addition to (i), condition $\mathrm{dom}(H^\alpha) \subseteq \mathrm{dom}(A^\alpha)$ are satisfied, then by Proposition 5.27

$$\eta(n) = c/n^{2\alpha-1}, \quad n \geqslant 2, \tag{6.182}$$

is an operator-norm error bound for the Trotter-Kato product formula for $\{f(tA)g(tB)\}_{t\geqslant 0}$ away from $t_0 > 0$. Therefore, applying Proposition 6.38 and proceeding as before, we prove the second estimate (6.180) of the proposition for $\Gamma_2 := 2^{2\alpha}c\,\|F^D(t_0)\|_\phi$.

Similarly, one proves the result for the families $\{g(tB)f(tA)\}_{t\geqslant 0}$, $\{F(t)\}_{t\geqslant 0}$ and $\{T(t)\}_{t\geqslant 0}$. □

Note that the rates of convergence $\ln(n)/n^{2\alpha-1}$ and $1/n^{2\alpha-1}$, in Proposition 6.40 coincide with the rates of convergence in the operator norm, cf. Corollary 6.39.

Now the following remarks are in order (see also notes in Section 5.6 and in Section 6.7):

Firstly, we remind that originally the operator-norm estimates have been proven in Proposition 5.8 under the assumption $A \geqslant \mathbb{1}$ and $B \geqslant \mathbb{1}$. A later revision of this proof shows that $B \geqslant 0$ is sufficient. Therefore, we use this observation in Proposition 6.40.

Secondly, we note that the fractional-power condition $\mathrm{dom}\, H^\alpha \subseteq \mathrm{dom}\, A^\alpha$ for some $\alpha \in (1/2, 1)$ is satisfied if $\mathrm{dom}\, A^\beta \subseteq \mathrm{dom}\, B^\alpha$ for some $\beta \in (0, \alpha)$, in particular, if $\mathrm{dom}\, A^{1/2} \subseteq \mathrm{dom}\, B^\alpha$ is valid.

6.7 Notes

Notes to Section 6.1. Here we follow the Introduction from [NZ99d]. The notions of *dominated* and *self-dominated* Kato functions (Definitions 6.1 and 6.2, or Appendix C) are also introduced in [NZ99d].

Notes to Section 6.2. Material concerning symmetrically-normed ideals can be found in [GK69] (Chapter III) and [Sim05] (Chapter 1). For the Ky Fan inequality, see [Fan51], or [GK69] (Chapter II).

Remarks 6.10, 6.12 and Example 6.13 serve as a preparation for the definition of the *Dixmier trace*, [Dix66]. We present the necessary details in Section 7.2.

Notes to Section 6.3. The material presented in this section is partially motivated by results and observations from the paper [Ara90] and from the book [Sim05], Chapter 1.

Notes to Section 6.4. The results of this section are essentially due to [NZ99d]. It turns out that assertions concerning the Gibbs semigroups (Chapter 5), [NZ90a], [NZ90b] (see also [Hia95]), and in general concerning the von Neumann-Schatten ideals, admit some extension to arbitrary symmetrically-normed ideals.

For convergence of the Trotter-Kato product formulae in the topology of symmetrically-normed ideals away from $t_0 > 0$ it is enough that the self-adjoint operator $F(t)$ (the *transfer matrix*) belongs to one of these ideals for $t = t_0$, provided the Kato functions f and g *behave well.* In particular, this is valid for the von Neumann-Schatten ideals and for dominated or self-dominated Kato functions.

These observations improve the results in [NZ90a], [NZ90b] or [Hia95], where stronger conditions were assumed. Namely, either an auxiliary operator $f_0(tA)$ for $t > 0$ or, respectively, $e^{-t_0 A}$ for some $t_0 > 0$, should belong to the von Neumann-Schatten ideal. Of course, these conditions imply that the operator $F(t)$ belongs to that ideal away from $t_0 > 0$.

Notes to Section 6.5. Proposition 6.34 is a generalisation of a result in the paper [NZ90a]. Proposition 6.35 is a generalisation of the main result [NZ99b].
Concerning the class of admissible Kato functions f and g, the present results are more restrictive than in [NZ90a], [NZ90b] since it is assumed that the Kato functions are regular, dominated or self-dominated and admit a certain decaying behaviour at infinity, see Appendix C for details. Only Propositions 6.23 and 6.28 allow arbitrary Kato functions g.

As a by-product, we also obtain the norm-convergence of the Trotter-Kato product formulae for the case when the operator $F(t)$ is *compact* away from $t_0 > 0$. This completes the results of Section 5.3, see also H. Neidhardt and V. A. Zagrebnov [NZ99b].

Notes to Section 6.6. Here we follow Section 5 of [NZ99d]. The general idea is the same as the *lifting* approach proposed in Chapter 5 (Section 5.4) that we considered above.

The results of this section easily yield the trace-norm convergence of the Trotter-Kato product formulae. For example, taking into account the operator-norm error bounds found in [DIT98], [IT98b] and applying Proposition 6.38, one immediately gets corresponding error bounds in the trace norm, as in Proposition 6.40.

Chapter 7

Product formulae in the Dixmier ideal

In this chapter we discuss product formulae in the weak-$\mathcal{C}_1$ ideal $\mathcal{C}_{1,\infty}(\mathcal{H})$ that we introduced in Section 6.2. Since this ideal is a natural domain for the Dixmier trace, it is also called the Dixmier ideal.

7.1 Ideals and singular traces

The Dixmier *trace* and *ideal* arose from the question of whether the algebra $\mathcal{L}(\mathcal{H})$ of all bounded linear operators on a Hilbert space $\mathcal{H}$ admits a unique nontrivial trace. This problem was solved in the negative. On $\mathcal{L}(\mathcal{H})$ there exist *singular* traces that vanish on the finite-rank operators, and consequently, on the ideal of trace-class operators.

Proposition 7.1. *The space $\mathcal{C}_{1,\infty}(\mathcal{H})$ endowed with the norm $\|\cdot\|_{1,\infty}$ (6.64) is a Banach space.*

The proof is quite standard, although tedious and long. For relevant references, see Section 7.4 (Notes to Section 7.1).

Proposition 7.2. *The space $\mathcal{C}_{1,\infty}(\mathcal{H})$ endowed with the norm $\|\cdot\|_{1,\infty}$ is a Banach ideal in the algebra of bounded operators $\mathcal{L}(\mathcal{H})$.*

Proof. It is sufficient to prove that if A and C are bounded operators, then $B \in \mathcal{C}_{1,\infty}(\mathcal{H})$ implies $ABC \in \mathcal{C}_{1,\infty}(\mathcal{H})$. Recall that the singular values of the operator

V. A. Zagrebnov, *Gibbs Semigroups*, Operator Theory: Advances and Applications 273, https://doi.org/10.1007/978-3-030-18877-1_7

ABC satisfy the estimate $s_j(ABC) \leqslant \|A\|\|C\|s_j(B)$. By (6.64), this yields

$$\begin{aligned} \|ABC\|_{1,\infty} &= \sup_{n\in\mathbb{N}} \frac{1}{1+\ln(n)} \sum_{j=1}^{n} s_j(ABC) \\ &\leqslant \|A\|\|C\| \sup_{n\in\mathbb{N}} \frac{1}{1+\ln(n)} \sum_{j=1}^{n} s_j(B) = \|A\|\|C\|\|B\|_{1,\infty}, \end{aligned} \tag{7.1}$$

which proves the proposition. □

Recall that for any $A \in \mathcal{L}(\mathcal{H})$ and all $B \in \mathcal{C}_1(\mathcal{H})$ one can define a linear functional on $\mathcal{C}_1(\mathcal{H})$ given by trace $\mathrm{Tr}_{\mathcal{H}}(AB)$, Definition 2.41. The set of these functionals $\{\mathrm{Tr}_{\mathcal{H}}(A\,\cdot)\}_{A\in\mathcal{L}(\mathcal{H})}$ is just the *dual* space $\mathcal{C}_1(\mathcal{H})^* := \mathcal{L}(\mathcal{C}_1(\mathcal{H}), \mathbb{C})$ of $\mathcal{C}_1(\mathcal{H})$, equipped with the operator-norm topology. In other words, $\mathcal{C}_1(\mathcal{H})^* = \mathcal{L}(\mathcal{H})$, in the sense that the map $A \mapsto \mathrm{Tr}_{\mathcal{H}}(A\,\cdot)$ is an isometric isomorphism of $\mathcal{L}(\mathcal{H})$ onto $\mathcal{C}_1(\mathcal{H})^*$.

Remark 7.3. Using the *duality* pairing

$$\langle A|B\rangle := \mathrm{Tr}_{\mathcal{H}}(AB), \tag{7.2}$$

one can also describe the space $\mathcal{C}_1(\mathcal{H})_*$, which is the *predual* of $\mathcal{C}_1(\mathcal{H})$, that is, its dual $\mathcal{L}(\mathcal{C}_1(\mathcal{H})_*, \mathbb{C}) = \mathcal{C}_1(\mathcal{H})$. To this aim, for each *fixed* $B \in \mathcal{C}_1(\mathcal{H})$ we consider the functionals $A \mapsto \mathrm{Tr}_{\mathcal{H}}(AB)$ on $\mathcal{L}(\mathcal{H})$. It is known that they are not *all* continuous linear functional on the bounded operators $\mathcal{L}(\mathcal{H})$, i.e., $\mathcal{C}_1(\mathcal{H}) \subset \mathcal{L}(\mathcal{H})^*$, but they yield the *entire* dual of the compact operators: $\mathcal{C}_1(\mathcal{H}) = \mathcal{C}_\infty(\mathcal{H})^*$. Hence, $\mathcal{C}_1(\mathcal{H})_* = \mathcal{C}_\infty(\mathcal{H}) \subset \mathcal{L}(\mathcal{H})$.

Now we note that under the duality relation (7.2) the Dixmier ideal $\mathcal{C}_{1,\infty}(\mathcal{H})$ is the dual of the Macaev ideal $\mathcal{C}_{1,\infty}(\mathcal{H}) = \mathcal{C}_{\infty,1}(\mathcal{H})^*$, where

$$\mathcal{C}_{\infty,1}(\mathcal{H}) = \{A \in \mathcal{C}_\infty(\mathcal{H}) : \sum_{n\geqslant 1} \frac{1}{n}\, s_n(A) < \infty\}, \tag{7.3}$$

see Example 6.13. By the same duality relation and by similar calculations one also concludes that the *predual* of $\mathcal{C}_{\infty,1}(\mathcal{H})$ is the ideal $\mathcal{C}_{\infty,1}(\mathcal{H})_* = \mathcal{C}^{(0)}_{1,\infty}(\mathcal{H})$, defined by

$$\mathcal{C}^{(0)}_{1,\infty}(\mathcal{H}) := \{A \in \mathcal{C}_\infty(\mathcal{H}) : \sum_{j\geqslant 1}^{n} s_j(A) = o(\ln(n)),\ n \to \infty\}. \tag{7.4}$$

By virtue of (6.63) (Remark 6.12), the ideal (7.4) is not self-dual, since

$$\mathcal{C}^{(0)}_{1,\infty}(\mathcal{H})^{**} = \mathcal{C}_{1,\infty}(\mathcal{H}) \supset \mathcal{C}^{(0)}_{1,\infty}(\mathcal{H}).$$

The problem that motivated the construction of the Dixmier trace was related to the desire of introducing general definition of the *trace*, i.e., a linear, positive, and unitarily invariant functional on a *proper* Banach ideal $\mathfrak{I}(\mathcal{H})$ of the unital algebra of bounded operators $\mathcal{L}(\mathcal{H})$, cf. Proposition 2.48. Since any proper two-sided ideal $\mathfrak{I}(\mathcal{H})$ of the ring $\mathcal{L}(\mathcal{H})$ is contained in the compact operators $\mathcal{C}_\infty(\mathcal{H})$ and contains the set $\mathcal{K}(\mathcal{H})$ of finite-rank operators (6.38), the *domain* of definition of the trace has to coincide with some ideal $\mathfrak{I}(\mathcal{H})$.

Remark 7.4. The *canonical* trace $\mathrm{Tr}_\mathcal{H}(\cdot)$ is nontrivial only on the trace-class ideal $\mathcal{C}_1(\mathcal{H})$, see Example 6.8. We recall that it is characterised by the property of *normality*: $\mathrm{Tr}_\mathcal{H}(\sup_\alpha B_\alpha) = \sup_\alpha \mathrm{Tr}_\mathcal{H}(B_\alpha)$, for every directed increasing bounded family $\{B_\alpha\}_{\alpha\in\Delta}$ of positive operators from the cone $\mathcal{C}_{1,+}(\mathcal{H})$ of positive operators. Since every nontrivial *normal* trace on $\mathcal{L}(\mathcal{H})$ is proportional to the canonical trace $\mathrm{Tr}_\mathcal{H}(\cdot)$, the Dixmier trace (6.67), $\mathcal{C}_{1,\infty} \ni X \mapsto \mathrm{Tr}_\omega(X)$, is *not* normal.

Definition 7.5. A *trace* on the proper Banach ideal $\mathfrak{I}(\mathcal{H}) \subset \mathcal{L}(\mathcal{H})$ is called *singular* if it vanishes on the set $\mathcal{K}(\mathcal{H})$.

Since by Proposition 2.59(b) a singular trace is defined up to trace-class operators $\mathcal{C}_1(\mathcal{H})$, then by Remark 7.4 it obviously is not normal.

7.2 Dixmier trace

Recall that only the ideal of trace-class operators has the property that on its positive cone $\mathcal{C}_{1,+}(\mathcal{H}) := \{A \in \mathcal{C}_1(\mathcal{H}) : A \geqslant 0\}$ the trace-norm is *linear*, since $\|A+B\|_1 = \mathrm{Tr}\,(A+B) = \mathrm{Tr}\,(A) + \mathrm{Tr}\,(B) = \|A\|_1 + \|B\|_1$ for $A, B \in \mathcal{C}_{1,+}(\mathcal{H})$, see Example 6.8. Then the uniqueness of the trace-norm allows one to extend the trace to the whole linear space $\mathcal{C}_1(\mathcal{H})$. This however *fails* for other symmetrically-normed ideals, either because of lack of linearity, or owing to lack of boundedness.

This problem motivates the Dixmier trace construction as a certain limiting procedure involving the $\|\cdot\|_{1,\infty}$-norm. Let $\mathcal{C}_{1,\infty,+}(\mathcal{H})$ be the positive cone of the Dixmier ideal. One can try to construct on $\mathcal{C}_{1,\infty,+}(\mathcal{H})$ a *linear*, *positive*, and *unitarily* invariant functional (called *trace* $\mathcal{T}$) via *extension* of the limit, called the Lim, of the sequence of the properly *normalised* finite sums of eigenvalues of the operator X:

$$\mathcal{T}(X) := \mathrm{Lim}_{n\to\infty} \frac{1}{1+\ln(n)} \sum_{j=1}^{n} \lambda_j(X), \quad X \in \mathcal{C}_{1,\infty,+}(\mathcal{H}). \tag{7.5}$$

First, keeping in mind that Lim is essentially the *limit*, we note that for any unitary operator $U : \mathcal{H} \to \mathcal{H}$, the eigenvalues of $X \in \mathcal{C}_\infty(\mathcal{H})$ are invariant: $\lambda_j(X) = \lambda_j(UXU^*)$. Hence, this is also true for the sequence

$$\sigma_n(X) := \sum_{j=1}^{n} \lambda_j(X), \quad n \geqslant 1, \tag{7.6}$$

and the Lim (7.5) (if it exists) inherits this property: $\mathcal{T}(UXU^*) = \mathcal{T}(X)$. Since for $\alpha \in \mathbb{R}$ one has $\lambda_j(\alpha X) = \alpha\lambda_j(X)$, the trace (7.5) is also homogeneous: $\mathcal{T}(\alpha X) = \alpha\mathcal{T}(X)$.

Similarly, we comment on positivity and on boundedness. Note that $X \geqslant 0$ implies the positivity of eigenvalues $\{\lambda_j(X)\}_{j\geqslant 1}$. Therefore, $\sigma_n(X) \geqslant 0$ and the Lim in (7.5) is a *positive* mapping: $X \mapsto \mathcal{T}(X) \geqslant 0$ for $X \in \mathcal{C}_{1,\infty,+}(\mathcal{H})$, which remains finite by virtue of the estimate $\mathcal{T}(X) \leqslant \|X\|_{1,\infty}$, (6.64).

The next issue with the formula (7.5) for $\mathcal{T}(X)$ is its *linearity*. To approach it, we need to understand better the meaning of Lim. To proceed, we recall that if $P : \mathcal{H} \to P(\mathcal{H})$ is an orthogonal projection on a finite-dimensional subspace with $\dim P(\mathcal{H}) = n$, then for any compact operator $X \geqslant 0$ formula (7.6) gives

$$\sigma_n(X) = \sup_P \{\mathrm{Tr}_{\mathcal{H}}(XP) : \dim P(\mathcal{H}) = n\} . \tag{7.7}$$

Using (7.7) and the inequality (3.17), Lemma 3.5, we obtain

$$\sigma_n(X + Y) \leqslant \sigma_n(X) + \sigma_n(Y), \quad n \in \mathbb{N}, \tag{7.8}$$

which is valid for any pair of *positive* compact operators $X, Y \in \mathcal{C}_{\infty,+}(\mathcal{H})$.

To estimate (7.8) from above we consider for the pair $X, Y \in \mathcal{C}_{\infty,+}(\mathcal{H})$ and for a given $\varepsilon > 0$ two orthogonal projections P_1 and P_2 satisfying the conditions

$$\begin{gathered}\dim P_1(\mathcal{H}) = \dim P_2(\mathcal{H}) = n, \\ \mathrm{Tr}_{\mathcal{H}}(XP_1) > \sigma_n(X) - \varepsilon \quad \wedge \quad \mathrm{Tr}_{\mathcal{H}}(YP_2) > \sigma_n(Y) - \varepsilon.\end{gathered}$$

Then by taking the linear span $P(\mathcal{H}) := P_1(\mathcal{H}) \vee P_2(\mathcal{H})$, that is, the orthogonal projection onto $P_1(\mathcal{H}) + P_2(\mathcal{H})$, we get

$$\begin{aligned}&\mathrm{Tr}_{\mathcal{H}}((X + Y)P) = \mathrm{Tr}_{\mathcal{H}}(XP) + \mathrm{Tr}_{\mathcal{H}}(YP) \\ &\geqslant \mathrm{Tr}_{\mathcal{H}}(XP_1) + \mathrm{Tr}_{\mathcal{H}}(YP_2) \geqslant \sigma_n(X) + \sigma_n(Y) - 2\,\varepsilon.\end{aligned} \tag{7.9}$$

Therefore, by virtue of $\dim P(\mathcal{H}) \leqslant 2n$ and of the arbitrariness of $\varepsilon > 0$, (7.7) and (7.9) yield

$$\sigma_n(X) + \sigma_n(Y) \leqslant \sigma_{2n}(X + Y), \quad n \in \mathbb{N}, \tag{7.10}$$

which is a kind of *anti-triangle* inequality.

Now to proceed with decoding the properties of (7.5) that would ensure the linearity of $\mathcal{T}(\cdot)$, we introduce

$$\xi_n(X) := \frac{1}{1 + \ln(n)}\sigma_n(X), \quad X \in \mathcal{C}_{1,\infty,+}(\mathcal{H}), \tag{7.11}$$

and a non-negative functional $\omega_{\lim} : l^\infty \to \mathbb{R}_0^+$, such that

$$\omega_{\lim}(\{\xi_n(X)\}_{n\geqslant 1}) := \mathrm{Lim}_{n\to\infty}\, \xi_n(X), \quad X \in \mathcal{C}_{1,\infty,+}(\mathcal{H}). \tag{7.12}$$

This functional is defined on the space of non-negative sequences $\{\xi_n(\cdot)\}_{n\geqslant 1} \in l^\infty$ generated by operators from the positive cone of the Dixmier ideal.

Note that by (7.11) the inequalities (7.8) and (7.10) yield for $n \in \mathbb{N}$

$$\xi_n(X) + \xi_n(Y) \geqslant \xi_n(X+Y), \tag{7.13}$$

$$\xi_n(X) + \xi_n(Y) \leqslant \frac{1+\ln(2n)}{1+\ln(n)}\, \xi_{2n}(X+Y). \tag{7.14}$$

Since by definition (7.12) the functional $\omega_{\lim}(\cdot)$ incorporates the limit $n \to \infty$, the inequalities (7.13), (7.14) *would* give a desired linearity of the trace $\mathcal{T}$:

$$\mathcal{T}(X+Y) = \mathcal{T}(X) + \mathcal{T}(Y), \tag{7.15}$$

if one can prove that for X, Y, as well as for $X+Y$, the $\mathrm{Lim}_{n\to\infty}$ in (7.13), (7.14) exist and the limits of the right-hand sides coincide. To this aim we have to clarify further the properties (a), (b) and (c), of the functional $\omega_{\lim}(\cdot)$.

Remark 7.6. Firstly, summarising our remarks above, we see that the functional $\omega_{\lim} : l^\infty \to \mathbb{R}$ is non-negative and homogeneous:

(a) $\omega_{\lim}(\alpha\,\eta) = \alpha\,\omega_{\lim}(\eta) \geqslant 0, \quad$ for $\alpha \geqslant 0$, $\forall \eta = \{\eta_n \geqslant 0\}_{n\in\mathbb{N}} \in l^\infty$.

Secondly, examining (7.5) and (7.12) we dediuce that the functional $\omega_{\lim}(\eta(\cdot)) = \mathrm{Lim}_{n\to\infty}\,\eta_n(\cdot)$ is completely determined by the "tail" behaviour of the sequences $\{\eta_n \geqslant 0\}_{n\geqslant 1} \in l^\infty$. For example, $\omega_{\lim}(\eta) = 0$ for all $\eta \in c_0$. Here $c_0 \subset l^\infty$ is the subspace of bounded sequences that converge to zero (Section 6.2). Therefore, $c_0 = \ker \omega_{\lim}$ and any $\xi \in \mathrm{dom}\,\omega_{\lim}$ is defined modulo $\eta \in c_0$, that is, $\omega_{\lim}(\xi + \eta) = \omega_{\lim}(\xi)$. Consequently,

(b) $\omega_{\lim}(\eta) = \lim_{n\to\infty} \eta_n, \quad$ if $\{\eta_n \geqslant 0\}_{n\in\mathbb{N}}$ is convergent.

By virtue of (a) and (b), the definitions (7.5) and (7.11) imply that for $X, Y \in \mathcal{C}_{1,\infty,+}(\mathcal{H})$ we have

$$\mathcal{T}(X) = \omega_{\lim}(\{\xi_n(X)\}_{n\in\mathbb{N}}) = \lim_{n\to\infty} \xi_n(X), \tag{7.16}$$

$$\mathcal{T}(Y) = \omega_{\lim}(\{\xi_n(Y)\}_{n\in\mathbb{N}}) = \lim_{n\to\infty} \xi_n(Y), \tag{7.17}$$

$$\mathcal{T}(X+Y) = \omega_{\lim}(\{\xi_n(X+Y)\}_{n\in\mathbb{N}}) = \lim_{n\to\infty} \xi_n(X+Y), \tag{7.18}$$

if the limits in the right-hand sides of (7.16)–(7.18) exist.

Now, to ensure that (7.15) holds one has to select the state $\omega_{\lim}$ in such a way that it allows to restore the equality of the right-hand sides of (7.13), (7.14), when $n \to \infty$. To this aim we require that the state $\omega_{\lim}$ be *dilation* $\mathfrak{D}_2$-invariant.

Definition 7.7. Define the dilation mapping $\mathfrak{D}_2 : l^\infty \to l^\infty$, $\eta \mapsto \mathfrak{D}_2(\eta)$ by the rule

$$\mathfrak{D}_2 : (\eta_1, \eta_2, \ldots \eta_k, \ldots) \to (\eta_1, \eta_1, \eta_2, \eta_2, \ldots \eta_k, \eta_k, \ldots), \quad \forall \eta \in l^\infty. \tag{7.19}$$

We say that $\omega_{\lim}$ is dilation $\mathfrak{D}_2$-invariant if for any $\eta \in l^\infty$

(c) $\omega_{\lim}(\eta) = \omega_{\lim}(\mathfrak{D}_2(\eta))$.

We shall discuss the question of *existence* of the dilation $\mathfrak{D}_2$-invariant states (also called the *invariant means*) on the Banach space l^∞ in Remark 7.9.

Let $X, Y \in \mathcal{C}_{1,\infty,+}(\mathcal{H})$. Then applying the property (c) to the sequence $\eta = (\xi_2, \xi_4, \xi_6, \ldots)$, where $\{\xi_{2n} := \xi_{2n}(X+Y)\}_{n=1}^\infty$, we obtain

$$\omega_{\lim}(\eta) = \omega_{\lim}(\mathfrak{D}_2(\eta)) = \omega_{\lim}(\xi_2, \xi_2, \xi_4, \xi_4, \xi_6, \xi_6, \ldots). \tag{7.20}$$

Note that for $\xi = \{\xi_n := \xi_n(X+Y)\}_{n=1}^\infty$ the difference

$$\mathfrak{D}_2(\eta) - \xi = (\xi_2, \xi_2, \xi_4, \xi_4, \xi_6, \xi_6, \ldots) - (\xi_1, \xi_2, \xi_3, \xi_4, \xi_5, \xi_6, \ldots), \tag{7.21}$$

is the sequence that converges to zero *if* for $n \to \infty$ one has $\xi_{2n} - \xi_{2n-1} \to 0$. Then $\mathfrak{D}_2(\eta) - \xi \in c_0$.

Since by Remark 7.6(b) the state $\omega_{\lim}$ is determined modulo sequences that belong to c_0, the relations (7.18), (7.20) and (7.21) *would* imply

$$\begin{aligned}&\omega_{\lim}(\{\xi_{2n}(X+Y)\}_{n\geqslant 1})\\ &= \omega_{\lim}(\mathfrak{D}_2(\{\xi_{2n}(X+Y)\}_{n\geqslant 1})) = \omega_{\lim}(\{\xi_n(X+Y)\}_{n\geqslant 1}).\end{aligned}$$

By (7.18), this *would* mean that

$$\lim_{n\to\infty} \xi_{2n}(X+Y) = \lim_{n\to\infty} \xi_n(X+Y),$$

which by estimates (7.13), (7.14) and $\lim_{n\to\infty}(1+\ln(2n))/(1+\ln(n)) = 1$, *would* yield

$$\lim_{n\to\infty} \xi_n(X+Y) = \lim_{n\to\infty} \xi_n(X) + \lim_{n\to\infty} \xi_n(Y). \tag{7.22}$$

Combining (7.16), (7.17), (7.18) and (7.22), we obtain the linearity (7.15) of the functional $\mathcal{T}$ on the positive cone $\mathcal{C}_{1,\infty,+}(\mathcal{H})$ *provided* that it is defined by the corresponding $\mathfrak{D}_2$-invariant state $\omega_{\lim}$, or by some equivalent dilation-invariant mean.

Therefore, to complete the proof of the linearity, it remains *only* to check that $\lim_{n\to\infty}(\xi_{2n} - \xi_{2n-1}) = 0$. To this end we note that, by definitions (7.6) and (7.11), one gets

$$\begin{aligned}\xi_{2n} - \xi_{2n-1} = &\left[\frac{1}{1+\ln(2n)} - \frac{1}{1+\ln(2n-1)}\right]\sigma_{2n-1}(X+Y)\\ &+ \frac{1}{1+\ln(2n)}\lambda_{2n}(X+Y).\end{aligned} \tag{7.23}$$

Since $X, Y \in \mathcal{C}_{1,\infty,+}(\mathcal{H})$, we have $\lim_{n\to\infty}\lambda_{2n}(X+Y) = 0$, and also $\sigma_{2n-1}(X+Y) \leqslant O(1+\ln(2n-1))$. Then taking into account that $1/(1+\ln(2n)) - 1/(1+\ln(2n-1)) = o(1/(1+\ln(2n-1))$ we conclude that the right-hand side of (7.23) converges to zero, as $n \to \infty$.

Concluding this construction of the trace $\mathcal{T}(\cdot)$ (7.5) and following definition (7.12), we denote it by

$$\mathrm{Tr}_\omega(X) := \omega_{\lim}(\xi(X)), \quad X \in \mathcal{C}_{1,\infty,+}(\mathcal{H}).$$

It is worth pointing out that by linearity (7.15) one can uniquely extend this functional from the positive cone $\mathcal{C}_{1,\infty,+}(\mathcal{H})$ to the real subspace of the Banach space $\mathcal{C}_{1,\infty}(\mathcal{H})$, and finally to the entire ideal $\mathcal{C}_{1,\infty}(\mathcal{H})$.

Definition 7.8. The *Dixmier trace* $\mathrm{Tr}_\omega(X)$ of the operator $X \in \mathcal{C}_{1,\infty,+}(\mathcal{H})$ is the value of the linear functional (7.5):

$$\mathrm{Tr}_\omega(X) = \mathrm{Lim}_{n\to\infty} \frac{1}{1+\ln(n)} \sum_{j=1}^{n} \lambda_j(X), \tag{7.24}$$

where by (7.6) and (7.11) $\mathrm{Lim}_{n\to\infty}\, \xi_n(\cdot)$ is defined as a positive dilation-invariant functional $\omega_{\lim}(\xi(\cdot))$ on l^∞ (7.12) that satisfies the properties (a), (b), and (c). Since any self-adjoint operator $X \in \mathcal{C}_{1,\infty}(\mathcal{H})$ has the representation $X = X_+ - X_-$, where $X_\pm \in \mathcal{C}_{1,\infty,+}(\mathcal{H})$, one gets $\mathrm{Tr}_\omega(X) = \mathrm{Tr}_\omega(X_+) - \mathrm{Tr}_\omega(X_-)$. Then for arbitrary $Z \in \mathcal{C}_{1,\infty}(\mathcal{H})$ the Dixmier trace is $\mathrm{Tr}_\omega(Z) = \mathrm{Tr}_\omega(\mathrm{Re}\, Z) + i\mathrm{Tr}_\omega(\mathrm{Im}\, Z)$.

Note that if $X \in \mathcal{C}_{1,\infty,+}(\mathcal{H})$, then definition (7.24) of $\mathrm{Tr}_\omega(\cdot)$ together with the definition of the norm $\|\cdot\|_{1,\infty}$ in (6.64), readily imply the estimate $\mathrm{Tr}_\omega(X) \leqslant \|X\|_{1,\infty}$, which in turn yields for arbitrary Z from the Dixmier ideal $\mathcal{C}_{1,\infty}(\mathcal{H})$ the inequality

$$|\mathrm{Tr}_\omega(Z)| \leqslant \|Z\|_{1,\infty}\,. \tag{7.25}$$

Remark 7.9. A decisive step in the construction of the Dixmier trace $\mathrm{Tr}_\omega(\cdot)$ (7.24) was the tacit assumption of the *existence* of the invariant mean (state) $\omega_{\lim}$. To make this existence evident the following remarks are in order:

(1) Although in our construction of $\omega_{\lim}$ we did not use this fact explicitly, we recall that the invariant functional can be included in the set $\mathcal{S}(l^\infty)$ of positive linear functionals (*states*) on the Banach space of bounded sequences l^∞. Note that $\mathcal{S}(l^\infty) \subset (l^\infty)^*$, where the space $(l^\infty)^*$ (of all linear functionals on l^∞) is the *dual* of the Banach space of bounded sequences. Recall that positivity and linearity make the state continuous (equivalently, bounded) on l^∞. So, one can consider the set of states $\omega \in \mathcal{S}(l^\infty)$ taking the positive linear functionals with unit norms.

(2) Now let $(l^\infty)^*$ be equipped with the weak*-topology. Then by the Banach-Alaoglu theorem, the convex set of states $\mathcal{S}(l^\infty)$ is *compact* in $(l^\infty)^*$ in the weak*-topology. For any state $\omega \in \mathcal{S}(l^\infty)$ the relation $\omega(\mathfrak{D}_2(\xi)) =: (\mathfrak{D}_2^*\omega)(\xi)$ defines dual $\mathfrak{D}_2^*$-dilation on the set of states. By definition (7.19) this map is such that $\mathfrak{D}_2^* : \mathcal{S}(l^\infty) \to \mathcal{S}(l^\infty)$, as well as continuous and affine (in fact linear). Then by the Markov-Kakutani theorem the dilation $\mathfrak{D}_2^*$ has a fix point

$$\omega_{\lim} = \mathfrak{D}_2^*\omega_{\lim}, \quad \omega_{\lim} \in \mathcal{S}(l^\infty).$$

This abstract observation justifies the existence of the invariant mean, or functional with the property (c) for $\mathfrak{D}_2$-dilation.

Note that Remark 7.9 has a straightforward extension to any $\mathfrak{D}_k$-dilation for $k > 2$, which is defined similarly to (7.19). Since dilations for different $k \geqslant 2$ *commute*, the extension of the Markov-Kakutani theorem shows that the commutative family $\mathcal{F} = \{\mathfrak{D}_k^*\}_{k\geqslant 2}$ has in $\mathcal{S}(l^\infty)$ a common fix point $\widehat{\omega} = \mathfrak{D}_k^*\widehat{\omega}$, which coincides with ω_{lim}. Therefore, Definition 7.8 of the Dixmier trace does not depend on the degree $k \geqslant 2$ of the dilation $\mathfrak{D}_k$. For more details about different constructions of *invariant* functionals and the corresponding Dixmier trace on $\mathcal{C}_{1,\infty}(\mathcal{H})$, see references in Section 7.4 (Notes to Section 7.2).

Proposition 7.10. *The Dixmier trace has the following properties:*

(a) *For any bounded operator* $B \in \mathcal{L}(\mathcal{H})$ *and any* $Z \in \mathcal{C}_{1,\infty}(\mathcal{H})$ *one has* $\mathrm{Tr}_\omega(ZB) = \mathrm{Tr}_\omega(BZ)$.

(b) *The Dixmier trace is singular:* $\mathrm{Tr}_\omega(C) = 0$ *for any operator* $C \in \mathcal{C}_1(\mathcal{H})$ *from the trace-class ideal, which is the closure of finite-rank operators* $\mathcal{K}(\mathcal{H})$ *in the* $\|\cdot\|_1$*-norm.*

(c) *The Dixmier trace* $\mathrm{Tr}_\omega : \mathcal{C}_{1,\infty}(\mathcal{H}) \to \mathbb{C}$ *is continuous in the* $\|\cdot\|_{1,\infty}$*-norm.*

Proof. (a) Since every operator $B \in \mathcal{L}(\mathcal{H})$ is a linear combination of four unitary operators, it is sufficient to prove the equality $\mathrm{Tr}_\omega(ZU) = \mathrm{Tr}_\omega(UZ)$ for a unitary operator U and moreover only for $Z \in \mathcal{C}_{1,\infty,+}(\mathcal{H})$. Then the corresponding equality follows from the unitary invariance: $s_j(Z) = s_j(ZU) = s_j(UZ) = s_j(UZU^*)$, of singular values of the positive operator Z for all $j \geqslant 1$.

(b) Since $C \in \mathcal{C}_1(\mathcal{H})$ yields $\|C\|_1 < \infty$, definition (7.6) for $\{\lambda_j(|C|)\}_{j\geqslant 1}$ implies $\sigma_n(|C|) \leqslant \|C\|_1$ for any $n \geqslant 1$. Then by Definition 7.8 one gets $\mathrm{Tr}_\omega(C) = 0$. The proof of the last part of the statement is standard.

(c) Since the ideal $\mathcal{C}_{1,\infty}(\mathcal{H})$ is a Banach space and $\mathrm{Tr}_\omega : \mathcal{C}_{1,\infty}(\mathcal{H}) \to \mathbb{C}$ a linear functional, it is sufficient to consider the continuity at $X = 0$. Hence, suppose the sequence $\{X_k\}_{k\geqslant 1} \subset \mathcal{C}_{1,\infty}(\mathcal{H})$ converges to $X = 0$ in $\|\cdot\|_{1,\infty}$-topology, that is, by (6.64)

$$\lim_{k\to\infty} \|X_k\|_{1,\infty} = \lim_{k\to\infty} \sup_{n\in\mathbb{N}} \frac{1}{1+\ln(n)} \sigma_n(X_k) = 0. \tag{7.26}$$

Since (7.25) implies $|\mathrm{Tr}_\omega(X_k)| \leqslant \|X_k\|_{1,\infty}$, the assertion follows from (7.26). □

Therefore, by Proposition 7.10(b) the Dixmier construction gives an example of a *singular* trace in the sense of Definition 7.5.

7.3 Product formulae

Consider first the trace-class ideal $\mathcal{C}_{\phi_1}(\mathcal{H}) = \mathcal{C}_1(\mathcal{H})$, where the symmetric norming function ϕ_1 is defined by (6.20), (6.21). Then under the assumptions of the general Proposition 6.22, the Trotter-Kato product formula convergence in the trace-norm:

$$\|\cdot\|_1\text{-}\lim_{r\to+\infty} F(t/r)^r = e^{-tH} , \tag{7.27}$$

locally uniformly away from zero for Kato functions of class $\mathcal{K}$ (Appendix C, Section C.1).

Since the trace Tr is a continuous functional on the ideal $\mathcal{C}_1(\mathcal{H})$, the limit (7.27) implies

$$\begin{aligned}\operatorname{Tr}(e^{-tH}) &= \lim_{r\to+\infty} \operatorname{Tr}(F(t/r)^r) = \lim_{r\to+\infty} \operatorname{Tr}(T(t/r)^r) \\ &= \lim_{n\to+\infty} \operatorname{Tr}((f(tA/n)g(tB/n))^n) = \lim_{n\to\infty} \operatorname{Tr}((g(tB/n)f(tA/n))^n),\end{aligned} \tag{7.28}$$

where the limits of approximants for the families $\{T(t)\}_{t\geqslant 0}$, $\{f(t)g(t)\}_{t\geqslant 0}$, and $\{g(t)f(t)\}_{t\geqslant 0}$ follow from the results of Section 6.4, Proposition 6.18.

Now we consider the Dixmier ideal: $\mathcal{C}_\phi(\mathcal{H})|_{\phi=(1,\infty)} = \mathcal{C}_{1,\infty}(\mathcal{H})$, (6.63).

Proposition 7.11. *Let A and B be non-negative self-adjoint operators on a Hilbert space $\mathcal{H}$ and let $f^D : \mathbb{R}_0^+ \to \mathbb{R}_0^+$ and $g^D : \mathbb{R}_0^+ \to \mathbb{R}_0^+$ be bounded Borel measurable functions such that $F^D(t_0)^p \in \mathcal{C}_\phi(\mathcal{H})$ for some $t_0 > 0$ and some integer $p \geqslant 1$.*

Suppose the Kato functions f and g are dominated by f^D and g^D, respectively, and the Trotter-Kato product formula converges in the operator-norm locally uniformly away from zero for self-adjoint approximants $\{F(t/r)^r\}_{r\geqslant 1}$.

Then for the Dixmier trace, similarly to (7.28), it holds that

$$\begin{aligned}\operatorname{Tr}_\omega(e^{-tH}) &= \lim_{r\to+\infty} \operatorname{Tr}_\omega(F(t/r)^r) = \lim_{r\to+\infty} \operatorname{Tr}_\omega(T(t/r)^r) \\ &= \lim_{n\to+\infty} \operatorname{Tr}_\omega((f(tA/n)g(tB/n))^n) = \lim_{n\to\infty} \operatorname{Tr}_\omega((g(tB/n)f(tA/n))^n).\end{aligned} \tag{7.29}$$

The limits are locally uniform away from pt_0 for all families of approximants generated by f and g.

Proof. The argument is close to that used for Proposition 6.22. Recall that, by Lemma 6.21 $e^{-tH} \in \mathcal{C}_\phi(\mathcal{H})$ for $t > pt_0$. Lemma 6.20 implies that for each bounded interval $[\tau_0, \tau] \subseteq (pt_0, \infty)$ there is an $r_0 \geqslant 1$ such that $F(t/r)^r \in \mathcal{C}_\phi(\mathcal{H})$ for $r \geqslant r_0$ and $t \in [\tau_0, \tau]$. By (6.113), we obtain $\sup_{r\geqslant r_0} \sup_{t\in[\tau_0,\tau]} \|F(t/r)^r\|_\phi \leqslant \|F^D(t_0)\|_\phi < \infty$, which shows that the sequence $\{F(t/r)^r\}_{r\geqslant 1}$ is bounded in $\mathcal{C}_\phi(\mathcal{H})$ locally uniformly away from pt_0. Now applying Proposition 6.19 we get that the Trotter-Kato product formula

$$\|\cdot\|_{1,\infty}\text{-}\lim_{r\to+\infty} F(t/r)^r = e^{-tH}, \tag{7.30}$$

converges locally uniformly away from pt_0. Since by Proposition 7.10(c) the Dixmier trace $\operatorname{Tr}_\omega : \mathcal{C}_{1,\infty}(\mathcal{H}) \to \mathbb{C}$ is continuous in the $\|\cdot\|_{1,\infty}$-norm topology, the limit (7.30) implies

$$\operatorname{Tr}_\omega(e^{-tH}) = \lim_{r\to+\infty} \operatorname{Tr}_\omega(F(t/r)^r). \tag{7.31}$$

The rest of (7.29) follows from (7.31) and Proposition 6.18 of Section 6.4. There it is shown that (7.30) yields the same limit in the symmetrically normed ideals $\mathcal{C}_\phi(\mathcal{H})$ for the families $\{T(t)\}_{t\geqslant 0}$, $\{f(t)g(t)\}_{t\geqslant 0}$, and $\{g(t)f(t)\}_{t\geqslant 0}$ and for the corresponding choice of $r \vee n$ in the product formulae. □

Different types of sufficient conditions that ensure assumptions of Proposition 7.11 are formulated in Proposition 6.28, Proposition 6.29, Proposition 6.30, Corollary 6.31, and Corollary 6.32.

Corollary 7.12. *The* $\|\cdot\|_\phi$*-norm estimates* $\eta_\phi(r)$ *and* $\eta_\phi(n)$ *of Proposition* 6.38 *determine the rate of convergence*

$$|\operatorname{Tr}_\omega(e^{-tH}) - \operatorname{Tr}_\omega(F(t/x)^x)| \leqslant C_\omega \eta_\omega(x), \tag{7.32}$$

of the Dixmier traces in (7.31) *for* $x = r \vee n$, *where* $r \in \mathbb{R}^+$ *or* $n \in \mathbb{N}$. *In fact, the rate is the same (modulo* C_ω *and* $r \vee n$*) for all Trotter-Kato approximants generated by* $\{F(t)\}_{t\geqslant 0}$, $\{T(t)\}_{t\geqslant 0}$, $\{f(t)g(t)\}_{t\geqslant 0}$, *and* $\{g(t)f(t)\}_{t\geqslant 0}$.

Consider for concreteness the approximants $\{(F(t/x))^x\}_{x=r\vee n}$. *Recall that by inequalities* (6.160) *and* (7.25) *we get the estimate*

$$\begin{aligned} |\operatorname{Tr}_\omega(e^{-tH}) - \operatorname{Tr}_\omega(F(t/x)^x)| &\leqslant \|e^{-tH} - F(t/x)^x\|_{1,\infty} \\ &\leqslant C\, \eta_{1,\infty}(x), \end{aligned} \tag{7.33}$$

for $t \in [\tau_0, \tau]$ *and* $x \geqslant x_0$, *where either* $x = r \in \mathbb{R}^+$, *or* $x = n \in \mathbb{N}$. *Then the rate* $\eta_\omega(\cdot) = \eta_{1,\infty}(\cdot)$.

Corollary 7.13. *Under assumptions of Proposition* 6.38 *the estimate of the rates of* $\|\cdot\|_{1,\infty}$*-convergence in* (7.33) *are entirely determined by the operator-norm error bound* η. *Hence, the lifting of operator-norm estimates to the error bound estimates in the* $\mathcal{C}_{1,\infty}(\mathcal{H})$*-topology for* $t > pt_0$ *coincide (up to factors) with* $\eta_\phi(r)$ *or* $\eta_\phi(n)$ *for* $\phi = (1,\infty)$. *Therefore, by Corollary* 7.12,

$$\begin{aligned} &\eta_\omega(r) := 2\,\|F^D(t_0)\|_{1,\infty}\,\eta(r/2), \ r \in \mathbb{R}^+, \quad \text{or} \\ &\eta_\omega(n) := \|F^D(t_0)\|_{1,\infty}\,\{\eta([n/2]) + \eta([(n+1)/2])\}, \quad n \in \mathbb{N}, \end{aligned} \tag{7.34}$$

where $[s]$ *denotes the entire part of* $s > 0$.

If the operator-norm estimate η *is optimal, then, similarly to* η_ϕ, *the rate* η_ω *is also optimal.*

Note that Proposition 6.40 gives the explicit expressions for the rates of convergence (7.34) and the Trotter-Kato formulae for all families $\{F(t)\}_{t\geqslant 0}$, $\{T(t)\}_{t\geqslant 0}$, $\{f(t)g(t)\}_{t\geqslant 0}$, and $\{g(t)f(t)\}_{t\geqslant 0}$.

Proposition 7.14. *Assume that the Kato functions* f *and* g *are dominated, respectively, by* f^D *and* g^D *and that the corresponding function* $F^D(t)$ *is such that* $F^D(t_0) \in \mathcal{C}_\phi(\mathcal{H})$ *for some* $t_0 > 0$. *Let for some* $\alpha \in (1/2, 1]$ *the Kato functions* f *and* g *belong to the class* $\mathcal{K}_\alpha$. *Suppose that*

$$b^{1/\alpha} C_{1/2\alpha}\, S_1 < 1.$$

Then the function (7.33)

$$\eta_{1,\infty}(n) \leqslant \Gamma_1 \ln(n)/n^{2\alpha-1},$$

for $n > 1$, *is the error bound in* $\mathcal{C}_{1,\infty}(\mathcal{H})$ *away from* $2t_0$ *for the Trotter-Kato product formulae for all families* $\{F(t)\}_{t\geqslant 0}$, $\{T(t)\}_{t\geqslant 0}$, $\{f(t)g(t)\}_{t\geqslant 0}$, *and* $\{g(t)f(t)\}_{t\geqslant 0}$. *Here the rate of convergence is* $\ln(n)/n^{2\alpha-1}$.

If in addition, besides $f, g \in \mathcal{K}_\alpha$ *and* (6.178), *for some* $\alpha \in (1/2, 1)$ *the condition* $\operatorname{dom} H^\alpha \subseteq \operatorname{dom} A^\alpha$ *is satisfied, then the function* (7.33)

$$\eta_{1,\infty}(n) \leqslant \Gamma_2/n^{2\alpha-1},$$

for $n > 1$, *is the error bound in* $\mathcal{C}_{1,\infty}(\mathcal{H})$ *away from* $2t_0$ *for the Trotter-Kato product formulae for all families* $\{F(t)\}_{t\geqslant 0}$, $\{T(t)\}_{t\geqslant 0}$, $\{f(t)g(t)\}_{t\geqslant 0}$, *and* $\{g(t)f(t)\}_{t\geqslant 0}$. *Here the rate of convergence is improved to* $1/n^{2\alpha-1}$.

More examples of explicit estimates for the rate of convergence in the Dixmier ideal can be found by lifting the operator-norm error bounds for the Trotter-Kato product formulae from Section 5.2.

7.4 Notes

Notes to Section 7.1. The contents of this section are standard. The section starts as a continuation of Section 6.2 by a discussion of two key definitions: the *normal* trace and the *singular* trace. Here we follow in part [Zag19].

For more details, including the proof of Proposition 7.1, the reader is referred to the book [LSZ12], which is a rather complete source on this topic.

Notes to Section 7.2. Recall that the *Dixmier trace* and *Dixmier ideal* arose from the question of whether the algebra $\mathcal{L}(\mathcal{H})$ of all bounded linear operators on a Hilbert space $\mathcal{H}$ admits a unique nontrivial trace. J. Dixmier solved this problem in the negative in a short note [Dix66]. He proved the existence on $\mathcal{L}(\mathcal{H})$ of (singular) traces that vanish on the ideal of trace-class operators.

The key steps of his construction are presented in Section 7.2. It obviously yields a singular trace in the sense of Definition 7.5. The example of construction is taken from the lecture notes [Su08]. We consider the simplest illustration based on the proof of the existence of the dilation $\mathfrak{D}_2$-invariant functional (mean) $\omega_{\lim}$. The reader can find in [Su08] examples of other constructions of invariant functionals based, e.g., on the *Banach limit*, or on the *Cesàro mean.*

Note that in our presentation we did not assume *a priori* that the functional $\omega_{\lim}$ is a state, i.e., that it is a positive linear functional with unit norm on the Banach space l^∞, see Remark 7.9. In fact, instead of continuity, the argument for proving of linearity needs only Remark 7.6(b) that the state $\omega_{\lim}$ is determined *modulo* sequences that belong to the set c_0, see [Zag19].

For more details and more examples we refer to the review article [CaSu06] and to the lecture notes [Su08].

Notes to Section 7.3. The Trotter-Kato product formulae in the Dixmier ideal $\mathcal{C}_{1,\infty}(\mathcal{H})$ were considered for the first time by H. Neidhardt and V. A. Zagrebnov in [NZ99d]. There under some sufficient conditions the convergence of Dixmier traces for the Trotter-Kato product formulae were announced without proof, as well as without hypothesis about estimates of the rate of convergence.

Here we developed indispensable arguments and made them explicit by proving in Proposition 7.11 that the Trotter-Kato product formulae convergence in the $\|\cdot\|_{1,\infty}$-topology. To establish these results with an estimate of the *rate* of convergence one needs more arguments.

In Proposition 7.14 the Trotter-Kato product formulae in Dixmier ideal $\mathcal{C}_{1,\infty}(\mathcal{H})$ with an estimate of the *rate* of convergence in the $\|\cdot\|_{1,\infty}$-topology is proved following a standard scheme of lifting arguments developed in Chapter 5 and Chapter 6. To this end we use the *lifting* of the estimates of the rate of convergence due to Proposition 6.38.

Appendix A. Spectra of closed operators

Although in this chapter we essentially focus on linear operators in a Hilbert space $\mathcal{H}$, the information below also concerns bounded and unbounded operators in a Banach space $\mathcal{B}$. We shall make clear if certain facts concern only a specific type of operators or a particular space. By $\mathcal{L}(\mathcal{H}_1, \mathcal{H}_2)$ we denote the Banach space of bounded linear operators from a Hilbert space $\mathcal{H}_1$ to a Hilbert space $\mathcal{H}_2$, and we set $\mathcal{L}(\mathcal{H}) := \mathcal{L}(\mathcal{H}, \mathcal{H})$. Similar notations will be used for the general case of Banach spaces $\mathcal{B}_1, \mathcal{B}_2$. Notice that we do not assume (*a priory*) that the domains of involved operators are dense, but do it whenever needed.

A.1 Resolvents and spectra

(a) Let A with domain $\operatorname{dom} A \subseteq \mathcal{H}$ be a *closed* linear operator in a Hilbert (or Banach) space $\mathcal{H}$ (or $\mathcal{B}$) and let $A_\zeta := A - \zeta \mathbb{1}$ for $\zeta \in \mathbb{C}$. Then the map $A_\zeta : \operatorname{dom} A \to \operatorname{ran} A_\zeta$ is said to be *continuous* in $\operatorname{dom} A$ if for any sequence $\{u_n\}_{n\geqslant 1} \subset \operatorname{dom} A$ such that $\lim_{n\to\infty} u_n = 0$, we have $\lim_{n\to\infty} A_\zeta u_n = 0$.

The $\operatorname{ran} A_\zeta$ of this closed map is in general not a *subspace*, but a *non-closed* linear subset (manifold) of $\mathcal{H}$ (or $\mathcal{B}$).

Since in $\mathcal{H}$ the orthogonal complement $(\operatorname{ran} A_\zeta)^\perp$ is closed (i.e., it is a *subspace* of $\mathcal{H}$), we can define the *deficiency* (or defect number) of the closed operator A_ζ by $\operatorname{def} A_\zeta := \dim(\operatorname{ran} A_\zeta)^\perp$.

(b) Below we would need the following version of the *closed graph theorem*: Let the operator $A \in \mathcal{C}(\mathcal{H})$ with domain $\operatorname{dom} A$ be closed in $\mathcal{H}$. Then, saying that $\operatorname{dom} A$ being a *subspace* of $\mathcal{H}$ is equivalent to saying that A is *continuous* on $\operatorname{dom} A$.

The latter is the same as the *boundedness* of A on $\operatorname{dom} A$: there exists $a > 0$ such that $\|Au\| \leqslant a\|u\|$, $u \in \operatorname{dom} A$. We denote this by $A \in \mathcal{L}(\operatorname{dom} A, \mathcal{H})$. Therefore, a closed *unbounded* operator $A \in \mathcal{C}(\mathcal{H})$ must have an *unclosed* domain.

(c) Note that the range $\operatorname{ran} A$ is a *subspace* of $\mathcal{H}$ if the operator A is closed and *injective*, that is, satisfies $\|Au\| \geqslant c\|u\|$, for some $c > 0$ and all $u \in \operatorname{dom} A$. In

V. A. Zagrebnov, *Gibbs Semigroups*, Operator Theory: Advances and Applications 273, https://doi.org/10.1007/978-3-030-18877-1

turn, this property of the range implies the existence of the *closed* inverse operator $A^{-1} : \operatorname{ran} A \to \operatorname{dom} A$, which is continuous: $A^{-1} \in \mathcal{L}(\operatorname{ran} A, \operatorname{dom} A)$, on this range. We recall that if in addition $\operatorname{ran} A = \mathcal{H}$, then A is *surjective* (*onto* map). Together with injectivity, this makes this map *bijective.*

Note that the general observations (b) and (c) are valid in Banach spaces.

(d) For a closed linear operator A in a Banach space $\mathcal{B}$ we call

$$\begin{aligned}\rho(A) &= \{\zeta \in \mathbb{C} : A_\zeta : \operatorname{dom} A \to \mathcal{B} \text{ is bijective}\} \\ &\Leftrightarrow \{A_\zeta : \operatorname{dom} A \to \mathcal{B} \text{ is injective}\} \wedge \{A_\zeta : \operatorname{dom} A \to \mathcal{B} \text{ is surjective}\},\end{aligned}$$

the *resolvent set.* Then, by definition, the *spectrum* of A is the set $\sigma(A) := \mathbb{C}\backslash\rho(A)$. By the closed graph theorem, for $\zeta \in \rho(A)$ the resolvent of A

$$R_A(\zeta) = (A_\zeta)^{-1} : \rho(A) \to \mathcal{L}(\mathcal{B}),$$

is a bounded operator-valued function, although with independent of ζ and (in general) unclosed range: $\operatorname{ran} R_A(\zeta) = \operatorname{dom} A$, see (c).

(e) Note that $\zeta_0 \in \rho(A)$ if and only if there exists $\delta_0 > 0$ such that $\|A_{\zeta_0} u\| \geqslant \delta_0\|u\|$ for any $u \in \operatorname{dom} A$. Then $\|R_A(\zeta_0)w\| \leqslant \delta_0^{-1}\|w\|$, $w \in \mathcal{B}$, and for other points $\zeta \in \rho(A)$ one obtains the *resolvent identity*

$$R_A(\zeta) - R_A(\zeta_0) = (\zeta - \zeta_0)\, R_A(\zeta)\, R_A(\zeta_0).$$

For a fixed $\zeta_0 \in \rho(A)$ this identity can be regarded as an *equation* for the unknown operator-valued function $\zeta \mapsto R_A(\zeta)$. It is solvable in the open disc $D_{\zeta_0}(\delta_0) := \{\zeta \in \mathbb{C} : |\zeta - \zeta_0| < \delta_0\}$. This solution is a uniformly operator-norm convergent in the disc $D_{\zeta_0}(\delta_0)$ power series

$$R_A(\zeta) = \sum_{n=0}^{\infty} (\zeta - \zeta_0)^n\, R_A(\zeta_0)^{n+1}.$$

Therefore, the resolvent is the piecewise holomorphic (in the *operator-norm* topology) function $R_A : \zeta \mapsto R_A(\zeta)$ defined in any connected component of the *open* resolvent set $\rho(A)$.

(f) Note that the power series representation for the resolvent implies for $\zeta \in \rho(A)$ the estimate from below $\|R_A(\zeta)\| \geqslant 1/\operatorname{dist}(\zeta, \sigma(A))$, i.e., the spectrum $\sigma(A)$ is a *closed* subset of $\mathbb{C}$ and its topological boundary $\partial\sigma(A) \subset \sigma(A)$.

(g) If the operator A is *bounded* with norm $\|A\|$, then the fact that the series

$$R_A(\zeta) = -\frac{1}{\zeta} \sum_{n=0}^{\infty} (A/\zeta)^n,$$

converges for $\{\zeta \in \mathbb{C} : |\zeta| > \|A\|\} \subseteq \rho(A)$ and the Liouville theorem for the operator-valued holomorphic functions in $\mathbb{C}$ show that the spectrum $\sigma(A) \neq \varnothing$. Note that it is compact and lies in the closed disc $\overline{\mathcal{D}_r}$ of radius $r = \|A\|$.

(h) Recall that $r(A) := \sup\{|\zeta| : \zeta \in \sigma(A)\}$ is the *spectral radius* of A, and one has $r(A) \leqslant \|A\|$. We note that $r(A) = \lim_{n\to\infty} \|A^n\|^{1/n}$. Hence, $r(A) = \|A\|$ if $A \in \mathcal{L}(\mathcal{H})$ is *self-adjoint.* In the opposite case it is possible that $r(A) = 0$ (e.g., for (*quasi*)-*nilpotent* operators), whereas $\|A\| \neq 0$. See below the example of the *Volterra* operator.

(i) Note that the assumption of *closeness* of the linear operator A makes the notion of spectrum of A *nontrivial.* Indeed, suppose A is *not* closed. To check whether the resolvent set of A is *non-empty,* suppose that for some $\zeta \in \mathbb{C}$ the operator A_ζ^{-1} is defined on $\mathcal{B}$ and that $A_\zeta^{-1} \in \mathcal{L}(\mathcal{B})$. Then A_ζ^{-1} is closed, which implies that A_ζ is also closed, and so is A. This contradiction shows that $\rho(A) = \varnothing$, that is, the spectral problem for nonclosed A is *trivial*: the solution is always $\sigma(A) = \mathbb{C}$.

However, there are examples of *closed* operators with spectrum $\sigma(A) = \mathbb{C}$ and by (g) they are *not* bounded operators.

(j) It is often useful to consider a *pseudo-resolvent* $\hat{R}(\zeta)$, which is a bounded operator-valued function on some domain $D \subset \mathbb{C}$. For motivation, note that the strong operator limit of the resolvents $\{R_k(\zeta)\}_{k\geqslant 1}$

$$\hat{R}(\zeta) = \operatorname*{s-lim}_{k\to\infty} R_k(\zeta), \quad \zeta \in D_s \subset \mathbb{C},$$

does *not* need to be the resolvent of an operator. Here D_s denotes the domain of *strong* convergence in the regular set for all $\{R_k(\zeta)\}_{k\geqslant 1}$. However, this strong limit satisfies the resolvent equation for $\zeta, \zeta_0 \in D_s$, and it is called *pseudo-resolvent.*

We recall that the pseudo-resolvent $\hat{R}(\zeta)$ is the *resolvent* of a closed operator A if and only if $\ker \hat{R}(\zeta) = \{0\}$ for any $\zeta \in D_s$.

Let D_s be non-empty. Then there is an alternative: either the limit $\hat{R}(\zeta)$ is not invertible for any $\zeta \in D_s$, or $\hat{R}(\zeta) = R_A(\zeta)$ is the resolvent of a unique closed operator A and one also gets that $D_s = \rho(A) \cap D_b$. Here D_b is the domain of *norm-boundedness* of the sequence $\{R_k(\zeta)\}_{k\geqslant 1}$.

A.2 Core $\widehat{\sigma}(A)$ of $\sigma(A)$

(a) By virtue of A.1(a), the deficiency of A_ζ is equal to $\operatorname{def} A_\zeta := \dim(\operatorname{ran} A_\zeta)^\perp$. If $\operatorname{dom} A$ is *dense* in $\mathcal{H}$, then the adjoint to A_ζ, $A_\zeta{}^* = A^* - \bar{\zeta}\mathbb{1}$, is well defined for any $\zeta \in \mathbb{C}$, and one has the orthogonal decomposition $\mathcal{H} = \overline{\operatorname{ran} A_\zeta} \oplus \ker A_\zeta{}^*$. Therefore, the deficiency of A_ζ can be expressed using the *closed* kernel $\ker A_\zeta{}^*$ by the *formula*: $\operatorname{def} A_\zeta = \dim(\ker A_\zeta{}^*)$.

Note that if A is *closed*, then the adjoint operator A^* is in turn densely defined in $\mathcal{H}$ and then one has that $A_\zeta{}^{**} = A_\zeta$.

(b) Let A be a *closed* linear operator in $\mathcal{H}$ and let $\operatorname{ran} A_\zeta \subset \mathcal{H}$ be closed for $\zeta \in \mathbb{C}$. Then one can define on the $\operatorname{ran} A_\zeta$ a continuous inverse $A_\zeta^{-1} \in \mathcal{L}(\operatorname{ran} A_\zeta, \mathcal{H})$. These points ζ constitute the set of *quasi-regular points* $\widehat{\rho}(A)$ of the operator A.

Note that the set $\widehat{\rho}(A)$ is *open* and that the integer-valued function $\zeta \mapsto \mathrm{def}\, A_\zeta$ is *constant* on each of the connected components of $\widehat{\rho}(A)$.

(c) Let $\rho(A) := \{\zeta \in \widehat{\rho}(A) : \mathrm{def}\, A_\zeta = 0\}$ denote the set of the *regular points* of the operator A. Then for $\zeta \in \rho(A)$ the map $A_\zeta : \mathrm{dom}\, A \to \mathcal{H}$ is *bijective* with $\mathrm{ran}\, A_\zeta = \mathcal{H}$ and by the closed graph theorem the inverse operator (resolvent) is bounded: $A_\zeta^{-1} \in \mathcal{B}(\mathcal{H})$. Therefore, by A.1(d) the set of regular points coincides with the resolvent set and $\rho(A) \subseteq \widehat{\rho}(A)$.

(d) The set $\sigma(A) := \mathbb{C}\backslash\rho(A)$ is known as the *spectrum* of the closed operator A, whereas the complement of the set of the quasi-regular points, $\widehat{\sigma}(A) := \mathbb{C}\backslash\widehat{\rho}(A)$, is called the *core* of the spectrum $\sigma(A)$. It is clear that $\widehat{\sigma}(A) \subseteq \sigma(A)$ and that both of these sets are closed.

A.3 Subsets of the spectrum $\sigma(A)$

(a) By A.1(d) the mapping $A_\zeta : \mathrm{dom}\, A \to \mathcal{B}$, for $\zeta \in \sigma(A)$, is not *bijective* with $\mathrm{ran}\, A_\zeta = \mathcal{B}$. This is equivalent to the statement:

$$\overline{\{A_\zeta : \mathrm{dom}\, A \to \mathcal{B} \ \text{ is injective}\}} \vee \overline{\{A_\zeta : \mathrm{dom}\, A \to \mathcal{B} \ \text{ is surjective}\}},$$

that is, *either* the mapping A_ζ is not one-to-one, *or* it is not onto.

Now note that the following three statements are equivalent:

$$\begin{aligned}\{A_\zeta : \mathrm{dom}\, A \to \mathcal{B} \text{ is surjective}\} &\Leftrightarrow \{\mathrm{ran}\, A_\zeta = \mathcal{B}\}\\ &\Leftrightarrow \{\mathrm{ran}\, A_\zeta \text{ is dense in } \mathcal{B}\} \wedge \{\mathrm{ran}\, A_\zeta \text{ is closed in } \mathcal{B}\}.\end{aligned}$$

Therefore, the statement $\{\zeta \in \sigma(A)\}$ is equivalent to the following *three-part* assertion:

$$\begin{aligned}\{\zeta \in \sigma(A)\} \Leftrightarrow &\{A_\zeta : \mathrm{dom}\, A \to \mathcal{B} \ \text{ is not injective}\}\\ &\vee \{\mathrm{ran}\, A_\zeta \text{ is not dense }\} \vee \{\mathrm{ran}\, A_\zeta \text{ is not closed }\}\ .\end{aligned}$$

Either of these three parts implies the statement $\{\zeta \in \sigma(A)\}$.

(b) Since the *first* part: $\{A_\zeta : \mathrm{dom}\, A \to \mathcal{B} \ \text{ is not injective}\}$, implies a non-trivial $\ker A_\zeta$, the *point* spectrum $\sigma_{\mathrm{p}}(A)$ of A is the subset of $\sigma(A)$ defined as

$$\sigma_{\mathrm{p}}(A) := \{\zeta \in \mathbb{C} : \ker A_\zeta \neq \{0\}\}\ .$$

This subset consists of eigenvalues of A, and $\ker A_\zeta$ contains the associated eigenvectors. In $\sigma_{\mathrm{p}}(A) \subseteq \sigma(A)$ we distinguish the subset of eigenvalues of *infinite* multiplicity:

$$\sigma_{\mathrm{p}}^{\infty}(A) := \{\zeta \in \mathbb{C} : \dim(\ker A_\zeta) = \infty\}.$$

(c) The *second* part: $\{\mathrm{ran}\, A_\zeta \text{ is not dense}\} \Leftrightarrow \{\overline{\mathrm{ran}\, A_\zeta} \neq \mathcal{B}\}$, implies that for the injective map A_ζ with closed range we have $(\mathrm{ran}\, A_\zeta)^\perp \neq \emptyset$. Hence, def $A_\zeta \neq 0$, and the operator A_ζ is boundedly invertible on $\operatorname{ran} A_\zeta$. Then by the definitions of the quasi-regular set $\widehat{\rho}(A)$ and the resolvent set $\rho(A)$ (see A.2) the corresponding set of ζ is

$$\sigma_{\text{res}}(A) := \widehat{\rho}(A)\backslash\rho(A) = \sigma(A)\backslash\widehat{\sigma}(A).$$

The subset $\sigma_{\text{res}}(A) \subset \sigma(A)$ is called the *residual* spectrum of A.

Since for $\zeta \in \sigma_{\text{res}}(A)$ the operator A_ζ remains injective, the point and the residual spectra are *disjoint*: $\sigma_{\text{p}}(A) \cap \sigma_{\text{res}}(A) = \varnothing$.

(d) The last remark has an important corollary: if a densely defined in a Hilbert space $\mathcal{H}$ operator A is self-adjoint, then $\sigma_{\text{res}}(A) = \varnothing$.

To show this we note that if $\zeta \in \sigma_{\text{res}}(A)$, that is $\{\overline{\operatorname{ran} A_\zeta} \neq \mathcal{H}\}$, then by the orthogonal decomposition and by *formula* in A.2(a) one gets $\ker A_\zeta{}^* \neq \varnothing$, which together with A.3(b) yields $\overline{\zeta} \in \sigma_{\text{p}}(A^*)$. Since $A = A^*$, the spectrum $\sigma(A) \subseteq \mathbb{R}$. Therefore, $\zeta \in \sigma_{\text{p}}(A)$ and consequently $\sigma_{\text{p}}(A) \cap \sigma_{\text{res}}(A) \neq \varnothing$ leads to a contradiction with the last remark in A.3(c).

(e) The *third* part: $\{\operatorname{ran} A_\zeta$ is not closed$\}$, motivates the definition of the *continuous* spectrum $\sigma_{\text{cont}}(A)$ of A as the subset

$$\sigma_{\text{cont}}(A) := \{\zeta \in \mathbb{C} : \ker A_\zeta = \{0\} \wedge \operatorname{ran} A_\zeta \neq \overline{\operatorname{ran} A_\zeta} = \mathcal{B}\}.$$

First we note that the intersection $\sigma_{\text{p}}(A) \cap \sigma_{\text{cont}}(A)$ can be *non-empty*, see examples below (Section A.4(f)). Next, we remark that $\sigma_{\text{p}}(A) \subset \widehat{\sigma}(A)$ and $\sigma_{\text{cont}}(A) \subset \widehat{\sigma}(A)$, that is, in fact one has

$$\widehat{\sigma}(A) = \sigma_{\text{p}}(A) \cup \sigma_{\text{cont}}(A).$$

Indeed, the inclusion $\sigma_{\text{p}}(A) \cup \sigma_{\text{cont}}(A) \subseteq \widehat{\sigma}(A)$ is obvious. Now suppose that $\zeta \notin \sigma_{\text{p}}(A) \cup \sigma_{\text{cont}}(A)$. Then by the definitions of the spectra $\sigma_{\text{p}}(A)$ and $\sigma_{\text{cont}}(A)$, the map A_ζ is injective and $\operatorname{ran} A_\zeta$ is a subspace. Therefore, $A_\zeta^{-1} \in \mathcal{L}(\operatorname{ran} A_\zeta, \mathcal{B})$. Then by A.2(b)-(d) the point ζ is quasi-regular: $\zeta \in \widehat{\rho}(A)$, that yields $\sigma_{\text{p}}(A) \cup \sigma_{\text{cont}}(A) \supseteq \widehat{\sigma}(A)$.

(f) The representation of the core $\widehat{\sigma}(A)$ found in (e) and the definition (c) of the residual spectrum $\sigma_{\text{res}}(A)$ yield $\sigma_{\text{res}}(A) \cap \widehat{\sigma}(A) = \varnothing$. Thus we infer that

$$\sigma(A) = \sigma_{\text{p}}(A) \cup \sigma_{\text{cont}}(A) \cup \sigma_{\text{res}}(A),$$

and that $\sigma_{\text{res}}(A) = \sigma(A) \cap \widehat{\rho}(A)$. Therefore, the residual spectrum is an *open* set, which consists of components of $\widehat{\rho}(A)$ with nontrivial deficiencies $\operatorname{def} A_\zeta \neq 0$. Equivalently, $\zeta \in \sigma_{\text{res}}(A)$ implies that the closed $\operatorname{ran} A_\zeta \neq \mathcal{B}$ and $A_\zeta^{-1} \in \mathcal{L}(\operatorname{ran} A_\zeta, \mathcal{B}))$.

(g) Suppose that $\widetilde{A} \supset A$ (i.e., $\operatorname{dom} A \subset \operatorname{dom} \widetilde{A}$) is a closed *extension* of a closed operator A. Then $\widehat{\sigma}(A) \subset \widehat{\sigma}(\widetilde{A})$. Indeed, by the representation of the core $\widehat{\sigma}(A)$ found in (e) one gets the following alternative:

$$\{\zeta \in \widehat{\sigma}(A)\} \Rightarrow \{\nexists\, A_\zeta^{-1}\} \vee \{A_\zeta^{-1} \notin \mathcal{L}(\mathcal{B})\}.$$

Since $A_\zeta \subset \widetilde{A}_\zeta$, the extension *cannot* reduce the core of the spectrum $\widehat{\sigma}(A) = \sigma_{\rm p}(A) \cup \sigma_{\rm cont}(A)$. Moreover, it is clear that $\sigma_{\rm p}(A) \subseteq \sigma_{\rm p}(\widetilde{A})$. On the other hand, the continuous spectrum $\sigma_{\rm cont}(\widetilde{A})$ may be *larger* or *smaller* than $\sigma_{\rm cont}(A)$. We note that the latter case is possible only if

$$\{\zeta \in \sigma_{\rm cont}(A)\backslash\sigma_{\rm cont}(\widetilde{A})\} \Rightarrow \{\zeta \in \sigma_{\rm p}^{\infty}(\widetilde{A})\}.$$

Now we recall that if $\zeta \in \sigma_{\rm res}(A)$, then there exists an extension $\widetilde{A} \supset A$ such that $\zeta \in \rho(\widetilde{A})$. On the other hand, if $\zeta \in \rho(A)$, then for this ζ and any nontrivial extension $\widetilde{A} \supset A$ one gets $\zeta \in \sigma_{\rm p}(\widetilde{A})$. Indeed, since $\operatorname{ran} A_\zeta = \mathcal{B}$, then for any $u \in \operatorname{dom} \widetilde{A}\backslash \operatorname{dom} A$ there exists vector $w \in \operatorname{dom} A$ such that $A_\zeta w = \widetilde{A}_\zeta u$. Hence, for $u \neq w$ one gets the non-zero element $u - w \in \ker \widetilde{A}_\zeta$.

A.4 Approximate and essential spectra

(a) Let $A : \operatorname{dom}(A) \to \mathcal{B}$, be a closed operator in a Banach space $\mathcal{B}$. Recall that if for $\zeta \in \mathbb{C}$ there exists a sequence $\{u_n\}_{n\geqslant 1} \subset \operatorname{dom} A$ such that $\|u_n\| = 1$ and $\lim_{n\to\infty} \|A_\zeta u_n\| = 0$, then ζ belongs to the set $\sigma_{\rm app}(A)$, which is called the *approximate spectrum* (or approximate point spectrum) of A.

(b) We note that according to this definition the approximate spectrum coincides with the core of the spectrum: $\sigma_{\rm app}(A) = \widehat{\sigma}(A)$. On the other hand, the definition is constructive and it helps to establish many results, especially in a Hilbert space. We start, however, with one general result valid in a Banach space.

If $A : \operatorname{dom} A \to \mathcal{B}$ is a closed operator, then the (topological) boundary of the operator spectrum belongs to the core of the spectrum: $\partial\sigma(A) \subseteq \widehat{\sigma}(A)$. Recall that the non-empty operator spectrum $\sigma(A)$ is a closed set. Let $\zeta_0 \in \partial\sigma(A) \subseteq \sigma(A)$. Then there exists a sequence $\{\zeta_n\}_{n\geqslant 1} \subset \rho(A)$ such that $\lim_{n\to\infty} \zeta_n = \zeta_0$. Using the estimate in A.1(f) one can find a vector $x \in \mathcal{B}$ such that $\lim_{n\to\infty} \|R_A(\zeta_n)x\| = \infty$. If we define $x_n := R_A(\zeta_n)x/\|R_A(\zeta_n)x\|$, then the sequence $\{x_n\}_{n\geqslant 1} \subset \operatorname{dom} A$. Moreover, it is such that $\|x_n\| = 1$ and $\|A_{\zeta_0} x_n\| = \|A_{\zeta_n} x_n - (\zeta_n - \zeta_0)x_n\| \to 0$ for $n \to \infty$. Hence, we get $\zeta_0 \in \sigma_{\rm app}(A)$.

(c) Now let $A : \operatorname{dom}(A) \to \mathcal{H}$, be a self-adjoint operator in the Hilbert space $\mathcal{H}$. Recall that $\sigma_{\rm res}(A) = \varnothing$, see A.2(d). Since by A.2(e),(f) one has $\sigma(A) = \widehat{\sigma}(A) \cup \sigma_{\rm res}(A)$, the spectrum of a self-adjoint operator A coincides with the core: $\sigma(A) = \widehat{\sigma}(A) = \sigma_{\rm p}(A) \cup \sigma_{\rm cont}(A)$. Hence, it is completely defined by the approximate spectrum as in (a). A *refinement* of definition (a) in a Hilbert space (Weyl's criterion) allows to distinguish in the spectrum of a self-adjoint operator an important subset $\sigma_{\rm ess}(A) \subseteq \sigma(A)$, called the *essential* spectrum of A.

(d) To this aim we define first the *discrete* spectrum $\sigma_{\rm d}(A) \subseteq \sigma_{\rm p}(A)$. It is the simplest subset of the point spectrum, which consists of *isolated* eigenvalues of *finite* multiplicity. Then the essential spectrum of a self-adjoint operator A is the complementary set of $\sigma_{\mathrm{d}}(A)$ in $\sigma(A)$:

$$\sigma_{\mathrm{ess}}(A) = \sigma(A)\backslash\sigma_{\mathrm{d}}(A) = (\sigma_{\mathrm{p}}(A)\backslash\sigma_{\mathrm{d}}(A)) \cup \sigma_{\mathrm{cont}}(A).$$

The *Weyl criterion.* Let $A : \mathrm{dom}\, A \to \mathcal{H}$, define a self-adjoint operator in $\mathcal{H}$. If for a real number $\lambda \in \mathbb{R}$ there exist an *orthonormal* sequence of vectors $\{e_n\}_{n\geqslant 1} \subset \mathrm{dom}(A)$, $\{\|e_n\| = 1 : e_n \perp e_{n+1}\}_{n\geqslant 1}$, such that $\lim_{n\to\infty} \|A_\lambda e_n\| = 0$, then λ belongs to $\sigma_{\mathrm{ess}}(A)$.

We note that by the Weyl criterion this subset of the spectrum $\sigma(A)$ is closed: $\sigma_{\mathrm{ess}}(A) = \overline{\sigma_{\mathrm{ess}}(A)}$. If A is self-adjoint, then $\lambda \in \sigma(A)$ belongs to the essential spectrum $\sigma_{\mathrm{ess}}(A)$ unless it lies in $\sigma_{\mathrm{d}}(A)$.

(e) Let A be a self-adjoint operator. Recall that a symmetric operator K is *A-compact* if there exists $\zeta \in \rho(A)$ such that $KR_A(\zeta) \in \mathcal{C}_\infty(\mathcal{H})$. In this case the operator $A + K$ with $\mathrm{dom}(A + K) = \mathrm{dom}\, A$ is self-adjoint and $\sigma_{\mathrm{ess}}(A + K) = \sigma_{\mathrm{ess}}(A)$. This stability of the *essential spectrum* is known as the Weyl theorem.

Moreover, if A and A' are two self-adjoint operators such that the difference of their resolvents, $(A - \zeta\mathbb{1})^{-1} - (A' - \zeta\mathbb{1})^{-1}$, is *compact* for some $\zeta \in \rho(A) \cap \rho(A')$, then $\sigma_{\mathrm{ess}}(A) = \sigma_{\mathrm{ess}}(A')$.

(f) To understand better the approximate and essential spectra we consider a few examples. We denote by $\sigma_{\mathrm{p}}^*(A) := \sigma_{\mathrm{p}}(A)\backslash\sigma_{\mathrm{d}}(A)$ the complement of a discrete spectrum in the point spectrum.

(1f) Let A be the self-adjoint operator given by $Ae_n = \lambda_n e_n$, $\lambda_n = n^{-1}$, on the orthonormal basis $\{e_n\}_{n\geqslant 1}$ in a Hilbert space $\mathcal{H}$. Then by (d) one gets that $\sigma(A)\backslash\{0\} = \sigma_{\mathrm{d}}(A)$ with multiplicity one and $\sigma_{\mathrm{ess}}(A) = \{0\}$. To elucidate the nature of the unique *non-isolated* point $\lambda = 0$ of $\sigma(A)$ one can either formally deduce from (d) and from $\sigma_{\mathrm{p}}^*(A) = \varnothing$ that $0 \in \sigma_{\mathrm{cont}}(A)$, or infer the *same* by referring to the approximate spectrum, see (a)–(c), and to the fact that non-isolated point $\lambda = 0$ belongs to the boundary $\partial\sigma(A) \subseteq \sigma_{\mathrm{p}}(A) \cup \sigma_{\mathrm{cont}}(A)$, but *not* to the point spectrum $\sigma_{\mathrm{p}}(A)$.

(2f) Let operator A_0 acts as $A_0 e_1 = 0$, $A_0 e_n = \mu_n e_n$, $\mu_n = (n-1)^{-1}, n \geqslant 2$, on the orthonormal basis $\{e_n\}_{n\geqslant 1}$ in a Hilbert space $\mathcal{H}$. Then by the definitions of the point and discrete spectra one obtains that $\sigma(A_0) = \sigma_{\mathrm{p}}(A_0) = \{0\} \cup \sigma_{\mathrm{d}}(A_0)$. Therefore, again by (d) the non-isolated point $\{0\} = \sigma_{\mathrm{ess}}(A_0)$, but now $0 \in \sigma_{\mathrm{p}}(A_0)$ and $\sigma_{\mathrm{cont}}(A_0) = \varnothing$.

(3f) Let the bounded self-adjoint operator A_1 acts as $A_1 e_n = r_n e_n$, $n \geqslant 1$, on the orthonormal basis $\{e_n\}_{n\geqslant 1}$ in a Hilbert space $\mathcal{H}$, where $\{r_n\}_{n\geqslant 1} \subset (0, 1)$ is the set of all *rational* numbers in the open interval $(0, 1)$. Then by the definitions of the point and discrete spectra we have $\sigma_{\mathrm{p}}(A_1) \neq \varnothing$, with non-isolated eigenvalues of multiplicity one and $\sigma_{\mathrm{d}}(A_1) = \varnothing$. Therefore, $\sigma(A_1) = \sigma_{\mathrm{ess}}(A_1) = \sigma_{\mathrm{p}}(A_1) \cup \sigma_{\mathrm{cont}}(A_1)$.

Note that for any $\lambda \in (0, 1)\backslash\sigma_{\mathrm{p}}(A_1)$ there exists a sequence $\{r_{n_k}\}_{k\geqslant 1} \subset \sigma_{\mathrm{p}}(A_1)$ such that $\lim_{k\to\infty} r_{n_k} = \lambda$. Then by the definition of the essential spectrum, $\|(A_1 - \lambda\mathbb{1})e_{n_k}\| \leqslant \|(A_1 - r_{n_k}\mathbb{1})e_{n_k}\| + \|(r_{n_k} - \lambda)e_{n_k}\|$. This yields that $\lambda \in \sigma_{\text{ess}}(A_1)\backslash\sigma_{\text{p}}(A_1) = \sigma_{\text{cont}}(A_1)$ and in accordance with (c) $\partial\sigma(A_1) \subseteq \sigma_{\text{p}}(A_1) \cup \sigma_{\text{cont}}(A_1)$, that is, we have $\sigma(A_1) = [0, 1]$.

A.5 Fredholm operators and essential spectrum

(a) Another approach to the essential spectrum is via Fredholm operators. Recall that an operator $A \in \mathcal{L}(\mathcal{B})$ is called *Fredholm* if $\ker A$ and the *quotient* space $\text{coker A} := \mathcal{B}/\operatorname{ran} A$, also known as the *cokernel* of A, are *finite*-dimensional. It is worth mentioning that cokerA "measures" the deficiency of surjection and in this way takes into account an eventual residual spectrum of A. Indeed, below we give an example of the Volterra operator V with non-empty $\sigma_{\text{res}}(V)$.

It is known that if A is a Fredholm operator, then $\operatorname{ran} A$ is a subspace of $\mathcal{B}$ and that if K is compact, then operator $\mathbb{1} + K$ is Fredholm.

(b) Let A be a closed unbounded operator with dense domain $\operatorname{dom} A \subset \mathcal{B}$ in a Banach space $\mathcal{B}$. Then A is a Fredholm operator if $\ker A$ is finite-dimensional, $\operatorname{ran} A$ is closed, and the cokernel coker A is also finite-dimensional. One defines the *essential* spectrum $\sigma_{\text{ess}}(A)$ of a closed operator A to be the set of $\zeta \in \mathbb{C}$ for which the operator $A_\zeta = A - \zeta\mathbb{1}$ is *not* Fredholm.

(c) This definition covers the one in A.4(c),(d) for the case of self-adjoint operators and allows a non-empty residual spectrum. Note that stability A.4(e) of the essential spectrum is valid for perturbations K, which are *A-compact*, i.e., $KR_A(\zeta) \in \mathcal{C}_\infty(\mathcal{B})$, in Banach spaces.

The nature of the essential spectrum of self-adjoint operator A in a Hilbert space is easy to describe: any $\lambda \in \sigma(A)$ lies in $\sigma_{\text{ess}}(A)$ unless it is a point of $\sigma_{\text{d}}(A)$.

A.6 Spectrum of compact operators

(a) Compact operators constitute an instructive example to explain basic notions concerning spectra of closed operators. Recall that a bounded operator on a Banach space $\mathcal{B}$ is *compact*, $K \in \mathcal{C}_\infty(\mathcal{B})$, if it maps *bounded* sets of $\mathcal{B}$ into *precompact* sets, cf. with the *completely continuous* operators, see Notes to Section 2.1. By A.1(h) the spectrum $\sigma(K)$ belongs to the disc in $\mathbb{C}$ with the (spectral) radius $r(A) \leqslant \|A\|$. Another general remark is that $\{0\} \subseteq \sigma(K)$. Otherwise, K would be boundedly invertible: $K^{-1}K = \mathbb{1}$, with a compact operator in the left-hand side, which is obviously impossible.

(b) Recall that by the *Fredholm alternative*: if the operator $K \in \mathcal{C}_\infty(\mathcal{B})$ and $\zeta \in \mathbb{C}\backslash\{0\}$, then $\zeta \in \rho(K) \cup \sigma_{\text{d}}(K)$. Moreover, the spectrum $\sigma_{\text{d}}(K)$ has no *accumulation* point different from $\zeta = 0$.

(c) As we learned from examples of Section A.4(f) the classification of the spectral point $\zeta = 0$ may be a subtle matter. We consider it first in a Hilbert space $\mathcal{H}$.

A general result for self-adjoint compact operators $K = K^*$ follows from A.4 (d),(e). Let $\mathcal{N} = \lambda\,\mathbb{1}$, for $\lambda = 0$, be the *zero-operator*. The spectrum of this bounded self-adjoint operator contains only one point, $\sigma(\mathcal{N}) = \{0\}$, which is an eigenvalue of infinite multiplicity: $\sigma_{\mathrm{p}}^{\infty}(\mathcal{N}) = \{0\}$. Hence, by A.4(d), $\sigma(\mathcal{N}) = \sigma_{\mathrm{ess}}(\mathcal{N})$. Since the stability of essential spectrum A.4(e) yields $\sigma_{\mathrm{ess}}(\mathcal{N}) = \sigma_{\mathrm{ess}}(\mathcal{N}+K)$, one obtains the accumulation point $\{0\} = \sigma_{\mathrm{ess}}(K)$. Note that this observation also agrees with the argument A.4(b),(c) since the accumulation point $\{0\}$ belongs to the topological boundary $\partial\sigma(K)$.

(d) We stress the point that the arguments in (c) do not imply that $\sigma_{\mathrm{ess}}(K)$ *coincides* with $\sigma_{\mathrm{p}}^{\infty}(\mathcal{N})$. Examples A.4(1f) and A.4(2f) of two self-adjoint compact operators show that the accumulation point $\{0\}$ does always belong to essential spectrum, but to its different parts. For the *non-self-adjoint* case, or for compact operators in Banach spaces, the classification of the spectral point $\{0\}$ gets even more complicated, since in general one has $\{0\} \subseteq \sigma_{\mathrm{ess}}(K) \cup \sigma_{\mathrm{res}}(K)$. To this aim we consider in subsection A.7 the example of the *Volterra* operator.

(e) We conclude by a remark concerning unbounded operators in a Banach space $\mathcal{B}$. Let A be a closed operator in $\mathcal{B}$ such that the resolvent $R_A(\zeta)$ exists and is compact for some $\zeta \in \mathbb{C}$. Then $R_A(\zeta) \in \mathcal{C}_\infty(\mathcal{B})$ for every $\zeta \in \rho(A)$, the spectrum $\sigma(A) = \sigma_{\mathrm{d}}(A)$, and the corresponding eigenvalues are such that $\lim_{k\to\infty} |\zeta_k| = \infty$. Since this implies that the spectral radius $r(A) = \infty$, the operator A is unbounded.

A.7 Example: the Volterra operator

(a) The Volterra operator $V : \mathcal{H} \to \mathcal{H}$ on the Hilbert space $\mathcal{H} = L^2[0,1]$ is the integral operator

$$(Vu)(t) := \int_0^t ds\, u(s), \qquad u \in \mathrm{dom}(V) = L^2[0,1].$$

To estimate the norm $\|V\|$ we note that

$$\begin{aligned}
\|Vu\|^2 &= \int_0^1 dt \Big| \int_0^t ds\, \sqrt{\cos(\pi s/2)}\, \frac{u(s)}{\sqrt{\cos(\pi s/2)}} \Big|^2 \\
&\leqslant \int_0^1 dt \left\{ \int_0^t ds_1 \cos(\pi s_1/2) \int_0^t ds_2 |u(s_2)|^2/\cos(\pi s_2/2) \right\} \\
&= \frac{2}{\pi} \int_0^1 ds \int_s^1 dt\, \sin(\pi t/2) |u(s)|^2/\cos(\pi s/2) = \left(\frac{2}{\pi}\right)^2 \|u\|^2 .
\end{aligned}$$

The norm $\|V\| = 2/\pi$, corresponds to the vector $u^*(s) := \cos(\pi s/2)$.

(b) Recall that the linear operator $A \in \mathcal{L}(\mathcal{H})$ belongs to the *Hilbert-Schmidt*-class if there exists an orthonormal basis $\{e_n\}_{n\in\mathbb{N}}$ such that

$$\|A\|_2^2 := \sum_{n=1}^{\infty} \|Ae_n\|^2 < \infty .$$

Here $\|A\|_2$ is the Hilbert-Schmidt *norm* of A, which (if finite) does not depend on the choice of the basis. Note that the set of Hilbert-Schmidt operators $\mathcal{C}_2(\mathcal{H})$ is a $*$-ideal in $\mathcal{L}(\mathcal{H})$ and that it is also a subset of the compact operators: $\mathcal{C}_2(\mathcal{H}) \subset \mathcal{C}_\infty(\mathcal{H})$. Moreover, if the operators A_1, A_2 belong to $\mathcal{C}_2(\mathcal{H})$, then the product A_1A_2 belongs to the *trace-class* $*$-ideal $\mathcal{C}_1(\mathcal{H}) \subset \mathcal{C}_2(\mathcal{H}) \subset \mathcal{C}_\infty(\mathcal{H})$.

Applying this definition to the Volterra operator on $L^2[0,1]$ for the orthonormal basis $\{e_n(x) = \sqrt{2}\sin(\pi n x)\}_{n\in\mathbb{Z}}$, one finds that $\|Ve_n\|^2 = 3/(n^2\pi^2)$. Therefore, the Volterra operator, together with its adjoint V^*, are Hilbert-Schmidt operators, and as such they are compact. Note that V is not normal and that by explicit calculations, the self-adjoint operator $V^*V \in \mathcal{C}_1(L^2[0,1])$, with $\sigma_{\mathrm{d}}(V^*V) = \{\lambda_k = 4/[(2k+1)^2\pi^2]\}_{k\geqslant 0}$ and $\sigma(V^*V) = \sigma_{\mathrm{d}}(V^*V) \cup \{0\}$. Since $\|V^*V\| = \|V\|^2$ and since for self-adjoint operators the spectral radius $r(V^*V) = \|V^*V\| = \max_{k\geqslant 0}\lambda_k$, we obtain $\|V\| = 2/\pi$, as in (a).

(c) The spectrum of the compact V has the canonical structure $\sigma(V) = \sigma_{\mathrm{d}}(V) \cup \{0\}$. To study the discrete part $\sigma_{\mathrm{d}}(V)$ one notes that

$$|V^n u|(t) \leqslant \int_0^t ds\, |u(s)| \frac{(t-s)^{n-1}}{(n-1)!} \leqslant \|u\| \frac{1}{(n-1)!} ,$$

which implies that $\|V^n\| \leqslant 1/(n-1)!$. Then the spectral radius $r(V) = \lim_{n\to\infty}(1/(n-1)!)^{1/n} = 0$. Hence, V is a *quasi-nilpotent* compact operator with $\sigma_{\mathrm{d}}(V) = \varnothing$.

Recall that if $r(A) = 0$ for a self-adjoint operator, then $A = 0$, cf. with (b) for $A = V^*V$, but this is not the case for the Volterra operator. Although V is quasi-nilpotent and its spectrum contains only zero point, $\sigma(V) = 0$, one has $\|V\| \neq 0$. Consequently, Proposition 2.45 (Lidskiĭ's spectral trace theorem) yields $\operatorname{Tr} V = \Lambda(V) = 0$, for a nontrivial set of matrix elements $\{(e_k, Ve_k)\}_{k\geqslant 1}$ and for any choice of the orthonormal basis $\{e_k\}_{k\geqslant 1}$ in $L^2[0,1]$.

(d) To study the zero spectral point of non-self-adjoint V we recall that $\{0\} \subseteq \sigma_{\mathrm{p}}^*(V) \cup \sigma_{\mathrm{cont}}(V) \cup \sigma_{\mathrm{res}}(V)$, see A.4(f), A.6(d), and note that according to the definition in (a) the range $\operatorname{ran} V = AC_0[0,1]$, is the set of *absolutely continuous* functions, which are zero at $t = 0$.

Now let $\zeta = 0 \in \sigma_{\mathrm{p}}^*(K)$ with the eigenvector u_0, i.e., one has $Vu_0 = 0$. Since $Vu_0 \in AC_0[0,1]$, this implies that $u_0 = 0$ in $L^2[0,1]$. Hence, $\sigma_{\mathrm{p}}^*(K) = \varnothing$.

Next we note that the closure of $\operatorname{ran} V$ in $L^2[0,1]$ gives $\overline{AC_0[0,1]} = L^2[0,1]$. So, the range of the Volterra operator $V_{\zeta=0}$ is dense in the Hilbert space $\mathcal{H}$. Therefore, $\sigma_{\mathrm{res}}(V) = \varnothing$ and thus $\sigma_{\mathrm{cont}}(V) = \{0\}$.

(e) The Volterra operator $V : \mathcal{B} \to \mathcal{B}$, on the Banach space $\mathcal{B} = C[0,1]$ with the sup-norm $\|\cdot\|_\infty$, is defined by the same integral operator as in (a). Then for any $f \in C[0,1]$ and for the *identity* function $[0,1] \ni t \mapsto \mathbb{1}(t) = 1$, we obtain the estimates

$$|(V^n f)(t)| \leqslant \frac{1}{n!}\|f\|_\infty \quad \text{and} \quad \|V^n\| \geqslant \|V^n \mathbb{1}\|_\infty = \frac{1}{n!}.$$

Hence, the operator norm $\|V^n\| = 1/n!$ and the spectral radius $r(V) = 0$. Therefore, the spectrum of this bounded operator with $\|V\| = 1$ is $\sigma(V) = \{0\}$.

Note that the range $\operatorname{ran} V$ of this operator is the set $AC_0[0,1]$ of absolutely continuous functions such that $(Vf)(t=0) = 0$. Since $|(Vf)(t)| \leqslant \|f\|_\infty t^{1/2}$ and $|(Vf)(t) - (Vf)(s)| \leqslant \|f\|_\infty |t-s|^{1/2}$, then by the *Arzela-Ascoli* theorem the image $V(B_1)$ of the unit ball $B_1 := \{f \in \mathcal{B} : \|f\|_\infty \leqslant 1\}$ is a compact set. So, V is a *compact* operator on $C[0,1]$ and $\sigma(V) = \{0\}$ implies that $\sigma_{\mathrm{d}}(V) = \varnothing$.

Now, to decide between the *three disjoint* possibilities: $\{0\} = \sigma_{\mathrm{p}}(V) \cup \sigma_{\mathrm{cont}}(V) \cup \sigma_{\mathrm{res}}(V)$, we claim first that $\{0\} \neq \sigma_{\mathrm{p}}(V)$ follows along the same line of reasoning as in (d). Second, note that $\operatorname{ran}(V_{\zeta=0}) = AC_0[0,1]$ is *not* dense in $C[0,1]$ in the $\|\cdot\|_\infty$-norm. Therefore, $\overline{\operatorname{ran} V_{\zeta=0}} \neq \mathcal{B}$, and consequently $\{0\} = \sigma_{\mathrm{res}}(V)$.

A.8 Example: unbounded operators

By the resolvent arguments in A.1(g) spectra of *bounded* operators are always non-empty. This is not the case for *unbounded* operators as it is shown below by two extreme examples of operators with $\sigma = \varnothing$ and with $\sigma = \mathbb{C}$.

(a) Let the unbounded operator A_0 be defined in the Hilbert space $\mathcal{H} = L^2[0,1]$ on $\operatorname{dom} A_0 = W_0^{1,2}(0,1)$ by

$$(A_0\, u)(x) := -i\, \partial_x u(x), \quad u \in \operatorname{dom} A_0\,.$$

The Sobolev space $W_0^{1,2}(0,1) = \{u \in AC[0,1] : \partial_x u \in L^2(0,1) \wedge u(0) = 0\}$. We note that $\operatorname{dom} A_0$ is dense in $L^2[0,1]$ and that the operator A_0 is closed. By inspection one sees that for any $\zeta \in \mathbb{C}$ the equation $(A_0 - \zeta\mathbb{1})u = 0$ for $u \in \operatorname{dom} A_0$, has only trivial solution, that is, $\ker(A_0 - \zeta\mathbb{1}) = \{0\}$. Therefore, $\sigma_{\mathrm{p}}(A_0) = \varnothing$ and $(A_0 - \zeta\mathbb{1}) : \operatorname{dom} A_0 \to \mathcal{H}$ is injective, see A.3(b).

Since for any $v \in L^2[0,1]$ there exists the vector $u_v \in W_0^{1,2}(0,1)$ (defined by the Volterra operator $V : v \mapsto u_v$), the map $A_0 - \zeta\mathbb{1}$ is *bijective*: $\operatorname{ran} A_\zeta = \overline{\operatorname{ran} A_\zeta} = \mathcal{H}$. By A.3(c)-(e), this yields $\sigma_{\mathrm{cont}}(A_0) = \varnothing$ and $\sigma_{\mathrm{res}}(A_0) = \varnothing$. Therefore, the spectrum $\sigma(A_0)$ is empty.

(b) Let the unbounded operator A_1 be defined in the Hilbert space $\mathcal{H} = L^2[0,1]$ by

$$(A_1 u)(x) = -i\partial_x u(x), \quad u \in \operatorname{dom} A_1 = W^{1,2}(0,1).$$

Here the domain is the Sobolev space $W^{1,2}(0,1) = \{u \in AC[0,1] : \partial_x u \in L^2(0,1)\}$. Note that the operator A_1 is *closed*. Since for any $\zeta \in \mathbb{C}$ we have

$$(A_1 - \zeta\mathbb{1})e_\zeta = 0 \quad \text{for } e_\zeta := e^{i\zeta x} \in \operatorname{dom} A_1\,,$$

$\ker(A_1 - \zeta\mathbb{1}) \neq \{0\}$ and by A.3(b) one gets that $\sigma(A_1) = \sigma_{\mathrm{p}}(A_1) = \mathbb{C}$.

A.9 Spectral mapping theorem for semigroups

(a) Let A be a closed unbounded operator in the Banach space $\mathcal{B}$, with dense domain $\operatorname{dom} A \subset \mathcal{B}$. Recall that if $\zeta \notin \sigma(A)$, then the resolvent *spectral mapping* theorem holds:

$$\sigma(R_A(\zeta)) = \{0\} \cup \{\lambda - \zeta)^{-1} : \lambda \in \sigma(A)\}.$$

(b) Let A be the generator of the strongly continuous semigroup $\{U_t(A)\}_{t\geqslant 0}$ on $\mathcal{B}$. Recall that the relationship between the spectra $\sigma(A)$ and $\sigma(U_t(A))$ is not as simple as in the resolvent spectral mapping theorem. In general, one can prove for the C_0-semigroups only the following form of this theorem:

$$\sigma(U_t(A)) \supseteq \{e^{-t\,z} : z \in \sigma(A)\}.$$

One can prove a *converse* statement under some additional *continuity* assumptions on the C_0-semigroups.

(c) To this aim we establish the following version of the spectral mapping theorem. Let the C_0-semigroup $\{U_t(A)\}_{t\geqslant 0}$ generated by the unbounded operator A be *quasi-bounded* of the *type* ω_0: $\|U_t(A)\| \leqslant Me^{\omega_0 t}, M \geqslant 1$. Let $\widehat{U}_{t,\alpha}(A) := U_t(A)R_A(\alpha)$, where $\alpha \notin \sigma(A)$. Then for $t > 0$ and $\Re\mathfrak{e}\, \alpha < -\omega_0$, we obtain

$$\sigma(\widehat{U}_{t,\alpha}(A)) = \{0\} \cup \{e^{-t\,z}(z-\alpha)^{-1} :\ z \in \sigma(A)\}.$$

(d) The next result is due to the improved *continuity* condition. Suppose the C_0-semigroup $\{U_t(A)\}_{t\geqslant 0}$ is *eventually norm-continuous*, i.e., there exists $t_0 > 0$ such that $t \mapsto U_t(A)$ is a norm-continuous function for $t > t_0$. Then $\sigma(U_t(A)) = \{e^{-t\,z} :\ z \in \sigma(A)\}$.

(e) Let us assume more: the C_0-semigroup $\{U_t(A)\}_{t\geqslant 0}$ is *eventually compact*, that is, there exists $t_0 > 0$ such that $t \mapsto U_t(A) \in \mathcal{C}_\infty(\mathcal{B})$ for $t > t_0$. Then there exists a direct sum decomposition $\mathcal{B} = \mathcal{B}_0 \oplus \mathcal{B}_1$ such that:

(1) The subspaces $\mathcal{B}_0$ and $\mathcal{B}_1$ are invariant under the semigroup $\{U_t(A)\}_{t\geqslant 0}$.

(2) $\dim(\mathcal{B}_0) < \infty$ and the restriction $S_t := U_t(A) \restriction \mathcal{B}_1$ satisfies the asymptotics $\|S_t\| = o(e^{-\beta t})$ as $t \to \infty$, for all $\beta > 0$.

(3) The spectrum $\sigma(A) = \sigma_{\mathrm{d}}(A)$, and if $\dim(\mathcal{B}) = \infty$, then

$$\sigma(U_t(A)) = \{0\} \cup \{e^{-t\,\sigma(A)}\}.$$

(4) The eigenvalues $\{z_k \in \sigma_{\mathrm{d}}(A)\}_{k\geqslant 1}$ are such that $\lim_{k\to\infty} \Re\mathfrak{e}\, z_k = \infty$.

For the proof one observes that the eventually compact C_0-semigroup $\{U_t\}_{t\geqslant 0}$ is in fact eventually norm-continuous. Indeed, let $U_{t_0} \in \mathcal{C}_\infty(\mathcal{B})$ and let $\mathfrak{B}_1 := \overline{U_{t_0}(B_1)}$ denote the closure of the compact image of the unit ball $B_1 \subset \mathcal{B}$. Since $(t, f) \mapsto U_t f$ is jointly continuous, for any given $\varepsilon > 0$ there exists $\delta > 0$ such that $\|U_t f - f\| < \varepsilon$ for all $f \in \mathfrak{B}_1$ and $0 \leqslant t < \delta$. Let $t_0 \leqslant t_1 \leqslant t < t_1 + \delta$ and $\|f\| < 1$. Then

$$\|U_t f - U_{t_1} f\| = \|U_{t_1 - t_0}(U_{t-t_1} - 1)U_{t_0} f\| \leqslant \|U_{t_1 - t_0}\| \, \varepsilon \, .$$

Therefore, $\|U_t - U_{t_1}\| \leqslant \|U_{t_1-t_0}\|\varepsilon$ and the C_0-semigroup $\{U_t\}_{t\geqslant 0}$ is norm-continuous for $t_0 < t < \infty$.

(f) Counterexamples show that the opposite inclusion to the inclusion in (b) does not hold in general. However, there is a partial result in this direction. Let $\{U_t(A)\}_{t\geqslant 0}$ be a quasi-bounded C_0-semigroup on a Banach space $\mathcal{B}$ generated by the unbounded operator A. Then the converse to inclusion (b) concerns only the *point* and the *residual* spectra and reads as

$$\sigma_{\mathrm{p}}(U_t(A)) = \{0\} \cup \{e^{-t\,z} : z \in \sigma_{\mathrm{p}}(A)\},$$
$$\sigma_{\mathrm{res}}(U_t(A)) = \{0\} \cup \{e^{-t\,z} : z \in \sigma_{\mathrm{res}}(A)\}.$$

(g) More versions of the spectral mapping theorem one finds in Section 4.1: Lemma 4.9 and Propositions 4.11, 4.12, and also in Section 4.2: Propositions 4.21, 4.22.

A.10 Notes

Notes to Section A.1–Section A.4 and to Section A.6. Although standard, this Appendix A recalls some useful key notations, definitions and facts from the spectral theory. This will facilitate understanding the text of the book.

In our presentation of Sections A.1–A.4 and partially of Section A.6 we followed essentially [BS87] (Chapters 3 and 9), as well as [HN01] (Chapter 9) and [Kat80] (Chapter 8, §1).

Notes to Section A.5, Section A.6 and to Section A.9.

In the presentation of Sections A.5, A.6, and A.9 we were motivated by Chapters 4 and 8 from [Dav07] and by [EN00] (Chapter IV).

Appendix B. More inequalities

B.1 The Araki inequality

(a) Recall that if $A \in \mathcal{L}(\mathcal{H})$ is a positive bounded self-adjoint operator on $\mathcal{H}$, then by the spectral representation of A and the elementary formula

$$s^\theta = \frac{\sin(\pi\theta)}{\pi} \int_0^\infty dt \, \frac{t^{\theta-1}\, s}{s+t}, \quad s > 0, \ 0 < \theta < 1,$$

the *fractional power* of A admits the representation

$$A^\theta = \frac{\sin(\pi\theta)}{\pi} \int_0^\infty dt \, \frac{t^{\theta-1}\, A}{A+t\mathbb{1}}, \quad 0 < \theta < 1, \tag{B.1}$$

where equality (B.1) is in the sense of sesquilinear forms and we have $(u, A^\theta u) = \|A^{\theta/2}u\|^2$, for $u \in \mathcal{H}$.

Lemma B.1 (Löwner-Heinz inequality). *Let $A, B \in \mathcal{L}(\mathcal{H})$ be positive operators such that $0 < B < A$. Then*

$$B^\theta < A^\theta \quad \textit{for any } 0 < \theta < 1. \tag{B.2}$$

Proof. The condition $0 < B < A$ implies that for $t > 0$ the operators: $A + t\mathbb{1} > 0$ and $B + t\mathbb{1} > 0$, and hence, $(A+t\mathbb{1})^{1/2}$ and $(B+t\mathbb{1})^{1/2}$, are invertible. Moreover, for any $u \in \mathcal{H}$ one also gets that $(u,(B + t\mathbb{1})u) < (u,(A + t\mathbb{1})u)$. Hence, for $u = (A+t\mathbb{1})^{-1/2}w$ and any $w \in \mathcal{H}$,

$$(w, (A+t\mathbb{1})^{-1/2}(B+t\mathbb{1})(A+t\mathbb{1})^{-1/2}w) < (w,w).$$

Therefore, the operator

$$C(t) := (A+t\mathbb{1})^{-1/2}(B+t\mathbb{1})u)(A+t\mathbb{1})^{-1/2} < \mathbb{1}.$$

is invertible. Then $C(t)^{-1} > \mathbb{1}$ and consequently

$$(A+t\mathbb{1})^{-1} < (B+t\mathbb{1})^{-1}. \tag{B.3}$$

V. A. Zagrebnov, *Gibbs Semigroups*, Operator Theory: Advances and Applications 273, https://doi.org/10.1007/978-3-030-18877-1

By (B.3) it follows that

$$A(A + t\mathbb{1})^{-1} = \mathbb{1} - t(A + t\mathbb{1})^{-1} > \mathbb{1} - t(B + t\mathbb{1})^{-1} = B(B + t\mathbb{1})^{-1}. \tag{B.4}$$

Therefore, representation (B.1) and inequality (B.4) give the Löwner-Heinz inequality (B.2) for positive bounded operators. □

Corollary B.2. *The inequality* (B.2) *is equivalent to* $\|B^{\theta/2}u\|^2 < \|A^{\theta/2}u\|^2$, $u \in \mathcal{H}$. *This yields that* $\|B^{\theta/2}u\| < \|A^{\theta/2}u\|$, *for* $0 < \theta < 1$.

Therefore, if the positive bounded operators A, B *satisfy* $\|Bu\| < \|Au\|$, *for all* $u \in \mathcal{H}$, *then* $\|Bu\|^2 < \|Au\|^2$, *i.e.* $B^2 < A^2$. *Now applying the inequality* (B.2) *one gets* $\|B^{\theta}u\| < \|A^{\theta}u\|$, *which is often also called the Löwner-Heinz inequality.*

(b) Let $X \in \mathcal{C}_\infty(\mathcal{H})$ be a compact operator and let $\sigma(X) = \{\lambda_j(X)\}_{j=1}^{\infty}$ be the corresponding sequence of eigenvalues, counting multiplicities. For $X \geqslant 0$ we assume that $\lambda_1(X) \geqslant \lambda_2(X) \geqslant \ldots \geqslant 0$. In this case the singular values $s(X) = \{s_j(X)\}_{j=1}^{\infty}$ coincide with $\sigma(X)$.

Recall that two positive compact operators X, Y are said to be *log-ordered*, and one writes $Y \prec_{\log} X$, if

$$\prod_{j=1}^{n} \lambda_j(Y) < \prod_{j=1}^{n} \lambda_j(X), \quad n = 1, 2, \ldots,$$

see Definition 6.14.

Proposition B.3 (Araki log-order inequality). *For every two positive compact operators* $A, B \in \mathcal{C}_\infty(\mathcal{H})$

$$(B^{1/2}AB^{1/2})^r \prec_{\log} B^{r/2}A^rB^{r/2}, \quad r > 1. \tag{B.5}$$

Proof. First, we define two invertible for $\varepsilon > 0$ operators, $A_\varepsilon = A + \varepsilon\mathbb{1}$ and $B_\varepsilon = B + \varepsilon\mathbb{1}$. The spectra $\sigma(A_\varepsilon)$ and $\sigma(B_\varepsilon)$ consist of isolated eigenvalues $\{\lambda_j(A_\varepsilon) > \varepsilon\}_{j=1}^{\infty}$ and $\{\lambda_j(B_\varepsilon) > \varepsilon\}_{j=1}^{\infty}$ of finite multiplicity and (eventually) the point $\varepsilon = \lambda_\infty(A_\varepsilon) = \lambda_\infty(B_\varepsilon)$, see Section A.6. Recall that $\varepsilon \mapsto \lambda_j(A_\varepsilon)$ and $\varepsilon \mapsto \lambda_j(B_\varepsilon)$ are continuous functions, including at $\varepsilon = +0$.

Second, we define for $r > 1$ two positive bounded operators with discrete spectra:

$$X_\varepsilon := B_\varepsilon^{r/2}A_\varepsilon^rB_\varepsilon^{r/2} \quad \text{and} \quad Y_\varepsilon := \left(B_\varepsilon^{1/2}A_\varepsilon B_\varepsilon^{1/2}\right)^r. \tag{B.6}$$

Note that the spectra $\{\lambda_j(X_\varepsilon)\}_{j=1}^{\infty}$ and $\{\lambda_j(Y_\varepsilon)\}_{j=1}^{\infty}$ are also continuous functions of $\varepsilon \geqslant 0$.

For each $n \in \mathbb{N}$ and bounded operator $Z \in \mathcal{L}(\mathcal{H})$ with a discrete spectrum $\{\lambda_j(Z)\}_{j=1}^{\infty}$ we introduce the n-fold tensor product space $\mathcal{H}^{\otimes n} = \mathcal{H} \otimes \mathcal{H} \otimes \cdots \otimes \mathcal{H}$ and denote by $Z^{\otimes n} = Z \otimes Z \otimes \cdots \otimes Z$ the corresponding tensor product of operators. Let $E^n : \mathcal{H}^{\otimes n} \to E^n\mathcal{H}^{\otimes n}$, be the projection onto the subspace of totally *antisymmetric* vectors. Then $Z^{\otimes_a n} := Z^{\otimes n}E^n$ is the restriction of the operator

$Z^{\otimes n}$ on this subspace. By this construction the largest eigenvalue $\Lambda(Z^{\otimes_a n})$ of the operator $Z^{\otimes_a n}$ is given by

$$\Lambda(Z^{\otimes_a n}) = \prod_{j=1}^{n} \lambda_j(Z). \tag{B.7}$$

Therefore, the log-order (B.5) is equivalent to

$$\Lambda(Y^{\otimes_a n}) \leqslant \Lambda(X^{\otimes_a n}), \tag{B.8}$$

or, by the continuity of the spectra $\sigma(X_\varepsilon^{\otimes_a n})$ and $\sigma(Y_\varepsilon^{\otimes_a n})$ at $\varepsilon = +0$, to inequality

$$\Lambda(Y_\varepsilon^{\otimes_a n}) \leqslant \Lambda(X_\varepsilon^{\otimes_a n}). \tag{B.9}$$

To prove (B.9) we introduce on $\mathcal{H}^{\otimes n}$ for $\varepsilon > 0$ two axillary invertible and positive operators

$$\alpha_\varepsilon := A_\varepsilon \otimes A_\varepsilon \otimes \cdots \otimes A_\varepsilon \quad \text{and} \quad \beta_\varepsilon := B_\varepsilon \otimes B_\varepsilon \otimes \cdots \otimes B_\varepsilon\,. \tag{B.10}$$

Since the tensor and operator products commute, restricting the operators (B.10) to the antisymmetric subspace $E^n\mathcal{H}^{\otimes n}$ one obtains the following two identities:

$$\beta_\varepsilon^{r/2}\alpha_\varepsilon^r\beta_\varepsilon^{r/2} = X_\varepsilon^{\otimes_a n} \quad \text{and} \quad (\beta_\varepsilon^{1/2}\alpha_\varepsilon\beta_\varepsilon^{1/2})^r = Y_\varepsilon^{\otimes_a n}\,. \tag{B.11}$$

Consequently, (B.11) yields for the corresponding largest eigenvalues,

$$\Lambda(\beta_\varepsilon^{r/2}\alpha_\varepsilon^r\beta_\varepsilon^{r/2}) = \Lambda(X_\varepsilon^{\otimes_a n}) \quad \text{and} \quad \Lambda((\beta_\varepsilon^{1/2}\alpha_\varepsilon\beta_\varepsilon^{1/2})^r) = \Lambda(Y_\varepsilon^{\otimes_a n}). \tag{B.12}$$

To proceed, we *rescale* the operator $\beta_\varepsilon^{r/2}\alpha_\varepsilon^r\beta_\varepsilon^{r/2}$ in such a way that

$$\Lambda(\beta_\varepsilon^{r/2}\alpha_\varepsilon^r\beta_\varepsilon^{r/2}) = 1. \tag{B.13}$$

We also note that the operators (B.10) are invertible and that (B.13) implies $\beta_\varepsilon^{r/2}\alpha_\varepsilon^r\beta_\varepsilon^{r/2} \leqslant \mathbb{1}$. Then one gets

$$\alpha_\varepsilon^r \leqslant \beta_\varepsilon^{-r}.$$

Since $1/r < 1$, the Löwner-Heinz inequality (B.2) for $\theta = 1/r$ yields $\alpha_\varepsilon \leqslant \beta_\varepsilon^{-1}$, whence

$$\beta_\varepsilon^{1/2}\alpha_\varepsilon\beta_\varepsilon^{1/2} \leqslant \beta_\varepsilon^{1/2}\beta_\varepsilon^{-1}\beta_\varepsilon^{1/2} = \mathbb{1}. \tag{B.14}$$

Since $r > 1$, definition (B.11) and inequality (B.14) yield

$$(\beta_\varepsilon^{1/2}\alpha_\varepsilon\beta_\varepsilon^{1/2})^r = Y_\varepsilon^{\otimes_a n} \leqslant \mathbb{1}. \tag{B.15}$$

Therefore, (B.12) together with (B.13) and (B.15) imply (B.9). Then by the continuity of the spectra $\sigma(X_\varepsilon^{\otimes_a n})$ and $\sigma(Y_\varepsilon^{\otimes_a n})$ at $\varepsilon = +0$ we deduce (B.8), which proves the Araki log-order (inequality) (B.5) for compact operators $A, B \in \mathcal{C}_\infty(\mathcal{H})$. □

Corollary B.4. *Note that inequality* (B.5) *is equivalent to the log-order in the form*

$$(B^{q/2}A^qB^{q/2})^{1/q} \prec_{\log} (B^{r/2}A^rB^{r/2})^{1/r}, \quad 0 < q \leqslant r\,. \tag{B.16}$$

B.2 The Araki-Lieb-Thirring inequality in symmetrically-normed ideals

Proposition B.5. *For every two positive compact operators $A, B \in \mathcal{C}_\infty(\mathcal{H})$ one has the inequality:*

$$\|B^{1/2}AB^{1/2})^r\|_\phi \leqslant \|B^{r/2}A^rB^{r/2}\|_\phi\,, \quad r > 1, \tag{B.17}$$

in the symmerically-normed ideals $\mathcal{C}_\phi(\mathcal{H})$.

Proof. The result (B.17) follows from the Araki inequality (B.5) and Lemma 6.15. □

Proposition B.6. *Let $\varphi : \mathbb{R}_0^+ \to \mathbb{R}_0^+$ be a monotone increasing function such that $\varphi(0) = 0$ and $\xi \mapsto \varphi(e^\xi)$ is convex in $\xi \in \mathbb{R}$. Then the inequality* (B.17) *can be generalised as follows:*

$$\|\varphi((B^{1/2}AB^{1/2})^r)\|_\phi \leqslant \|\varphi(B^{r/2}A^rB^{r/2})\|_\phi\,, \quad r > 1. \tag{B.18}$$

Now setting $\varphi(t) = t^q$ for $q > 0$ one obtains the Araki-Lieb-Thirring inequality

$$\|(B^{1/2}AB^{1/2})^{rq}\|_\phi \leqslant \|(B^{r/2}A^rB^{r/2})^q\|_\phi\,, \quad r > 1, \tag{B.19}$$

in the symmetrically-normed ideals $\mathcal{C}_\phi(\mathcal{H})$.

B.3 Notes

Notes to Section B.1. The log-order inequality (B.5) was first proved by E. H. Lieb and W. Thirring in [LiTh76] for matrices as an inequality between traces. Using Araki's arguments ([Ara90], Theorem 1) one can generalise the log-order inequality (B.5) to positive compact operators ([Hia95], Proposition 2.1), see Proposition B.3 and Corollary B.4.

Similarly to [Ara90], our proof is based on the Löwner-Heinz inequality [BS87], Ch.10, and the arguments are close to the one developed in [LiSe10], Sec.4.5.

Notes to Section B.2 Another generalisation due to Araki ([Ara90], Theorem 2) is the trace inequality

$$\operatorname{Tr}\varphi((B^{1/2}AB^{1/2})^r) \leqslant \operatorname{Tr}\varphi(B^{r/2}A^rB^{r/2}), \quad r > 1. \tag{B.20}$$

Recall that here the operator-valued function $X \mapsto \varphi(X) \in \mathcal{C}_1(\mathcal{H})$ (in general for unbounded $X \geqslant 0$) is such that $\varphi(0) = 0$ and $\xi \mapsto \varphi(e^\xi)$ is convex in $\xi \in \mathbb{R}_0^+$.

In Propositions B.5 and B.6 we extend the Araki-Lieb-Thirring inequality (B.20) from the trace norm to symmetrically-normed ideals $\mathcal{C}_\phi(\mathcal{H})$.

Appendix C. Kato functions

The functions named above were introduced by T. Kato to extend the Trotter product formula to non-exponential approximants $\{(f(tA/n)g(tB/n))^n\}_{n\geqslant 1}$. Here the Borel measurable functions $f, g : [0, \infty) \to [0, 1]$ are assumed to imitate the exponential function $t \mapsto \mathrm{e}^{-t}$ for $t \geqslant 0$ in the vicinity of zero and monotonically decrease to zero at infinity. Soon, it became clear that to prove even the strong operator convergence of *non-exponential* Trotter product formulae one has to impose additional conditions on this natural and simple class. For example, to ensure the convergence for generators A and B subordinated by a relative smallness condition one has to *tune* correspondingly the properties of the generic Kato functions $f, g \in \mathcal{K}$.

For the reader's convenience Section C.1 collects a list of the key Kato functions, which are mentioned in the book, with comments and proofs that were omitted in the text.

We note that the conditions for the convergence of Trotter-Kato product formulae are often formulated in the text not directly for the Kato functions f, g, but after mapping them into *auxiliary* functions f_0, g_0. The constructions and properties of the latter are presented in Section C.2.

There are two kinds of properties to distinguish in the description of the Kato and auxiliary functions: the *global* behaviour, including at infinity, and the *local* behaviour in the vicinity of $t = +0$. To this aim we introduce in Section C.3 the concepts of *regularity* and *domination*. There we define also the corresponding classes of Kato functions.

In the Notes in Section C.4, one finds more details about the origin of particular classes of the Kato functions, as well as further references.

C.1 Classes: $\mathcal{K}$, $\mathcal{K}_\alpha$, $\widehat{\mathcal{K}}_\beta$

Since monomials in the product formulae are defined by at least *two* Kato functions, in definition below we consider a pair of them.

Definition C.1. Two Borel measurable functions f, g, defined on $\mathbb{R}_0^+ = [0, \infty)$ and

V. A. Zagrebnov, *Gibbs Semigroups*, Operator Theory: Advances and Applications 273, https://doi.org/10.1007/978-3-030-18877-1

which satisfy the conditions

$$0 \leqslant f(x) \leqslant 1, \qquad f(0) = 1, \qquad f'(+0) = -1, \tag{C.1}$$
$$0 \leqslant g(x) \leqslant 1, \qquad g(0) = 1, \qquad g'(+0) = -1, \tag{C.2}$$

are called *generic Kato functions*. We denote the ring of monomials generated by f, g, including their fractional powers, by $\mathcal{K}$.

Elementary examples of the *individual* Kato functions and the *monomials* that, in turn, satisfy conditions (C.1) (or (C.2)) are: $f(x) = \mathrm{e}^{-x}$, $f(x) = (1 + qx)^{-1/q}$, $g(x) = (1 - |\sin x|)$ and $f(x)g(x) = (1 + qx)^{-1/q}(1 - |\sin x|)$, $f(x)g(x) = \mathrm{e}^{-x}(1 + qx)^{-1/q}$, $f(x)^{1/2}g(x)f(x)^{1/2} = \mathrm{e}^{-x/2}(1 + qx)^{-1/q}\mathrm{e}^{-x/2}$, etc., for $q > 0$.

We introduced this class of Kato functions $\mathcal{K}$ in Section 5.2 preparing its specification $\mathcal{K}_{\alpha=1}$ for the case of pairs A and B subordinated by the Kato smallness condition of the operator B with respect to the operator A, with the relative bound $b < 1$. Therefore, we recall first the definition of the class $\mathcal{K}_\alpha$ for $\alpha \in (1/2, 1]$.

Definition C.2. For each $\alpha \in (1/2, 1]$ we denote by $\mathcal{K}_\alpha$ the class of Kato functions $f, g \in \mathcal{K}$ defined by the following conditions:

$$C_{1/2\alpha} := \sup_{x>0} \frac{x f(x)^{1/2\alpha}}{1 - f(x)} < +\infty, \tag{C.3}$$
$$C_1 := \sup_{x>0} (1 - f(x)) \frac{1}{x} < +\infty, \tag{C.4}$$
$$C_2 := \sup_{x>0} \left| \left(f(x) - \frac{1}{1+x} \right) \frac{1}{x^2} \right| < +\infty, \tag{C.5}$$
$$S_1 := \sup_{x>0} (1 - g(x)) \frac{1}{x} < +\infty, \tag{C.6}$$
$$S_2 := \sup_{x>0} \left| \left(g(x) - \frac{1}{1+x} \right) \frac{1}{x^2} \right| < +\infty. \tag{C.7}$$

To link the relative Kato smallness of B ($b < 1$) with properties of the Kato functions $f, g \in \mathcal{K}_\alpha$ we consider also the additional condition

$$b^{1/\alpha} C_{1/2\alpha} S_1 < 1, \tag{C.8}$$

which is indispensable for Proposition 5.8 ($\alpha = 1$) and for Proposition 5.27 ($\alpha \in (1/2, 1)$).

Comments:

1. If $f \in \mathcal{K}_\alpha$, then one evidently has $C_{1/2\alpha} \geqslant 1$ and $S_1 \geqslant 1$. Then the left-hand side of (C.8) $0 \leqslant b < (C_{1/2\alpha} S_1)^{-1} \leqslant 1$. For large $C_{1/2\alpha} \geqslant 1$ and $S_1 \geqslant 1$ the condition (C.8) is nontrivial for b. On the other hand, for $f(x) = g(x) = \mathrm{e}^{-x}$ one gets $C_{1/2\alpha} S_1 = 1$.

2. The proofs of the statements in Proposition 5.8 and Proposition 5.27 are valid for *all* Kato functions from $\mathcal{K}_{\alpha=1}$ and respectively from $\mathcal{K}_{1/2<\alpha<1}$ if the relative bound for the generator B is $b < 1$.

3. The operator-norm convergence in the case $\alpha = 1/2$ requires the relative A-compactness of the operator B and Kato functions from the class $\mathcal{K}'$, see Proposition 5.30 and Notes to Section C.4.

4. Note that (C.3) and, hence the tuning condition (C.8), exclude from the class $\mathcal{K}_\alpha$ such elementary function as $f(x) = (1+x)^{-1}$, because in this case $C_{1/2\alpha} = \infty$. This is a consequence of the subordination of smallness between the generators A and B.

Now let us consider the opposite case, when there is no subordination between A and B. Suppose that A, B are non-negative self-adjoint operators in $\mathcal{H}$ such that the operator sum $C := A + B$ is *self-adjoint* on the domain $\operatorname{dom} C := \operatorname{dom} A \cap \operatorname{dom} B$.

Recall (see Proposition 5.25) that the Trotter-Kato product formulae converge in the operator norm for the Kato functions from the class $\widehat{\mathcal{K}}_{\beta=2}$.

Definition C.3. We say that $h \in \widehat{\mathcal{K}}_\beta$ if:

(i) $h : [0, \infty) \to [0, 1]$ is a Borel-measurable function such that $h(0) = 1$, and $h'(+0) = -1$;

(ii) there exist $\varepsilon > 0$ and $\delta(\varepsilon) < 1$, such that $h(s) \leqslant 1 - \delta(\varepsilon)$ for $s \geqslant \varepsilon$, and

$$[h]_\beta := \sup_{s>0} \frac{|h(s) - 1 + s|}{s^\beta} < \infty, \quad \text{for } 1 < \beta \leqslant 2.$$

The standard examples of $h \in \widehat{\mathcal{K}}_\beta$ are $h(x) = e^{-x}$ and $h(x) = (1 + a^{-1}x)^{-a}$, for $a > 0$, where h denotes either f or g. The class $\widehat{\mathcal{K}}_\beta$ is *larger* than $\mathcal{K}_\alpha$.

Comments:

5. In contrast to $\mathcal{K}_\alpha$, one has $f(x) = (1 + x)^{-1} \in \widehat{\mathcal{K}}_\beta$.

6. For $\alpha = 1$ Definition C.2 yields as a local condition for f (and the same for g) the estimate $|f(x) - 1 + x| \leqslant \tilde{C}_2 x^2$. This estimate coincides with that for $[h]_{\beta=2}$.

C.2 Auxiliary functions and $\mathcal{K}_*$

For many reasons, see Section 5.3, it is convenient (and in fact necessary) to use certain auxiliary functions f_0, g_0, associated with Kato functions $f, g \in \mathcal{K}$. To this aim and to study the properties of f_0, g_0 we recall first their construction, see Definition 5.34.

Definition C.4. First we associate with the Kato functions $f, g \in \mathcal{K}$ the pair of functions

$$\varphi(x) := x^{-1}\left(\frac{1}{f(x)} - 1\right) \quad \text{and} \quad \psi(x) := x^{-1}(1 - g(x)). \tag{C.9}$$

Here we assume that $f(x) > 0$. Using (C.9), we define a pair of auxiliary functions by

$$0 \leqslant \varphi_0(x) := \inf_{0<s\leqslant x} \varphi(s) = \inf_{0<s\leqslant x} s^{-1}\left(\frac{1}{f(s)} - 1\right) \leqslant 1, \tag{C.10}$$

$$0 \leqslant \psi_0(x) := \inf_{0<s\leqslant x} \psi(s) = \inf_{0<s\leqslant x} s^{-1}(1 - g(s)) \leqslant 1. \tag{C.11}$$

Here we make the convention that $0^{-1} := +\infty$. Finally, we set for $x \in \mathbb{R}_0^+$:

$$f_0(x) := \begin{cases} 1, & \text{if } x = 0, \\ (1 + x\varphi_0(x))^{-1}, & \text{if } x > 0, \end{cases} \tag{C.12}$$

and

$$g_0(x) := \begin{cases} 1, & \text{if } x = 0, \\ (1 - x\psi_0(x)), & \text{if } x > 0. \end{cases} \tag{C.13}$$

Comments:

1. Since $f \in \mathcal{K}$, definitions (C.10) and (C.12) yield

$$0 \leqslant \varphi_0(x) = t^{-1}\left(\frac{1}{f_0(t)} - 1\right) \leqslant t^{-1}\left(\frac{1}{f(t)} - 1\right) \leqslant 1,$$

which implies $0 \leqslant f(x) \leqslant f_0(x) \leqslant 1$, $x \in \mathbb{R}_0^+$. Similarly (C.11) and (C.13) yield $0 \leqslant g(x) \leqslant g_0(x) \leqslant 1$, $x \in \mathbb{R}_0^+$.
Therefore, since for $f \in \mathcal{K}$ it holds that $\lim_{x\to+0} f(x) = 1$, we get also that $\lim_{x\to+0} f_0(x) = 1$ and similarly $\lim_{x\to+0} g_0(x) = 1$.

2. Note that (C.10) implies for $\delta > 0$ that

$$\varphi_0(x) := \inf_{0<s\leqslant x} \varphi(s) \geqslant \inf_{0<s\leqslant x+\delta} \varphi(s) = \varphi_0(x+\delta),$$

i.e., the function $\varphi_0(x)$ is monotone *non-increasing* on $\mathbb{R}_0^+$. Moreover,

$$\lim_{x\to+0} \varphi_0(x) = \lim_{x\to+0} \inf_{0<s\leqslant x} \varphi(s) = \lim_{x\to+0} \varphi(x) = 1, \tag{C.14}$$

where $\lim_{x\to+0} \varphi(x) = \lim_{x\to+0}(1 - f(x))/x = 1$ by (C.9) and because of $f \in \mathcal{K}$. By *Comment* 1, (C.14), and definition (C.10), we obtain for the derivative

$$\begin{aligned} f_0'(+0) &= \lim_{x\to+0} x^{-1}\left(\frac{1}{f_0(x)} - 1\right) \\ &= -\lim_{x\to+0} \frac{\varphi_0(x)}{1 + t\varphi_0(x)} = -\lim_{x\to+0} \varphi_0(x)\, f_0(x) = -1. \end{aligned}$$

3. Arguing in much the same way for the function g, we deduce that the function $\psi_0(x)$ is monotone *non-increasing* on $\mathbb{R}_0^+$ and the right derivative $g_0'(+0) = -1$.

4. Summarising *Comments* 1.–3., we conclude that the auxiliary functions f_0, g_0 associated with $f, g \in \mathcal{K}$ are themself *Kato functions.* Moreover, the corresponding functions φ_0, ψ_0 are monotone *non-increasing,* whereas the functions φ, ψ (C.9) *a priori* are not.
Consider for example $f(x) = \mathrm{e}^{-x}$, $x \in \mathbb{R}_0^+$. Then $\varphi(x) = (\mathrm{e}^x - 1)/x$, $\varphi_0(x) = 1$, and $f_0(x) = (1 + x)^{-1}$.

5. Besides the inequalities in *Comment* 1 that involve the functions f and f_0, we use in the text many other properties of them. For example:
   - $0 \leqslant \varphi_0(x) \leqslant 1$ (C.10) implies that $1 + x\varphi_0(x) \leqslant 1 + x$, for $x \in \mathbb{R}_0^+$, or by definition of f_0, that *always* $f_0(x) \geqslant (1 + x)^{-1}$.
   - For any $f \in \mathcal{K}$ one has $\lim_{n\to\infty} f(x/n)^n = \mathrm{e}^{-x}$.

6. In order to prove Proposition 5.61 we restricted the set of Kato functions $\mathcal{K}$ by a condition, that is a modification of (C.9). This defines a news class of Kato functions $\mathcal{K}_*$, as below.

Definition C.5. We say that $f, g \in \mathcal{K}$ belong to the class $\mathcal{K}_*$, if the the pair of functions φ, ψ associated with $f, g \in \mathcal{K}$ are such that

$$\varphi(x) := s^{-1}\left(\frac{1}{f(x)} - 1\right) \quad \text{and} \quad \psi(x) := x^{-1}(1 - g(x))$$
$$\text{are monotone } \textit{non-increasing} \text{ for } x \in \mathbb{R}^+ . \tag{C.15}$$

7. As noted in *Comment* 4. by adding the condition (C.15) one eliminates, for example, the exponential Kato function $f(x) = \mathrm{e}^{-x}$, whereas $f(x) = (1 + x/\kappa)^{-\kappa} \in \mathcal{K}_*$ for $0 < \kappa \leqslant 1$.

C.3 Regularity, domination: $\mathcal{K}_r$, $\mathcal{K}_D$, $\mathcal{K}_{s\text{-}d}$, $\mathcal{K}'$

As we have observed (for example in Chapter 5) the important part of conditions ensuring the convergence of Trotter-Kato product formulae in norm topologies are formulated using *auxiliary* functions f_0, g_0. These conditions have an impact (restriction) on the choice of the set of original Kato functions $f, g \in \mathcal{K}$ admissible for these assertions.

An example dealing with this issue is dealt within Lemma 5.44 of Section 5.4.

Let $f \in \mathcal{K}$ and A be a non-negative self-adjoint operator in a (infinite dimensional) Hilbert space $\mathcal{H}$. Let $(\mathbb{1} + A)^{-1}$ be compact. Is the operator $f_0(A)$ also compact ?

Note that since $f_0(x) \geqslant (1+x)^{-1}$ (Section C.2, *Comment* 5) and, hence, $f_0(A) \geqslant (1+A)^{-1}$, the converse statement follows easily. It turns out the affirmative answer to the question is possible for the class of *regular* Kato functions $\mathcal{K}_r \subset \mathcal{K}$, see Lemma 5.44.

Definition C.6. Let f be a Kato function from $\mathcal{K}$. We set $b(x) := \sup_{0 \leqslant s \leqslant x} sf(s)$ and $r(x) := \sup_{s \in [x,\infty)} f(s)$, for $x \in \mathbb{R}^+$. The Kato function f is called *regular* if $\lim_{x \to +\infty} b(x)/x = 0$ and $0 \leqslant r(x) < 1$.

Comments:

1. Note that, for instance, $f(x) = e^{-x}$ and $f(x) = 1/(1+x)$ are regular Kato functions, whereas $f(x) = 1 - |\sin(x)|$ is obviously *not* regular. Therefore, *regularity* controls the *global* behaviour of the Kato functions, including at infinity: $x\varphi_0(x) \geqslant (1 - r(\delta))x/b(x) \to +\infty$, $x \to \infty$, and $\delta > 0$.

2. It is instructive to compare this control at infinity with the "tail" control of $f \in \mathcal{K}_\alpha$ (*Comment* 4. Section C.1) and of $f \in \mathcal{K}_*$ (*Comment* 7. Section C.2)

One of the key ingredients of the *lifting method* (see Sections 6.1 and 6.4) is a control of the local uniform boundedness (6.6) of the sequence of Trotter-Kato product approximants in an appropriate norm-topology. To develop this observation, we introduce the notions of *dominated* and *self-dominated* functions. This concern f, g as well as the auxiliary functions f_0, g_0.

Definition C.7. We say that two generic Kato functions $f, g \in \mathcal{K}$ are *dominated* by the Borel measurable functions $f^D : \mathbb{R}_0^+ \to \mathbb{R}_0^+$ and $g^D : \mathbb{R}_0^+ \to \mathbb{R}_0^+$, respectively, if

$$f(qx)^{1/q} \leqslant f^D(x) \quad \text{and} \quad g(qx)^{1/q} \leqslant g^D(x), \quad 0 < q \leqslant 1. \tag{C.16}$$

We denote the sub-class of *dominated* Kato functions by $\mathcal{K}_D$.

Comments:

3. It is trivial, but useful to note that any Kato function $f \in \mathcal{K}$ is dominated by $f^D(x) = 1$ for $x \geqslant 0$.

A concept related to domination is *self-domination.*

Definition C.8. The Kato functions f and g with

$$f(qx)^{1/q} \leqslant f(x) \quad \text{and} \quad g(qx)^{1/q} \leqslant g(x), \quad 0 < q \leqslant 1, \tag{C.17}$$

for $x \geqslant 0$, are called *self-dominated* if

$$C := \sup_{x>0} \frac{xf(x)}{1 - f(x)} < +\infty \quad \text{and} \quad S := \sup_{x>0} \frac{xg(x)}{1 - g(x)} < +\infty. \tag{C.18}$$

We denote this class of Kato functions by $\mathcal{K}_{s\text{-}d}$.

Comments:

4. Note that $C \leqslant C_{1/2\alpha}$ (C.3). Therefore, in contrast to *Comment* 4 in Section C.1, one has $f(x) = (1+x)^{-1} \in \mathcal{K}_{s\text{-}d}$. Hence, for self-dominated Kato functions the "tail" control is weaker than in $\mathcal{K}_\alpha$, see *Comment* 2.

5. By *Comment* 5 in Section C.1, and by (C.16) one obtains that $\mathrm{e}^{-x} = \lim_{q\to 0} f(qx)^{1/q} \leqslant f^D(x)$. Since $\mathrm{e}^{-x} = (\mathrm{e}^{-qx})^{1/q}$, this means that if f^D dominates the Kato function f, then it also dominates the exponential Kato function $\hat{f}(x) = \mathrm{e}^{-x}$, $x \geqslant 0$.

6. Similarly, if $f \in \mathcal{K}_{s\text{-}d}$, then by (C.17) it dominates $\hat{f}(x)$.

7. By Lemma 6.24, any function $f \in \mathcal{K}_{s\text{-}d}$ has a unique *representation* as $f(x) = \mathrm{e}^{-xh(x)}, x \geqslant 0$, where $h : \mathbb{R}_0^+ \to \mathbb{R}_0^+$ is non-increasing and $\lim_{x\to+0} h(x) = 1$.

8. By this representation and by virtue of definition (C.10)

$$\varphi_0(x) = \inf_{0<t\leqslant x} \frac{1}{t}\left(\frac{1}{f(t)} - 1\right) = \inf_{0<t\leqslant x} \frac{1}{t}\left(e^{th(t)} - 1\right),$$

which yields for $x \geqslant 0$

$$f_0(x) = \frac{1}{1 + x\varphi_0(x)} \leqslant \frac{1}{1 + xh(x)} = \frac{1}{1 - \ln(f(x))} \leqslant 1.$$

Therefore, combining these estimates with *Comment* 1 in Section C.2, we obtain for any $f \in \mathcal{K}_{s\text{-}d}$ the following chain of inequalities:

$$0 \leqslant f(x) \leqslant f_0(x) \leqslant \{1 - \ln(f(x))\}^{-1} \leqslant 1. \tag{C.19}$$

9. Recall an important property of self-dominated Kato functions, see Lemma 6.26.

Let A be a non-negative self-adjoint operator in a Hilbert space $\mathcal{H}$ and let $f \in \mathcal{K}_{s\text{-}d}$. Then $f(t_0 A) \in \mathcal{C}_\phi(\mathcal{H})$ for some $t_0 > 0$, implies $f_0(t_0 A) \in \mathcal{C}_\infty(\mathcal{H})$.

10. Let $s \in \mathcal{K}_{s\text{-}d}$ be self-dominated. Then $f(x) \leqslant s(x)$ implies $f(qx)^{1/q} \leqslant (s(x))^{1/q} \leqslant s(x)$, $0 < q \leqslant 1$, that is, the function f is dominated by s.

Definition C.9. The class of Kato functions f and g obeying for $x \geqslant 0$ the conditions

$$C := \sup_{x>0} \frac{xf(x)}{1 - f(x)} < +\infty \quad \text{and} \quad S := \sup_{x>0} \frac{xg(x)}{1 - g(x)} < +\infty, \tag{C.20}$$

will be denoted by $\mathcal{K}'$.

Comments:

11. Functions of the class $\mathcal{K}'$ appeared in the formulation of sufficient conditions for the operator-norm convergence of Trotter-Kato product formulae in Proposition 5.30 and Proposition 5.46.

12. Since these statements concern only the operator-norm topology, one does not need the first part of the assumptions of the dominance in Definition C.8.

C.4 Notes

Notes to Section C.1. Definition C.1 of the class $\mathcal{K}$ is due to T. Kato. It appeared for at the first time in [Kat74] and then in his seminal paper [Kat78] about what is now called the strong operator convergent *Trotter-Kato* product formulae for an arbitrary pair of non-negative self-adjoint generators.

To *lift* this convergence to the *trace-norm* topology we introduced in [NZ90a], [NZ90b] Definition C.2 for $\alpha = 1$.

In the framework of the class $\mathcal{K}_{\alpha=1}$ the *operator-norm* convergent Trotter-Kato product formulae with error bound estimate for the rate of convergence were proved in [NZ98]. The *operator-norm* convergence of the Trotter-Kato product formulae and the rate estimate in the class $\mathcal{K}_{1/2<\alpha<1}$ were established in [NZ99a]. It was subsequently shown that for the Trotter product approximants this estimate is optimal [Tam00].

The optimal rate of convergence for $\alpha = 1$ was established in [ITTZ01] for the class $\widehat{\mathcal{K}}_\beta$ (Definition C.3), proposed by Takashi Ichinose and Hideo Tamura in [IT01].

For $\alpha = 1/2$ the conditions for the operator-norm convergence of the Trotter-Kato product formulae were formulated in [NZ99b], see also Proposition 5.30. The class of admissible Kato functions considered there coincides with $\mathcal{K}'$, Section C.3, but there are no results about the error bound estimate.

Counterexamples show that for $\alpha < 1/2$ the operator-norm convergence, in general, does not hold.

Notes to Section C.2. The auxiliary functions φ and ψ (C.9) appeared already in the first paper [Kat74] by Kato on this subject. There he proved the *strong operator* convergence only for the class $\mathcal{K}_*$, Definition C.5, which unfortunately excludes the exponential function. In a subsequent paper [Kat78], Kato proposed a completely different proof that improves his result for the strong topology by extending the class $\mathcal{K}_*$ to the class $\mathcal{K}$.

Under the conditions of Proposition 5.61 we need to restrict to the class $\mathcal{K}_*$ for the proof of the trace-norm convergence, see [NZ90a], Lemma 3.2. In [NZ90b] and then in [NZ99d] we developed further the auxiliary functions control of $\mathcal{K}$ to produce other sufficient conditions: $\mathcal{K}_r$, $\mathcal{K}_D$, $\mathcal{K}_{s\text{-}d}$, $\mathcal{K}'$, permitting the exponential function. Another elegant way to do this is to consider the class $\widehat{\mathcal{K}}_\beta$ proposed in [IT01]. This way is closer to Kato's original approach [Kat78] to the problem.

Notes to Section C.3. The concepts of regularity (Definition C.6), domination (Definition C.7), and self-domination (Definition C.8) originated in the analysis of sufficient conditions for the convergence of the Trotter-Kato product formulae in symmetrically-normed ideals by means of *auxiliary* functions in [NZ99d].

The leading idea is to *optimise* the choice of the admissible/appropriate Kato functions by taking into account the conditions on the pairs of generators A and B involved into the Trotter-Kato approximants. The ultimate aim is to ensure in this way the convergence in the corresponding norm topology. For example, in the definition of the class $\mathcal{K}'$ (Definition C.9) one does *not* need the conditions corresponding to self-dominated functions (Definition C.8) since the convergence in Proposition 5.30 and Proposition 5.46 concerns *only* the operator-norm topology.

Appendix D. Lie-Trotter-Kato product formulae: comments on the bibliography

D.1 From the strong to the norm convergence

We recall that since the work of Sophus Lie (1875) one knows that for any pair of finite square matrices $A, B \in \mathcal{M}(\mathbb{R}^d)$ on the d-dimensional Euclidean space $\mathbb{R}^d$, one has for $n \to \infty$ the estimates

$$\left\|\left(\mathrm{e}^{-tA/n}\mathrm{e}^{-tB/n}\right)^n - \mathrm{e}^{-t(A+B)}\right\| \leqslant O(1/n), \tag{D.1}$$

$$\left\|\left(\mathrm{e}^{-tA/2n}\mathrm{e}^{-tB/n}\mathrm{e}^{-tA/2n}\right)^n - \mathrm{e}^{-t(A+B)}\right\| \leqslant O(1/n^2). \tag{D.2}$$

Here $\|\cdot\|$ is *any* norm on the matrix space $\mathcal{M}(\mathbb{R}^d)$. Note that the power $\left(\mathrm{e}^{-tA/n}\mathrm{e}^{-tB/n}\right)^n$ is called the *Lie product* (or the product *approximant*), and the power $\left(\mathrm{e}^{-tA/2n}\mathrm{e}^{-tB/n}\mathrm{e}^{-tA/2n}\right)^n$ is called the *symmetric* (or *symmetrised*) *Lie product* (or *approximant*). For the proof see Section 5.1 to understand the reason of the difference in the convergence rate between (D.1) and (D.2).

Generalisations of the Lie-type product formula to infinite-dimensional spaces appeared in several papers by Yu. L. Daletskiĭ (see, e.g., [Dal60], [Dal61]) in connection with a path-integral representation (known as the Feynman-Kac formula) for solutions of operator differential evolution equations. But the first abstract result was due to H. Trotter [Tro59], who extended the Lie product formula to strongly continuous contraction semigroups on a Banach space $\mathcal{B}$. The convergence of the Lie-Trotter product formula was proved in the *strong* operator topology.

Let A and B be generators of contraction semigroups in a Banach space $\mathcal{B}$; if the closure C of the operator sum $A + B$ is the generator of a contraction

V. A. Zagrebnov, *Gibbs Semigroups*, Operator Theory: Advances and Applications 273, https://doi.org/10.1007/978-3-030-18877-1

semigroup, then

$$\operatorname*{s-lim}_{n\to\infty}\left(e^{-tA/n}e^{-tB/n}\right)^n = e^{-tC}, \quad t \geqslant 0. \tag{D.3}$$

In [Nel64] E. Nelson pointed out the importance of the Trotter product formula for semigroups generated by Schrödinger operators, and for the proof of the Feynman-Kac formula. To him also belongs a simple proof of the Trotter theorem, in the case where A and B are two non-negative self-adjoint operators in a separable Hilbert space $\mathcal{H}$ such that the operator sum $A + B$ is also self-adjoint [Nel64]. It is striking that much later it was proved in [ITTZ01] that Nelson's conditions ensure convergence of the product formula in the *operator-norm* topology, with an error bound estimate for the rate of convergence that is *optimal*, since it coincides with the rate in (D.1).

A beautiful and elegant way to treat the Lie-Trotter product formula in the framework of the general theory of strongly continuous contractions is due to P. Chernoff [Che68, Che74]. In particular, he proposed a new way to consider perturbations (sums) of linear operators in a Hilbert space. Since the right-hand side of the Trotter product formula defines a semigroup, the corresponding generator can be associated with a "sum" of the generators in the left-hand side, see [Che74], and [Far67a, Far67b], where perturbations of linear propagators were also considered.

Further progress was achieved by T. Kato [Kat78]. In 1978 he obtained the following important result:

Let $A \geqslant 0$ and $B \geqslant 0$ be non-negative self-adjoint operators in a separable Hilbert space $\mathcal{H}$. Denote by $\mathcal{H}_0$ the subspace

$$\mathcal{H}_0 := \overline{\operatorname{dom} A^{1/2} \cap \operatorname{dom} B^{1/2}}\,.$$

It may happen that $\operatorname{dom} A \cap \operatorname{dom} B = \{0\}$, but the form-sum $H = A \dotplus B$ is well defined on the subspace $\mathcal{H}_0$. Under these conditions the Trotter product formula converges strongly to the *degenerate* semigroup $\{e^{-tH}P_0\}_{t>0}$, locally uniformly away from zero. That is, one has

$$\operatorname*{s-lim}_{n\to\infty}\left(e^{-tA/n}e^{-tB/n}\right)^n = e^{-tH}P_0, \quad t > 0, \tag{D.4}$$

uniformly in $t \in [a, b]$, for any $[a, b] \subset (0, +\infty)$. Here P_0 denotes the orthogonal projection from $\mathcal{H}$ onto $\mathcal{H}_0$. Further, T. Kato [Kat74, Kat78] discovered that the Trotter product formula is valid not only for the exponential function e^{-x}, $x \geqslant 0$, but also for a whole class of Borel measurable functions f and g that are defined on $[0, \infty)$ and satisfy the conditions

$$0 \leqslant f(x) \leqslant 1, \qquad f(0) = 1, \qquad f'(+0) = -1, \tag{D.5}$$

$$0 \leqslant g(x) \leqslant 1, \qquad g(0) = 1, \qquad g'(+0) = -1. \tag{D.6}$$

Kato proved that in this case one always has

$$\operatorname*{s-lim}_{n\to\infty}\left(f(tA/n)g(tB/n)\right)^n = e^{-tH}P_0, \quad t > 0, \tag{D.7}$$

uniformly in $t \in [a,b]$, for any $[a,b] \subset (0,+\infty)$. These product formulae for pairs f,g are known as the *Trotter-Kato* product formulae for *generic Kato functions* $f,g \in \mathcal{K}$, [NZ90a, NZ90b]. More details about classes of Kato functions are provided in Appendix C.

The next important step concerning the Trotter product formula is related to the following result by Dzh. L. Rogava in 1993 [Rog93]:
Let A and B be two bounded from below self-adjoint operators in a separable Hilbert space $\mathcal{H}$. If $\operatorname{dom} A \subset \operatorname{dom} B$ and the operator $C = A + B$ is self-adjoint, then

$$\left\| \left(\mathrm{e}^{-tA/n}\mathrm{e}^{-tB/n}\right)^n - \mathrm{e}^{-tC} \right\| \leqslant \frac{c\ln(n)}{\sqrt{n}}, \quad n > 1, \tag{D.8}$$

locally uniformly in $t \in [0,T]\,, 0 < T < \infty$. This *operator-norm* convergence of the Trotter product formula with error bound estimate (D.8) is uniform in $t \geqslant 0$ if the operators A and B are positive.

Since Rogava discovered that the Trotter product formula may converge in the operator norm topology, the question of the error bound estimate and its dependence on the pair A, B becomes an interesting problem. Trying to understand the Rogava statement, the following result was established in [Zag95]: let A and B be two positive self-adjoint operators such that B is *bounded*, $B \in \mathcal{L}(\mathcal{H})$, then for self-adjoint $C = A + B$

$$\left\| \left(\mathrm{e}^{-tA/n}\mathrm{e}^{-tB/n}\right)^n - \mathrm{e}^{-tC} \right\| \leqslant \frac{c\,(\ln(n))^2}{n}, \quad n > 1, \tag{D.9}$$

uniformly in $t \geqslant 0$. It was conjectured that the improved error bound estimate (D.9) is owing to the boundedness of B.

In 1997, T. Ichinose and Hideo Tamura [IT97] found that if A and B are two non-negative self-adjoint unbounded operators in a separable Hilbert space $\mathcal{H}$ such that

$$\operatorname{dom} A^\alpha \subset \operatorname{dom} B, \quad 0 \leqslant \alpha < 1, \tag{D.10}$$

then for self-adjoint $C = A + B$ the error bound for the rate of convergence for the Trotter product formula is of the order $O(n^{-2/(3+\alpha)})$, when $n \to \infty$.

Then in 1998, H. Neidhardt and V. A. Zagrebnov [NZ98] relaxed the Ichinose-Tamura operator smallness condition (D.10) and generalised the Trotter product formula to the Trotter-Kato product formulae for *unbounded* operator B with a better error bound estimate than (D.9).

Namely, suppose $A \geqslant \mathbb{1}$ and $B \geqslant \mathbb{1}$ be self-adjoint operators such that unbounded operator B, with $\operatorname{dom} A \subset \operatorname{dom} B$, is Kato-small with respect to A for the relative A-bound $b < 1$, that is,

$$\|Bu\| \leqslant a\,\|u\| + b\,\|Au\|\,, \quad u \in \operatorname{dom} A,\ a,b \geqslant 0. \tag{D.11}$$

Then for self-adjoint $C = A + B$ we proved that

$$\left\| \left(f(tA/n)g(tB/n)\right)^n - \mathrm{e}^{-tC} \right\| \leqslant \frac{c\,\ln(n)}{n}, \quad n > 1, \tag{D.12}$$

uniformly in $t \geqslant 0$, where the Borel measurable functions $f, g \in \mathcal{K}_\alpha$ satisfy, besides the generic Kato conditions (D.5) and (D.6), some additional requirements (5.15)–(5.20). For details about the sub-class $\mathcal{K}_\alpha$ of the generic Kato functions $\mathcal{K}$, see Appendix C.

Note that the Ichinose-Tamura condition (D.10) on A and B implies that the relative bound $b > 0$ is *infinitesimally* small, that is, $B \in \mathcal{P}_{0+}$, see Section 4.4. In the subsequent paper [IT98a], T. Ichinose and H. Tamura improved their estimate to $(c\ln(n))/n$, and extended it to the case where the generator $B = B(t)$ is *time-dependent*. Their condition (D.10) has to be uniform in $t \in [0, T], 0 < T < \infty$, and in addition they require that

$$\left\|A^{-\alpha}(B(t) - B(s))A^{-\alpha}\right\| \leqslant d\,|t - s|\,, \quad d > 0, \tag{D.13}$$

for positive operators A and $B(\cdot)$, when $t, s \in [0, T]$. This was the first Trotter-type result about the operator-norm convergent product formula for *propagators*. The paper [IT98a] generalises a previous result in [Far67b] about the *strong operator* topology convergent product formula for propagators to the *operator-norm* topology under a time-dependent condition (D.10):

$$\operatorname{dom} A^\alpha \subset \operatorname{dom} B(t), \quad 0 \leqslant \alpha < 1 \text{ for } t \in [0, T]\,.$$

D.2 Norm convergence: optimal rate

In 1999, H. Neidhardt and V. A. Zagrebnov [NZ99a] considered, instead of (D.11), the optimal fractional power conditions on the pair of self-adjoint non-negative generators A and B for the Trotter-Kato product formula (D.12). In particular, they proved the following assertion:

Let $A \geqslant \mathbb{1}$ and $B \geqslant 0$ be two self-adjoint operators such that B^α is Kato-small with respect to A^α, with the relative bound $b < 1$:

$$\|B^\alpha u\| \leqslant b\,\|A^\alpha u\|\,, \quad u \in \operatorname{dom} A^\alpha,\ b \geqslant 0, \tag{D.14}$$

and

$$\operatorname{dom} H^\alpha \subset \operatorname{dom} A^\alpha, \tag{D.15}$$

for some $\alpha \in (1/2, 1)$, where $H = A \dot{+} B$. Then

$$\left\|\left(f(tA/n)g(tB/n)\right)^n - \mathrm{e}^{-tH}\right\| \leqslant \frac{c}{n^{2\alpha-1}}, \quad n > 1, \tag{D.16}$$

uniformly in $t \geqslant 0$.

Hiroshi Tamura [Tam00] proved that this estimate is *optimal.* More precisely, there exist operators A and B satisfying the statement above, such that one also gets

$$\left\|\left(\mathrm{e}^{-tA/n}\mathrm{e}^{-tB/n}\right)^n - \mathrm{e}^{-tH}\right\| \geqslant \frac{c_t}{n^{2\alpha-1}}, \quad \text{for } \alpha \in (1/2, 1), \tag{D.17}$$

for large n, whereas

$$\left\|\left(e^{-tA/n}e^{-tB/n}\right)^n - e^{-tH}\right\| \geqslant d_t\,, \tag{D.18}$$

if $\alpha \in (0,1/2]$. Here c_t and d_t are positive continuous functions for $t > 0$.

Note that the case $\alpha = 1/2$ is an example where the Trotter product formula does not hold in the operator norm, while it still converges strongly by Kato's theorem [Kat78]. On the other hand, inequality (D.18) proves a conjecture in [NZ99a] that the operator norm convergence of (D.16), and of all other Trotter-Kato formulae, cannot be extended to the case $\alpha \in (0,1/2]$ without additional conditions. Some of these (sufficient) conditions were found by H. Neidhardt and V. A. Zagrebnov in [NZ99b, NZ99c]. Their first result concerns a generalisation of the main proposition by Chernoff (Theorem 1.1 in [Che74]) to the operator norm topology. This result gives a criterion for the convergence of the Trotter-Kato product formulae in the operator norm, which implies the following assertion:

Let $A \geqslant 0$ and $B \geqslant 0$ be two self-adjoint operators in a separable Hilbert space $\mathcal{H}$. The Trotter-Kato product formulae converge in the operator norm topology if one of the following conditions is satisfied [NZ99b]:

(i) $(\mathbb{1} + A)^{-1} \in \mathcal{C}_\infty(\mathcal{H})$;

(ii) $A \geqslant \mathbb{1}$ and $B^{1/2}A^{-1/2} \in \mathcal{C}_\infty(\mathcal{H})$;

(iii) $(\mathbb{1} + A)^{-1}(\mathbb{1} + B)^{-1} \in \mathcal{C}_\infty(\mathcal{H})$.

In particular, condition (ii) ensures the operator norm convergence of the Trotter-Kato formula (D.16) for $\alpha = 1/2$.

The abstract condition (iii) covers the case of Schrödinger semigroups on $\mathcal{H} = L^2(\mathbb{R}^d)$ with potentials increasing at infinity considered in the papers [Hel95, DS97, ITak97, DIT98, Tak97, ITak98, ITak00], giving a rate not better than order $O(1/n)$, see review in [IT09]. Note that in these references neither $(\mathbb{1} + A)^{-1}$, nor $(\mathbb{1} + B)^{-1}$ are compact, but their *product* is. Note also that the operator norm convergence was proven in [NZ99b] without error bound estimate.

The optimality of the convergence rate was elucidated by T. Ichinose, Hideo Tamura, Hiroshi Tamura, and V. A. Zagrebnov in [ITTZ01]. This was the next after [NZ99a] result about the optimal convergence rate for the Trotter-Kato product formulae. In [ITTZ01] we proved the following statement:

Let A and B be non-negative self-adjoint operators in a Hilbert space $\mathcal{H}$ such that $C = A + B$ is also self-adjoint. Suppose, in addition to (D.5) and (D.6), that the Borel functions f and g are decreasing and satisfy the conditions

$$\sup_{x>0} \frac{|f(x) - 1 + x|}{x^\beta} < \infty, \tag{D.19}$$

$$\sup_{x>0} \frac{|g(x) - 1 + x|}{x^\beta} < \infty, \tag{D.20}$$

for $\beta = 2$, that is, $f, g \in \widehat{\mathcal{K}}_{\beta=2}$. For details about the Ichinose-Tamura class $\widehat{\mathcal{K}}_\beta$ of the Kato-functions, see Appendix C. Then the Trotter-Kato product formulae

(D.12) converges in the operator norm topology uniformly in $t \in [0,T]$, for $0 < T < \infty$, with the *upper* error bound estimate $O(1/n)$, instead of $O(\ln(n)/n)$. The optimality of the rate $O(1/n)$ is shown in [ITTZ01] via simple noncommutativity of generators A and B, which gives $O(1/n)$ for the estimate of the Trotter-Kato product formulae (D.12) from *below*, this means that the rate $O(1/n)$ is *sharp*.

Similarly to [NZ99a], the same conclusion is also valid (sic!) for symmetric Trotter-Kato approximants with the same rate of convergence $O(1/n)$. Based on an example from Tamura's optimal theorem [Tam00], it was shown in [ITTZ01] that symmetrisation *itself* does not improve the rate of convergence because there are operators A, B and (exponential) functions $f, g \in \widehat{\mathcal{K}}_2$ satisfying the conditions of [ITTZ01] such that for $t > 0$ and $n > 1$ one has

$$\left\|(g(tB/2n)f(tA/n)g(tB/2n))^n - \mathrm{e}^{-tC}\right\| \geqslant L(t)/n\,, \quad L(t) > 0. \tag{D.21}$$

This is an unexpected result if one compares it with the Lie product formula (D.2), which improves the rate of convergence after symmetrisation. Therefore, the rate of convergence of the Trotter-Kato product formulae under the [ITTZ01] conditions is *optimal.*

This abstract result covers almost all cases of the Trotter-Kato product formulae for self-adjoint semigroups mentioned above. The only exception is the one where the Neidhardt-Zagrebnov fractional power conditions are involved [NZ99a].

We note also that condition [ITTZ01]: the operator $C := A + B$ is *self-adjoint* if A and B are self-ajoint, is quite subtle and cannot be relaxed. In the paper [Tam00], Proposition 3, Hiroshi Tamura constructed example in which the operator-norm convergence of the Trotter product formula does not hold (cf.(D.18)):

$$\left\|(g(tB/2n)f(tA/n)g(tB/2n))^n - \mathrm{e}^{-tC}\right\| \geqslant D_t\,, \quad D_t > 0, \tag{D.22}$$

even though the operator sum $C = A + B$, is *essentially* self-adjoint and the self-adjoint operator B is A-form bounded with a relative bound less than one.

Of course, in this case the product formula (D.22) still converges strongly according to Kato's theorem [Kat78], or by Trotter's theorem [Tro59] for exponential functions f, g.

Another result by T. Ichinose, H. Neidhardt, and V. A. Zagrebnov [INZ03, INZ04] (INZ) is also proved with the optimal error bound.

Let A and B be non-negative self-adjoint operators on a Hilbert space such that their densely defined form-sum $H = A \dot{+} B$ obeys the condition $\operatorname{dom} H^\alpha \subseteq \operatorname{dom} A^\alpha \cap \operatorname{dom} B^\alpha$, for some $\alpha \in (1/2, 1)$. If in addition A and B satisfy $\operatorname{dom} A^{1/2} \subseteq \operatorname{dom} B^{1/2}$, then the nonsymmetric and symmetric Trotter product formulae converge in the operator norm:

$$\left\|\left(\mathrm{e}^{-tB/2n}\mathrm{e}^{-tA/n}\mathrm{e}^{-tB/2n}\right)^n - \mathrm{e}^{-tH}\right\| = O(n^{-(2\alpha-1)}), \tag{D.23}$$

$$\left\|\left(\mathrm{e}^{-tA/n}\mathrm{e}^{-tB/n}\right)^n - \mathrm{e}^{-tH}\right\| = O(n^{-(2\alpha-1)}), \tag{D.24}$$

uniformly in $t \in [0,T]$, $0 < T < \infty$ as $n \to \infty$, both with the same optimal error bound. The same holds true if one replaces the exponential function in the products (D.23), (D.24) by functions from the Kato class $\hat{\mathcal{K}}_\beta$, see Appendix C.

The INZ result improves the previous one in [NZ99a] by relaxing the assumption of smallness of B^α with respect to A^α to the *milder* assumption $\operatorname{dom} A^{1/2} \subseteq \operatorname{dom} B^{1/2}$ and by extending the *admissible* class of the Kato functions from $\mathcal{K}_\alpha$ to $\hat{\mathcal{K}}_\beta$.

If instead of $\operatorname{dom} A^{1/2} \subseteq \operatorname{dom} B^{1/2}$ one has a stronger condition: $B^{1/2}$ is relatively compact with respect to $A^{1/2}$ (condition (ii) of [NZ99b], see above), then we have operator norm convergence of the Trotter-Kato product formulae, but without any error bound estimate. In [NZ99b] we conjectured that instead of the relative compactness the relative form boundedness with the relative bound $b = 0^+$, is sufficient for this convergence.

The optimality problem was treated by Ichinose and Tamura in [IT04] for Schrödinger C_0-semigroups on $\mathcal{H} = L^2(\mathbb{R}^d)$. This was an important step in the resolution of the [ITTZ01] *no-go* problem (D.21) concerning the improvement of the convergence rate by symmetrisation of the *Trotter* product formula. In [IT04] it is proved that for generators $A = -\Delta$ and $B = V$, where V is a non-negative function $\mathbb{R}^d \ni x \mapsto V(x)$ growing polynomially at infinity and hence verifying the conditions of Proposition 5.19, the *symmetric* Trotter product formula converges in the operator norm with the rate $O(n^{-2})$. This result advises against a prejudice that for *any* pair of noncommuting unbounded operators A and B verifying the conditions of Proposition 5.19 the rate of convergence $O(n^{-1})$ is necessarily the best possible for the symmetric Trotter product formula. In fact, if one imposes on A, B *additional* conditions (here $(\mathbb{1} + A)^{-1}(\mathbb{1} + B)^{-1} \in \mathcal{C}_\infty(L^2(\mathbb{R}^d)))$ compared with the [ITTZ01] Proposition 5.19 (subsection 5.2.4), then the symmetrisation could improve the rate of convergence up to the asymptotic $O(n^{-2})$.

Moreover, in [AzIch08], taking a simple example of the one-dimensional harmonic oscillator: $H = A + B$, with $A = -\partial_x^2$, $B = V(x) = x^2$ in $\mathcal{H} = L^2(\mathbb{R})$, it is proved by explicit calculations that the estimate of the symmetric Trotter formula from *below* is also of the order $O(n^{-2})$:

$$\frac{c}{n^2} \leqslant \left\| \left(\mathrm{e}^{-tB/2n}\mathrm{e}^{-tA/n}\mathrm{e}^{-tB/2n}\right)^n - \mathrm{e}^{-tH} \right\| \leqslant \frac{C}{n^2}\,.$$

Here $0 < c < C$, locally uniformly for $t \geqslant 0$. Hence the rate $O(n^{-2})$ is *sharp*, and so optimal. A natural hypothesis is that the same rate $O(n^{-2})$ is optimal for any non-negative growing polynomially at infinity potential $V(x)$. For more details see review in [IT09].

D.3 Norm convergence: non-self-adjoint semigroups and Banach spaces

On the other hand, we are actually rather far from the optimal error bound estimates for the rate of convergence of product formulae in the case of non-self-adjoint semigroups, or of semigroups on a Banach space.

In paper [CZ01a] V. Cachia and V. A. Zagrebnov proved the following result: Let A generate a holomorphic contraction semigroup on a Banach space $\mathcal{B}$ and let B be the generator of a contraction semigroup such that for some $\alpha \in [0,1)$,

$$\operatorname{dom} A^{\alpha} \subseteq \operatorname{dom} B \quad \text{and} \quad \operatorname{dom} A^{*} \subseteq \operatorname{dom} B^{*}. \tag{D.25}$$

Then operator $C = A + B$ is generator of the holomorphic semigroup and the Trotter product formula converges in the operator norm with the error bound estimates

$$\left\| \left(\mathrm{e}^{-tA/n}\mathrm{e}^{-tB/n}\right)^n - \mathrm{e}^{-tC} \right\| \leqslant M_{\alpha>0} \, \frac{\ln(n)}{n^{1-\alpha}}, \quad n > 1, \tag{D.26}$$

$$\left\| \left(\mathrm{e}^{-tA/n}\mathrm{e}^{-tB/n}\right)^n - \mathrm{e}^{-tC} \right\| \leqslant M_{\alpha=0} \, \frac{(\ln(n))^2}{n}, \quad n > 1, \tag{D.27}$$

locally uniformly in $t \geqslant 0$.

We note that in the case $\alpha = 0$ the operator B in (D.25) is by definition *bounded*: $B \in \mathcal{L}(\mathcal{B})$. The asymptotics $O((\ln(n))^2/n)$ in (D.27) was established by the same argument as in [Zag95] for a Hilbert space. This argument was not based on the self-adjointness as the proof in [NZ98], but on analytic extension of the semigroup $\{\mathrm{e}^{-tA}\}_{t \geqslant 0}$. So, it was not unexpected that the rates of convergence in (D.9) and in (D.27) coincide.

If $\alpha > 0$, then (D.25) (similar to the Ichinose-Tamura condition (D.10)) implies that the operator B is *infinitesimally* A-small, i.e., $B \in \mathcal{P}_{0+}$. But in contrast to the improved self-adjoint estimate $(c\ln(n))/n$, [IT98a], the rate of convergence (D.26) in the non-self-adjoint setting rests less sharp and it is α-dependent.

In [CZ99] the same authors showed that some results of [NZ99a], [NZ99b], and [NZ99c] can be extended to m-sectorial generators A and B, but *without* error bound estimates. In [CZ01b] they introduced the notion of *quasi-sectorial* contractions and have shown that for them the Chernoff approximation theorem [Che68] is valid in the operator norm topology.

The Trotter product formulae for a class of non-self-adjoint contraction semigroups with error bound estimate was proved by V. Cachia, H. Neidhardt, and V. A. Zagrebnov in [CNZ01, CNZ02]. In [CNZ01], they proved the following generalisation of [NZ98]:

Let A be a positive self-adjoint operator, and let B be an m-accretive (i.e., closed and maximal accretive) operator such that B and B^* are Kato-small with respect to A, with relative bounds less than 1. Then the Trotter product formulae converge in the operator norm topology with error bound estimate $O(\ln(n)/n)$, uniformly in $t \geqslant 0$.

In [CNZ02] it is proved that if one replaces the above conditions on the m-accretive operator B by the fractional conditions:

$$\operatorname{dom}(H^*)^\alpha \subseteq \operatorname{dom} A^\alpha \cap \operatorname{dom}(B^*)^\alpha \neq \{0\}, \tag{D.28}$$

for some $\alpha \in (0,1]$, then the Trotter product formula converges in the operator norm with error bound estimate $O(n^{-\alpha}\ln(n))$ uniformly in $t \geqslant 0$. A similar estimate is also true for $H^* = (A+B)^*$, that is, for adjoint semigroups generated by $A = A^*$, B^*, and H^*. Note that in the latter case we do not require the smallness of B^* with respect to A as in [CNZ01], but the weaker condition (D.25).

We must stress the point that just the A-smallness of B (even with relative bound less than 1) is not sufficient for the existence of a nontrivial operator $A+B^*$. The paper [CNZ02] starts with an example in which for these conditions one gets $\operatorname{dom} A \cap \operatorname{dom} B^* = \{0\}$. So, in this case $H^* \neq A + B^*$. For more details see a nice short book [Ca10].

Recently we revised the operator-norm convergence (D.26), (D.27) of the Trotter product formula on a Banach space [NSZ18a]. The operator-norm convergence holds true if the dominating operator A generates a holomorphic contraction semigroup and B is an infinitesimally small generator of a contraction semigroup (D.26), in particular, if B is a bounded operator (D.27). Inspired by our studies of the evolution semigroups in Banach spaces, we show that in general the operator-norm convergence of the Trotter product formula *fails*, even for bounded operators B, if A is *not* generator of a holomorphic semigroup. We proposed examples that in this case by varying B the rate of the operator-norm convergence of the Trotter product formula can be made arbitrary slow. For further reading we suggest a short review [NSZ18b].

D.4 Trace-norm convergence

Since the present book is about the Gibbs semigroups, which are trace-norm continuous away from $t_0 \geqslant 0$, the bibliographic comments about this topology are dispersed throughout the Notes at the end of Chapters 4–6 and Chapter B.3. So, I would like to add here only few more comments on the question of convergence of semigroups and product formulae in the trace-norm topology.

As far as I know, it was D. Uhlenbrock, [Uhl71], who introduced for the first time the notion the *Gibbs semigroups* (Definition 4.1). He was motivated by the one-parameter Gibbs exponential for the density-matrix operator in Quantum Statistical Mechanics. Then this concept was supported in papers [ANB75] and [Zag80].

The problem of the possible convergence of the Trotter product formula for the Gibbs semigroups in the trace-norm topology was suggested to the author by Robert A. Minlos. A positive answer in the quantum statistical mechanics setup for the Schrödinger (Gibbs) semigroups was obtained in [Zag88]. The first abstract result was due to H. Neidhardt and V. A. Zagrebnov [NZ90a, NZ90b]. We proved the following assertion:

Let $A \geqslant 0$ and $B \geqslant 0$ be two non-negative self-adjoint operators in a separable Hilbert space $\mathcal{H}$. Let $\mathcal{H}_0$ denote the subspace

$$\mathcal{H}_0 := \overline{\operatorname{dom} A^{1/2} \cap \operatorname{dom} B^{1/2}} \, .$$

Further, let f, g be the Borel functions defined on $[0, \infty)$ satisfying conditions:

$$\begin{aligned} &0 \leqslant f(x) \leqslant 1, \qquad && f(0) = 1, \qquad && f'(+0) = -1, \\ &0 \leqslant g(x) \leqslant 1, && g(0) = 1, && g'(+0) = -1, \end{aligned}$$

that is, they are generic Kato functions: $f, g \in \mathcal{K}$, Appendix C. Then the Trotter-Kato product formula

$$\| \cdot \|_1\text{-}\lim_{n\to\infty} (f(tA/n)g(tB/n))^n = \mathrm{e}^{-tH} P_0 \, , \quad t > 0, \tag{D.29}$$

converges locally uniformly away from zero in the *trace-norm* topology on $\mathcal{C}_1(\mathcal{H})$ if

$$f_0(tA) \in \mathcal{C}_p(\mathcal{H}), \quad \text{for } t > 0, \text{ and } 1 \leqslant p < \infty.$$

Here P_0 is the orthogonal projection from $\mathcal{H}$ onto $\mathcal{H}_0$ and the function $f_0 : \mathbb{R}^+ \to \mathbb{R}^+$ is completely determined by f in (C.12), Appendix C.

For example, if $f(x) = \mathrm{e}^{-x}$, then $f_0(x) = (1 + x)^{-1}$. Therefore, the operator A is a p-generator since $(1 + tA)^{-1} \in \mathcal{C}_p(\mathcal{H})$ for $t > 0$, Definition 4.26. We remaind that, incidentally, (D.29) implies the operator-norm convergence. Thus, the operator-norm convergence of the product formulae for Gibbs semigroups was establishes before [Rog93].

The trace-norm Trotter product formula with *error bound* estimates of the convergence rate appeared first for the case of the Schrödinger (Gibbs) semigroups in [IT98b]. Then estimates of the rate of trace-norm convergence for the Trotter-Kato product formulae and abstract Gibbs semigroups were obtained in [NZ99d] and in [CZ01c]. They were established by *lifting* the corresponding operator-norm error bound estimates. We consider this lifting method in Chapter 5.

The study of the convergence of the Trotter-Kato product formulae in general framework of symmetrically-normed ideals was initiated by the paper [NZ99d]. Therein a programme for treating the case of the Dixmier ideal was mentioned in concluding remarks. Chapter 7 is a partial realisation of this programme, see also [Zag19].

D.5 Unitary product formulae

This book is about the Gibbs semigroups, which are trace-norm continuous away from $t_0 \geqslant 0$. Therefore we do not discuss product formulae that involve, or have as their limits, the *unitary groups* of operators. Thus, we conclude this appendix by only few remarks to indicate some known results about the unitary product formulae. The first one is due to H. Trotter [Tro59].

Proposition D.1 (Unitary Trotter product formula). *Let A, B and C be self-adjoint generators of unitary groups on $\mathcal{H}$. Suppose that algebraic sum*

$$Cu = Au + Bu, \tag{D.30}$$

is valid for all $u \in \mathcal{D}$, where $\mathcal{D} \subset (\operatorname{dom} A \cap \operatorname{dom} B)$ is a core of the operator C, i.e., the operator $A + B$ is essentially self-adjoint. Then the unitary group $\{U_C(t) = \mathrm{e}^{-itC}\}_{t\in\mathbb{R}}$ can be approximated in the strong operator topology on $\mathcal{H}$ by the Trotter product formula

$$\mathrm{e}^{-itC}\ u = \lim_{n\to\infty} (\mathrm{e}^{-itA/n}\mathrm{e}^{-itB/n})^n\ u, \quad u \in \mathcal{H}, \tag{D.31}$$

for all $t \in \mathbb{R}$. Here generator $C := \overline{(A + B) \restriction \mathcal{D}}$ is the closure of the algebraic sum (D.30).

Note that similar to the case of *semigroup* Trotter product formula (D.3), the *unitary* product formula (D.31) generally does not hold in the operator norm. There are examples showing that conditions of Proposition D.1 are neither *sufficient*, nor *necessary* for convergence of the unitary product formula in the operator-norm topology, see [Ich03], [Che74] §5. Since the unitary groups involved in (D.31) do not belong to the trace-class $\mathcal{C}_1(\mathcal{H})$, this note is also valid for convergence in the trace-norm topology.

Recall that a systematic treatment of convergence of the unitary Trotter product formula in the operator-norm topology can be based on the *commutator conditions* [IT04c].

Let A and B be self-adjoint operators in a Hilbert space $\mathcal{H}$ with domains $\operatorname{dom} A$ and $\operatorname{dom} B$ such that the operator sum $A + B$ is essentially self-adjoint on $\operatorname{dom} A \cap \operatorname{dom} B$. Both A and B may not be semi-bounded, that is, neither bounded form below, nor from above. We denote the unique self-adjoint extension, which is the closure of $A + B$, by $C := \overline{A + B}$. Here the case when the $A + B$ itself is already self-adjoint is included.

Assume that there exists a dense set $\mathcal{D} \subset \operatorname{dom} A \cap \operatorname{dom} B$ such that

(d1) $\mathrm{e}^{-itA}, \mathrm{e}^{-itB} : \mathcal{D} \to \mathcal{D}$ for $t \in \mathbb{R}$.

(d2) There exists a symmetric operator E on $\mathcal{D}$ such that

$$i\big((Bu, Av) - (Au, Bv)\big) = (Eu, v),$$

for every u, $v \in \mathcal{D}$.

Remark D.2. The operator E is formally represented in the form of the commutator $E = i[A, B] = i(AB - BA)$.

Proposition D.3. *Let A and B be self-adjoint operators with domains* $\operatorname{dom} A$ *and* $\operatorname{dom} B$, *respectively. Suppose that operator $A+B$ is essentially self-adjoint on* $\operatorname{dom} A \cap \operatorname{dom} B$ *and denote by the C its unique self-adjoint extension.*

If conditions (d1) *and* (d2) *are satisfied and operator E in* (d2) *admits a bounded extension on $\mathcal{H}$, then*

$$\left\|\left(\mathrm{e}^{-itA/n}\mathrm{e}^{-itB/n}\right)^n - \mathrm{e}^{-itC}\right\| = O(n^{-1}), \tag{D.32}$$

$$\left\|\left(\mathrm{e}^{-itB/2n}\mathrm{e}^{-itA/n}\mathrm{e}^{-itB/2n}\right)^n - \mathrm{e}^{-itC}\right\| = O(n^{-1}), \tag{D.33}$$

as $n \to \infty$, locally uniformly in $t \in \mathbb{R}$.

Since the unitary groups involved in (D.32), (D.33) do not belong to trace-class $\mathcal{C}_1(\mathcal{H})$, the Proposition D.3 does not allow us to use the *lifting* method of Section 5.2 for improving the operator-norm to the trace-norm convergence.

To proceed further with analysis of unitary versus semigroup product formulae we recall the Kato result (D.4) for $\mathcal{H}_0 = \mathcal{H}$ and $H = C$,

$$\mathrm{e}^{-tC} = \operatorname*{s-lim}_{n\to\infty}\left(\mathrm{e}^{-tA/n}\mathrm{e}^{-tB/n}\right)^n, \quad t \in \mathbb{R}^+. \tag{D.34}$$

In this case intersection $\operatorname{dom} A^{1/2} \cap \operatorname{dom} B^{1/2}$ is dense in $\mathcal{H}$ and self-adjoint operator $C := A \dot{+} B$ is the *form-sum* of A and B.

Now one can poses a question, whether the Trotter product formula (D.34) remains valid for the *imaginary* parameter it, $t \in \mathbb{R}$, see [Che74, Remarks pp. 90-91] and [JL00, Problem 11.3.9]. Note that if A and B are non-negative self-adjoint operators in $\mathcal{H}$ and the limit in the right-hand side of (D.31) exists for all $t \in \mathbb{R}$, then $\operatorname{dom} A^{1/2} \cap \operatorname{dom} B^{1/2}$ is dense in $\mathcal{H}$, see [JL00, Proposition 11.7.3]. Moreover, applying Proposition D.1 we find that formula (D.31) is valid *if* the operator $C = \overline{A+B}$, i.e. the closure is self-adjoint. However, if $A+B$ is *not* essentially self-adjoint, then all attempts to verify the Trotter product formula (D.34) for $t \geqslant 0$ substituted by it, $t \in \mathbb{R}$, have failed so far.

One of the ways out was proposed by T. Ichinose in [Ich80]. His main idea was to find for product formulae some appropriate *non-exponential* approximants à la Trotter-Kato. In [ENZ11] we generalised this scheme and proved that for non-negative self-adjoint operators A and B, such that the intersection $\operatorname{dom}(A^{1/2}) \cap$ *dom*$(B^{1/2})$ is dense in $\mathcal{H}$ one gets

$$\mathrm{e}^{-itC} = \operatorname*{s-lim}_{n\to\infty}\left(\phi(tA/n)\psi(tB/n)\right)^n, \tag{D.35}$$

locally uniformly in $t \in \mathbb{R}$, where ϕ and ψ are some *admissible functions* such that $\mathfrak{Re}(\phi(x)) \geqslant 0$, $\mathfrak{Im}(\phi(x)) \leqslant 0$ and $\mathfrak{Im}(\psi(x)) \leqslant 0$ for $x \in \mathbb{R}^+$.

The admissible functions are defined by the properties

$$|\varphi(x)| \leqslant 1, \quad x \in [0,\infty), \quad \varphi(0) = 1, \quad \text{and} \quad \varphi'(+0) = -i,$$

cf. Kato functions $\mathcal{K}$ in Appendix C (Section C.1). These functions were introduced in [EINZ07] in the context of the *Zeno product formula*, which is tightly related with the general unitary product formulae. The typical examples are: $\varphi(x) = \mathrm{e}^{-ix}$ and $\varphi(x) = (1 + ix)^{-1}$, for $x \in \mathbb{R}$.

Another way to *legitimate* the unitary product formulae is to consider their convergence in topology, which is *different* than the operator strong, or norm, topology. In [ENZ11] we show that the unitary Trotter product formula makes sense in the L^2-topology, that is, it holds

$$\lim_{n\to\infty} \int_{-T}^{T} \left\| \left(\mathrm{e}^{-itA/n}\mathrm{e}^{-itB/n}\right)^n h - \mathrm{e}^{-itC} h \right\|^2 dt = 0 \tag{D.36}$$

for $h \in \mathcal{H}$ and any $T > 0$, and for the standard conditions: A, B are non-negative self-adjoint operators in a Hilbert space $\mathcal{H}$ such that the form-sum $A \dotplus B$ is a densely defined operator and self-adjoint operator C is its closure.

Using the concept of the *holomorphic* Kato functions [ENZ11] we also proved the Trotter-Kato product formula in the L^2-topology. For example, under the same conditions on the operators, we obtain

$$\lim_{n\to\infty} \int_{-T}^{T} \left\| \left(\widetilde{f}(itA/n)\widetilde{g}(itB/n)\right)^n h - \mathrm{e}^{-itC} h \right\|^2 dt = 0 \tag{D.37}$$

for $h \in \mathcal{H}$ and any $T > 0$. Here f, g are holomorphic Kato functions and $\widetilde{f}, \widetilde{g}$ are Borel measurable extensions of f and g on the imaginary axis, see [ENZ11] for details. Similarly one gets the Trotter-Kato product formulae for other approximants.

Bibliography

[ANB75] N. Angelescu, G. Nenciu, and M. Bundaru, *On the perturbation of Gibbs semigroups*, Commun. Math. Phys. **76** (1975), 29–30.

[Ara90] H. Araki, *On an inequality of Lieb and Thirring*, Lett. Math. Phys. **19** (1990), 167–170.

[ABHN01] W. Arendt, C. Batty, M. Hieber, and F. Neubrander *Vector-valued Laplace Transforms and Chauchy Problems*, Birkhäuser Verlag, Basel-Berlin, 2001.

[AzIch08] Y. Azuma and T. Ichinose *Note on norm and pointwise convergence of exponential products and their integral kernels for the harmonic oscillator*, Integral Equations Oper. Theory **60** (2008), 151–176.

[Bal76] A. V. Balakrishnan, *Applied Functional Analysis*, Springer-Verlag, Berlin, 1976.

[Bern88] D. S. Bernstein, *Inequalities for the trace of matrix exponentials*, SIAM J. Martix Anal. Appl. **9** (1988), 156–158.

[BS87] M. S. Birman and M. Z. Solomyak, *Spectral Theory of Selfadjoint Operators in Hilbert Space*, Mathematics and its Applications (Soviet Series), D. Reidel Publishing Co., Dordrecht, 1987.

[BEH94] J. Blank, P. Exner, and M. Havlíček, *Hilbert Space Operators in Quantum Physics*, Amer. Inst. Phys. Press, New York, 1994.

[BBZKT81] N. N. Bogolyubov (jr), J. G. Brankov, V. A. Zagrebnov, A. M. Kurbatov and N. S. Tonchev, *The Approximating Hamiltonian Method in Statistical Physics*, Bulgarian Acad. Sci., Sofia, 1981.

[BBZKT84] N. N. Bogolyubov (jr), J. G. Brankov, V. A. Zagrebnov, A. M. Kurbatov and N. S. Tonchev, *Some classes of exactly soluble models of problems in quantum statistical mechanics: the method of the approximating Hamiltonian*, Russian Math. Surv. **39** (1984), 1–50.

[BR79] O. Bratteli and D. W. Robinson, *Operator Algebras and Quantum Statistical Mechanics 1*, Springer, Berlin, 1979.

V. A. Zagrebnov, *Gibbs Semigroups*, Operator Theory: Advances and Applications 273, https://doi.org/10.1007/978-3-030-18877-1

[BR96] O. Bratteli and D. W. Robinson, *Operator Algebras and Quantum Statistical Mechanics 2*, Springer, Berlin, 1996.

[BG72] M. Breitenecker and H.-R. Grümm, *Note on trace inequalities*, Commun. Math. Phys. **26** (1972), 276–279.

[CNZ01] V. Cachia, H. Neidhardt, and V. A. Zagrebnov, *Accretive perturbations and error estimates for the Trotter product formula*, Integral Equations Oper. Theory **39** (2001), 396–412.

[CNZ02] ———, *Comments on the Trotter product formula error-bound estimates for nonself-adjoint semigroups*, Integral Equations Oper. Theory **42** (2002), 425–448.

[CZ99] V. Cachia and V. A. Zagrebnov, *Operator-norm convergence of the Trotter product formula for sectorial generators*, Lett. Math. Phys. **50** (1999), 203–211.

[CZ01a] ———, *Operator-norm convergence of the Trotter product formula for holomorphic semigroups*, J. Oper. Theory **46** (2001), 199–213.

[CZ01b] ———, *Operator-norm approximation of semigroups by quasi-sectorial contractions*, J. Funct. Anal. **180** (2001), 176–194.

[CZ01c] ———, *Trotter product formula for nonself-adjoint Gibbs semigroups*, J. London Math. Soc. **64** (2001), 436–444.

[Ca10] V. Cachia, *La formule de Trotter-Kato*, Editions universitaires européennes, Saarbrücken, 2010.

[CaSu06] A. L. Carey and F.A. Sukochev, *Dixmier traces and some applications in non-commutative geometry*, Russian Math. Surv. **61**:6 (2006), 1039-1099.

[Car09] E. A. Carlen, *Trace Inequalities and Quantum Entropy: An introductory course*, in: Entropy and the Quantum (R. Sims and D. Ueltschi, Eds.), Contemporary Mathematics v.529, 73–140, AMS Providence 2010.

[Che68] P. R. Chernoff, *Note on product formulas for operator semigroups*, J. Funct. Anal. **2** (1968), 238–242.

[Che74] ———, *Product formulas, nonlinear semigroups and addition of unbounded operators*, Memoirs Amer. Math. Soc. **140** (1974), 1–121.

[Con94] A. Connes, *Noncommutative Geometry*, Academic Press, London, 1994.

[Dal60] Yu. L. Daletskiĭ, *Representatability of solutions of operator equations in the form of contunual path integrals*, Doklady Akad. Nauk USSR **134** (1960), 1013–1016, in Russian.

[Dal61] ———, *Contunual integrals and characteristics related to a group of operators*, Doklady Acad. Nauk USSR **141** (1961), 1290–1293, in Russian.

[Dav69] E. B. Davies, *Quantum stochastic processes* Commun. Math. Phys. **15** (1969), 277–304.

[Dav80] E. B. Davies, *One-parameter Semigroups*, Academic Press, London, 1980.

[Dav07] E. B. Davies, *Linear Operators and their Spectra*, Cambridge University Press, Cambridge, 2007.

[dellA67] G. F. Dell'Antonio, *On the limits of sequences of normal states*, Commun. Pure Applied. Math. **20** (1967), 413–429.

[DS97] B. O. Dia and M. Schatzmann, *An estimate on the transfer operator*, J. Funct. Anal. **145** (1997), 108–135.

[Dix66] J. Dixmier, *Existence des traces non normales*, C. R. Acad. Sci. Paris, Sér. A **262** (1966), 1107–1108.

[DIT98] A. Doumeki, T. Ichinose, and Hideo Tamura, *Error bounds on exponential product formulas for Schrödinger operators*, J. Math. Soc. Japan **50** (1998), 359–377.

[DLP79] W. Dressler, L. Landau, and J. F. Perez, *Estimates of critical length and critical temperatures for classical and quantum lattice systems*, J. Stat. Phys. **20** (1979), 123–162.

[DLS78] F. Dyson, E. H. Lieb, and B. Simon, *Phase transitions in quantum spin systems with isotropic and nonisotropic interactions*, J. Stat. Phys. **18** (1978), 335–383.

[EN00] K.-J. Engel and R. Nagel, *One-parameter Semigroups for Linear Evolution Equations*, Springer-Verlag, Berlin, 2000.

[EINZ07] P. Exner, T. Ichinose, H. Neidhardt, and V. A. Zagrebnov, *Zeno product formula revisited*, Integral Equations Oper. Theory **57** (2007), 67–81.

[ENZ11] P. Exner, H. Neidhardt, and V. A. Zagrebnov, *Reamrks on the Trotter-Kato product for unitary groups*, Integral Equations Oper. Theory **69** (2011), 451–478.

[Fan49] K. Fan, *On a theorem of Weyl concerning eigenvalues of linear transformations* I , Proc. Nat. Acad. Sci. USA **35** (1949), 652–655.

[Fan51] K. Fan, *Maximum properties and inequalities for the eigenvalues of completely continuous operators*, Proc. Nat. Acad. Sci. USA **37** (1951), 760–766.

[Far67a] W. G. Faris, *The product formulas for semigroups defined by Friedrichs extensions*, Pacific J. Math. **22** (1967), 47–70.

[Far67b] ______, *Product formulas for perturbations of linear propagators*, J. Funct. Anal. **1** (1967), 93–108.

[Fr76] J. Fröhlich, *Phase transitions, Goldstone bosons and topological superselection rules*, Acta Phys. Austriaca Suppl. **XV** (1976), 133–269.

[FSS76] J. Fröhlich, B. Simon, and T. Spencer, *Infrared bounds, phase transitions and continuous symmetry breaking*, Commun. Math. Phys. **50** (1976), 79–85.

[FrLi78] J. Fröhlich and E.H. Lieb, *Phase transitions in anisotropic lattice spin models*, Commun. Math. Phys. **60** (1978), 233–268.

[FILS78] J. Fröhlich, R. Israel, E. H. Lieb, and B. Simon, *Phase transitions and reflection positivity. I. General theory and long range lattice models*, Commun. Math. Phys. **62** (1978), 1–34.

[GG69] J. Ginibre and C. Gruber, *Green functions of the anisotropic Heisenberg model*, Commun. Math. Phys. **11** (1969), 198–213.

[GK69] I. C. Gohberg and M. G. Kreĭn, *Introduction to the Theory of Linear Nonselfadjoint Operators*, Transl. Math. Monographs, vol. 18, Amer. Math. Soc., Providence, R. I., 1969.

[Gol85] J. A. Goldstein, *Semigroups of Operators and Applications*, Oxford Univ. Press, Oxford, 1985.

[Gold65] S. Golden, *Lower bounds for the Helmholtz function*, Phys.Rev. **137** (1965), B1127–B1128.

[Gra10] M. Gramegna, *Serie di equazioni differenziali lineari ed equazioni integro-differenziali*, Atti Reale Accad. Sci. Torino **45** (1910), 291–313.

[Grü73] H.-R. Grümm, *Two theorems about $\mathcal{C}_p$*, Commun. Math. Phys. **4** (1973), 211–215.

[Hel95] B. Helffer, *Around the transfer operator and the Trotter-Kato formula*, Oper. Theory Adv. Appl. **78** (1995), 161–175.

[Hia95] F. Hiai, *Trace norm convergence of exponential product formula*, Lett. Math. Phys. **33** (1995), 147–158.

[Horn50] A. Horn, *On the singular values of a product of completely continuous operators*, Proc. Nat. Acad. Sci. USA **36** (1950), 374–375.

[HP57] E. Hille and R. S. Phillips, *Functional Analysis and Semigroups*, vol. 31, Amer. Math. Soc. Coll. Publ., Providence, R.I., 1957.

[HN01] J. K. Hunter and B. Nachtergaele, *Applied Analysis*, World Scientific, Singapore, 2001.

[Ich80] T. Ichinose, *A product formula and its application to the Schrödinger equation*, Publ. RIMS, Kyoto Univ., **16**(1980), 585–600.

[Ich03] T. Ichinose, *Time-Sliced Approximation to Path Integral and Lie-Trotter-Kato Product Formula*, in: A Garden of Quanta: Essays in Honor of Hiroshi Ezawa (J.Arafune et al, Eds.), pp. 77–93, World Scientific Publishing Co. Pte. Ltd., 2003.

[INZ03] T. Ichinose, H. Neidhardt, and V. A. Zagrebnov, *Operator norm convergence of the Trotter-Kato product formula*, in: Proceedings of the International Conference on Functional Analysis, Kiev (August 22–26, 2001) 2003, pp. 100–106.

[INZ04] ———, *Trotter-Kato product formula and fractional powers of self-adjoint generators*, J. Funct. Anal. **207** (2004), 33–57.

[ITak97] T. Ichinose and S. Takanobu, *Estimate of the difference between the Kac operator and the Schrödinger semigroup*, Commun. Math. Phys. **27** (1997), 167–197.

[ITak98] ———, *The norm estimate of the difference between the Kac operators and the Schrödinger semigroup: A unified approach to the non-relativistic and relativistic cases*, Nagoya Math. J. **149** (1998), 53–81.

[ITak00] ———, *The norm estimate of the difference between the Kac operators and the Schrödinger semigroup II: The general case including the relativistic case*, Electron. J. Probab. **5** (2000), 1–47.

[IT97] T. Ichinose and Hideo Tamura, *Error estimate in operator norm for Trotter-Kato product formula*, Integral Equations Oper. Theory **27** (1997), 195–207.

[IT98a] ———, *Error estimate in operator norm of exponential product formulas for propagators of parabolic evolution equations*, Osaka J. Math. **35** (1998), 751–770.

[IT98b] ———, *Error bound in trace norm for Trotter-Kato product formula of Gibbs semigroups*, Asymptotic Anal. **17** (1998), 239–266.

[IT01] ———, *The norm convergence of the Trotter-Kato product formula with error bound*, Commun. Math. Phys. **217** (2001), 489–502.

[IT04] ———, *Sharp error bound on norm convergence of exponential product formula and approximation to kernels of Schrödinger semigroups*, Commun. Partial Diff. Equations **29** (2004), 1905–1918.

[IT04c] ———, *Note on the norm convergence of the unitary Trotter product formula*, Lett. Math. Phys. **70** (2004), 65–81.

[IT09] ———, *Results on convergence in norm of exponential product formulas and pointwise of the corresponding integral kernels*, Operator Theory: Advances and Applications, **190**, (2009) 315327.

[ITTZ01] T. Ichinose, Hideo Tamura, Hiroshi Tamura, and V. A. Zagrebnov, *Note on the paper "The norm convergence of the Trotter-Kato product formula with error bound" by Ichinose and Tamura*, Commun. Math. Phys. **221** (2001), 499–510.

[JL00] G. W. Johnson and M. L. Lapidus. *The Feynman integral and Feynman's operational calculus.* Oxford Mathematical Monographs. The Clarendon Press Oxford University Press, Oxford 2000.

[Kat74] T. Kato, *On the Trotter-Lie product formula*, Proc. Japan Acad. **50** (1974), 694–698.

[Kat78] ———, *Trotter's product formula for an arbitrary pair of self-adjoint contraction semigroups*, Topics in Funct. Anal., Adv. Math. Suppl. Studies (I. Gohberg and M. Kac, eds.), vol. 3, Acad. Press, New York, 1978.

[Kat80] ———, *Perturbation Theory for Linear Operators*, Springer-Verlag, Berlin, 1980, Corrected Printing of the Second Edition.

[Kle31] O. Klein, *Zur quantenmechanischen begrundung des zweiten hauptsatzes der warmelehre*, Z. Physik **72** (1931), 767–775.

[KF99] A. N. Kolmogorov and S. V. Fomin, *Elements of the Theory of Functions and Functional Analysis*, Dover Pub., Mineola, New York, 1999.

[Kon94] Yu. G. Kondratiev, *Phase transitions in quantum models of ferroelectrics*, in: Stochastic Processes, Physics and Geometry, World Scientific, Singapore, 1994, pp. 465–475.

[Kre71] S. G. Kreĭn, *Linear Differential Equations in Banach Spaces*, Transl. Math. Monogr., vol. 29, Amer. Math. Soc., Providence, R. I., 1971.

[Kva56] I. A. Kvasnikov, *Application of variational principle to the Ising model of ferromagnetism*, Doklady Akad. Nauk USSR **110** (1956), 755–757.

[Lid59] V. B. Lidskiĭ, *Nonselfadjoint operators having a trace*, Doklady Akad. Nauk USSR **125** (1959), 485-487.

[LiSe10] E. H. Lieb and R. Seiringer, *The Stability of Matter in Quantum Mechanics*, Cambridge University Press, Cambridge, 2010.

[LiTh76] E. Lieb and W. Thirring, *Inequalities for the moments of the eigenvalues of the Schrödinger hamiltonian and their relation to Sobolev inequalities*, in: Studies in Mathematical Physics (Eds. E. Lieb, B. Simon and A. Wightman), pp. 301–302, Princeton Univ.Press, 1976.

[LSZ12] S. Lord, F. Sukochev and D. Zanin, *Singular Traces: Theory and Applications*, De Gruyter Studies in Mathematics 46. De Gruyter, Berlin. 2012.

[Mai71] H. D. Maison, *Analyticity of the partition function for finite quantum systems*, Commun. Math. Phys. **22** (1971), 166–172.

[MVZ00] R.A.Minlos, A.Verbeure, and V.A.Zagrebnov, *A quantum crystal model in the light-mass limit: Gibbs states*, Rev. Math. Phys., **12** (2000), 981–1032.

[NZ90a] H. Neidhardt and V. A. Zagrebnov, *The Trotter product formula for Gibbs semigroups*, Commun. Math. Phys. **131** (1990), 333–346.

[NZ90b] H. Neidhardt and V. A. Zagrebnov, *On the Trotter product formula for Gibbs semigroups*, Ann. der Physik **47** (1990), 183–191.

[NZ98] ———, *On error estimates for the Trotter-Kato product formula*, Lett. Math. Phys. **44** (1998), 169–186.

[NZ99a] ———, *Fractional powers of self-adjoint operators and Trotter-Kato product formula*, Integral Equations Oper. Theory **35** (1999), 209–231.

[NZ99b] ———, *Trotter-Kato product formula and operator-norm convergence*, Commun. Math. Phys. **205** (1999), 129–159.

[NZ99c] ———, *On the operator-norm convergence of the Trotter-Kato product formula*, Operator Theory: Advances and Applications, **108** (1999), 323–334.

[NZ99d] ———, *Trotter-Kato product formula and symmetrically normed ideals*, J. Funct. Anal. **167** (1999), 113–167.

[NSZ18a] H. Neidhardt, A. Stephan and V. A. Zagrebnov, *Remarks on the operator-norm convergence of the Trotter product formula*, Integral Equations Oper. Theory **90:15**(2) (2018), pp.1–14.

[NSZ18b] H. Neidhardt, A. Stephan and V. A. Zagrebnov, *Operator-norm convergence of the Trotter product formula on Hilbert and Banach spaces: a short survey*, in: Current Research in Nonlinear Analysis. Springer Optimization and Its Applications vol. 135, Springer, Berlin 2018, pp. 229–247.

[Nel64] E. Nelson, *Feynman integral and the Schrödinger equation*, J. Math. Phys. **5** (1964), 332–343.

[PK87] L. A. Pastur and V. A. Khoruzhenko, *Phase transition in quantum models of rotators and ferroelectrics*, Theor. Math. Phys. **73** (1987), 111–124.

[Paz83] A. Pazy, *Semigroups of Linear Operators and Applications to Partial Differential Equations*, Springer-Verlag, Berlin, 1983.

[Peie38] R. Peierls, *On a minimum property of the free energy*, Phys. Rev. **54** (1938), 918–920.

[Petz83] D. Petz, *A survey of certain trace inequalities*, in: Functional Analysis and Operator Theory, Banach Center Publications, v. 30, pp. 287–298, Institute of Mathematics PAS, Warszawa, 1994.

[Piet14] A. Pietsch, *Traces of operators and their history*, Acta et Comment Univer. Tartuensis de Math. **18** (2014), 51–64.

[RS80] M. Reed and B. Simon, *Methods of Modern Mathematical Physics,* I, *Functional Analysis*, (Revised and Enlarged Edition) Academic Press, New York, 1980.

[RS75] ———, *Methods of Modern Mathematical Physics* II, *Fourier-analysis, Self-Adjointness*, Academic Press, New York, 1975.

[RS78] ———, *Methods of Modern Mathematical Physics* IV, *Analysis of Operators*, Academic Press, New York, 1978.

[RR93] M. Renardy and R. C. Rogers, *An Introduction to Partial Differential Equations*, Springer-Verlag, Berlin, 1993.

[Rog93] Dzh. L. Rogava, *Error bounds for Trotter-type formulas for self-adjoint operators*, Funct. Anal. Appl. **27** (1993), 217–219.

[Rue69] D. Ruelle, *Statistical Mechanics. Rigorous Results*, W. A. Benjamin, Inc., Amsterdam, 1969.

[Sad91] V. A. Sadovnichiĭ, *Theory of Operators*, Consultants Bureau, New York, 1991.

[Sch70] R. Schatten, *Norm Ideals of Completely Continuous Operators*, Springer-Verlag, Berlin, 1970.

[Sim05] B. Simon, *Trace Ideals and their Applications*, Second Edition, Mathematical Surveys and Monographs Vol.120, AMS, 2005.

[Su08] F. A. Sukochev, *Dixmier Traces: Analytical Approach and Applications* (Lectures Noncom. Int. Week) Leiden Univ., 2008.

[Suz96] M. Suzuki, *Exponential Product Formulas and Quantum Analysis*, University of Tokyo, Tokyo, 1996.

[Tak97] S. Takanobu, *On the error estimate of the integral kernel for Trotter product formula operators for Schrödinger operators*, Ann. Probab. **25** (1997), 1895–1952.

[Tam00] Hiroshi Tamura, *A remark on operator-norm convergence of Trotter-Kato product formula*, Integral Equations Oper. Theory **37** (2000), 350–356.

[Thom65] C. J. Thompson, *Inequality with applications in statistical mechanics*, J.Math.Phys. **6** (1965), 1812–1813.

[Tro59] H. F. Trotter, *On the products of semigroups of operators*, Proc. Amer. Math. Soc. **10** (1959), 545–551.

[Tya67] S. V. Tyablikov, *Methods in the Quantum Theory of Magnetism*, Plenum Press, New York, 1967.

[Uhl71] D. A. Uhlenbrock, *Perturbation of statistical semigroups in quantum statistical mechanics*, J. Math. Phys. **12** (1971), 2503–2512.

[Ves96] E. Vesentini, *Introduction to Continuous Semigroups*, Scuola Normale Superiore Pisa, Pisa, 1996.

[Voi77] J. Voigt, *On the perturbation theory for strongly continuous semigroups*, Math. Ann. **229** (1977), 161–171.

[Wey49] H. Weyl, *Inequalities between the two kinds of eigenvalues of a linear transformation*, Proc. Nat. Acad. Sci. USA **35** (1949), 408–411.

[Yos65] K. Yosida, *Functional Analysis*, Springer-Verlag, Berlin, 1965.

[You92] P. You, *Characteristic conditions for a C_0-semigroup with continuity in the uniform operator topology for $t > 0$ in Hilbert space*, Proc. Amer. Math. Soc. **112** (1992), 991–997.

[Zag80] V. A. Zagrebnov, *On the families of Gibbs semigroups*, Commun. Math. Phys. **76** (1980), 269–276.

[Zag88] ______, *The Trotter-Lie product formula for Gibbs semigroups*, J. Math. Phys. **29** (1988), 888–891.

[Zag89] ______, *Perturbations of Gibbs semigroups*, Commun. Math. Phys. **120** (1989), 653–664.

[Zag95] ______, *On the operator-norm error-bound estimate for the Trotter product formula: Comments on Rogava's paper*, Preprint CPT-95/P.3177, CNRS-Luminy, 1995.

[Zag03] ______, *Ultimate optimal error bound for the Trotter product formula: Gibbs semigroups*, in: Proceedings of the International Conference on Functional Analysis, Kiev (August 22–26, 2001), Naukova Dumka, Kiev 2003, pp. 94–99.

[Zag03b] ______, *Topics in the Theory of Gibbs semigroups*, Leuven Notes in Mathematical and Theoretical Physics, vol. 10 (Series A: Math. Phys.), Leuven University Press, 2003.

[Zag05] ______, *Trotter-Kato product formula: some recent results*, in: Proceedings of the XIVth International Congress on Mathematical Physics, Lisbon (July 28–August 02, 2003), World Scientific, Singapore 2005, pp. 634–641.

[Zag17] ———, *Comments on the Chernoff $\sqrt{n}$-lemma*, in: Functional Analysis and Operator Theory for Quantum Physics (The Pavel Exner Anniversary Volume), European Mathematical Society, Zürich, 2017, pp. 565–573.

[Zag19] ———, *Trotter-Kato product formulae in Dixmier ideal*, in: Analysis and Operator Theory. Dedicated in Memory of Tosio Kato's 100th Birthday. Springer Optimization and Its Applications vol. 146, Springer, Berlin 2019, pp. 395–416.

[ZBT75] V. A. Zagrebnov, J. G. Brankov, and N. S. Tonchev, *A rigorous statement about systems interacting with a Boson field*, Doklady Akad. Nauk USSR **225** (1975), 71–73.

Index

V. A. Zagrebnov, *Gibbs Semigroups*, Operator Theory: Advances and Applications 273, https://doi.org/10.1007/978-3-030-18877-1

List of Symbols

V. A. Zagrebnov, *Gibbs Semigroups*, Operator Theory: Advances and Applications 273, https://doi.org/10.1007/978-3-030-18877-1

www.ingramcontent.com/pod-product-compliance
Ingram Content Group UK Ltd.
Pitfield, Milton Keynes, MK11 3LW, UK
UKHW021919270726
14059UKWH00002B/99